Chemical Kinetics

Chemical Kinetics

Edited by **Sarah Gates**

Willford Press
New York

Published by Willford Press,
118-35 Queens Blvd., Suite 400,
Forest Hills, NY 11375, USA
www.willfordpress.com

Chemical Kinetics
Edited by Sarah Gates

International Standard Book Number: 978-1-68285-245-3 (Hardback)

Printed in China.

Contents

Preface

Chemical kinetics is an emerging field of chemistry that is concerned with rates of chemical reactions as well as factors and mechanisms involved in it. The study of chemical kinetics includes the determination of rate constants, rate laws and order of a reaction. The topics covered in this extensive book deal with the core concepts of chemical kinetics. The objective of this book is to give a comprehensive overview of the different aspects of homogenous catalysis, free radical chemistry and kinetic models. The content of this book provides the readers with a thorough understanding of the subject.

Significant researches are present in this book. Intensive efforts have been employed by authors to make this book an outstanding discourse. This book contains the enlightening chapters which have been written on the basis of significant researches done by the experts.

Finally, I would also like to thank all the members involved in this book for being a team and meeting all the deadlines for the submission of their respective works. I would also like to thank my friends and family for being supportive in my efforts.

Editor

Investigation into the kinetics of biodegradation of crude oil in different soils

C. P. Ukpaka

Department of Chemical/Petrochemical Engineering, Rivers State University of Science and Technology, Nkpolu, P. M. B. 5080, Port Harcourt, Nigeria. E-mail: chukwuemeka24@yahoo.com.

An experimental study was conducted to examine crude oil degradation in different soil (clay, sandy loamy and swampy soil) of 1, 2, 3, 4 and 5 ml × 10 kg^{-1} pollution. The polluted crude oil and soil samples were obtained from Niger Delta area of Nigeria and microbial analysis were conducted on the samples for the purpose of identification, isolation and characterization of possible microbes with a view to determine the kinetics of the reaction (microbial growth rate). Model equations were developed to simulate the degradation and microbial growth rate kinetics of different soil polluted with crude oil as a function of volume, time, substrate concentration and specific rate. The increase in volume of crude oil applied in the soil samples also induced the microbial activities and the two microbial species discussed in this paper is applicable to the study of the kinetics of biodegradation and the characterization of the product, heat generated and new biomass produced in the system.

Key words: Investigation, kinetics, biodegradation, crude oil, different soil.

INTRODUCTION

Through the years, many experimental and theoretical studies have been done on the kinetics of biodegradation of crude oil (James and David, 1977; Alkinson and Marintuna, 1983; Barelay et al., 1990). The subject is of fundamental importance in the comprehension of any phenomenon involving crude oil degradation. The microbial activities enhance environmental cleanup (Antal and Mgbomo 1973), the release of gases, carbon dioxide, heat and new biomass on surface as it occurs in the bioremediation process (Milkin and Steart, 1974; Ogoni and Gumus, 2001; Ogoni, 2002), characterization of micro-organism (Alkinson and Marintuna, 1983; Sterling et al., 2002; Fingas, 2000; Lee et al., 2002; Lessard and Demarco, 2000; Page et al., 2000; Ukpaka, 2010, 2011), determination of total aerobic heterotrophic bacteria and fungi (Ronald 1992; Ukpaka et al., 2005) and enumeration of hydrocarbon utilizing bacteria and fungi (Antal and Mgbomo 1973); Folsom et al., 1990; Lodaya et al., 1991). The activities of the petroleum industry on the environment have attracted the attention of environmentalists on the effluent discharge into the environment emanating from exploration, production, refining and utilization. Recent investigation carried out by various research groups reveal that due to the high level of petroleum activities in Nigeria, and other parts of the world, there is a high concentration of the petroleum hydrocarbon contaminants, traceable to both upstream and downstream sectors of the economy (Gandy and Gandy, 1988). This has resulted in environmental pollution, which affects the ecological system (Schmidt and Gier, 1990; Parson et al., 1990; Phelps et al., 1990; Bashir et al., 1990; Scow et al., 1990; Kistner and Kornelius, 1990; Deweerd et al., 1991; Schnell et al., 1991; Zheng and Yapa, 2000; Ukpaka, 2011a; Ukpaka and David, 2010).

Although, biodegradation is an important process used in minimizing potentially adverse impacts on environmental system, traditionally, it has not been considered quantitatively in environmental assessments. Efforts to use the kinetic of biodegradation system to judge the qualitative evaluation of microbial activities, its biodegradation properties (for example, “fast” or “slow”) or rigorous quantitative models, to predict the kinetics of biodegradation and microbial activities on the different types of the soil used for this investigation (Wami and Ogoni, 1997; Parsons and Govers, 1990; Ronald, 1992).

However, scientific investigation on the effect of oil pollution in Nigeria only began recently after the shell-BP Bomu 11 blows out in July, 1970. The Texaco blow out in 1980 and the Agip Oyakame pipeline leakage of 1980. The Safram (now elf) Obagi 21 blow of 1970 has all without any doubt resulted in disastrous effects on land. Since after the blow out, studies have been carried out on the effect of crude oil on the soil and aquatic environment with special emphasis in its degradation by micro-organisms, since the recognition of the interaction of micro-organisms and petroleum (Antal and Mgbomo 1973). They have been used greatly in solving the environmental problems.

In this research work, however, investigation was conducted in identification, isolation, characterization of micro-organisms and the determination of the physicochemical properties of the crude oil and the soil samples (sandy loam, swampy soil and clay soil). In this case heat substrate reaction kinetics, microbial growth kinetics and product kinetics modeled equations were developed to study the principle behind the bioremediation process. Bioremediation processes are used to treat contaminated soil, success of such treatment processes lies in degrading the organic contaminants and reducing both the toxicity as well as the migration potential of the hazardous constituents (Irwin, 1975; Alkinson and Marintuna, 1983; Barelay et al., 1990; Digrazia et al., 1990; Dasappa and Loehr, 1991; Miller and Alexander, 1991). The role of microbes in the corrosion of metals is due to the chemical activities (metabolism) associated with microbial growth under favourable condition (James and David, 1977; Goldsmith et al., 1989; Arvin et al., 1991). This microbial activities associated with chemical reaction can accelerate exponentially the rate of degradation. Although these activities may be suppressed by mechanical techniques, biological activities often grow again when favourable conditions are restored (Awwa, 2000; Guan, 2001; Kakporbia, 2001; Ukpaka et al., 2010; Ukpaka, 2009).

In fact, it confirms that no comprehensive and feasible models have been developed on effect of different soil samples in biodegradation kinetics of crude oil in batch reactors. However, it is the purpose of this research work to correlate the biodegradation kinetics system of the different soil samples as a function of timer, volume of crude oil to soil ratio, microbial population, substrate concentration and then determination of Line Weaver Bulk Plot parameters (Vmax or μ_{max} and Ks or Km) since the microbial degradation of crude oil under environmental conditions in soil phase using aerobic, anaerobic and facultative anaerobic condition is influenced by volume of crude oil deposited, the soil texture and structure, the type of micro-organism present and the composition of both the crude oil and the soil (Ukapa, 2010; Antunes-do et al., 2010; Scbastiao and Guides, 2006; Brekke and Solberg, 2005; Jha et al., 2008; Antunes-do and Coasta, 2000; Guo and Wang, 2009).

THE MODEL

Biomass growth rate model

When an organic waste (petroleum hydrocarbon) is discharged into the environment, the organic content of the effluent undergoes biochemical reactions as reported (Irwin, 1975). The rate of biodegradation is influenced by the concentration of the substrates and the product inhibition (James and David, 1977). The biodegradation is generally classified as aerobic, anaerobic and facultative anaerobic process taking place under the same environmental condition, resulting in the production of new biomass, carbon dioxide, water and product.

$$[\text{Soil + crude oil}]_{mixed} + \text{micro-organism} \xrightarrow{K} \text{gases} + CO_2 = \text{heat release + new biomass} \quad (1)$$

Where, $[\text{soil + crude oil}]_{mixed}$ = A, micro-organism = E, product [gas and CO_2] = P heat generated = Q, and other hydrocarbon gases.

Therefore Equation (1) can be written as A+E $\rightarrow$ P+E

(2)

The microbial growth rate kinetics can be experienced at the lag, progressive, stationary and decline phase and for dynamic studies, the general conservation equation for a steady state must be modified to give the following unsteady state mass balance;

$$\frac{d}{dt}(\text{biomass in the system}) = (rate\ of\ addition\ of\ micro-organism\ to\ system) - (rate\ of\ removal\ of\ micro-organism\ from\ system) + (rate\ of\ production\ within\ system) \quad (3)$$

Assuming the rate of addition and removal of micro-organism from the system is equal. Therefore equation (3) becomes;

$$\frac{d}{dt}(biomass\ in\ the\ sytem) = (rate\ of\ production\ within\ system) \quad (4)$$

Since the rate of growth of biomass is proportional to the initial content or number of micro-organism (biomass growth of particular microbial species presented), the time required for the biomass increase was determined by the application of the differential.

$$\frac{d\mu}{dt} = \phi\mu \quad (5)$$

On rearranging Equation (5) yields

$$\frac{d\mu}{dt} - \phi\mu = 0 \qquad (6)$$

Where $\frac{d\mu}{dt}$ change in biomass per unit time is, μ is the micro organism and ϕ is the proportionality constant.

Therefore Equation (6) can be written in terms of bacteria and fungi species present in the system.
hence,

$$\frac{d\mu_B}{dt} - \phi_B\mu_B = 0 \qquad (7)$$

and

$$\frac{d\mu_F}{dt} - \phi_F\mu_F = 0 \qquad (8)$$

where $\frac{d\mu_B}{dt}$ and $\frac{d\mu_F}{dt}$ are changes in biomass per unit time for bacteria and fungi species, ϕ_B and ϕ_F are proportional constant for bacteria and fungi species, μ_B and μ_F are population of microorganism for bacteria and fungi species in the system.

Equation (6) was solved by considering the necessary boundary conditions such as; t = 0, $\mu_{(0)} = \mu_0$.
Therefore application of mathematical approach (Laplace transfer) to Equation (6) yielded

$$S\mu_o - \mu(0) - \phi\mu_{(s)} = 0 \qquad (9)$$

Substituting the boundary condition at, t = 0, $\mu(0) = \mu_o$ into equation (9) and rearranging, it becomes

$$\mu_{(s)} = \frac{\mu_o}{S - \phi} \qquad (10)$$

Therefore Laplace inverse of equation (10) with respect to time gives

$$\mu_{(t)} = \mu_o e^{\phi t} \qquad (11)$$

Therefore the value of ϕ was determined by rearranging Equation (11) thus:

$$\frac{\mu_t}{\mu_0} = e^{\phi t} \qquad (12)$$

Simplifying Equation (12), it becomes

$$In\frac{\mu_t}{\mu_o} = \phi t \qquad (13)$$

$$\therefore \ \phi = \frac{1}{t} In \frac{\mu_{(t)}}{\mu_o} \qquad (14)$$

Where t = time per week.

Considering the initial statement at the progressive phase for one week interval neglecting the effect of temperature and nutrient, the following expressions will be obtained as shown in Table 1, for the determination of various ϕ_B and ϕ_F. Using Equation (14), at t = 1, yields the different values in Table 1.

Where μ is specific rate, μ_{max} is maximum specific rate, [S] is substrate concentration and k_m is the equilibrium constant.

Monod equation model

The Monod equation for microbial kinetic which rates the specific growth rate of the micro-organism and the limiting components. The general form of this equation is presented by Alkinson and Marintuna (1983) as;

$$\mu = \frac{\mu_{max}[S]}{K_m + [S]} \qquad (15)$$

Model for correlation of Monod equation and biomass growth rate

Considering the stage at which the specific rate μ is a function of time (t) or time dependent, therefore equation (15) can be written as:

$$\mu_o e^{\phi t} = \frac{[\mu_o e^{\phi t}]_{max}[S]}{K_m + [S]} \qquad (16)$$

Substituting the values obtained in Table 1 into Equation (12) yields the result in the Table 2.

Table 1. Determination of the various proportionality constants for ϕ_B and ϕ_F.

Oil applied ml x 10 kg^{-1} (volume)	Proportionality constants ϕ_B and ϕ_F					
	$[\phi]_{clay\ soil}$		$[\phi]_{sandy\ loamy}$		$[\phi]_{swamp\ soil}$	
	ϕ_B	ϕ_F	ϕ_B	ϕ_F	ϕ_B	ϕ_F
0	$3/4\,\mu_{OB}$	$33/40\,\mu_{OF}$	$16/25\,\mu_{OB}$	$57/100\,\mu_{OF}$	$3/5\,\mu_{OB}$	$9/50\,\mu_{OF}$
1	$39/50\,\mu_{OB}$	$33/50\,\mu_{OF}$	$57/100\,\mu_{OB}$	$11/25\,\mu_{OF}$	$83/100\,\mu_{OB}$	$3/10\,\mu_{OF}$
2	$22/25\,\mu_{OB}$	$3/4\,\mu_{OF}$	$21/25\,\mu_{OB}$	$37/50\,\mu_{OF}$	$3/4\,\mu_{OB}$	$9/25\,\mu_{OF}$
3	$7/10\,\mu_{OB}$	$83/100\,\mu_{OF}$	$17/20\,\mu_{OB}$	$9/10\,\mu_{OF}$	$22/25\,\mu_{OB}$	$33/50\,\mu_{OF}$
4	$73/100\,\mu_{OB}$	$43/50\,\mu_{OF}$	$43/50\,\mu_{OB}$	$14/25\,\mu_{OF}$	$39/50\,\mu_{OB}$	$19/25\,\mu_{OF}$
5	$34/50\,\mu_{OB}$	$4/5\,\mu_{OF}$	$39/50\,\mu_{OB}$	$1/2\,\mu_{OF}$	$4/5\,\mu_{OB}$	$73/100\,\mu_{OF}$

Michael's Menten model

The Michael's Menten equation for substrate kinetic which gives the specific rate of substrate concentration and limiting component; the general form of this equation is presented by Irwin (1975) as:

$$V = \frac{V_{max}[S]}{K_S + [S]} \quad (16a)$$

Enzyme/substrate reaction model (Figure 1), considering the following reaction mechanism;

Where A is substrate, E is enzyme, P is product EA is enzyme-substrate complex, K_1, K_2, K_3 and K_4 are the various equilibrium constants.

$$A + E \xrightarrow{K_1} P + E \quad (17)$$

Rate of reaction model

With the rate equation given as

$$-\gamma_A = \frac{dC_A}{dt} = K \quad (18)$$

Where $K = KC_{EO}C_{AO}$ = constant, integrating (18) between the limits

$C_A = CAO$ at $t = 0$ and $C_A = C_A$ at $t = t$ yields

$$C_A = CAO = -K_t \quad (19)$$

Still considering the reaction given as:

$$A + E \xrightarrow{K_1} P + E$$

With the rate equation given as:

$$-\gamma_A = \frac{dC_A}{dt} = KC_{EO}C_A \quad (20)$$

Where K is kinetic rate constant, C_{Eo} is initial concentration of the enzyme, C_A is concentration of the substrate, r_A is rate of reaction, $\frac{dC_A}{dt}$ is the rate of change of substrate concentration per unit time. Since one of the characteristics of enzyme or catalyst is that, their concentration is constant in any reacting system

Thus,

$$-\gamma_A = \frac{dC_A}{dt} = KC_A \quad (21)$$

Where $K = KC_{EO}$ = constant, integrating Equation (21) at $C_A = C_{Ao}$ and at t = t yields

$$\ln C_A - \ln C_{AO} = -Kt \quad (22)$$

Substrate concentration model

Let us consider the saturation kinetics

$$A + E \underset{K_3}{\overset{K_2}{\rightleftarrows}} EA \xrightarrow{K_e} P + E$$

Table 2. Equations that led to the determination of specific growth, maximum specific growth rate and equilibrium constants for different soil samples.

Oil applied ml x 10kg^{-1}	Specific rate equations					
	Clay soil		Sandy loamy soil		Swampy soil	
	Bacterial (μ_B)	Fungi (μ_F)	Bacterial (μ_B)	Fungi (μ_F)	Bacterial (μ_B)	Fungi (μ_F)
0	$\left[\mu_o e^{(In\frac{3}{4})t}\right] = \frac{\left[\mu_0 e^{(In\frac{3}{4})t}\right]_{max}[S]_B}{K_{mB}+[S]_B}$	$\mu = \frac{\left[\mu_0 e^{(In\frac{33}{50})t}\right]_{max}[S]_F}{K_{mF}+[S]_F}$	$\mu = \frac{\left[\mu_0 e^{(In\frac{16}{25})t}\right]_{max}[S]_B}{K_{mB}+[S]_B}$	$\mu = \frac{\left[\mu_0 e^{(In\frac{57}{100})t}\right]_{max}[S]_F}{K_{mF}+[S]_F}$	$\mu = \frac{\left[\mu_0 e^{(In\frac{3}{5})t}\right]_{max}[S]_B}{K_{mB}+[S]_B}$	$\mu = \frac{\left[\mu_0 e^{(In\frac{9}{50})t}\right]_{max}[S]_F}{K_{mF}+[S]_F}$
1	$\left[\mu_o e^{(In\frac{39}{50})t}\right] = \frac{\left[\mu_0 e^{(In\frac{39}{50})t}\right]_{max}[S]_B}{K_{mB}+[S]_B}$	$\mu = \frac{\left[\mu_0 e^{(In\frac{33}{50})t}\right]_{max}[S]_F}{K_{mF}+[S]_F}$	$\mu = \frac{\left[\mu_0 e^{(In\frac{57}{100})t}\right]_{max}[S]_B}{K_{mB}+[S]_B}$	$\mu = \frac{\left[\mu_0 e^{(In\frac{11}{25})t}\right]_{max}[S]_F}{K_{mF}+[S]_F}$	$\mu = \frac{\left[\mu_0 e^{(In\frac{83}{100})t}\right]_{max}[S]_B}{K_{mB}+[S]_B}$	$\mu = \frac{\left[\mu_0 e^{(In\frac{3}{10})t}\right]_{max}[S]_F}{K_{mF}+[S]_F}$
2	$\left[\mu_o e^{(In\frac{22}{25})t}\right] = \frac{\left[\mu_0 e^{(In\frac{22}{25})t}\right]_{max}[S]_B}{K_{mB}+[S]_B}$	$\mu = \frac{\left[\mu_0 e^{(In\frac{3}{4})t}\right]_{max}[S]_F}{K_{mF}+[S]_F}$	$\mu = \frac{\left[\mu_0 e^{(In\frac{21}{25})t}\right]_{max}[S]_B}{K_{mB}+[S]_B}$	$\mu = \frac{\left[\mu_0 e^{(In\frac{37}{50})t}\right]_{max}[S]_F}{K_{mF}+[S]_F}$	$\mu = \frac{\left[\mu_0 e^{(In\frac{3}{4})t}\right]_{max}[S]_B}{K_{mB}+[S]_B}$	$\mu = \frac{\left[\mu_0 e^{(In\frac{9}{25})t}\right]_{max}[S]_F}{K_{mF}+[S]_F}$
3	$\left[\mu_o e^{(In\frac{7}{10})t}\right] = \frac{\left[\mu_0 e^{(In\frac{7}{10})t}\right]_{max}[S]_B}{K_{mB}+[S]_B}$	$\mu = \frac{\left[\mu_0 e^{(In\frac{83}{100})t}\right]_{max}[S]_F}{K_{mF}+[S]_F}$	$\mu = \frac{\left[\mu_0 e^{(In\frac{17}{20})t}\right]_{max}[S]_B}{K_{mB}+[S]_B}$	$\mu = \frac{\left[\mu_0 e^{(In\frac{9}{50})t}\right]_{max}[S]_F}{K_{mF}+[S]_F}$	$\mu = \frac{\left[\mu_0 e^{(In\frac{22}{25})t}\right]_{max}[S]_B}{K_{mB}+[S]_B}$	$\mu = \frac{\left[\mu_0 e^{(In\frac{33}{50})t}\right]_{max}[S]_F}{K_{mF}+[S]_F}$
4	$\left[\mu_o e^{(In\frac{73}{100})t}\right] = \frac{\left[\mu_0 e^{(In\frac{73}{100})t}\right]_{max}[S]_B}{K_{mB}+[S]_B}$	$\mu = \frac{\left[\mu_0 e^{(In\frac{43}{50})t}\right]_{max}[S]_F}{K_{mF}+[S]_F}$	$\mu = \frac{\left[\mu_0 e^{(In\frac{43}{50})t}\right]_{max}[S]_B}{K_{mB}+[S]_B}$	$\mu = \frac{\left[\mu_0 e^{(In\frac{14}{25})t}\right]_{max}[S]_F}{K_{mF}+[S]_F}$	$\mu = \frac{\left[\mu_0 e^{(In\frac{39}{50})t}\right]_{max}[S]_B}{K_{mB}+[S]_B}$	$\mu = \frac{\left[\mu_0 e^{(In\frac{19}{25})t}\right]_{max}[S]_F}{K_{mF}+[S]_F}$
5	$\left[\mu_o e^{(In\frac{34}{50})t}\right] = \frac{\left[\mu_0 e^{(In\frac{34}{50})t}\right]_{max}[S]_B}{K_{mB}+[S]_B}$	$\mu = \frac{\left[\mu_0 e^{(In\frac{4}{5})t}\right]_{max}[S]_F}{K_{mF}+[S]_F}$	$\mu = \frac{\left[\mu_0 e^{(In\frac{39}{50})t}\right]_{max}[S]_B}{K_{mB}+[S]_B}$	$\mu = \frac{\left[\mu_0 e^{(In\frac{1}{2})t}\right]_{max}[S]_F}{K_{mF}+[S]_F}$	$\mu = \frac{\left[\mu_0 e^{(In\frac{4}{5})t}\right]_{max}[S]_B}{K_{mB}+[S]_B}$	$\mu = \frac{\left[\mu_0 e^{(In\frac{73}{100})t}\right]_{max}[S]_F}{K_{mF}+[S]_F}$

where $\mu = \mu_0 e^{In\phi t}$

Recalling the Michael's – Menten equation for such reaction gives

$$-\gamma_A = \frac{K_2 C_A}{K_m + C_A} \qquad (23)$$

Where $K_m = K_2 / K$

Substituting Equation (21) into Equation (23) and integrating between the limits yields:

$$\frac{In\, C_A / C_{AO}}{C_A - C_{AO}} = \frac{K_2 t}{K_m (C_{AO} - C_A)} - \frac{1}{K_m} \qquad (24)$$

Hence,

$$In\frac{C_A / C_{AO}}{(C_A - C_{AO})} \quad Vs \quad \frac{t}{(C_{AO} - C_A)}$$

Product kinetic model

The rate at which crude oil is being used up to produce hydrocarbon gases, CO_2, H_2O, heat and biomass is given as:

$$\frac{dC_{CO2}}{dt} = -\gamma_{CO2} \qquad (25)$$

$$\frac{dC_{H_2O}}{dt} = -\gamma_{H_2O} \qquad (26)$$

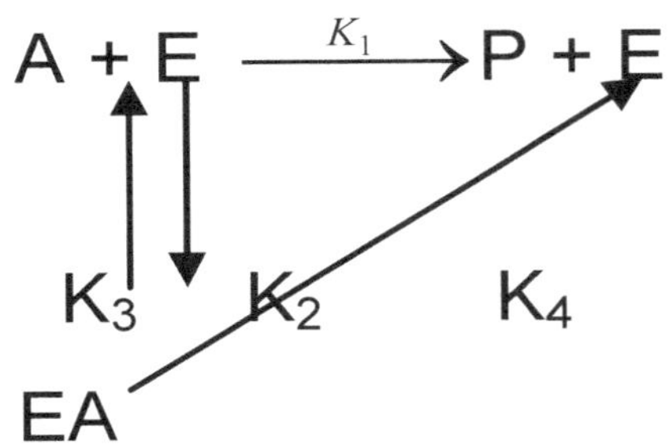

Figure 1. The enzyme-substrate reaction mechanisms.

$$\frac{dC_{Heat}}{dt} = -\gamma_{Heat} \quad (27)$$

$$\frac{dC_b}{dt} = -\gamma_{biomass} \quad (28)$$

The rate law for Equations (25), (26), (27) and (28) is given as:

$$-\gamma_{CO_2} = \frac{V_{\max(CO_2)}C_{CO2}}{K_{CO2} + C_{CO2}} \quad (29)$$

$$-\gamma_{H_2O} = \frac{V_{\max(H_2O)}C_{H_2O}}{K_{H_2O} + C_{H_2O}} \quad (30)$$

Therefore the overall rate of the product produce can be expressed as:

$$\frac{dC}{dt} = \frac{dC_{CO2}}{dt} + \frac{dC_{H_2O}}{dt} + \frac{dQ_{Heat}}{dt} + \frac{dC_{biomass}}{dt} \quad (31)$$

$$-\gamma_{overall} = (-\gamma_{CO2}) + (-\gamma_{H2O}) + (-\gamma_{Heat}) + (-\gamma_{biomass}) \quad (32)$$

where $-\gamma_{overall}$ is the overall rate of reaction, $-\gamma_{CO2}$ is the rate of production of carbon dioxide, $-\gamma_{H2O}$ is the rate of production of water, $-\gamma_{Heat}$ is the rate of heat generated during the reaction, $-\gamma_{biomass}$ is the rate of production of biomass in the system, $\frac{dc}{dt}$ is the change in substrate concentrate of the reaction system per unit time, $\frac{dc_{CO_2}}{dt}$ is the change in carbon dioxide concentration per unit time, $\frac{dc_{H_2O}}{dt}$ is the change in water concentration per unit time, $\frac{dQ_{Heat}}{dt}$ is the change in quantity of heat generated per unit time and $\frac{dc_{biomass}}{dt}$ is the change in biomass built up per unit time

Relating in terms of Michael's Menten equation yields

$$\gamma_{overall} = \frac{V_{\max(CO_2)}}{K_{CO_2} + C_{CO_2}} + \frac{V_{\max(H_2O)}C_{H_2O}}{K_{H_2O} + C_{H_2O}} \quad (33)$$

Where ($-\gamma_{Heat}$) and ($-\gamma_{biomass}$) is negligible

Model for the determination of fractional conversion

Substituting equation (25) into equation (29) and then rearranging and integrating yields

$$t = \int_{C_{CO_2}}^{C_{CO_2}} \frac{dC_{CO_2}}{-\gamma_{CO_2}} \quad (34)$$

$$t = \int_{C_{CO_2}}^{C_{CO_2(0)}} \frac{K_{CO_2} + C_{CO_2}}{V_{\max(CO_2)}C_{CO_2}} dC_{O_2} \quad (35)$$

Writing equation (35) in terms of fractional conversion yields

$$C_{CO2} = C_{CO2(0)}(1 - x) \quad (36)$$

Therefore substituting Equation (36) into Equation (35) yields

$$t = \frac{K_{CO_2}}{V_{\max(co_2)}} In\frac{1}{1-x} + \frac{C_{CO_2}}{V_{\max}C_{O_2}} \quad (37)$$

Similarly for H_2O: $t = \frac{K_{H_2O}}{V_{\max(H_2O)}} In\frac{1}{1-x} + \frac{C_{H_2O}}{V_{\max}C_{H_2O}}$ (38)

$$CH_U : t = \frac{K_{CH_U}}{V_{\max(CH_U)}} In \frac{1}{1-x} + \frac{C_{CH_U}}{V_{\max} C_{CH_U}} \quad (39)$$

Where X = fractional conversion, K_{H2O}, K_{CO2}, K_{CH4} are equilibrium constants for water, carbon dioxide and methane respectively. V_{max} (H_2O), $V_{max(CO2)}$, $V_{max(CH4)}$, are maximum specific rate values for water, carbon dioxide and methane respectively.

Heat kinetic model

Since the instantaneous microbial heat generated rate Q_{gr} is given as

$$Q_{gr} = U_{reactor}\beta\mu \, {}^{1}\!/_{Y_A} \quad (40)$$

Where Q_{gr} is heat generated during the reaction process, $V_{reactor}$ is the reactor volume, β is a constant, μ is the microbial growth rate, Y_A is yield factor of component A, therefore Equation (40) can be rearranged and written as:

$$\mu = \frac{Q_{gr} Y_A}{V_{reactor}} \beta \quad (41)$$

Recalling the Monod equation

$$\mu = \frac{\mu_{max} S}{K_m + S} \quad (42)$$

Therefore substituting Equation (42) into Equation (41) yields

$$\frac{Q_{gr} Y_A}{V_{reactor} \beta} = \frac{\mu_{max}[S]}{K_m + [S]} \quad (43)$$

where $-\gamma_{Heat} = \mu$ and $\gamma = \mu = V$

Hence

$$Q_{gr} = \left[\frac{\mu_{max}[S]}{K_m + [S]}\right]\left[\frac{V_{reactor}\beta}{Y_A}\right] \quad (44)$$

Similarly,

$$-\gamma_{biomass} = \frac{\mu_{max\,(biomass)} \, C_{biomass}}{K_{biomass} + C_{biomass}} \quad (45)$$

Where μ is specific rate of microbes, μ_{max} is maximum specific growth rate of microbes, [S] is substrate concentration, $V_{reactor}$ is volume reactor, Y_A is yield factor, is yield factor, Y_A $C_{biomass}$ is biomass concentration

MATERIALS AND METHODS

Sample collection

Soil samples were collected without bias, the surface soil of about 0 to 10 cm depth. Each soil sample was collected from six different locations and then mixed together to obtain a gross representation of each sample. The cap of the sterile tube is used to collect soil sample from each location for microbial study of the unpolluted soil. And they are then stored in a refrigerator before analysis is carried out on it, which is by preparing a solution of 14g of nutrient agar and 500ml of distilled water with a pelt of 7.4 in a conical flask. Sample location: Middle Belt-soil in Kaima, clay texture mangrove – swampy – Eagle Island and coast plain Terrac-sandy soil.

Determination of total aerobic heterotrophic bacteria and fungi

The variable plate count method, using a surface spreading technique was used to determine the number of soil aerobic bacteria and fungi. Serial dilution was prepared using soil suspension, which is obtained by adding one gram (1 g) of soil sample obtained from the centre of soil cores into 100 m of sterile distilled water. The suspension was shaken vigorously and allowed to settle before 1 ml is used for serial dilution. The 1 ml dilution is plated on triplicate agar supplement at 28°C for 24 h (1day) (Atlas 1981).

Aerobic fungi was determined also by surface spreading technique, using serial dilution, 1 ml of each dilution was plated on malt extract agar into which 100 µg/l of streptomycin and 15 mg of penicillin G has been incorporated in plates triplicate were incubated at 25°C for 72 h (3days) (Atlas 1981).

Enumeration of hydrocarbon utilizing bacteria and fungi

It involves the viable count method using surface spreading technique. Serial dilution of the soil samples was prepared form 10^{-1} to 10^{-15}. One ml of soil dilution was plated in triplicate with agar and 50 µg/l of nystatin to inhibit fungi for bacteria count. After then the polluted soil is put in the Petri dishes, the plates was wrapped with masking tape and made airtight. This is aimed at supplying hydrocarbon as the sole source of carbon energy for growth of the utilizer through vapor phase transfer. The plates were incubated at 37°C for seven days before enumeration.

RESULTS AND DISCUSSION

Biodegradation of Nigeria crude oil in different soil were studied to determine the microbial and substrate kinetics and the Monod's constants (V_{max}, μ_{max} K_S and K_m) for both species used during the investigation. Also the microbial numbers of the micro-organism (Bacteria and fungi) per week activities are shown in Tables 3, 4, and 5 for the different soil types (clay soil, sandy soil and swampy soil). The initial microbial number used for this investigation is 5cfug^{-1} for the different experimental set-up.

Similarly, the experimental results obtained from the

Table 3. Microbial number (x $10^5 kg^{-1}$) in top 10cm of polluted soil; clay soil (Bacteria and fungi).

Oil applied ml x 10 kg^{-1}	**Total microbial number introduced = 5cfug^{-1} for each of the species (microorganism)**													
	1st week		**2nd week**		**3rd week**		**4th week**		**5th week**		**6th week**		**7th week**	
	Ba_1	**Fu_1**	**Ba_2**	**Fu_2**	**Ba_3**	**Fu_3**	**Ba_4**	**Fu_4**	**Ba_5**	**Fu_5**	**Ba_6**	**Fu_6**	**Ba_7**	**Fu_7**
0	9	7	12	6	10	9	14	10	20	16	15	16	13	12
1	11	10	14	15	12	17	16	20	28	23	22	16	14	6
2	16	14	18	12	14	16	19	23	32	27	29	22	18	10
3	19	16	20	10	16	12	23	18	50	27	32	20	24	5
4	21	24	24	20	19	23	26	37	64	51	36	69	28	30
5	25	24	27	30	20	46	29	60	83	74	40	38	32	21

Table 4. Microbial number (x $10^5 kg^{-1}$) in top 10cm of polluted soil, farm land soil (sandy, loamy) bacterial and fungi.

Oil applied ml x 10 kg^{-1}	**Total microbial number introduced = 5 cfug^{-1} for each of the specie (microorganism)**													
	1st week		**2nd week**		**3rd week**		**4th week**		**5th week**		**6th week**		**7th week**	
	Ba_1	**Fu_1**	**Ba_2**	**Fu_2**	**Ba_3**	**Fu_3**	**Ba_4**	**Fu_4**	**Ba_5**	**Fu_5**	**Ba_6**	**Fu_6**	**Ba_7**	**Fu_7**
0	11	8	17	14	18	16	20	20	30	26	20	39	14	10
1	16	14	28	32	36	40	48	60	100	65	80	68	78	47
2	22	14	26	22	38	30	80	71	96	83	70	64	40	25
3	23	20	27	22	51	47	78	61	125	117	92	77	40	38
4	25	20	29	36	70	68	91	135	173	206	82	104	50	70
5	25	25	32	50	88	90	101	200	250	310	83	112	50	80

Table 5. Microbial number (x 10^5 kg^{-1}) in top 10 cm of polluted soil, swamp soil (Bacterial and fungi).

Oil applied ml x 10 kg^{-1}	**Total microbial number introduced = 5cfug^{-1} for each of the species (microorganism)**													
	1st week		**2nd week**		**3rd week**		**4th week**		**5th week**		**6th week**		**7th week**	
	Ba_1	**Fu_1**	**Ba_2**	**Fu_2**	**Ba_3**	**Fu_3**	**Ba_4**	**Fu_4**	**Ba_5**	**Fu_5**	**Ba_6**	**Fu_6**	**Ba_7**	**Fu_7**
0	3	5	5	11	6	19	7	28	8	53	7	24	6	10
1	5	8	6	27	8	44	9	63	11	108	12	63	11	41
2	6	11	8	30	10	34	14	40	16	53	17	23	15	17
3	7	16	8	24	12	27	14	30	18	35	24	10	20	5
4	7	23	9	16	12	21	14	23	19	23	28	9	25	4
5	8	30	10	22	12	30	16	30	19	17	32	8	28	8

investigations for the consumption rate of substrate concentration on the different types of soil used is shown in Table 4.

Where A_{sa}, A_{sw} A_{cl} = 0ml x $10kg^{-1}$ (Zero pollution)
B_{sa}, B_{sw} B_{cl} = 1 ml × 10 kg^{-1} (1 ml pollution)
C_{sa}, C_{sw} C_{cl} = 2 ml × 10 kg^{-1} (2 ml pollution)
D_{sa}, D_{sw} D_{cl} = 0 ml × 10 kg^{-1} (3 ml pollution)
E_{sa}, E_{sw} E_{cl} = 0 ml × 10 kg^{-1} (4 ml pollution)
F_{sa}, F_{sw} F_{cl} = 5 ml × 10 kg^{-1} (5 ml pollution)

Also, the computational result that led to the determination of specific rate (V) for the different soil types is shown in Table 7.

Microbial kinetic characteristics

The results obtained from the investigation shows lag phase, progressive phase, stationary phase and death or decline phase was used in the determination of the microbial number (μ) cfu/kg at each phase. It was observed that for the lag phase, there was a decrease in the initial microbial number and increase in biomass at

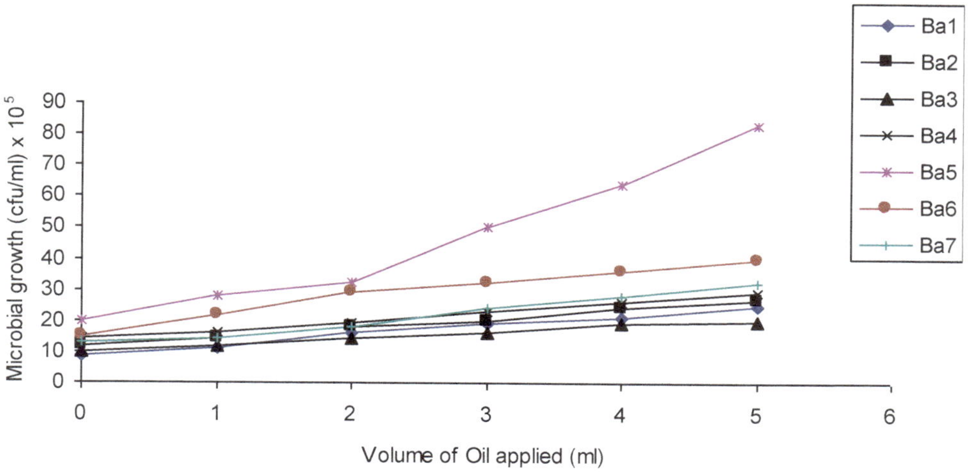

Figure 2. Microbial growth vs volume of oil applied for clay soil (bacteria sp).

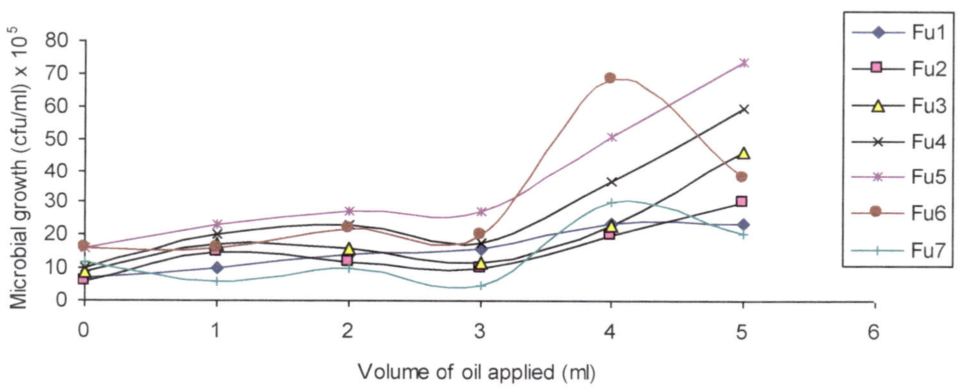

Figure 3. Microbial growth vs volume of oil applied for sandy loamy (fungi sp).

the progressive phase as shown in Tables 3, 4 and 5 for the different soil samples.

As shown in Tables 1, 2 and 3, it was discovered, from analysis, that growth rate of bacteria and fungi species at the progressive phase increased linearly until the whole substrate concentration was reduced to the minimum level when the process became stagnant (stationary phase). Similarly, the decay rate increases as the substrate consumption decreases. Results obtained as reflected in this paper indicates that the growth rate of the bacteria and fungi species increased with increase in time until the stationary phase was attained. Similarly, the decay rate of the bacteria and fungi species increased with increase in time until the whole substrate concentration was reduced to the minimum level.

In modeling, the rate of microbial growth of bacteria and fungi species in degrading crude oil was influenced by the degree of environmental factors such as temperature. (20 to 38°C), pH (sandy-loamy = 6.9, swampy soil = 9.7 and clay soil = 7.1) and electrical conductivity (sandy-loamy = 120 × 10^{-6}mol/cm, swampy soil = 498 × 10^{-6} mol/cm and clay soil = 260 × 10^{-6} mol/cm).

Effect of volume of crude oil applied on the biodegradation process

The effect of the increase in volume of crude oil applied on the soil samples to accompanying biodegradation are shown in Figures 2, 3, 4, 5, 6 and 7 for crude oil concentration of 1, 2, 4 and 5 ml. The substrate concentration at which the comparison for sandy loamy is 1 < l < 3 < 5 < 4 ml, clay is 1 < 2 < 3 < 4 < 5 ml only at 28 and 35 days that 5 < 4 ml for clay and sandy loamy soil. The higher the volume of crude oil concentration, the higher the microbial growth rate and the higher the estimated degradation period.

At 1 ml, volume of crude oil applied to the different soil samples yields substrate concentration of the following order $S_{sa} < S_{cL} < S_{sw}$ (S is the substrate concentration of

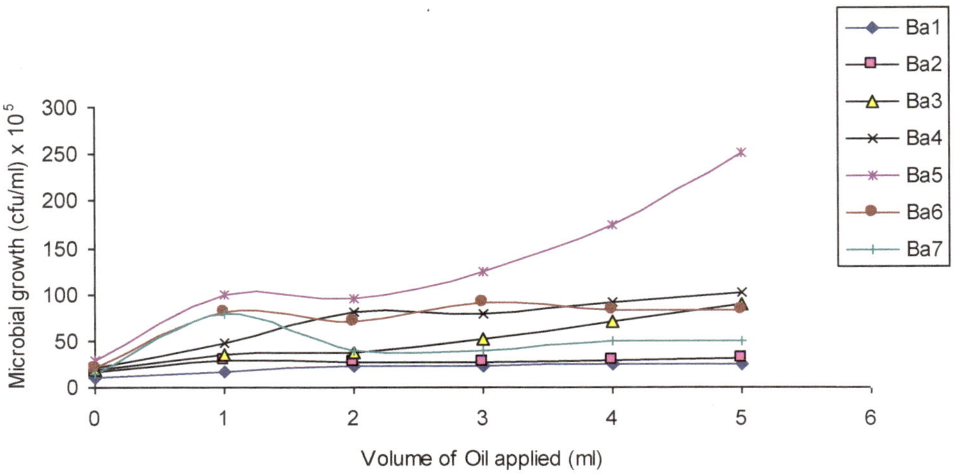

Figure 4. Microbial growth vs volume of oil applied for sandy loamy soil (bacteria sp.).

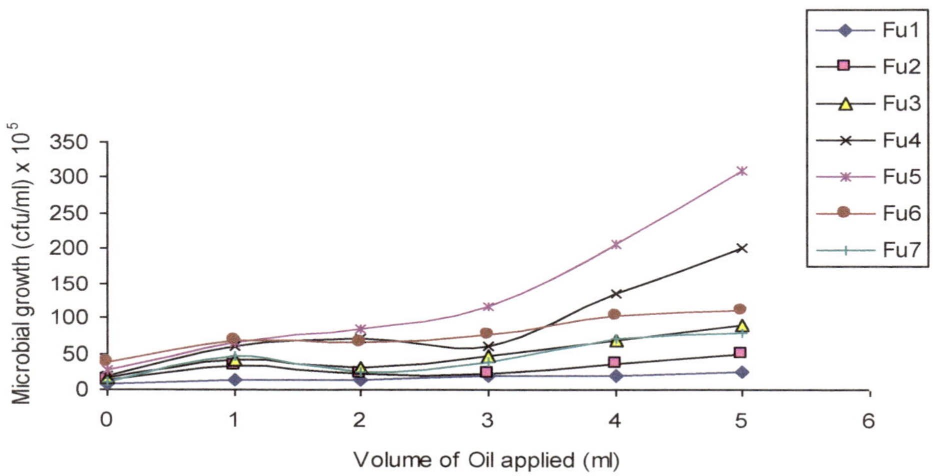

Figure 5. Microbial growth vs volume of oil applied for sandy loamy soil (fungi sp.).

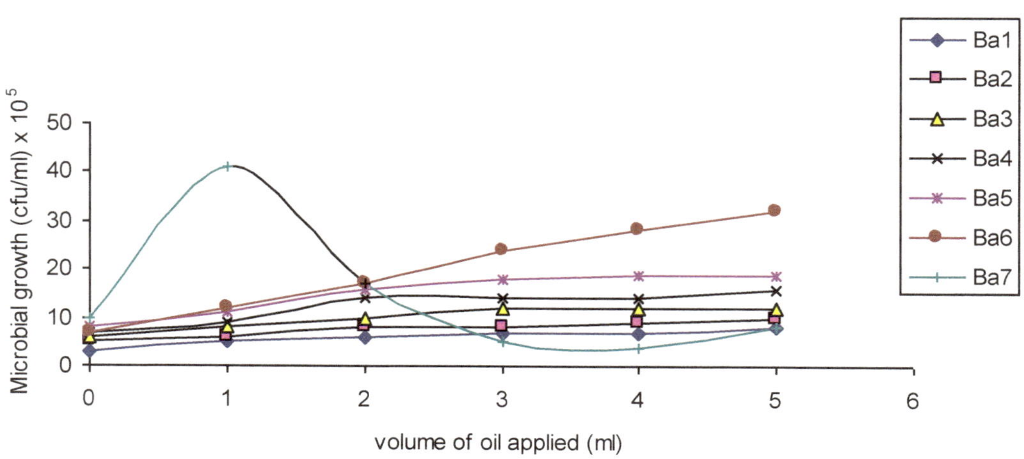

Figure 6. Microbial growth vs volume of oil applied for swampy soil (bacteria sp.).

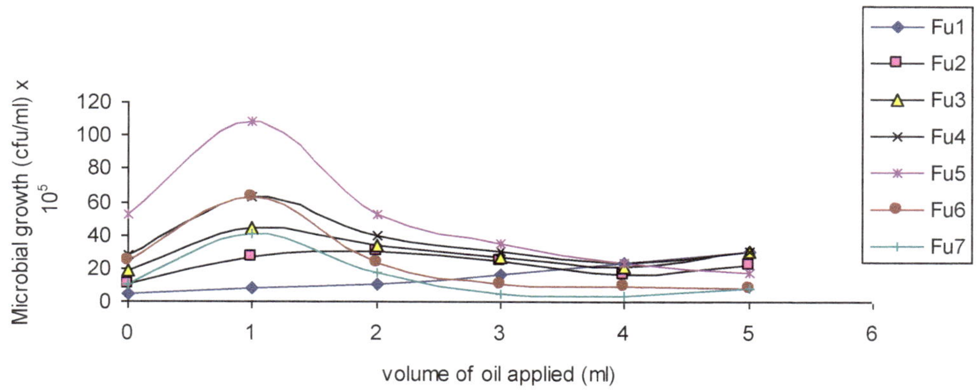

Figure 7. Microbial growth vs volume of oil applied for swampy soil (fungi sp.).

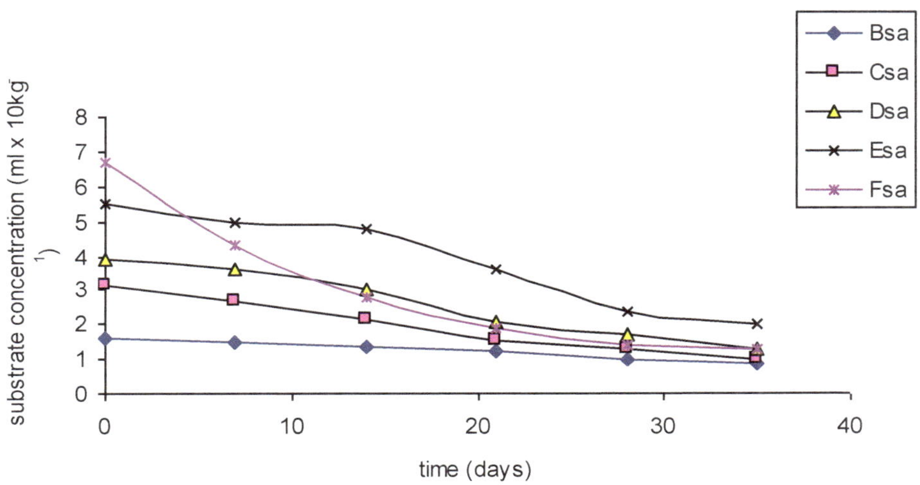

Figure 8. Substrate concentration vs time for sandy loamy soil.

1.45 < 1.92 < 1.98); for 2 ml, $S_{sa} < S_{sw} < S_d$ (2.64 < 3.60 < 3.966); for 3 ml, $S_{sa} = S_{sw} < S_{cL}$ (3.60 = 3.60 < 3.96); for 4 ml, $S_{sa} < S_{sw} < S_{cL}$ (5.00 < 5.08 < 5.70) and 5 ml, $S_{Sq} < S_{cL} < S_{sw}$ (4.32 < 5.46 < 5.76) at 7 days observation. This could be attributed to the physicochemical properties of the sample such as, pH (sandy loamy = 6.9, swampy soil = 9.7, and clay soil = 7.1) and electrical conductivity (sandy loamy = 120 × 10^{-6} mol/cm, swampy soil = 498 × 10^{-6}mol/cm and clay soil = 260 × 10^{-6} mol/cm).

Effect of time on the substrate concentration

Figures 8, 9 and 10 show the variation of the substrate concentration with the time for 1, 2, 3, 4 and 5 ml × 10 kg^{-1}. On the different soil samples, the biodegradation pattern of this reaction is slightly different as most of the substrate concentrations increases. The rate of change in substrate concentration varies with time and is different for different volume of crude oil applied. These results were used for determining the substrate concentration gradient. The investigation result shows that the values for concentration gradient depends on initial substrate concentration, composition of the crude oil, spreading rate, environmental temperature and temperature of the system, mechanical weathering which includes the rate of diffusion of the crude oil, evaporation of the lighter components (gases, and CO_2), microbial activity in the system and absence of microbial inhibitors.

The results of the theoretical model show a clear behaviour similar with those of the experimental values. The behaviour suggests that there is a difference in the substrate concentration gradient, as evaluated by the theoretic model.

Table 6 illustrates the substrate concentration at zero pollution, 1, 2, 3, 4 and 5 ml pollution for sandy loam, swampy and clay environment. Decrease in substrate concentration was experienced with increase in time for

Table 6. Experimentally determining the substrate concentration for the different types of soil.

Time (Day)	Substrate concentration (ml x 10kg^{-1})																	
	Sandy loam						Swampy						Clay					
	A_{sa}	B_{sa}	C_{sa}	D_{sa}	E_{sa}	F_{sa}	A_{sw}	B_{sw}	C_{sw}	D_{sw}	E_{sw}	F_{sw}	A_c	B_{cl}	C_{cl}	D_{cl}	E_{cl}	F_{cl}
0	Tr	1.58	3.16	3.90	5.50	6.72	Tr	2.06	3.65	4.81	5.10	6.10	Tr	1.96	4.16	4.83	5.77	5.96
7	Tr	1.46	2.64	3.60	5.00	4.32	Tr	1.98	3.60	4.71	5.08	5.76	Tr	1.92	3.96	4.50	5.70	5.46
14	Tr	1.36	2.16	3.00	4.80	2.80	Tr	1.92	3.54	4.53	5.00	5.12	Tr	1.86	3.60	4.42	4.90	4.63
21	Tr	1.26	1.56	2.10	3.60	1.92	Tr	1.86	3.48	4.43	4.96	5.00	Tr	1.52	3.48	3.92	4.50	3.93
28	Tr	1.02	1.32	1.72	2.40	1.40	Tr	1.82	3.38	4.09	4.92	4.86	Tr	1.46	3.40	3.73	4.00	3.55
35	Tr	0.90	1.00	1.31	2.00	1.31	Tr	1.79	3.27	4.03	4.80	4.73	Tr	1.38	3.30	3.70	3.93	2.97

Table 7. Computation of 1/S and 1/V for sandy loamy, clay, and swamp.

Time	Sandy loamy Substrate concentration (ml x 10kg^{-1})					Specific rate					Reciprocal substrate					Reciprocal of specific rate				
	S_{Fsa}	S_{Bsa}	S_{Csa}	S_{Dsa}	S_{Esa}	V_{Fsa}	V_{Bsa}	V_{Csa}	V_{Dsa}	V_{Esa}	$1/S_{Fsa}$	$1/S_{Bsa}$	$1/S_{Csa}$	$1/S_{Dsa}$	$1/S_{Esa}$	$1/V_{Fsa}$	$1/V_{Bsa}$	$1/V_{Csa}$	$1/V_{Dsa}$	$1/V_{Esa}$
0	6.72	1.58	3.16	3.90	5.50	0.34	0.02	0.07	0.04	0.07	0.15	0.63	0.32	0.26	0.18	-	-	-	-	-
7	4.32	1.46	2.64	3.60	5.00	0.22	0.01	0.07	0.09	0.03	0.23	0.68	0.38	0.28	0.20	2.94	50.00	14.29	25.00	14.29
14	2.80	1.36	2.16	3.00	4.80	0.13	0.01	0.09	0.13	0.17	0.36	0.74	0.46	0.33	0.21	4.55	100.00	14.29	11.11	33.33
21	1.92	1.26	1.56	2.10	3.60	0.07	0.03	0.03	0.05	0.17	0.52	0.79	0.64	0.48	0.28	7.69	100.00	11.11	7.69	5.88
28	1.40	1.02	1.32	1.72	2.40	0.01	0.02	0.05	0.06	0.06	0.71	0.98	0.76	0.58	0.42	14.29	33.33	33.33	20.00	5.88
35	1.31	0.90	1.00	1.31	2.00						0.76	1.11	1.00	0.76	0.50	100	50.00	20.00	16.67	16.67

CLAY SOIL

	S_{Fsa}	S_{Bsa}	S_{Csa}	S_{Dsa}	S_{Esa}	V_{Fsa}	V_{Bsa}	V_{Csa}	V_{Dsa}	V_{Esa}	$1/S_{Fsa}$	$1/S_{Bsa}$	$1/S_{Csa}$	$1/S_{Dsa}$	$1/S_{Esa}$	$1/V_{Fsa}$	$1/V_{Bsa}$	$1/V_{Csa}$	$1/V_{Dsa}$	$1/V_{Esa}$
0	5.96	1.96	4.16	4.83	5.77	-	-	-	-	-	0.17	0.51	0.24	0.21	0.17	-	-	-	-	-
7	5.46	1.92	3.96	4.50	5.70	0.07	0.01	0.03	0.05	0.01	0.18	0.52	0.25	0.22	0.18	14.27	100.00	33.33	20.00	100.00
14	4.63	1.86	3.60	4.42	4.90	0.12	0.01	0.05	0.01	0.11	0.22	0.54	0.28	0.23	0.20	8.33	100.00	20.00	100.00	9.09
21	3.93	1.52	3.48	3.92	4.50	0.10	0.05	0.02	0.07	0.06	0.25	0.66	0.29	0.26	0.22	10.00	20.00	50.00	14.29	16.67
28	3.55	1.46	3.40	3.73	4.00	0.05	0.01	0.01	0.03	0.07	0.28	0.68	0.29	0.27	0.25	20.00	100.00	100.00	33.33	14.29
35	2.97	1.38	3.30	3.70	3.93	0.08	0.01	.01	0.01	0.01	0.34	0.72	0.30	0.27	0.25	12.5	100.00	100.00	100.00	100.00

SWAMP

	S_{Fsa}	S_{Bsa}	S_{Csa}	S_{Dsa}	S_{Esa}	V_{Fsa}	V_{Bsa}	V_{Csa}	V_{Dsa}	V_{Esa}	$1/S_{Fsa}$	$1/S_{Bsa}$	$1/S_{Csa}$	$1/S_{Dsa}$	$1/S_{Esa}$	$1/V_{Fsa}$	$1/V_{Bsa}$	$1/V_{Csa}$	$1/V_{Dsa}$	$1/V_{Esa}$
0	6.10	2.06	3.65	4.81	5.10	-	-	-	-	-	0.16	0.49	0.27	0.21	0.20	-	-	-	-	-
7	5.76	1.98	3.60	4.71	5.08	0.05	0.01	0.01	0.01	0.01	0.17	0.51	0.28	0.21	0.20	20.0	100.00	100.00	100.00	100.00
14	5.12	1.92	3.54	4.53	5.00	0.09	0.01	0.01	0.03	0.01	0.20	0.52	0.28	0.22	0.20	11.11	100.00	100.00	33.33	100.00
21	5.00	1.86	3.48	4.43	4.96	0.02	0.01	0.01	0.01	0.01	0.20	0.53	0.29	0.23	0.20	50.00	100.00	100.00	100.00	100.00
28	4.80	1.82	3.38	4.09	4.92	0.03	0.01	0.01	0.05	0.01	0.21	0.55	0.30	0.24	0.20	33.33	100.00	100.00	20.00	100.00
35	4.73	1.79	3.27	4.03	4.80	0.01	0.01	0.02	0.01	0.02	0.21	0.56	0.31	0.25	0.20	100.00	100.00	50.00	100.00	50.00

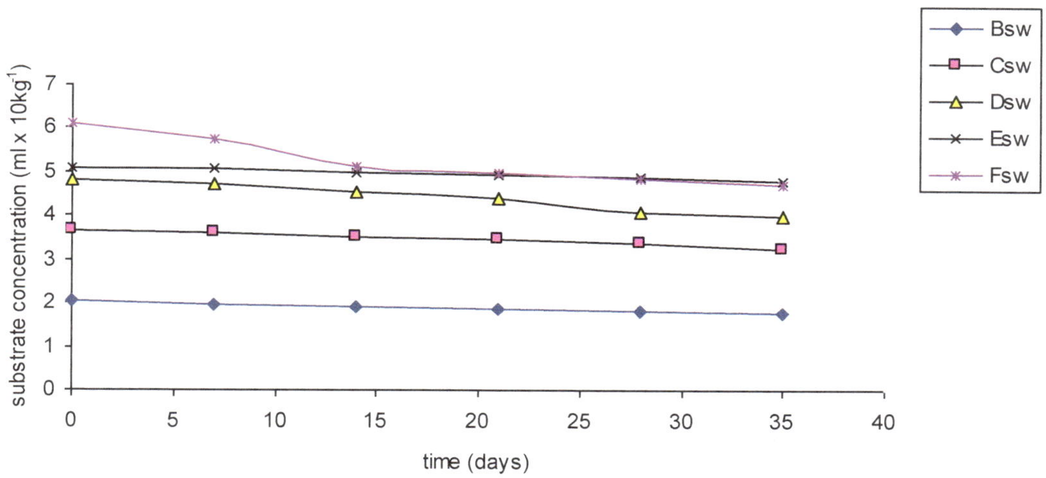

Figure 9. Substrate concentration vs time for swampy soil.

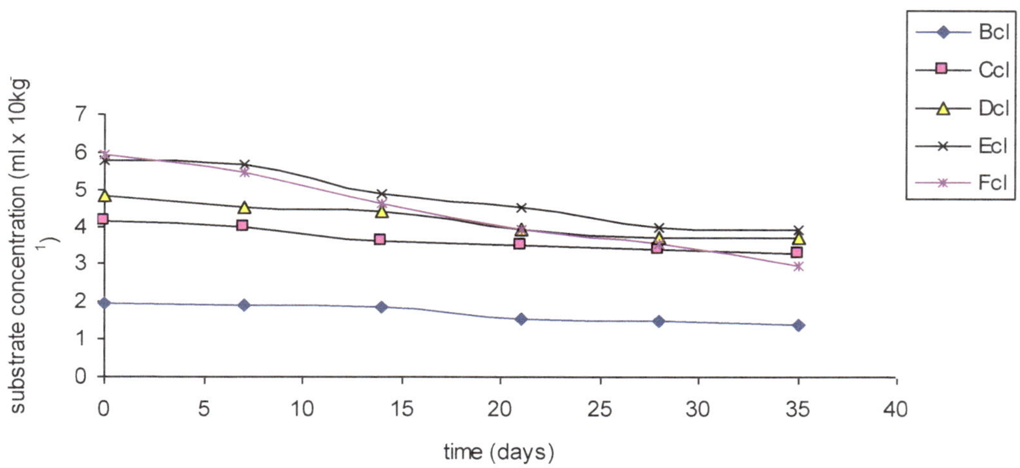

Figure 10. Substrate concentration vs time for clay soil.

1, 2, 3, 4 and 5 ml in the different soil environment.

Specific rate (Vor μ) maximum specific rate and (V_{max} or μ_{max}) and dissociation rate constant (K_m or K_s)

The specific rate of degradation was determined only for sandy loam, clay and swampy soil polluted with crude oil. The values of the specific rate for each of the experimental condition were obtained using the procedures outlined in this research investigation, the specific rates were determined by plotting substrate concentration against time (slope of the curve) for sandy loamy, clay and swampy soil mixtures (Nml × 10/kg, where N = 1,2,3,4 and 5).

The values of the maximum specific rate of substrate concentration and Michael's Menten constant is obtained by rearranging the equation obtained. Line-Weaver Bulk plot for the various experimental conditions, the results of the theoretical computed values of the reciprocal of the various values of the specific rate of substrate concentration are presented in Figures 8, 9 and 10. The values of maximum specific rate of substrate concentration and Michael's Menten constants were computed from the plots in Figures 11 and 12 using the equation $\left(\frac{K_D}{V_{max}} = slope\right)$

Conclusion

Models for the simulation of specific growth rate of bacteria and fungi species for 1, 2, 3, 4 and 5 ml × 10 kg^{-1} of different soil samples and degradation kinetic

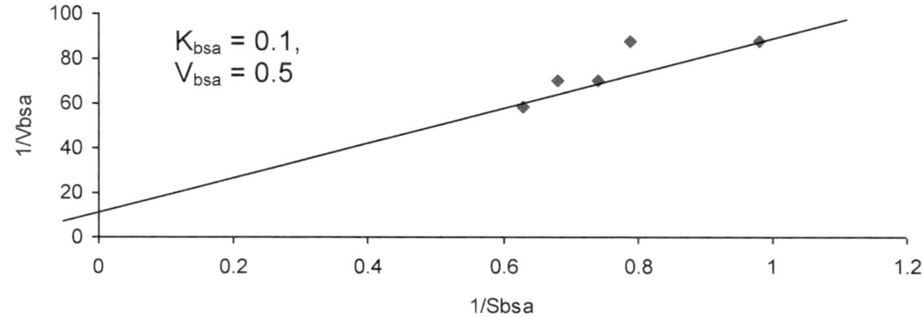

Figure 11. 1/Vbsa vs 1/Sbsa for C_8 degredation on sandy loamy soil.

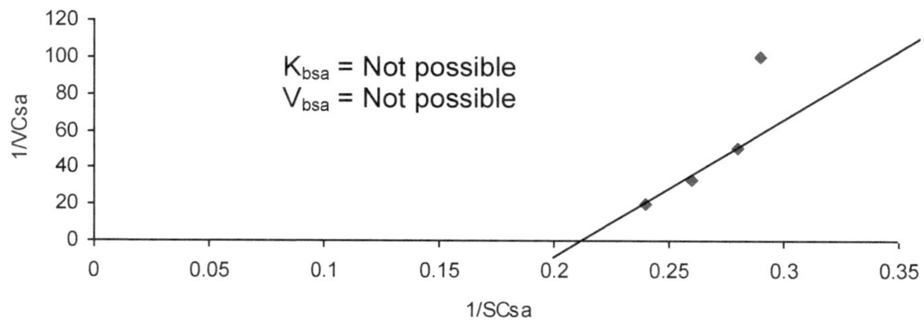

Figure 12. 1/VCsa vs 1/Scsa for C_8 degredation on clay soil.

equations developed and equation for the determination of heat generated during the enzyme – catalysed reaction has been proposed. They are based on the description of the simultaneous microbial growth rate and substrate utilization inside the reactor, on the characterization of the microbes and the substrate concentration.

The proposed models were assessed using data obtained from the experimental investigation. In the present experimental study, investigation into the kinetics of biodegradation of Nigeria crude oil in different soil was examined. The following results were obtained:

1) For the microbial growth rate of the species (bacteria and fungi) was sandy loamy > day > swamp soil.
2) The degradation rate for the process to enhanced environmental clean up was faster in the following order; sandy loamy > clay > swamp soil in bioremediation process.
3) Substrate concentration decreases with increase in microbial activities and time.
4) For the three soil sample and two microbial species, models were developed to predict the microbial growth, substrate concentration specific growth rate, specific rate of degradation, maximum specific growth rate, maximum specific rate of degradation, model for carbon dioxide, water, biomass and gas production.
5) Effect of volume increase of crude oil application on the various soil samples. The developed models were found useful in monitoring and predicting the biodegradation of Nigeria crude oil in different soil.

The bio-production of carbon dioxide and heat evolution on the kinetics of biodegradation of crude oil in different soil samples will appear in the next paper.

NOMENCLATURE

S_{Bsa} = substrate concentration of sandy loamy soil for 1 ml x 10/kg
S_{Dsa} = substrate concentration of sandy loamy soil for 2 ml x 10/kg
S_{Esa} = substrate concentration of sandy loamy soil for 3 ml x 10/kg
S_{Fsa} = substrate concentration of sandy loamy soil for 4 ml x 10/kg
V_{Rsa} = specific rate of sandy loamy soil degradation for 1 ml x 10/kg per week
V_{Csa} = specific rate of sandy loamy soil degradation for 2 ml x 10/kg per week
V_{Dsa} = specific rate of sandy loamy soil degradation for 3 ml x 10/kg per week

V_{Esa} = specific rate of sandy loamy soil degradation for 4 ml x 10/kg per week
V_{Fsa} = specific rate of sandy loamy soil degradation for 5 ml x 10/kg per week
A = substrate concentration (ml/g)
E = enzyme concentration (cfu/g) per week
P = product concentration (ml/g)
Q_{gr} = quantity of heat generated (J/kg.k)
S = substrate concentration (ml/g)
K = kinetic rate constant (dimensionless)
C_{EO} = initial concentration of the enzyme (cfu/g)
C_A = final concentration of the substrate (ml/g)
γY_A = yield factors
$C_{biomass}$ = biomass concentration (cfu/g) per week
$V_{reactor}$ = volume of reactor (m^3)
C_{Co2} = carbon oxide concentration (ml/g)
C_{H2o} = water concentration (ml/g)
$V_{max(H2o)}$ = maximum specific rate of carbon dioxide produced (ml/g)
$V_{max(H2o)}$ = maximum specific rate of water produced (ml/g)
$K_{biomass}$ = equilibrium constant of biomass (dimensionless)
K_{Co2} = equilibrium constant of carbon dioxide (dimensionless)
K_{Ho2} = equilibrium constant of water (dimensionless)
t = time per week
X = fractional conversion (%)
K_S = equilibrium constant for substrate (dimensionless)
K_m = equilibrium constant of carbon dioxide (dimensionless)
V = specific rate of enzyme catalysed reaction (kg/ml)

Greek symbols

ϕ = proportionality constant
ϕ_β = proportionality constant of bacteria
ϕ_F = proportionality constant of fungi
μ_0 = initial inoculate of microorganism (cfu/g)
μ = final biomass concentration (cfu/g)
μ_S = biomass concentration at steady state condition (cfu/g)
μ_t = biomass concentration with respect time (cfu/g) per week
β = constant (dimensionless)
$\gamma_{biomass}$ =rate of production of biomass (cfu/g) per week
γ_{CO2} = rate of production of carbon dioxide (ml/g) per week
γ_{H2O} = rate of production of water (ml/g) per week
γ_{QHeat} = rate of production of heat (J/g.K) per week
γ_{gases} = rate of production of various gases (ml/g) per week
$\mu_{max\,(biomass)}$ = maximum specific rate of biomass (cfu/g) per week

Subscripts

B = Bacteria
F = Fungi
t = time
S = laplace sign
Sa = sandy loamy soil
Cl = clay soil
Sw = swampy soil

REFERENCE

Alkinson B, Marintuna J (1983). Biochemical Engineering and Biotechnology Hand Book, the Nature press, pp. 205-207.
Antal SP, Mgbomo H (1973). Distribution of hydrocarbon utilizing bacteria in oil spill area, Lewis Publishers, Boca Ratan, pp. 368-371.
Antunes-do C, Costa M (2000). Modelling of oil spill evolution in estuaries and coastal region. In proceeding of 2nd International Conference on oil spill, pp. 63-72.
Antunes do C, Pinbo JL,Vieira JP (2010). Oil spills in coastal zones: predicting slick transport and Wealthering processes. Open ocean engineering J., 3: 129-142.
Arvin E, Jensen BK, Gundersen AT (1991). "Biodegradation Kinetics of Phenols in as aerobic biofilm at low concentrations. J. Water Sci. Technol., 23(7-9):1375-1384.
Atlas RM (1981). Microbial Degradation of Petroleum Hydrocarbon, J. Environ. Perspective Microbiol., 45(1):180-209.
AWWA C209-00 (2000). Cold-applied tape coating for the exterior of special section, connections and fittings for steel water pipelines, Awwa Denver Co. p. 88.
Barelay CD, Moore, DM. Lander SR, Legge RL (1990). Heat – denaturation Kinetics of Lignin Perioxidases from phanerochaete chrysosporium, J. Biochem. Eng. Group, 23(2):16-23.
Bashir B, Cseh T, Leduc R, Samson R (1990). Effect of soil/contaminant interaction on the biodegradation of naphthalene in flooded soil under denitrifying conditions, Appl. Microbiol. Biotechnol., 34(3): 414-419.
Brekke C, Solberg AHS (2005). Oil spill detection by satellite remote sensing. Remote sensing Environ., 95(1): 1-13.
Dasappa SM, Loehr RC (1991). Toxicity reduction in contaminated soil bioremediation processes, J. Water Res. Oxford, 25(9): 1121 – 1130.
Deweerd KA, Concannon F, Suflita JM (1991). Reduction of sulfur oxyamions by Desulfomonile tiedjei, Appl. Environ. Microbiol., 57(7): 1929-1934.
Fingas M (2000). Use of surfactants for environmental applications. In: Surfactants fundamentals and applications in the petroleum industry Schramm L, I, ed. Cambridge University Press, Cambridge U.K. pp. 461-539.
Folsom BR, Chapman PJ, Pritchard PH (1990). Phenol an trichloroethylene degradation by pseudomonas cepacia C4, kinetics and interactions between substrates, J. Appl. Environ. Microbial., 56(5): 279-1285.
Gandy AF, Gandy ET (1988). Element of Bio-environmental engines press Inc. Sam Jose Califonia, pp. 78-79.
Goldsmith CDJ, Balderson PJ (1989). Biokinetic constants of a fixed microbial culture with model diesel fuel. J. Hazard. Waste Hazard Mater., 6(2): 145-154.
Guan SW (2001). 100% Solid polyurethare coating technology for corrosion protection in water and waste water system, the 9th middle east corrosion conference Bahraim, 9(12): 72.
Guo WJ, Wang YX (2009). A numberical oil spill model based on a hybrid method. Marine pollution bulletin, 58(5): 726-734.
Irwin HS (1975). Enzyme Kinetics behavior and analysis of rapid equilibrium and steady state enzyme system, Interscience publisher, pp. 205-251.

James EB, David FO (1977). Biochemical Engineering fundamentals, 2nd edition, Kingsport press in Publishers, pp.10-780.

Jha MN, Levy J, Gao Y (2008). Advances in remote sensing for oil spill disaster management; state of the Art sensors technology for oil spill surveillance. Sensors, 8, pp. 230-255.

John TC Jr. (1999). Bioremediation Engineering design and application. McGraw-Hill Inc. 202-214.

Kakporbia T (2001). Cathodic protection survey techniques, a training manual by pipeline integrity Nigeria Limited plot 25, Trans-Amadi Industrial layout, Port Harcourt, p.18.

Kistner A, Kornelius JH (1990). Studies of plant fibre degradation by anaerobic bacteria, J. Microbiol. Methods, 12(3-4): 173-182.

Lee DG, Bonner JS, Garton IS, Ernest ANS, Autenrieth RL (2002). Modelling coagulation kinetics incorporating fractual theories comparison with observed data. Water Research, 36(4):1056-1066.

Lessard RR, Demarco G (2000). The Significance of oil spill dispersants spill sceicne and Technology Bulltin, 6(1): 59-68.

Lodaya M, Lakhwala F, Rus E, Manjari S, Lewandowski G, Sofer S, Singh V (1991). Biodegradation of benzene and a BTX mixture using immobilized activated sludge J. Environ. Sci. Health, part as Environ. Sci. Eng., 26(1):121-137.

Milkin PGT, Steart JE (1974). Effects of Environmental Parameter on bacterial degradation of bulk oil, crude oil and hydrocarbon application, J. Microbial., 6(2): 915-722.

Miller ME, Alexander M (1991). Kinetics of bacterial degradation of benzylamine in a montmorillinite suspension, J. Environ. Sci. Technol., 25(2): 250-245.

Ogoni HA, Gumus RH (2001). Development of kinetic model of product inhibition in Biodegradation of Petroleum hydrocarbon, Afr. J. Environ.Studies, 2(1): 431-437.

Ogoni HA (2002). Topics in Biochemical Engineering, 1st ed. Pearl Publishers, 17-59.

Page CA, Bonner JS, Sumner OL, McDonald TJ, Autenrieth RL, Fuller CB (2000). Behaviour of a chemically-dispersed oil and a whole oil in a near-shore environment. Water Res., 34(9): 2507-2516.

Parsons JR, Govers HAJ (1990). Quantitative Structure-activity relationships for biodegradation, Ecotoxicol. Environ. Safety, 19(2): 212-227.

Phelps JJ, Niedzielski JJ, Schram RM, Herbes SE, White DC (1990). Biodegradation of trichloroethylene in continuous-recycle expanded-bed bioreactors, Appl. Environ. Microbiol., 56(6): 1702-1709.

Ronald LD (1992). Theory and practice of water and wastewater treatment, John Wiley and Sons, Inc. New York, pp. 87-146.

Scbastiao P, Guedes SC (2006). Uncertainty in production of oil spill trajectories in a coastal zone, J. Marine syst., 63(3-4):257-269.

Schmidt SK, Gier MJ (1990). Coexisting bacterial populations responsible for multiphasic mineralization kinetics in soil, Appl. Environ. Microbiol., 56(9):2692-2697.

Schnell S, Brune A, Shink B (1991). Degradation of hydroxyhydroquinone by the strictly anaerobic fermenting bacterium pelobacter massiliensis sp. Nov., J. Archives Microbiol., 155(5): 511-516.

Scow KM, Merica RR, Alexander M (1990). Kinetic analysis of enhanced biodegradation of carbofuran, J. Agric. Food Chem., 38(3): 908-912.

Sterling MC, Ojo Jr T, Autenrieth RL, Bonner JS, Page CA, Ernest ANS (2002). Coalescence Kinetics of Dispersed crude oil in a laboratory reactor, From proceedings of the twenty-fifty Artice and marine oilspill program (AMOP June 11-14) Technical Seminar: Environment Canada, Ottawa, Canada, 2: 741-753.

Ukpaka CP, Ogoni HA, Amadi SA, Adebayo TA (2005). Mathematical modeling of the microbial growth and decay rate of pseudomonas species on Biodegradation of Bonny light crude oil, Global J. Pure Appl. Sci, 11(3): 423-431.

Ukpaka CP (2009). Development of mathamtical correlative model equation for the microbial growth in biodegradation of Benzylehloride in a CSTR. Knoweldge Rev., A multi-disciplinary J, 19(2): 86-98.

Ukpaka CP (2010). Evaluation of Biokinetic model for the prediction of biochemcail oxygen demand in a pond system for wet season degradation of petroleum hydrocarbon. Int. J. Phama world Res., 2(2): 1-26.

Ukpaka CP (2011). The effect of some physicochemical parameters in aromatic compound degradation in salt water medium ICASTOR J. Eng., 4(2):177-200.

Ukpaka CP (2011a). Biodegradation model on effect of some physicochemical parameters on aromatic compounds in fresh water medium, J. Bacteriol. Res., 3(3): 42-55.

Ukpaka CP, Davis DD (2010). Studying the biodegrdation of petroleum hydrocarbon in soil using *pseudomonas sp.* in Niger Delta Area of Nigeria. Multidisciplinary J. Acad.Excellence, 1(2): 1-15.

Ukpaka CP, Ogoni HA, Odharo J (2010). Evaluation of temperature impact on the characteristics of sewage sludge for Agro—soil/water application. Global Res. Eng. Technol., 3(2): 59-588.

Wami EN, Ogoni HA (1997). Kinetic model fro Biodegradation of Petroleum hydrocarbon mixture Bony light crude oil without activator. J. Nig.Society Chemical Engr., 16(2):72-78.

Zheng L, Yapa PD (2000). Buoyant velocity of spherical and nonspherical bubbles/droplets. J. Hydraulic Eng. – ASCE, 126(11): 852-854.

Numerical study of heat transfer in pulsating turbulent air flow

E. A. M. Elshafei, M. Safwat Mohamed, H. Mansour and M. Sakr*

Mechanical Power Department, Faculty of Engineering, Mansoura University, El-Mansoura 35516, Egypt.

A numerical investigation of heat transfer characteristics of pulsating turbulent flow in a circular tube is carried out. The flow is thermally and hydrodynamically fully developed and the tube wall is subjected to a uniform heat flux. The flow inlet to the pipe consists of fixed component and pulsating component that varies sinusoidally with time. The flow and temperature fields are computed numerically using computational fluid dynamics (CFD) Fluent code. Prediction of heat transfer characteristics is performed over a range of $10^4 \leq Re \leq 4\times10^4$ and $0 \leq f \leq 70$ Hz are observed. Results showed little reduction in the mean time-averaged Nusselt number with respect to that of steady flow. However, in the fully developed established region, the local Nusselt number either increases or decreases over the steady flow-values depending on the frequency parameter. These noticed deviations are rather small in magnitude for the computed parameter ranges. The characteristics of heat transfer are qualitatively consistent with the available experimental and numerical predictions.

Key words: Nusselt number, heat transfer, computational fluid dynamics (CFD), pulsating flow in pipe.

INTRODUCTION

The study of pulsating flow in pipe has been a subject of interest among many researches. The problem is important in biological applications in relation to blood flow and in industrial applications related to heat exchange efficiency, cavitation in hydraulic pipe lines, pressure surges and reciprocating machines. In general, pulsating flow filed is assumed to consist of a steady poiseuille flow part and purely oscillating part. Many experimental, analytical and numerical works studied the heat transfer in pulsating laminar flow but little in turbulent flow. One of the key issues concerning pulsating convection heat transfer in tubes is whether a superposed flow pulsation effects on heat transfers in the original steady flow. The answer to this question in the previous studies can be classified into four different opinions: (1) flow pulsation enhances heat transfer (Zheng et al., 2004; Xuefeng and Nengli, 2005; Faghri et al., 1979); (2) it deteriorates heat transfer (Mostafa, 2005a, b; Hemeada et al., 2002; Guo and Sung, 1997); (3) it has no effect on heat transfer (Chattopadhyay et al., 2006); and (4) it either enhances or deteriorates heat transfer, depending on flow parameters (Al-Haddad and Al-Binally, 1989; Martineelli et al., 1943; Cho and Hyun 1990; Gbadebo et al., 1999; Gupa et al., 1982; Moschandreou and Zamir, 1997; Said et al., 2003; Scotti and Piomelli, 2001; Zhao and Cheng, 1998; Zohir et al., 2005).

Numerical study presented by (Xuefeng and Nengli, 2005) showed that in a pulsating turbulent flow there is an optimum Womersley number at which heat transfer is maximally enhanced. Experimental studies by (Al-Haddad and Al-Binally, 1989; Habib et al., 1999; Barid et al., 1996; Habbib et al., 2002; Liao and Wang, 1985; Martineelli et al., 1943) showed that increment and reductions in mean Nusselt number depends on pulsation frequency, turbulent bursting frequency, amplitude, axial location, Reynolds number and Prandtl number. In order to understand these phenomena and to resolve the contradictory results, different models of turbulence for pulsating flows were considered. Two of these models are well known, the quasi-steady flow model (Shemer, 1985) and the bursting model discussed by (Liao and Wang, 1985; Genin et al., 1992; Havemann and Rao, 1954; Martineelli et al., 1943).

Due to its complexity, studies on heat transfer in

*Corresponding author. E-mail: moh_saker1981@yahoo.com or moh_saker@mans.edu.eg.

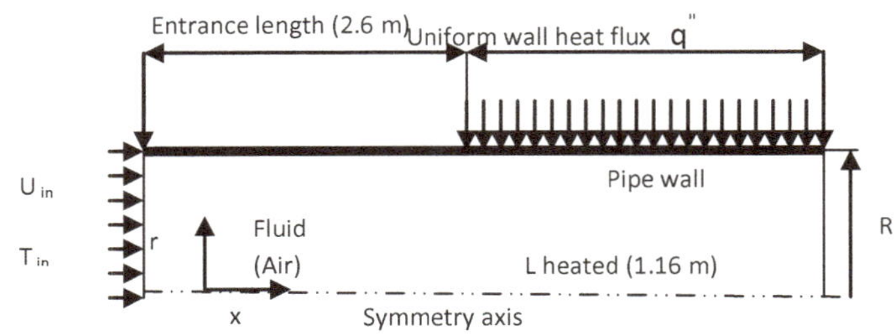

Figure 1. Schematic diagram for the studied problem.

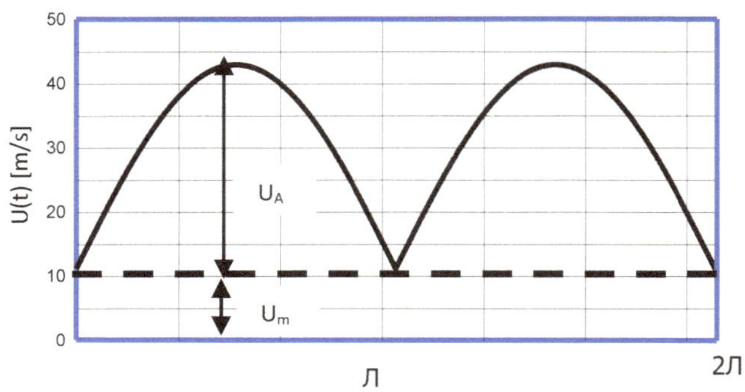

Figure 2. Variations of inlet velocity.

work aims to numerically and experimentally investigate the characteristics of heat transfer in pulsating pipe flow and how it is affected by pulsation frequency as well as the value of Reynolds number.

In the present study, a pipe subjected to a constant heat flux from outside is considered. The governing equations are solved with the aid of CFD code (Fluent 6·1, 2003). The obtained numerical results can be served as a useful source material against which the outcome of future analytical and experimental investigations can be checked.

MATHEMATICAL FORMULATIONS

Pipe geometry

In the present study, the numerical solution of the pulsating turbulent flow through an externally heated pipe is analyzed. Air is selected as a working fluid. The radius and length of the pipe are R and L respectively. The thickness of the pipe is neglected and the thermal boundary condition on the pipe wall is assumed to be a uniform heat flux. The two-dimensional axisymmetric model of the considered pipe is shown in Figure 1.

The pipe geometry at the inlet is long enough to assure a hydrodynamically fully developed flow at the heated section and simulated to the experiential facility of Huseyin et al. (2005). Since the pipe cross-section is circular, it is assumed that the flow is axisymmetric. In cylindrical polar coordinates, this means that the flow variables depend only on the axial coordinate x and radial coordinate r. The pipe wall section is modeled as: unheated section of 2.6 m length and a heated one of 1.16 m long. Sinusoidal pulsating flow is assumed to be entering to the pipe of fixed period (the velocity is only oscillates) as can be seen in Figure 2.

The governing conservation equations

The governing equations for transient turbulent incompressible flow and heat transfer in the flow region ($0 \le x \le L$ and $0 \le r \le R$) can be written as follows (Huseyin et al., 2005):

Continuity:

$$\frac{\partial u}{\partial x}+\frac{v}{r}+\frac{\partial v}{\partial r}=0 \tag{1}$$

Momentum in x-direction:

$$\begin{aligned}&\rho\left[\frac{\partial u}{\partial t}+\frac{1}{r}\frac{\partial}{\partial x}(ruu)+\frac{1}{r}\frac{\partial}{\partial r}(rvu)\right]\\&=-\frac{\partial p}{\partial x}+\frac{1}{r}\frac{\partial}{\partial x}\left\{r\mu\left[2\frac{\partial u}{\partial x}-\frac{2}{3}\left(\frac{\partial u}{\partial x}+\frac{\partial v}{\partial r}+\frac{v}{r}\right)\right]\right\}\\&+\frac{1}{r}\frac{\partial}{\partial r}\left[r\mu\left(\frac{\partial u}{\partial r}+\frac{\partial v}{\partial x}\right)\right].\end{aligned} \tag{2a}$$

Momentum in r-direction:

$$\begin{aligned}&\rho\left[\frac{\partial v}{\partial t}+\frac{1}{r}\frac{\partial}{\partial x}(ruv)+\frac{1}{r}\frac{\partial}{\partial r}(rvv)\right]\\&=-\frac{\partial P}{\partial r}+\frac{1}{r}\frac{\partial}{\partial r}\left[r\mu\left(\frac{\partial v}{\partial x}+\frac{\partial u}{\partial r}\right)\right]\\&+\frac{1}{r}\frac{\partial}{\partial r}\left\{r\mu\left[2\frac{\partial v}{\partial r}-\frac{2}{3}\left(\frac{\partial u}{\partial x}+\frac{\partial v}{\partial r}+\frac{v}{r}\right)\right]\right\}\\&-2\mu\frac{v}{r^2}+\frac{2}{3}\frac{\mu}{r}\left(\frac{\partial u}{\partial x}+\frac{\partial v}{\partial r}+\frac{v}{r}\right).\end{aligned} \tag{2b}$$

Energy:

$$\begin{aligned}&\rho Cp\left(\frac{\partial T}{\partial t}+u\frac{\partial T}{\partial x}+v\frac{\partial T}{\partial r}\right)\\&=\lambda_{eff}\left[\frac{\partial^2 T}{\partial x^2}+\frac{1}{r}\frac{\partial}{\partial r}\left(r\frac{\partial T}{\partial r}\right)\right]+\mu\varphi\end{aligned} \tag{3a}$$

Where φ is viscous dissipation term, which is

$$\varphi = 2\left[\left(\frac{\partial u}{\partial x}\right)^2 + \left(\frac{\partial v}{\partial r}\right)^2 + \left(\frac{v}{r}\right)\right] + \left(\frac{\partial v}{\partial x} + \frac{\partial u}{\partial r}\right)^2, \quad (3b)$$

and λ_{eff} is the effective thermal conductivity, and for standard *k*-ε model, is given by

$$\lambda_{eff} = \lambda + \lambda_t \quad (3c)$$

$$\lambda_t = C_p\, \mu_t / Pr_t \quad (3d)$$

Turbulence model

Among the available turbulent models, those based on Eddy viscosity are widely applied to the turbulent kinetic energy – turbulent energy dissipation rate (*k*-ε). Two additional equations for the standard *k*-ε turbulence model: the turbulence kinetic energy, *k,* and the dissipation rate, *ε,* are determined using the following transport equations, respectively.

k – Equation:

$$\rho\frac{Dk}{Dt} = \frac{\partial}{\partial x}\left(\left(\mu + \frac{\mu_t}{\sigma_k}\right)\frac{\partial k}{\partial x}\right) + \frac{1}{r}\frac{\partial}{\partial r}\left(r\left(\mu + \frac{\mu_t}{\sigma_k}\right)\frac{\partial k}{\partial r}\right) + G - \rho\varepsilon \quad (4a)$$

Where

$$G = \mu_t\left(2\left(\frac{\partial u}{\partial x}\right)^2 + 2\left(\frac{\partial v}{\partial r}\right)^2 + 2\left(\frac{v}{r}\right)^2 + \left(\frac{\partial u}{\partial r} + \frac{\partial v}{\partial x}\right)^2\right) \quad (4b)$$

ε- equation:

$$\rho\frac{D\varepsilon}{Dt} = \frac{\partial}{\partial x}\left(\left(\mu + \frac{\mu_t}{\sigma_\varepsilon}\right)\frac{\partial \varepsilon}{\partial x}\right) + \frac{1}{r}\frac{\partial}{\partial r}\left(r\left(\mu + \frac{\mu_t}{\sigma_\varepsilon}\right)\frac{\partial \varepsilon}{\partial r}\right) + \frac{\varepsilon}{k}(C_{1\varepsilon}G - C_{2\varepsilon}\rho\varepsilon) \quad (5)$$

The turbulent viscosity, μ_t is computed by combining *K* - ε as follows:

$$\mu_t = \rho C_\mu (k^2/\varepsilon) \quad (6)$$

Where C_μ, $C_{1\varepsilon}$, and $C_{2\varepsilon}$ are the model constant; σ_k and σ_ε are turbulent Prandtl numbers for k and ε, respectively. These constants have the following default values (Launder and Spalding, 2006): σ_k =1.0, $C_{1\varepsilon}$= 1.44, $C_{2\varepsilon}$= 1.92, σ_ε =1.3, and C_μ= 0.09, respectively.

Heat transfer coefficient and Nusselt number

Due to the unique characteristics of the pulsating flow, several heat transfer coefficients have to be carefully defined.

The instantaneous local heat transfer coefficient is given by

$$h(x,t) = \frac{q}{T_w - T_{av}} \quad (7a)$$

and the local heat transfer coefficient, defined as the average of heat transfer coefficient over a cycle, is formulated as

$$h(x) = \frac{1}{2\pi}\int_0^{2\pi} h\, d(\omega t) \quad (7b)$$

The instantaneous heat transfer coefficient, defined as the average of heat transfer coefficient over the whole length of pipe at a given moment, is expressed as

$$h(t) = \frac{1}{L}\int_0^L h\,dx. \quad (7c)$$

Subsequently, several relevant Nusselt numbers are rationally introduced. The instantaneous local Nusselt number is

$$Nu(x,t) = \frac{hD}{K} \quad (7d)$$

The overall Nusselt number; Nu is given by

$$Nu = \frac{1}{2\pi L}\int_0^L\int_0^{2\pi} Nu_{x,t}\, d(\omega t)\,dx. \quad (7e)$$

To ensure sufficient temporal resolution of the numerical solution, five to seven cycles are required to obtain a steady periodic solution with an acceptable accuracy.

Boundary conditions

The solution domain of the considered 2D, axisymmetric pipe flow is geometrically quite simple, which is a rectangle on the *x-r* plane, enclosed by the inlet, outlet, symmetry and wall boundaries. On walls no-slip condition are assumed for the momentum equations. The inlet velocity values have been derived from given Reynolds numbers. Inaccuracies due to an uncertainty in the shape of the inlet velocity profiles are not expected to play an important role, since all calculations were carried out at places sufficiently downstream from the inlet, so that quite fully-developed conditions should be expected at the calculating section. For pulsating cases, the inlet velocity is varied with time. The outlet boundary condition is called "outflow", which implies zero-gradient condition at the outlet.

At the pipe inlet, (x=0):

For steady flow condition, U (t) = U_{in}, and for $\frac{\partial \phi}{\partial x} = 0$ sinusoidal pulsating flow, U (t) = $U_m + U_A \sin(\omega t)$

At the pipe exit plane, (x = L):

At the pipe axis, (r = 0): $\frac{\partial \phi}{\partial r} = 0$

At the pipe wall, (r = R): (u, v k, ε) = 0, a uniform heat flux is imposed, q" = q_o for heated section and q" = 0 for the adiabatic wall.

Values of k and ε are not known at the inlet, but, if they are not given by experimental data, some reasonable assumptions can be made. The kinetic energy of turbulence according to a certain value of the square of the average inlet velocity is expressed by Launder and Spalding (2006) as

$k = 0.03\ U_{in}^2$

In addition, dissipation is calculated according to the equation

$$\varepsilon = C_\mu \frac{k^{\frac{3}{2}}}{0.03R}$$

Simulation values

In the analysis, the following simulation values are taken as:

U_{in} = (12.7 to 42.8) m/s, U_m =11.07 m/s, U_A = (1.63 to 31.81) m/s, ω = (41.9 to 427.25) rad/s, and q = 922 W/m^2.

COMPUTATIONAL PROCEDURE

Calculation tools

Even the difficult general differential equations now yield to the approximating technique known as numerical analysis, whereby the derivates are simulated by algebraic relations between a finite numbers of grid points in the flow field which are then solved on a digital computer. A suitable CFD computer code is used to solve the governing equations numerically along with the boundary and the initial conditions. The CFD program is a process by which the fluid flow characteristics can be predicted through arbitrary geometries giving such information as: flow speed, pressure, residence time, flow patterns, etc. The Fluent 6·1 (Fluent, 2003) program is chosen as the CFD computer code for this work because of the ease with which the analysis model can be created, and because the software allows users to modify the code for special analysis conditions through the use of user subroutines. The Fluent computer code uses a finite-volume procedure to solve the governing equations of fluid flow in primitive variables such as velocity components in axial and radial directions and pressure. A variety of turbulence models is offered by the Fluent computer code. The standard *k*–ε model was used as a turbulence model in this study. This model is a semi-empirical one, based on model transport equations for the turbulent kinetic energy (*k*) and its dissipation rate (ε). The model transport equation for *k* is derived from the exact equation while the model transport equation for ε is obtained using physical reasoning and bears little resemblance to its mathematically exact counterpart. In the derivation of the *k*–ε model, it was assumed that the flow is fully turbulent, and the effects of molecular viscosity are negligible. The standard *k*–ε model is, therefore, valid only for fully turbulent flows. A detailed description of turbulence models and its applications to turbulence can be found in "Fluent (2003) (Launder and Spalding, 2006). In the case of the standard *k*–ε models, two additional transport equations, (for the turbulent kinetic energy and the turbulence dissipation rate) are solved, and turbulent viscosity, μ_t, is computed as a function of *k* and ε. The solution method for this study is axisymmetric. In order to define the pulsating inlet velocities in all cases, a UDF (User-Defined Function) file was introduced to the prepared fluent case file. By using the results of the calculations performed with the fluent code. This program, written in FORTRAN 77 language, calculates numerically the local and average Nusselt number.

Grid size

Grid-independent tests were carried out to ensure grid independence of the calculated results; consequently, the grid size and the grid orientation giving grid independent results were selected, and thus a total cell number of 45000 cells (300 × 150) were adopted. The grid points are clustered in the radial direction so that the finer spacing is formed near the wall and at the inlet. Figure 3 shows the grid.

RESULTS AND DISCUSSION

Steady flow

To validate the numerical calculations, the computational results for the fully developed steady turbulent pipe flow are compared with the experimental data observed in (Elshafei et al., 2006). For two different values of Reynolds number, the numerical results are in consistent with the experimental results; within about ±5% as can be noticed in Figure 4. The flow was observed to be thermally fully developed for X/D ≥ 20.

Pulsating flow

In the present work, investigations are performed within the ranges of $10^4 \leq Re \leq 4 \times 10^4$, and $0 \leq f \leq 70$ Hz.

The effect of pulsating frequency on the local time average Nusselt number at fixed Reynolds number of 16.8×10^3 is shown in Figure 5. It can be seen that, as

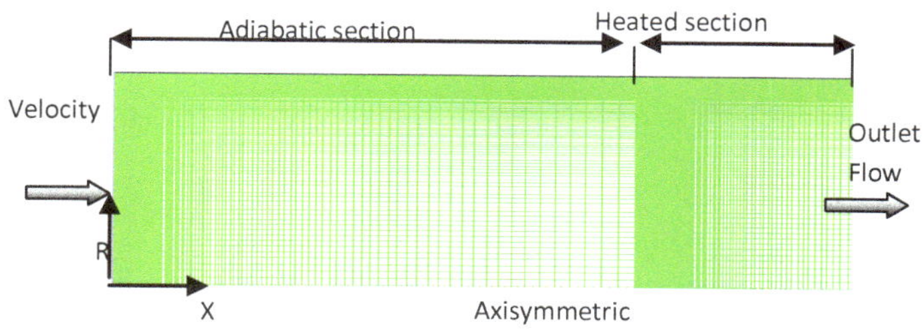

Figure 3. Computational grid for pipe.

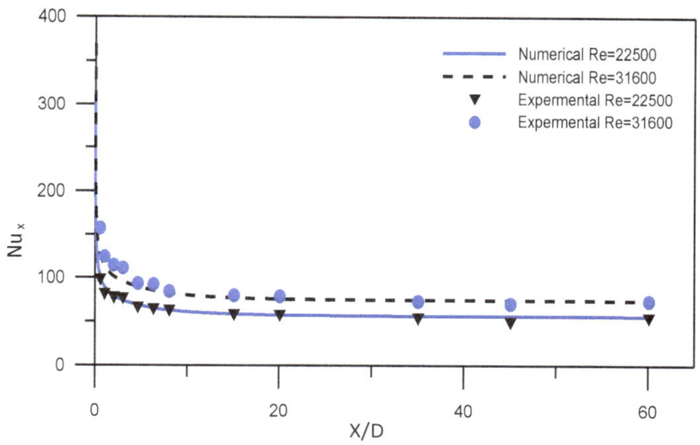

Figure 4. Variation of local Nusselt number at different Reynolds number.

the axial location moves downstream, the time average local Nusselt number for both pulsating and steady flows decays rapidly from a very large value near the entrance and the axial decrease becomes less step in the bulk of the pipe soon after leaving this entrance region.

The variation of the time averaged local Nusselt number for pulsating turbulent air flow exhibits a similar trend to that of the steady air flow case, but generally with lower values than that of the steady flow for f < 61.13 Hz. The maximum reduction of about 6% in the local Nusselt number was predicted in the thermal entrance region. For $f \geq 61.13$ Hz, little enhancement in the local time average Nusselt number of about 8% for X/D ≥ 50 was detected.

Figure 6 illustrates the variation of the ratio of the local Nusselt number value for pulsating flow to that of the steady flow value; η with the axial distance at $Re = 22.5 \times 10^3$. For $f \leq 39.3$ Hz, it can be observed from Figure 6a that the ratio η starts rising rapidly in the entrance zone of the thermally developed region (X/D<20) and becomes nearly fixed below unity where its value in the bulk of the pipe soon after leaving the entrance length (X/D≥20) implies that the effect of pulsation on the local heat transfer diminishes.

For $f \geq 42.5$ Hz, the value of η close to the end of the pipe enhanced by about 4% as shown in Figure 6b.

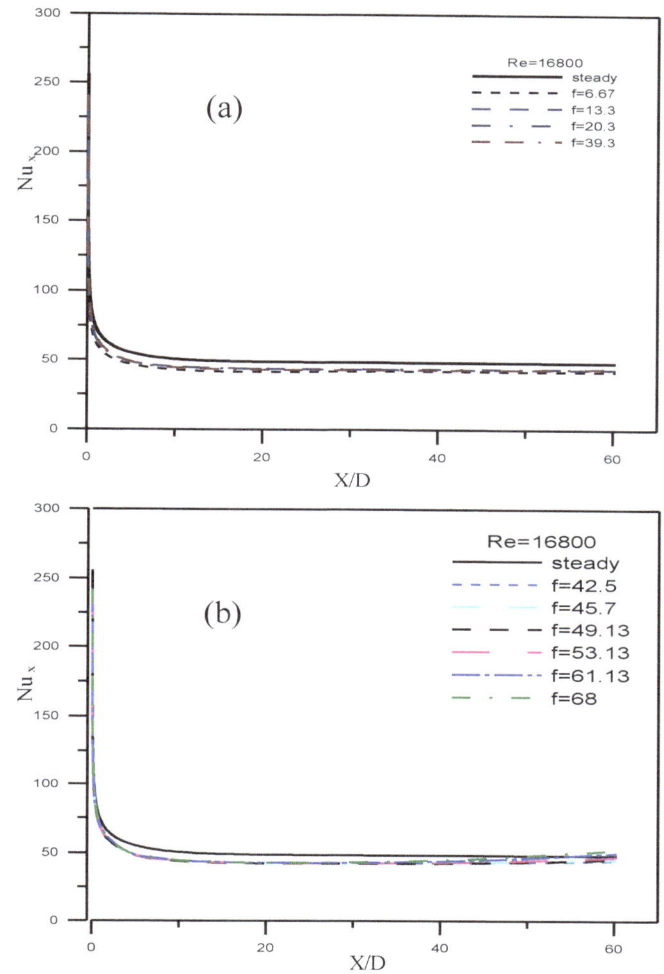

Figure 5. Distrbution of Nu_x at different frequency, Re=16800.

The impacts of other relevant parameters on heat transfer are illustrated in Figure 7a and b. The detected values of η due to the variation of pulsation frequency is at higher value of Reynolds number ($Re = 31.6 \times 10^3$). At lower frequency ($f \leq 39.3$ Hz), a small change in heat transfer ratio is visible near the entrance region. At f=13.3 Hz, a rapid increases in η can be observed up to X/D =20, after which it seems to be unchanged and generally, the value of η is less than unity. The same trends can be seen for f = 28.3 Hz and f = 39.3 Hz. However, in the downstream region the value of η reaches about unity.

The effect of changing pulsation frequency on the ratio η seems to be negligible specially, for higher values of pulsation frequency as can be noticed in Figure 7b.

At high values of Reynolds number; $Re \geq 37 \times 10^3$, certain qualitative trends can be shown in Figure 8a and b, which are similar to those described in the previous figures. For all studied values of pulsation frequency, the values of the ratio η are always less than unity.

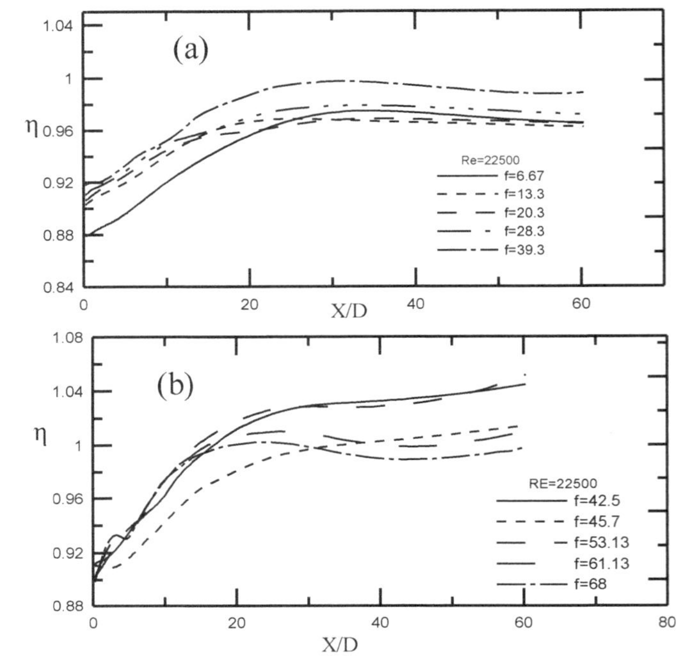

Figure 6. Distrbution of η at different frequency, Re=22500.

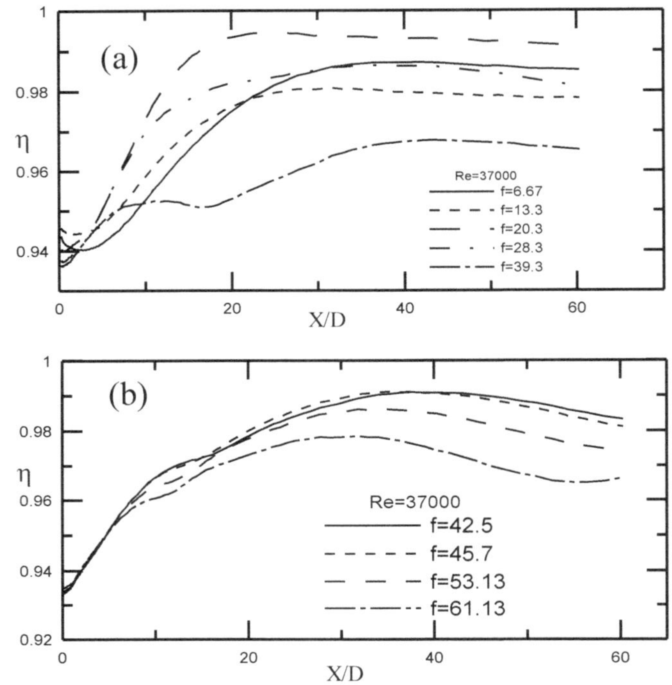

Figure 8. Distribution of η at different frequency, Re=37000.

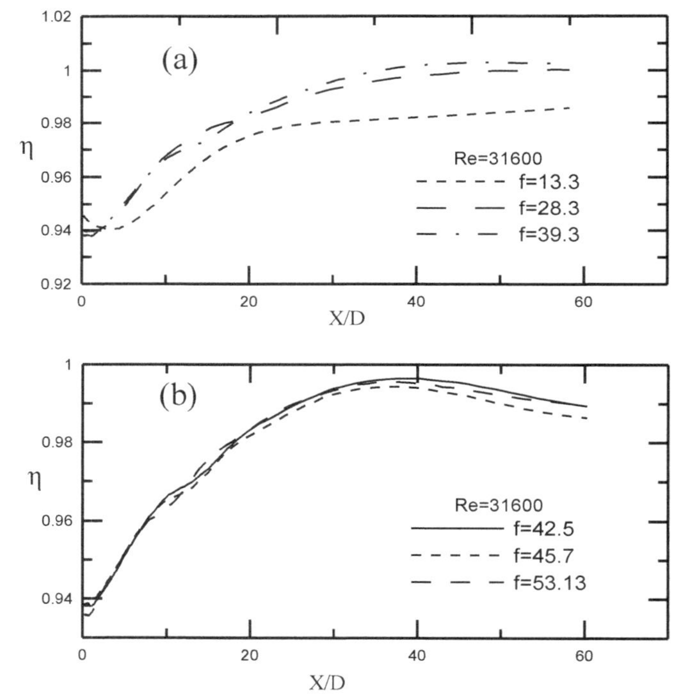

Figure 7. Distribution of η_x at different frequency, Re=31600.

This global behavior of pulsating flow is consistent with the analytical and numerical results of (Mostafa, 2005b; Chattopadhyay et al., 2006; Lee et al., 1998; Shemer, 1985; Kim et al., 1993; Jie-Chang et al., 2004; Elshafei et al., 2006). The overall patterns of η curves are qualitatively similar as discussed earlier.

Number and pulsation frequency

The third set of results pertains to the effect of the Reynolds number on the relative mean time- averaged Nusselt; η_m in case of pulsating flow in pipes. The calculated data for η_m versus Re are described in Figure 9a and b. In the range of pulsating frequency; $f \leq 39.3$ Hz, it is found that Reynolds number strongly affects the heat transfer ratio; η_m when Reynolds number is relatively low; Re < 25×10^3 as can be noticed in Figure 9a. The heat transfer ratio sharply increases with Reynolds number up to nearly unity. However, when Reynolds number exceeds this value, the effect of increasing Reynolds number diminishes and the value of η_m remains nearly constant. It is also noticed that the variation of the imposed pulsating frequency has a relatively small effect on η_m. These computed results are in consistent with that reported in (Xuefeng and Nengli, 2005).

For higher pulsation frequency; $f \geq 42.5$ Hz, the value of the ratio η_m increases also with the increase of Reynolds number. As η_m reaches to its peak value at Re $\approx 22.5 \times 10^3$, it then smoothly decreases with increasing Reynolds number up to 33×10^3 where η_m value is held nearly constant for higher values of Reynolds number as

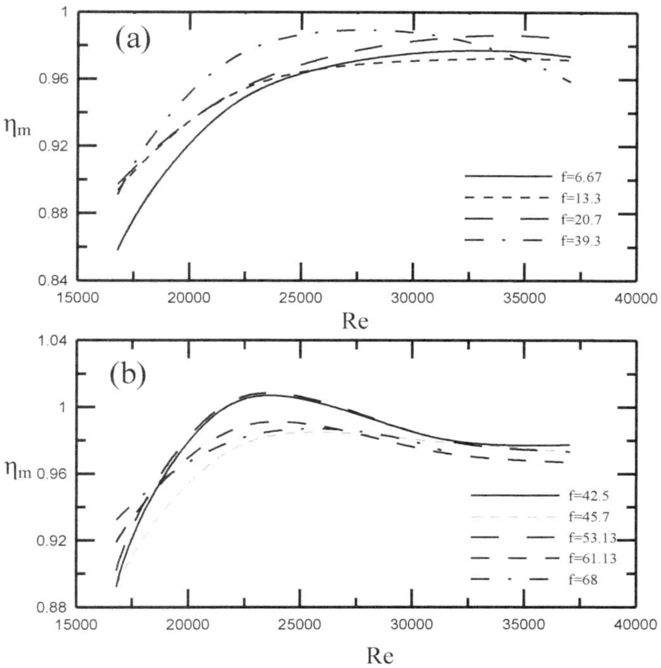

Figure 9. Heat transfer ratio for different Reynolds.

can be noticed in Figure 9b.

The behavior of Nusselt number corresponding to the variation in both Reynolds number and pulsation frequency can also be explained in view of the bursting phenomena (Liao and Wang, 1985; Genin et al., 1992). The bursting model defines certain regions on frequency-Reynolds number plane, where, the bursting frequency can be dependent or independent of the pulsation frequency.

The present results showed that the heat transfer coefficient may be increased or decreased with the change of both Reynolds number and frequency. The mean bursting frequency may be subdued to the pulsation frequency leading to the occurring of resonance that is dependent only on the pulsation frequency. Accordingly, the heat transfer process is expected to be affected resulting in either reduction or enhancement in the rate of heat transfer. These results are consistent with those reported by Habbib et al. (2004), and Liao and Wang (1985).

A comparison between the present numerical results with the experimental ones [32] over a ranges of $16 \times 10^3 < Re < 38 \times 10^3$ and $6.67 \leq f \leq 68$ Hz are described in Figure 10. The data are presented in the form of η_m as a function of bursting frequency ω^*. It can be observed that for all values of Reynolds number, the numerical predictions of η_m as a function of ω^* have the same trend as those for the experimental data. The discrepancies between the experimental data and the computed ones at low bursting turbulent frequency are of about 10%. As ω^*

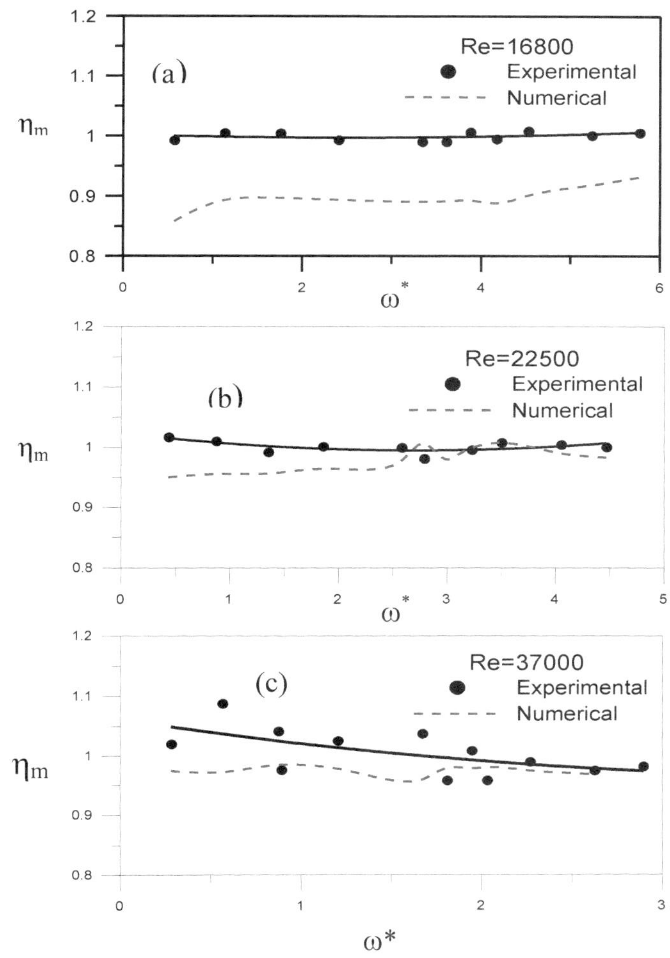

Figure 10. Comparison of the experimental data with numerical data at different (ω^*).

increases, the computed results closes to the experimental ones, specially, at higher value of Reynolds number.

In summary, the present elaborated numerical results for heat transfer characteristics of turbulent pulsating flow are useful. It can serve as a source of materials against which further analytical data may be checked for consistency.

Conclusion

Simulation was performed to describe heat transfer in pulsating turbulent air flow in a pipe using CFD code. The simulation results on the fully developed pulsating turbulent flow and heat transfer are compared with the available experimental data. The findings suggest that the heat transfer augmentation in the established flow regions takes place only for a certain band of frequencies. For extreme low and high frequencies, heat transfer is actually reduced below that of steady flow. The following may be concluded from this study:

i) At all values of pulsation frequency, the variations in the local time-averaged Nusselt number exhibit similar trends to the steady flows.
ii) The local time-averaged Nusselt number either increases or decreases than steady flow values depending on the frequency and Reynolds number.
iii) Frequency has little effect on the values of mean time-averaged Nusselt number especially at higher values of pulsation frequency.
iv) The presence of any pulsation frequency has a notable effect on the heat transfer coefficient in comparison with that of steady flow, even at a very small value of pulsation frequency. However, the change of the value of pulsation frequency has relativity small effect on the heat transfer rates.
v) No increase in the mean time averaged-Nusselt number for frequency $f \leq 39.34$ Hz. For $f \geq 42.5$ Hz, little increase in time averaged-Nusselt number is detected at Re ≈ 22.5×10^3.

Nomenclature

C_p: Specific heat at constant pressure (J/Kg K)
C_1, C_2, C_μ: Empirical constant of *k*-ε model.
D: Diameter of pipe (m)
G: The production of turbulent kinetic energy (Kg/m s^3)
h(x,t): Instantaneous local heat transfer coefficient (W/m^2 K)
h(t): Instantaneous heat transfer coefficient which is the average of heat transfer coefficient over the whole length of pipe at given moment (W/m^2 K)
h(x): Local heat transfer coefficient which is the average of heat transfer coefficient over a cycle (W/m^2 k).
p: Pressure (Pa)
K: Thermal conductivity (W/m K)
***k*:** Turbulent kinetic energy (m^2/s^2)
L: Length of pipe (m)
Nu: Overall Nusselt number
Nu_t: Instantaneous Nusselt number
Nu_x: Local Nusselt number
Pr: Prandtl number (υ/α)
q: Heat flux per unit area (W/m^2)
r: Radial coordinate (m)
R: Radius of pipe (m)
Re: Reynolds number (UD/υ)
T_{av}: Average bulk temperature along cross-sectional area of the pipe (°C)
T_{in}: Inlet temperature of the pipe (°C)
U_{in} Average of mean velocity at x=0 (m/s) $U_{in} = U_m + U_A$
U_m: Mean velocity of pulsating component, (m/s)
U_A: Amplitude velocity of pulsating component, (m/s)
UDF: User defied function
u: Velocity component in the axial direction (m/s)
v: Velocity component in the radial direction (m/s)
x: Axial distance(m)
t: Time (s).

Greek symbols

α: Thermal diffusivity of fluid $K/\rho C_p$ (m^2/s)
ε: Turbulent energy dissipation rate (W/kg)
φ: Viscous dissipation
μ: Dynamic viscosity (kg/m s)
μ_t: Eddy (or turbulent) viscosity (kg/m s)
υ: Kinematics viscosity (m^2/s)
ρ: Density (Kg/m^3)
σ_k: Turbulent Prandtl number for k
σ_T: Turbulent Prandtl number for T
σ_ε: Turbulent Prandtl number for ε
ω: Angular frequency (1/s)
ω*: Dimensionless frequency; (ω D/U^*), U* is the fraction velocity described as $U^* = 0.199\ Um / Re^{-0.125}$
η: Relative local Nusselt number = (Nu_{px}/Nu_{sx}).

Subscripts

s: Steady state
p: Pulsation
m: Mean.

REFERENCES

Al-Haddad A, Al-Binally N (1989). "Prediction of Heat Transfer Coefficient in Pulsating Flow". Int. J. Heat Fluid Flow, 10(2): 131-133.
Barid MHI, Duncan GJ, Smith JI, Taylor J (1996). Heat Transfer in pulsed Turbulent Flow. Chem. Eng. Sci. 21: 197-199.
Chattopadhyay H, Durst F, Ray S (2006). "Analysis of heat transfer in simultaneously developing pulsating laminar flow in a pipe with constant wall temperature". J. Heat Mass Transf. 33: 475-481.
Cho HW, Hyun JM (1990). Numerical solutions of pulsating flow and heat transfer characteristics in a pipe. Int. J. Heat Fluid Flow, 11(4): 321-330.
Elshafei EAM, Safwat Mohamed M, Mansour H, Sakr M (2006). Experimental Study of Heat Transfer in Pulsating Turbulent Flow in a Pipe". Publications in process (8th International Congress of Fluid Dynamics and Propulsion.
Erdal OK, Gainer JL (1979). The effect of pulsation on heat transfer. Int. Eng. Chem. Fundam. 18(10): 11-15.
Faghri M, Javdani K, Faghri A (1979). Heat Transfer with Laminar Pulsating Flow in Pipe". Lett. Heat Mass Transf. 6(4): 259–270.
Gbadebo SA, Said SAM, Habbib MA (1999). Average Nusselt number correlation in the thermal entrance region of steady and pulsating turbulent pipe flows. J. Heat Mass Transf. 35: 377–381.
Genin LG, Koval AP, Manachkha SP, Sviridov VG (1992). Hydrodynamics and heat transfer with pulsating fluid flow in tubes". Therm. Eng. 39950: 30-34.
Guo Z, Sung HJ (1997). Analysis of the Nusselt Number in pulsating Pipe Flow". J. Heat Mass Transf. 40(10): 2486-2489.

Gupa SK, Patel RD, Ackerberg RC (1982). Wall heat /Mass Transfer in pulsating Flow. Chem. Eng. Sci. 37(12): 1727-173.
Habbib MA, Attya AM, Said SAM, Eid Al, Aly AZ (2004). Heat transfer characteristics and Nusselt number correlation of turbulent pulsating air flows". J. Heat Mass Transf. 40: 307-318.
Habbib MA, Attya AM,.Said SAM, Eid Al, Ally AZ (2002). Convective Heat transfer characteristics of laminar pulsating air flow. J. Heat Mass Transf. 38: 221-232.
Habib MA, Said SAM, Al-Dini SA, Asghar A, Gbadebo SA (1999)."Heat transfer characteristics of pulsated turbulent pipe flow". J. Heat Mass Transf. 34(5): 413–421.
Havemann HA, Rao NN (1954). Heat transfer in pulsating flow. Nature, 7: 4418.
Hemeada HN, Sabry MN, Mansour A (2002). Theoretical analysis of heat transfer in laminar pulsating Flow". J. Heat Mass Transf. 45: 1767-1780.
Huseyin Y, Gamze B, Nesrin K, Şenay Y (2005). Numerical Study on Transient Local Entropy Generation in Pulsating Turbulent Flow through an Externally Heated Pipe". Sădhană, 30(5): 625–648.
Jie-Chang Yu, Zhi-Xin Li, Zhaoan TS (2004). Analytical study of pulsating laminar heat convection in a circular tube with constant heat flux. J. Heat Mass Transf. 47: 5297-5301.
Kim SY, Kang BH, Hyun JM (1993). Heat Transfer in The thermally Developed Region of a Pulsating Channel Flow. J. Heat Mass Transf. 36(17): 4257-4266.
Launder BE, Spalding DB (2006). Mathematical Models of Turbulence, London/New York 1972. Academic Press.
Lee DY, Park SJ, Ro ST (1998). Heat transfer in the thermally region of a laminar oscillating pipe flow. Ctyogenics, 38: 585-594.
Liao NS, Wang CC (1985). An investigation of the heat transfer in pulsating turbulent pipe flow". Fundamentals of forced and Mixed Convection. The 23th National Heat Transfer Conference Denver. Co. U.S.A. Aug. 4-7, pp. 53-59.
Martineelli RC, Boeleter LMK, Weinberg EB, Yakahi S (1943). Heat transfer to a fluid flowing periodical at low frequencies in a vertical tube. Trans. ASME, pp. 789-798.
Moschandreou T, Zamir M (1997). Heat transfer in a tube with pulsating flow and constant heat flux". Int. J. Heat Mass Transf. 40(10): 2461-2466.
Mostafa HM, Torki AM, Abd-Elsalam KM (2005a). Experimental Study for Forced Convection Heat Transfer of Pulsating Flow Inside Horizontal Tube". 4thInternational Conference on Heat Transfer, Fluid Mechanics and Thermodynamics, Cario, Ak3.
Mostafa HM, Torki AM, Abd-Elsalam KM (2005b). Theoretical study of laminar pulsating flow in pipe". 4th International Conference on Heat Transfer, Fluid Mechanics and Thermodynamics, Cario.
Said SAM, Habib MA, Iqal MO (2003). Heat transfer to pulsating flow in an abrupt pipe expansion. Int. J. Num. Heat Fluid Flow, 13(3): 286,308.
Scotti A, Piomelli U (2001). Numerical simulation of pulsating turbulent channel flow". Phys. Fluid, 13(5): 1367-1384.
Shemer L (1985). Laminar-turbulent transition in a slowly pulsating pipe flow. Phys. Fluids, 28(12): 3506-3509.
Xuefeng W, Nengli Z (2005). Numerical Analysis of Heat Transfer in Pulsating Turbulent Flow in a Pipe". J. Heat Mass Transf. 48: 3957-3970.
Zhao TS, Cheng P (1998). Heat Transfer in Oscillatory Flows. Ann. Rev. Heat Transf. 9(7): 1-64.
Zheng J, Zeng D, Ping W, Hong G (2004). An Experimental Study of Heat Transfer Enhancement With Pulsating Flow". Heat Transf.Asian Res. 23(5): 279-286.
Zohir AE, Habbi MMA, Attya AM, Eid Al (2005)."An Experimental Investigation of Heat Transfer to Pulsating Pipe Air Flow with Different Amplitudes". J. Heat Mass Transf. 42(7): 625-635.

3

Sodium chloride concentration on the rheological and dynamic properties of aloe vera gel

Ukpaka C. P.

Department of Chemical/Petrochemical Engineering, Rivers State University of Science and Technology, Nkpolu, P. M. B. 5080, Port Harcourt, Nigeria. E-mail: chukwuemeka24@yahoo.com.

Aloe vera gel is a highly medicinal gel which is in high demand in the cosmetics and pharmaceutical industry. Mathematical models were developed to simulate the rheological and dynamic properties of aloe vera gel in relation to the addition of sodium chloride (NaCl – common salt) at various concentrations. The result from the investigation shows the effect of sodium chloride concentration on the density, viscosity, shear rate, shear stress and flow concentration of the aloe vera gel; hence, giving the manufacturers of aloe vera products an insight and awareness of aloe vera gel rheology and dynamic properties. The mucilaginous fluid samples were obtained from aloe vera bardadensis miller leaves, washed and wiped dry and mechanical analysis were conducted on the samples for the purpose of separation of aloe juice and aloe gel from the sample. Increase in the concentration of sodium chloride result to a change in the concentration of aloe vera gel mixture (viscosity and density). The result obtained from the analysis revealed the sudden change in aloe versa gel concentration alongside with rheological and dynamic characteristics of the aloe vera gel mixture were attributed to increase in salt added. The developed models obtained from the investigation were found useful in monitoring and predicting the quality of aloe vera products.

Key words: Modeling, sodium chloride concentration, rheology, dynamic property, aloe vera products.

INTRODUCTION

The study of modeling the effect of sodium chloride concentration on the rheological and dynamic properties of aloe vera gel is of great concern to the engineers, as a result of the effect of rheological and dynamic characteristics on the quality of aloe vera products by the cosmetics and pharmaceutical industry (Perry and Maloney, 1997; Douglas, 2004). Gels are among the numerous fluids that are of great importance to mankind. Gel is define as a colloidal dispersion that may be jelly-like (for example, gelatin) or not (for example, silica gel), usually formed by coagulation for example, by cooling or evaporation (Arends, 1965; Fujioka, 1969; Beck and Muttzall, 1980; Osipora, 1987; Kapel, 1988). Among the gels of importance to man is aloe vera gel. The name aloe vera is derived from the Arabic work "Aloe" meaning shinning bitter substance. Aloe vera is a species of the genus "Aloe", a member of the lilianceae family of which over two hundred species exist of all these aloe vera Barbadensis miller is the most potent and is widely used in commercial quantity. Aloe vera is a succulent with green, fleshy, lanceolated spiked leaves which grow from the base in a rosette pattern. And it is best known for its healing properties, which is most evident when the plant is at least four years old. Its leaves can be separated to give two basic products such as aloe gel and aloe juice (Hoffman, 1974; Schangraw and Mattocks, 1961; Ukpaka and Oboho, 2006).

The science that deals with the mechanical behaviour of fluids is called rheology. The rheology of fluid is usually described by its measure of the fluid resistance to flow and hence, quantifying this parameter is of great importance to assessing or predicting the function at press drop in various parts of a flowing system (Osipora, 1987; Casson, 1959; Goodwin, 1975; Herschel and Bulkley, 1926a). All fluids exhibit a certain viscosity (Beck and Muttall, 1980). A fluid is often termed as being either thick (having a high viscosity) or thin (having a low viscosity) (Kapel, 1988). For more quantitative description of viscosity two more physical parameters – shear stress and shear rate have to be defined "shear stress (τ) is the force required to overcome a fluid resistance to flow divided by the area that the force is acting upon" or "force per unit area" and is expressed in Newton's per meter square (N/m^2) or (dynes/cm^2).

Several research conducted (Dennis, 1982; Perry and Maloney, 1997; Donald 1998; Douglas et al., 2004; Herschel, Bulkley, 1926b; Pal and Maliyah, 1990. On flow behaviour index of power law fluid reveal that when n < 1 the liquid is said to be pseudoplastic, n = 1 Newtonian liquid and > 1 the liquid is said to be dilatant (Radford and Kichardson, 1974; Grimm, 1976). Many liquids of interest to the process and chemical engineers do not exhibit Newtonian behaviour, that is strict proportionality between shear stress and shear rate (rate of strain) (Grimm et al., 1976; Ukpaka, 2005). Investigation conducted reveals that for a fluid to enter the pump, it must have sufficient energy to force itself towards the inlet, but in this research work such system is been influenced by the sodium chloride concentration (Ukpaka, 2005). Similarly investigation carried out by John on engineering materials showed that the shear rate, shear stress, velocity distribution, density, viscosity and flow, head and power coefficients of aloe vera gel and sodium chloride concentration affect the quality of the product.

MATHEMATICAL MODEL FORMULATION

For fluid obeying Newton's law of viscosity, taking the direction of motion as the x-direction and V_x as the velocity of the fluid in the x-direction at a distance y from the boundary, the shear stress in the X-direction is given as

$$\tau_x = \mu \frac{dv_x}{dy} \tag{1}$$

Let F_g = mass flow rate
L = distance

Therefore $\mu\left(\frac{Kg}{ms}\right) = \frac{mass\ flow\ rate\ (kg/s)}{\text{distance (m)}}$ (2)

$$\frac{dVx}{dy} = \frac{I}{\mu}\tau \tag{3}$$

Also in terms of species involved in the reaction (i) the above equation is written as

$$\frac{dVx}{dy} = \frac{I}{\mu}\tau_i \tag{4}$$

Model I

The expression for the velocity distribution of fluid flow through a long tube of radius R, considering steady state condition and the force balance on a cylindrical element of aloe vera gel mixture (aloe vera gel/sodium chloride) interms of radius is given as

$$2\pi r L\tau_x = (p_1 - p_2)\pi r^2 \tag{5}$$

Rearranging Equation (5) yields

$$\tau_x = \frac{(p_1 - p_2)r}{2\pi L} \tag{6}$$

Model II

The rheological test on aloe vera gel mixture (aloe vera gel/Sodium chloride) provide data which indicate that it shear stress/shear rate characteristic can be represented accurately by the equation

$$\tau = -\phi\left|\frac{dv}{dy}\right|^{\lambda-1}\frac{dv}{dy} \tag{7}$$

where ϕ and λ are constants

Resolving Equation (7) using the necessary mathematical technique yields

$$\tau = \phi\left[\frac{-dv}{dy}\right]^{\lambda} \bullet \left[\frac{dv}{dy}\right]^{-1} \bullet \left[\frac{dv}{dy}\right] \tag{8}$$

Similarly, Equation (8) can be written as

$$\tau = \phi\left[\frac{-dv}{dy}\right]^{\lambda} \tag{9}$$

Equation (9) can be expressed as

$$\frac{\tau}{\phi} = \left[\frac{-dv}{dy}\right]^{\lambda} \tag{10}$$

Multiplying both side of Equation (10) with power index of $\frac{1}{\lambda}$ and rearranging yields

$$\frac{dv}{dy} = -\tau^{1/\lambda} \Big/ \phi^{1/\lambda} \qquad (11)$$

Where

$$dv = -\ \tau^{1/\lambda} / \phi^{1/\lambda}\ dy \qquad (12)$$

integrating Equation (12) yields

$$\int_o^v dv = -\ \tau^{1/\lambda} / \phi^{1/\lambda} \int_o^y dy \qquad (13)$$

$$[V]_0^V = -\tau^{1/\lambda} / \phi^{1/\lambda}\ [Y]_0^y \qquad (14)$$

$$V = -\ \tau^{1/\lambda} / \phi^{1/\lambda}\ y_o + c \qquad (15)$$

since V = 0 for y = y_o, then substituting the boundary conditions into Equation (15) and rearranging yields

$$0 = -\tau^{1/\lambda} / \phi^{1/\lambda}\ y_o + c \qquad (16)$$

Therefore

$$C = \frac{\tau^{1/\lambda}\ y_o}{\phi^{1/\lambda}} \qquad (17)$$

substituting Equation (17) into Equation (15) yields

$$V = -\frac{\tau^{1/\lambda}}{\phi^{1/\lambda}}\ y + \frac{\tau^{1/\lambda}}{\phi^{1/\lambda}}\ y_0 \qquad (18)$$

$$V = -\frac{\tau^{1/\lambda}}{\phi^{1/\lambda}}\ (y - y_o) \qquad (19)$$

Equation (19) can also be written as

$$\tau^{\frac{1}{-1}} = -\frac{V\,\phi^{1/\lambda}}{(y - y_o)} \qquad (20)$$

therefore substituting Equation (6) into Equation (20) yields

$$\left[\frac{(p_1 - p_2)r}{2\pi\ L}\right]^{1/\lambda} = -\frac{V\,\phi^{1/\lambda}}{(y - y_o)} \qquad (21)$$

where

$$\phi = \tau_x^{1-\lambda}\left[\frac{\mu\ \infty}{1 - \dfrac{4\ \beta}{3\ \tau_x}\ 1 + \left(\dfrac{\beta}{\tau_x}\right)^{4/3}}\right]^{\lambda} \qquad (22)$$

$$\lambda = \left[1 - \frac{4\ \beta}{3\ \tau_x} + 1 - \left(\frac{\beta}{\tau_x}\right)^{4/3}\right] \Big/ 1 - \left(\frac{\beta}{\tau_x}\right)^{4} \qquad (23)$$

Hence rearranging Equation (21) yields

$$V = -\frac{(p_1 - p_2)}{2\pi\,L\,\phi^{\frac{1}{\lambda}}}\ r^{1/\lambda}\ (y - y_o) \qquad (24)$$

Since

$$\text{viscosity}\ \mu = \frac{shear\ \ stress}{shearrate}\ \frac{\tau_x}{\beta} \qquad (25)$$

Also assuming that in a ideal situation where $\phi\ and\ \dfrac{1}{\lambda} = 1$ Equation (24) reduces to

$$V = -\ \frac{(p_1 - p_2)\,r\,(y - y_o)}{2\pi\ L} \qquad (26)$$

Dividing through Equation (26) by (y-yo) yields

$$\frac{V}{y_1} = -\frac{(p_2 - p_1)\,r}{2\pi\ L} \qquad (27)$$

where $y_1 = y - y_0$

Hence Equation (27) can be written as

$$\tau = \frac{(p_2 - p_1)r}{2\pi L} \quad (28)$$

Where $\upsilon = (p_2 - p_1)r$ (28a)

For υ = flow concentration

$V_R = 2\pi L$ (28b)

But in terms of space time and space velocity

$$\tau_t = \frac{\upsilon}{V_R} \quad (28c)$$

$$\tau_v = \frac{V_R}{\upsilon} \quad (28d)$$

And since $\tau_t = \frac{1}{\tau_v}$ (28e)

Therefore $\tau_v \; \dfrac{2\pi L}{(p_2 - p_1)r}$ (28f)

where τ_t = space time, τ_v = space velocity since D = 2r, hence r = D/2. Substituting r = D/2 into Equation (28) yields.

$$\tau = \frac{(p_2 - p_1)D}{4\pi L} \quad (29)$$

Models III

Since the flow rate characteristics for the flow of a viscous liquid through a tube can be predicted from the rheogram for that liquid using the expression as presented by Arcadia et al. (1991).

$$Q = \frac{-\pi R^3}{\tau_w^3} \int_0^{\tau_w} \tau_x^2 \, f(\tau_x) d\tau_x \quad (30)$$

where $f(\tau_x) = -\dfrac{\tau_x}{\mu}$ (31)

therefore substituting Equation (31) into Equation (30), yields.

$$Q = \frac{-\pi R^3}{\tau^3_w} \int_0^{\tau_w} \tau^2_x \, f\left(\frac{-\tau_x}{\mu}\right) d\tau_x \quad (32)$$

resolving Equation (32) mathematically yields

$$Q = \frac{-\pi R^3}{\tau^3_w} \left(\frac{\tau_x^4}{4\mu}\right) \tau_w \quad (33)$$

$$= \frac{\pi R^3 \tau_w^3}{4\mu} \quad (34)$$

But $\tau_w = R/2 \, \dfrac{\Delta P}{\Delta L}$ (35)

Therefore $Q = \dfrac{\pi R^4 \Delta p}{8\mu \Delta L}$ (36)

But pressure can be express mathematically as

$p = \rho g L$ (37)

Therefore

$\Delta p = \rho g \Delta L$ (38)

Substituting Equation (38) into Equation (36), becomes

$$Q = \frac{\pi R^4 \rho g \Delta^L}{8\mu \Delta L} \quad (39)$$

$$Q = \frac{\pi R^4 \rho g}{8\mu} \quad (40)$$

$$\therefore Q = \frac{\pi D,^4 \rho g}{128\mu} \quad (41)$$

Similarly, expressing Equation (40) in terms of $\rho = m/v,$ hence Equation (30) becomes

$$Q = \frac{\pi R^4 mg}{8\mu V} \quad (42)$$

Table 1. Experimentally determined physical properties of aloe vera gel + sodium chloride mixture of two different samples collected in Niger Delta area.

Mass of sodium chloride (common salt)	Viscosity (cp)		Mean viscosity (cp kg/m.h)	Mass of sample + salt (mg)	Volume of sample + salt (ml)	Density of sample + salt (mg/ml)		Mean density (mg/ml)
	A_1	A_2				A_1	A_2	
0	57.4	58.6	58.0	96.23	100	0.9623	0.9273	0.9146
10	43.6	44.4	44.0	109.10	102	1.0696	1.1002	1.0834
30	44.5	57.8	46.2	115.00	106	1.0848	1.9973	1.5067
60	55.2	57.3	56.3	151.59	112	1.3535	1.3865	1.3729
90	53.2	56.8	55.0	188.59	120	1.5716	1.4996	1.5986
100	44.9	44.5	44.7	194.49	122	1.5902	5.843	1.6175

$$\therefore Q = \frac{\pi D^4 mg}{128 \mu V} \quad (43)$$

EXPERIMENT AND METHODS

Sampling procedure/collection of samples

Aloe vera (Barbadensis miller) leaves samples were collected from two communities (Obite and Obagi) in Niger Delta Area of Nigeria. The leaves of the Aloe vera (Barbadensis miller) were transferred into glass bottles for transportation to the laboratory for analysis.

The research work was carried out by using the following materials and equipment such as, aloe vera (Barbadensis miller) leaves, sodium chloride (common/table salt) beakers, measuring cylinder, spatula, chemical balance, Brookfield, viscometer, spoon and knife.

Procedure

Aloe vera (Barbadensis miller) leaves were harvested, washed and wiped dry. The leaves were then slashed with a sharp knife to expose the gel. The gel, which is the dear mucilaginous fluid in the central part of the leaf, was then scooped with a spoon into a sample container. It is important to note that the mechanical separation of aloe juice and aloe gel is not always complete so aloe juice can be found in aloe gel. The temperature of the aloe vera gel in the sample container was measured and recorded. Six empty beakers were weighed with a chemical balance and their weight were recorded and 100 ml of aloe vera was measured and put into each of the six beakers. The following mass of sodium chloride (common salt) were measured 10,30,60,90 and 100 g using spatula and chemical balance each of these salt quantities was added to the five other samples containing 100 ml aloe vera gel. The viscosity of each of the sample was then determined using a Brookfied viscometer at a speed of 100 RMP. The viscosity of the sample was determined after the salt and aloe vera gel in each beaker was properly stirred for some time to allow mixing of the two substances. Similarly, the densities of the aloe vera gel were measured with the aid of hygrometers.

RESULTS AND DISCUSSION

Physical properties of aloe vera gel mixture (aloe vera gel/sodium chloride) samples obtained from a mixture of 100 ml of aloe vera gel added with 10,30,60,90 and 100 g of sodium chloride (common salt) in Nigeria environment as listed in Table 1 to confirm their effect on the viscosity and density. The viscosities and densities were measured with the aid of Brookfield viscometer and hydrometer respectively.

The developed models were simulated by using the following parameters change in pressure ($p_1 - p_2$) = 172 N/m^2, length (L) = 5 to 25 m, radius (r) = 0.3 m, acceleration due to gravity (g) = 9.81 m/s^2.

The data obtained were fed into the model computer using the developed mathematical expression in Equations (6), (28), (28d) and (40) by considering various incremental steps as specified in Table 1.

Rheology of aloe vera gel with sodium chloride

Figures 1 and 2 shows the variation of the viscosity and density with increase in the mass of sodium chloride concentrations. It should be noted that viscosity increase from 10 to 60%, and the mass of sodium chloride added are 10, 30 and 60 g and decreases from 90 to 100%, and the mass of sodium chloride added are 90 and 100 g for both A_1 and A_2 samples. Whereas density increases from 10 to 100%, and the sodium added are 10, 30, 60, 90 and 100 g. The aloe vera gel flow concentration increases as the mass of sodium chloride concentration increase ranging from 10 to 60% and decreases within the range of 90 to 100% as shown in Figure 6. It is evident that the addition of sodium chloride to the aloe vera gel increases the density and viscosity (10 to 60%, for 10, 30 and 60 g) and decreases (90 to 100%, for 90 and 100 g) as shown in Figures 1 and 2. Figure 1 shows that after a rapid increase in sodium chloride added, the viscosity decrease with increasing shear stress (Figures 12 and 13). Close examination of Figures 12 and 13 reveals that after decreasing to a minimum, the viscosity begins to increase with the shear stress, the aloe vera

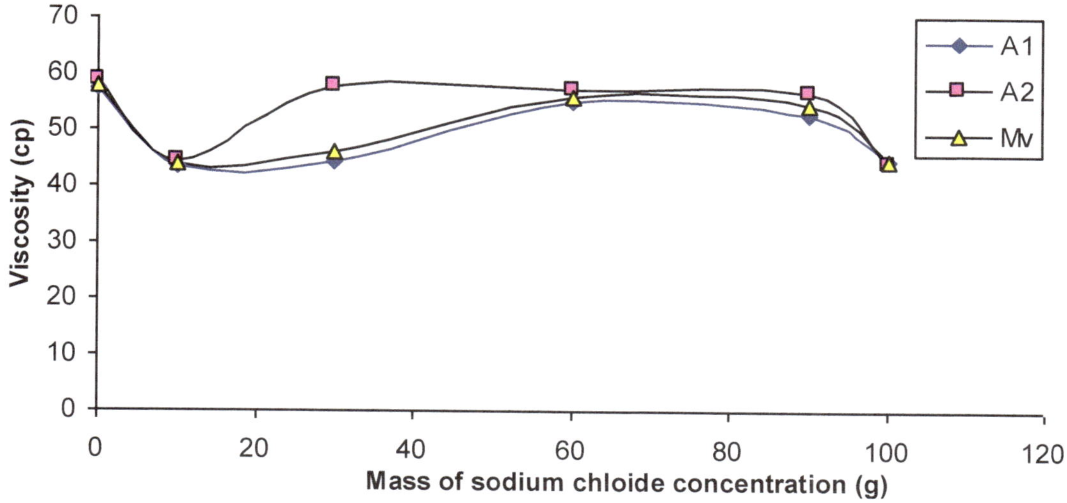

Figure 1. Viscosity versus mass of sodium chloride concentration.

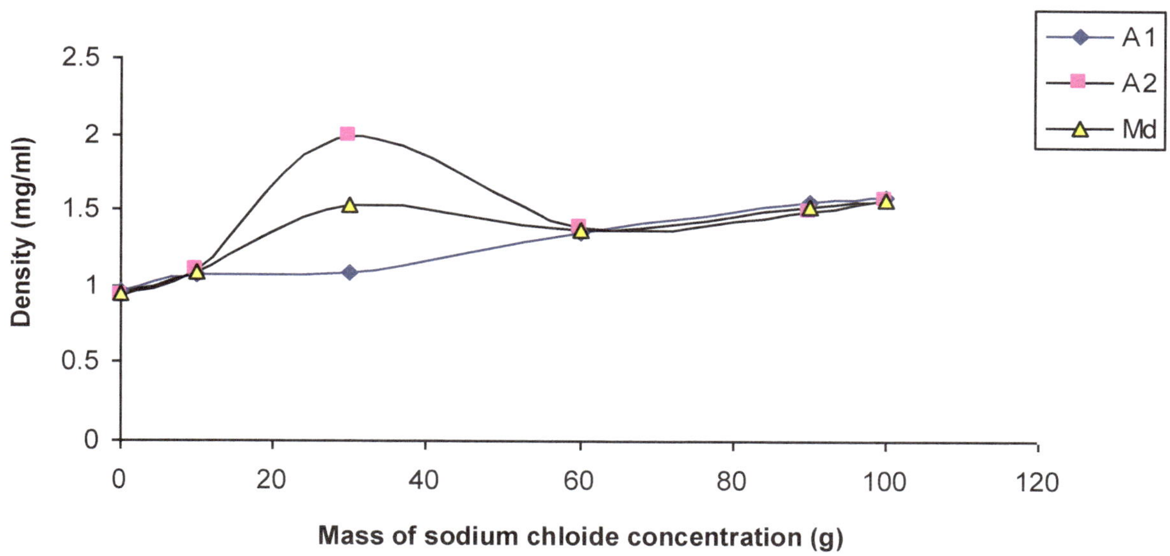

Figure 2. Density versus mass of sodium chloride concentration.

gel/sodium chloride mixture exhibit dilatant behaviour. It is to be noted from Figures 12 and 13 that the dilatant behaviour becomes more noticeable as the mass of sodium chloride concentration increases. Comparison of Figures 12 and 13 indicates that both aloe gel of sample A_1 and A_2 exhibit the dilatant behaviour.

Rheology of aloe vera gel without sodium chloride

Figure 10 shows the variation of the viscosity (A_1 = 57.4cp, A_2 = 58.6cp and M_v = 0.9448 mg/ml) of the same species of aloe vera bardadensis miller leaves samples collected in two areas of Niger Delta. It can be seen that the viscosity of the aloe vera gel decreases without sodium chloride concentration. The aloe vera gel exhibit shear thinning behaviour at low shear stress. At high shear stress, the aloe vera gel becomes essentially Newtonian with a slight increase in the viscosity with shear stress. The aloe gel used by this investigation was of high softening characteristics, such that at room temperature the aloe vera gel flows were essentially rigid. The aloe vera gel used in the present study is similar in terms of physical and chemical properties.

In the absence of sodium chloride concentration, the rheological and dynamic behaviour of the sample becomes rigid, as shown in Table 1 with viscosity values of A_1 = 57.7 cp, A_2 = 58.6 cp and m_v = 58.0 cp, with density values of 0.9623, 09273 and 0.9448 mg/ml, in which the viscosity values for A_1, A_2 and m_v are higher than others when sodium chloride is added into the system. The flow characteristic decreases with increase in viscosity and decrease in density for both A_1 and A_2 samples.

Effect of flow concentration on the viscosity of aloe vera gel/sodium chloride mixture

Figures 7 and 8 shows a comparison of the viscosity of aloe vera gel/sodium chloride mixture containing sample A_1 and A_2. Figures 7 and 8 show the flow concentration increases with viscosity and decreases with increase in the tube length Figure 9. This process is characterized by a decrease in the viscosity as the shear rate is increased. When the shear rate increases up to a critical value, the flow concentrations begin to disrupt and flow concentration is disturbed and the product quality as well being reduced (Table 3). Thus, the viscosity begins to increase with the shear rate (that is, dilatant behaviour).

There are various explanations in the literature concerning shear rate, shear stress, which all tend to rely on some kind of change base on the type of samples. According to Goodwin, (1975) and Hoffman, (1974), a change in the flow pattern takes place when dilatancy occurs. From Figure 3, it shows that the shear rate decreases with increase in length; this therefore shows that there is a very low shear rate, indicating that the viscosity is high as shown in Figure 4. Similarly, in Figure 10, it is noted that the shear stress decreases with increase in tube length as shown in Figure 10. Consequently, the aloe vera gel and the sodium chloride concentrations are additive for the evaluation of the viscosity of the aloe vera gel plus sodium chloride mixture (Table 4). Therefore, care must be taken when one calculates the viscosity of an aloe vera get plus sodium chloride mixture.

Effect of tube length on the shear rate and flow concentration of aloe vera gel plus sodium chloride

Figure 3 shows the variation of shear rate with increases in tube length with variation in viscosity. From Figure 3, it is noted that shear rate decreases with increase in tube length, therefore affecting the viscosity of the aloe vera gel/sodium chloride mixture as well as the quality of the product produced.An analysis of the data obtained from the application of varying the length highlighted different length-dependent behaviour for the system examined. Figures 3 and 10 shows different stress by system A_1 and A_2 samples at a given viscosity.

Several rheological models (Herschel and Bulkley, 1926a, 1926b; Casson, 1959; Shangraw et al., 1961; Robertson and Stiff, 1976) have been taken into consideration for the correlation of the experimental shear-stress-shear rate data with the purpose of selecting one equation suitable for describing the shear dependent behaviour of all the system examined. Accordingly, the comparison of the different systems (A_1 and A_2 sample) and evaluation of the aloe vera gel/sodium chloride mixture, the concentration dependence of the rheological properties can be based upon a limited set of rheologica parameter like the yield stress and the infinite – shear rate viscosity μ_∞.

Effect of tube length on the shear stress

In Figure 11, it is noted that shear stress decreases with increase in tube length. The shear stress can be used in predicting the rheological and dynamic properties of aloe vera gel; and the quality of the product obtained after production. Also, the manufacturer needs to acknowledge the effect of shear stress on the compatibility of the product (Table 2).

Figure 5 shows the variation of shear rate with increase in shear stress. From Figure 5, it is noted that shear stress increases with increase in shear rate. Several literature (Goodwin, 1975; Hoffman 1974; Grimm et al., 1976; Douglas et al., 2004) reveals that the characteristics of shear rate and shear stress affects the flow pattern when dilatancy occurs.

Conclusion

The rheological and dynamic properties of aloe vera gel are studied by the experimental determinations of the viscosity and density of the gel. Different quantities of common salt were added separately to a given volume of aloe gel. The viscosity of the pure gel dropped on addition of salt. Increasing the quantity of salt added to the aloe gel led to an increase in the viscosity of the gel. The viscosity of the gel increases, gets to a maximum and declines. There was an observed continuous increase in the density of the gel with increase in the quantity of salt added.

Generally, the viscosity of aloe vera is high. This is attributed to the presence of high molecular weight long chain compound (shears). It is important to note that the essence of the addition of salt is to thin down the gel to have a certain viscosity for easy pumping or flow characteristics during aloe vera processing. The addition of salt to aloe vera gel can be used to control and monitor the quality of the gel based on the characteristics of the rheological and dynamics properties of the samples under investigation.

The following conclusion was drawn from the research work:

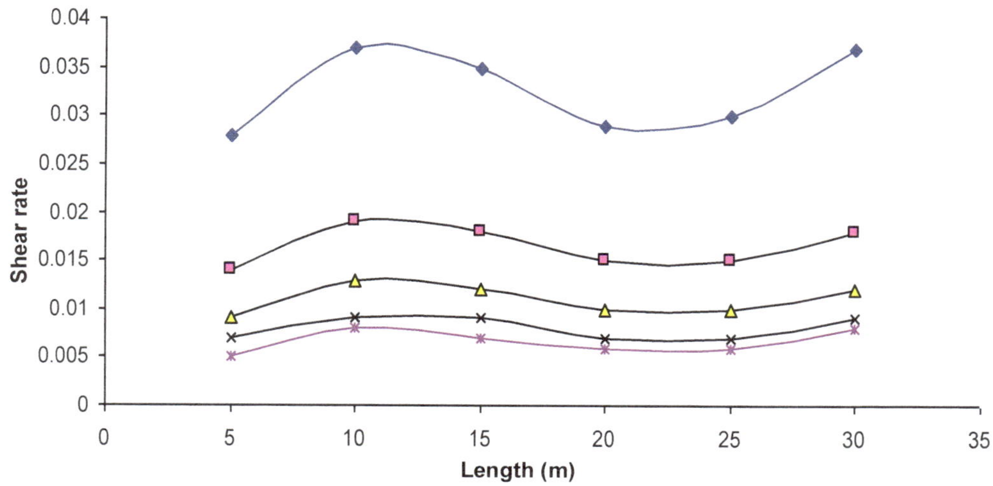

Figure 3. Shear rate versus length.

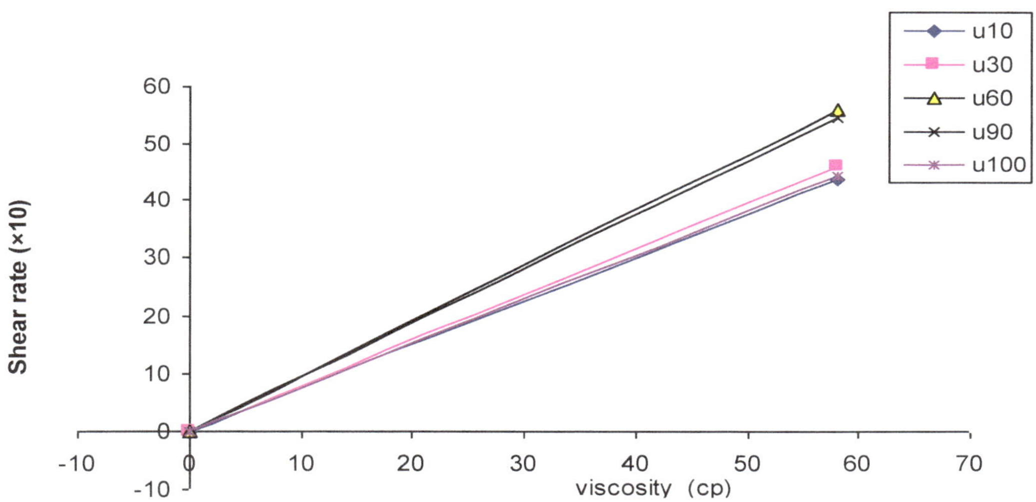

Figure 4. Shear rate versus Viscosity,

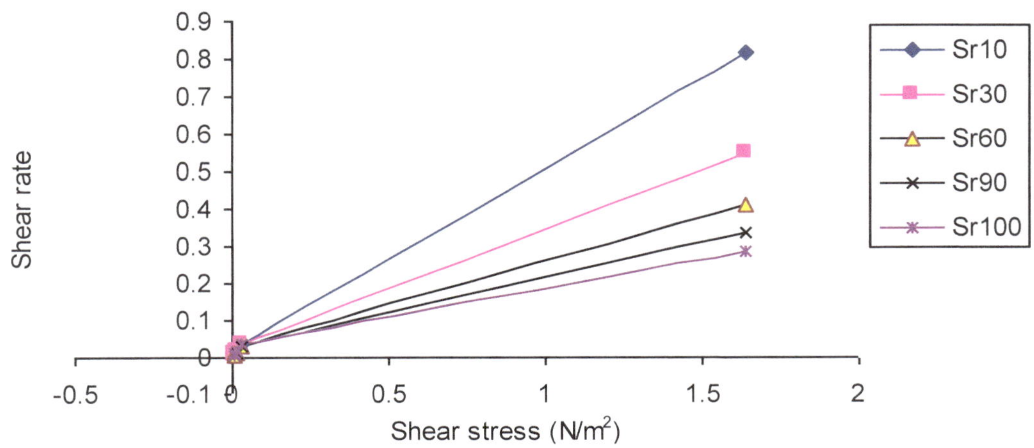

Figure 5. Shear rate versus shear stress.

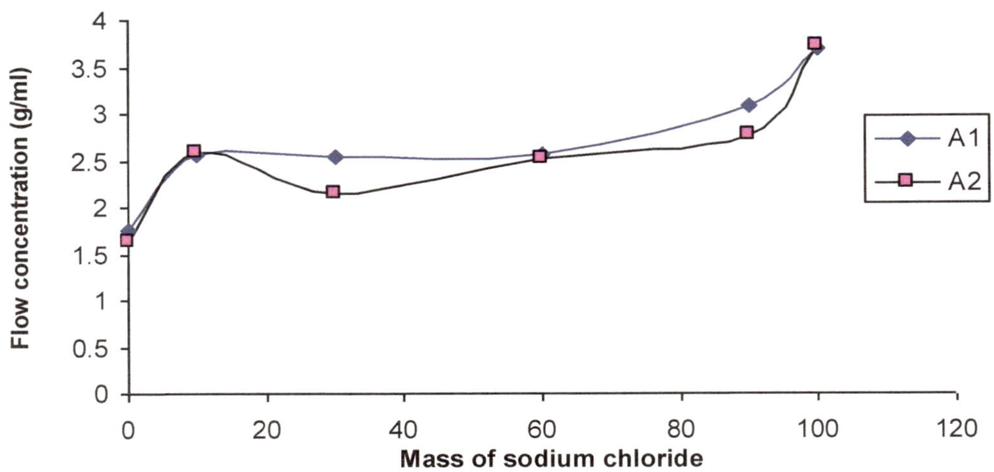

Figure 6. Flow concentration versus mass of sodium chloride.

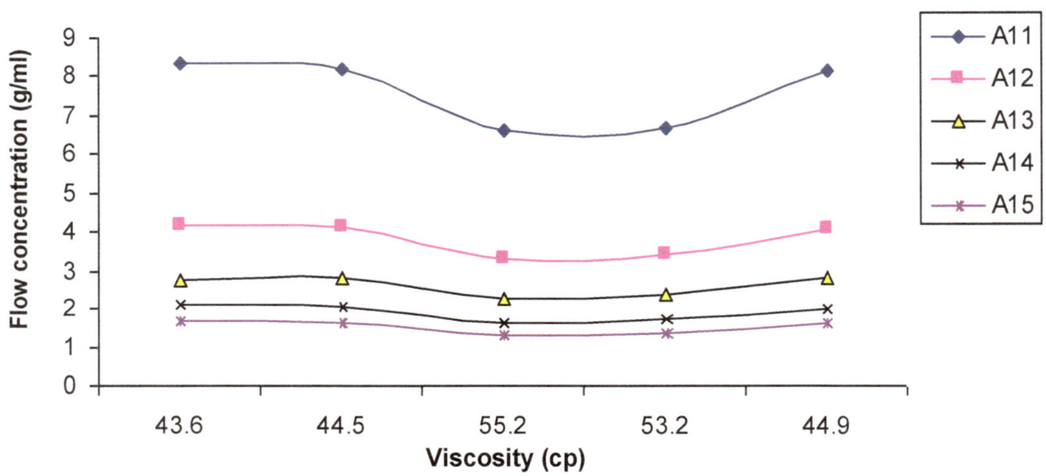

Figure 7. Flow concentration versus viscosity.

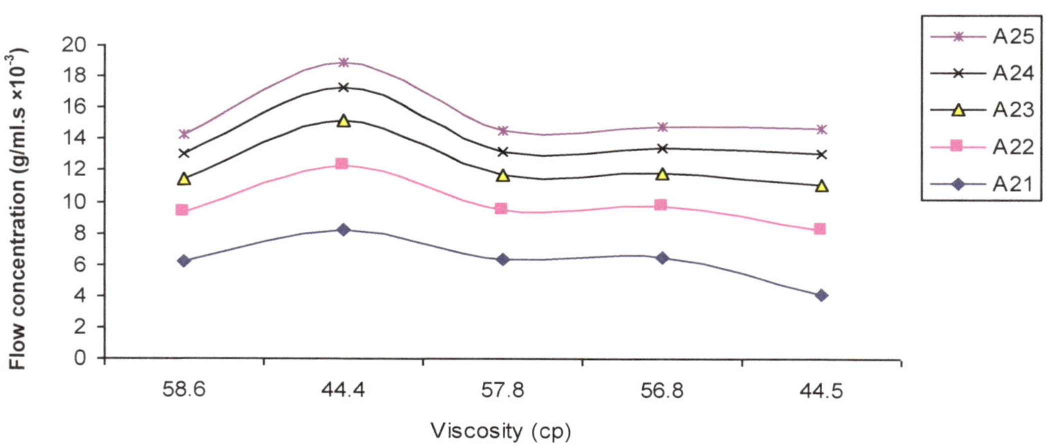

Figure 8. Flow concentration versus viscosity for A2.

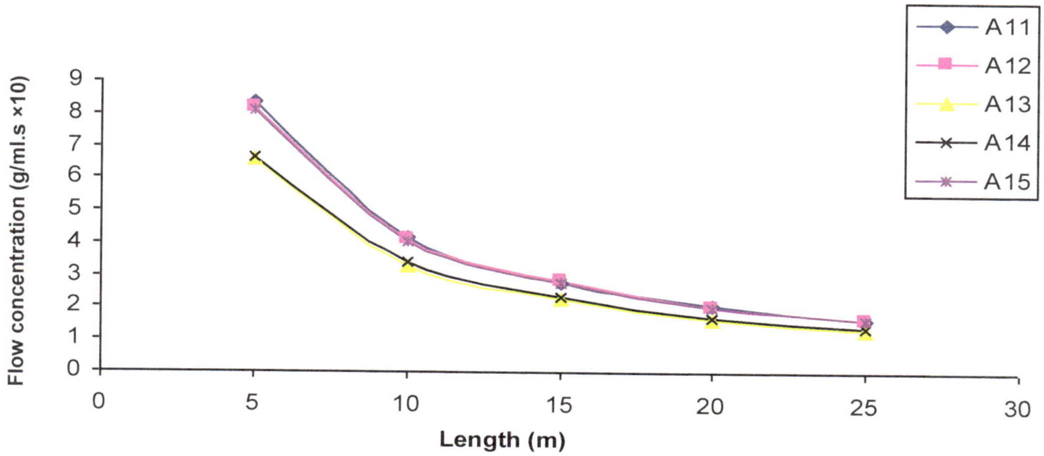

Figure 9. Flow concentration versus length for A1.

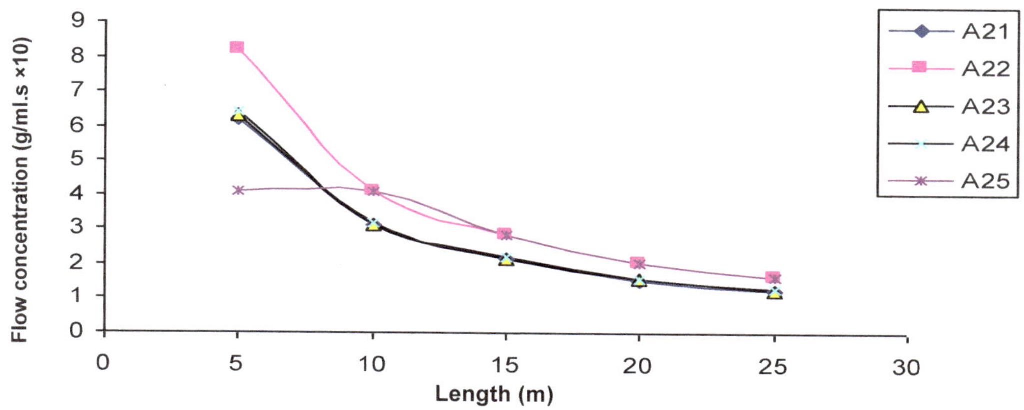

Figure 10. Flow concentration versus length A2.

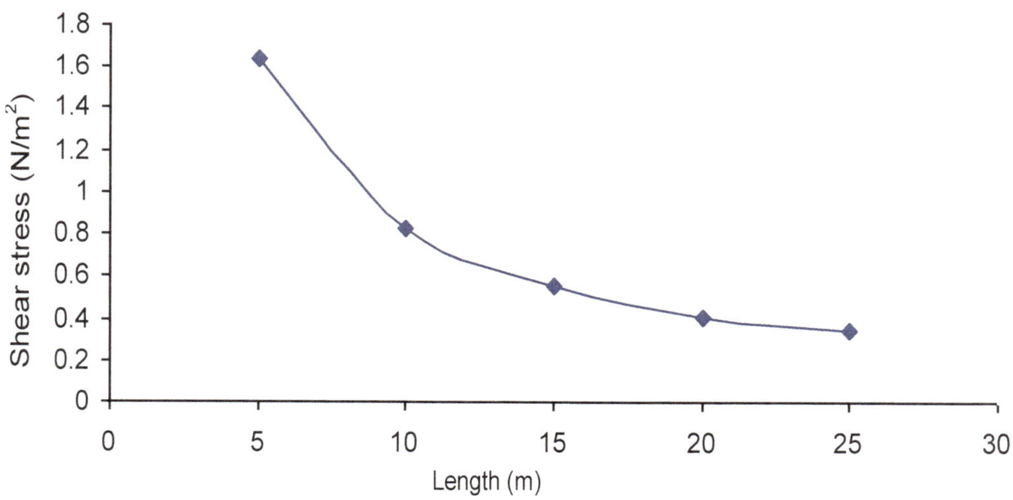

Figure 11. Shear stress versus length.

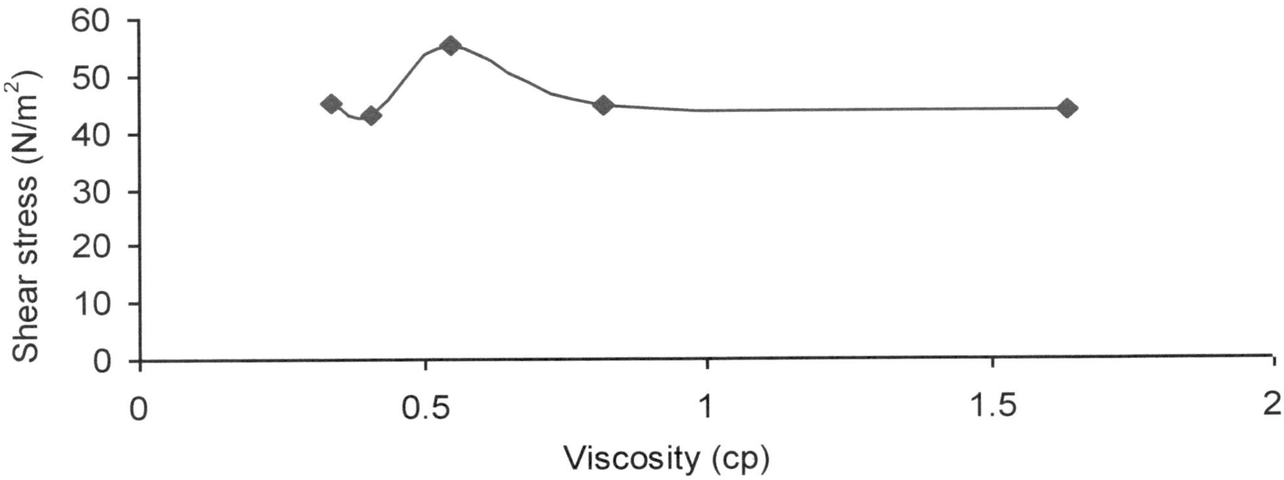

Figure 12. Shear stress versus viscosity for A1.

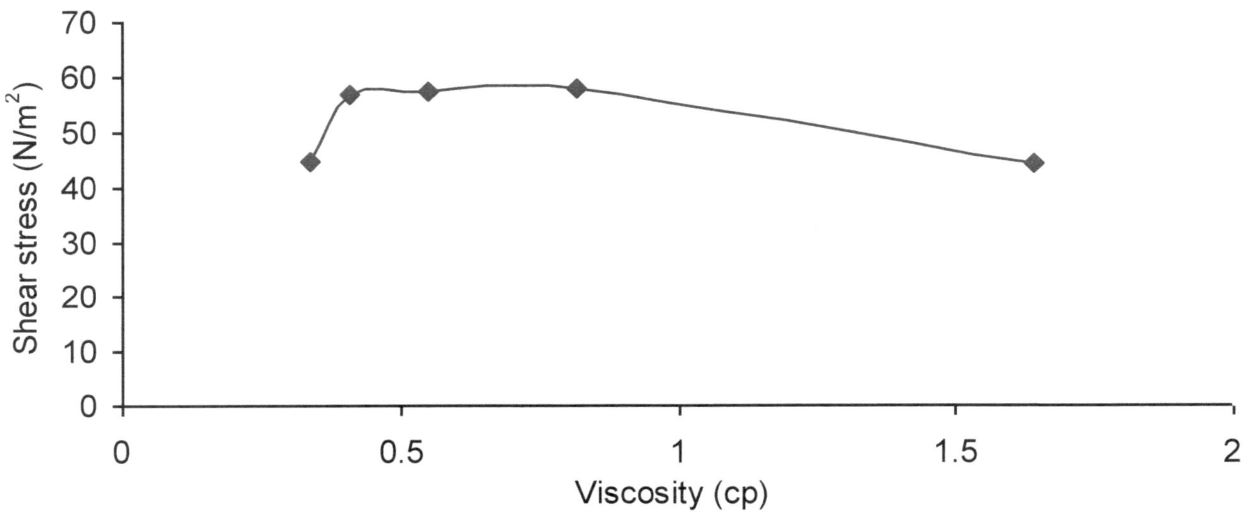

Figure 13. Shear stress versus viscosity for A2.

Table 2. Determination of shear rate at a given length and viscosity.

Length (m)	**Shear stress** $\tau_x = \frac{(p_1 - p_2)}{2\pi L}$	**Viscosity (cp)** $\mu_0 = 58.0$	$\mu_{10} = 44.0$	$\mu_{30} = 46.2$	$\mu_{60} = 56.3$	$\mu_{90} = 55.0$	$\mu_{100} = 44.7$
	2π	τ_{R0}	τ_{R10}	τ_{R30}	τ_{R60}	τ_{R90}	τ_{R100}
5	1.64	0.028	0.037	0.035	0.029	0.030	0.037
10	0.82	0.014	0.019	0.018	0.015	0.015	0.018
15	0.55	0.009	0.013	0.012	0.010	0.010	0.012
20	0.41	0.007	0.009	0.009	0.007	0.007	0.009
25	0.34	0.005	0.008	0.007	0.006	0.006	0.008

Table 3. Determination of the flow concentration at different mass of sodium chloride, viscosity and density.

Mass of sodium chloride common salt (g)	Viscosity reading (cp)		Density reading (mg/L)		Flow concentration (g/ml.s)	
	A_1	A_2	A_1	A_2	A_1	A_2
0	57.4	58.6	0.9623	0.9273	1.7440×10^{-3}	1.6457×10^{-3}
10	43.6	44.4	1.0696	1.100	2.5513×10^{-3}	2.5766×10^{-3}
30	44.5	57.8	1.0848	1.1973	2.5354×10^{-3}	2.1543×10^{-3}
60	55.2	57.3	1.3535	1.3865	2.5500×10^{-3}	2.5165×10^{-3}
90	43.2	56.8	1.5716	1.4996	3.0723×10^{-3}	2.7457×10^{-3}
100	44.9	44.5	1.5902	1.5843	3.6883×10^{-3}	3.7026×10^{-3}

Table 4. Determination of flow concentration of the aloe vera gel + sodium chloride at a given mass and at a different length.

Length of tube (m)	Flow concentration (g/ml.s)										
	u = 57.4	μ=43.6	μ=44.5	μ=55.2	μ=53.2	μ=44.9	μ_{PH}=58.6	μ_{PH}=44.4	μ_{PH}=57.8	μ=56.8	μ=44.5
5	$6.3552 x10^{-3}$	$8.3665 x10^{-3}$	$8.1973 x10^{-3}$	$6.6083 x10^{-3}$	$6.6568 x10^{-3}$	$8.1243 x10^{-3}$	$6.2249 x10^{-3}$	$8.2157 x10^{-3}$	$6.3111 x10^{-3}$	$6.4222 x10^{-3}$	$4.0987 x10^{-3}$
10	$3.1775 x10^{-3}$	$4.1833 x10^{-3}$	$4.0987 x10^{-3}$	$3.3042 x10^{-3}$	$3.4284 x10^{-3}$	$4.0621 x10^{-3}$	$3.1125 x10^{-3}$	$4.1079 x10^{-3}$	$3.1555 x10^{-3}$	$3.2111 x10^{-3}$	$4.0987 x10^{-3}$
15	$2.1183 x10^{-3}$	$2.7888 x10^{-3}$	$2.8299 x10^{-3}$	$2.2813 x10^{-3}$	$2.3671 x10^{-3}$	$2.8047 x10^{-3}$	$2.1490 x10^{-3}$	$2.8363 x10^{-3}$	$2.1787 x10^{-3}$	$2.2171 x10^{-3}$	$2.8299 x10^{-3}$
20	$1.5888 x10^{-3}$	$2.0916 x10^{-3}$	$2.0493 x10^{-3}$	$1.6521 x10^{-3}$	$1.7142 x10^{-3}$	$2.0311 x10^{-3}$	$1.5562 x10^{-3}$	$2.0539 x10^{-3}$	$1.5778 x10^{-3}$	$1.6055 x10^{-3}$	$2.0493 x10^{-3}$
25	$1.2710 x10^{-3}$	$1.6733 x10^{-3}$	$1.6395 x10^{-3}$	$1.3217 x10^{-3}$	$1.3714 x10^{-3}$	$1.6249 x10^{-3}$	$1.2450 x10^{-3}$	$1.6431 x10^{-3}$	$1.2622 x10^{-3}$	$1.2844 x10^{-3}$	$1.6395 x10^{-3}$

1. Sodium chloride (NaCl) concentrations influence the rheological behaviours of the aloe vera gel.
2. The density and viscosity is the functional parameters that influence the quality product obtained by using aloe vera gel for the product of aloe vera soap, cream etc.
3. The effectiveness of the aloe vera gel dependents on its maturity for medical purpose.
4. The shear stress and the shear rate which is a function of Young modulus influence the characteristics of aloe vera gal in process plant.
5. The flow characteristic is also other parameters that influence the biochemical composition of the aloe vera gel in term of input and output composition.
6. The aloe vera gel obtained from the two communities from Egi clan in Niger Delta area of Nigeria can be used in the manufacturing of aloe vera cream, soap, used in it natural state for medical herbs and treatment of wounds.
7. The additional of the sodium chloride (NaCl) increases the rheoloical characteristics of the aloe vera gel as well as an activator to the microorganisms presented in it.

Nomenclature: $\mathbf{P_1}$, Initial pressure (N/m^2); $\mathbf{P_2}$, final pressure (N/m^2); $\mathbf{V_R}$, reactor volume (m^3); $\mathbf{\Delta L}$, change in length (m); **G**, acceleration due to gravity (m/s^2); **D**, diameter (m); **R or r**, radius (red); $\mathbf{m_v}$, mean viscosity (cp); $\mathbf{m_{ad}}$, mean density (mg/ml); $\mathbf{A_1}$, aloe vera samples collected from Obite in Niger Delta Area; $\mathbf{A_2}$, aloe vera samples collected from Obagi in Niger Delta Area; **L**, distance (m); **C**, integration constant; **Q**, flow concentration (g/ml.s); **V**, volume (m^3); μ, viscosity (cp); τ_x, shear stress (N/m^2); τ_R, shear rate; τ_t, space time (s); τ_v, space velocity (1/s); τ_w, shear stress at the wall (N/m^2); β, rheological constant (dimensionless); ϕ, constant (dimensionless); υ, flow concentration or flow rate (g/ml.s); ρ, density (mg/ml); λ, constant (dimensionless); **X**, direction (dimensionless); **T**, time (s); 1, 2…, etc. = numbers.

REFERENCES

Arends CB (1965). Impact Testing Methods. J. Appl. Polym. Sci. 9(3):531.

Beck WJ, Muttzall KMK (1980). Transport Phenomena, Chic ester New York. Brisbane Toronto: A Wiley – Interscience pp. 227-290.
Casson NI (1959). Rheology of Disperse System, C.C. Mill. Ed. Pergamon Press. Oxford pp. 84-104.
Dennis GZ (1982). First Course in Differential Equation with Application, 2nd Edition pp. 280-310.
Donald QK (1998). Process Heat Transfer. McGraw-Hill, Kogakusha Ltd. New York pp. 3-93.
Douglas JF, Gasiorek JM, Swaffield JA (2004). Fluid mechanics. 2nd ed. Pearson Education Ltd. India pp. 11-14.
Fujioka K (1969). Materials Characterization Instrumented Impact Testing. El Du pont de Nemoars and Co. Inc. Wilmington USA. J. Appl. Polym. Sci. 13:1421.
Goodwin JW (1975). The Rheology of Dispersions, In Colloid Science 2. Royal Society f Chemistry, Special Periodical Reports. London 310
Grimm RJ (1976). Sequeezing Flows of Newtonian Liquid Film. Naperville, Illinois 60540. Appl. Sci. Res. 32:149.
Grimm RJ, Davies JM, Brindley G, Walter K (1976) Sequeezing Flows of Polymeric Liquids. Ache J. 24(3):435.
Herschel W, Bulkley R (1926a). Consistenzmessungeon von Gummy – Benzol Losingen. Proc. Am. Soc. Test. Matts 26:621.
Herschel WH, Bulkley R (1926b). Consistenzmessungeon von Gummy – Benzol Losingen, Kolloid – Z 39 : 291.
Hoffman RL (1974). Discontinuous and Dilatant Viscosity Behaviour in Concentrated Suspensions. II Theory and Experimental Tests. J. Coll. Interf. Sci. 46:491–506.
Kapel EV (1998). Drilling Fluid Manual. SIPM EP. D/24, Nigeria pp. 1-325.
Osipora M (1987). Heat Transfer. Mir Publication. Moscow pp. 40-440.
Pal R, Maliyah J (1990). Rheology of oil in water emulsions with added solids. Can. J. Chem. Eng. 68:24-28.
Perry HR, Maloney JO (1997). Chemical Engineering Handbook. 7th Edition, McGraw-Hill Companies. Inc. Publisher pp. 2-195.
Radford JD, Richardson DB (1974). Production Engineering Technology. 3rd ed. McGraw-Hill Companies UK p. 273.
Schangraw RW, Grim, Mattocks AM (1961). An Equation for Non-Newtonian Flow. Trans. Soc. Rheol. 5:247–260.
Ukpaka CP (2005). Modeling solid-liquid separation on a rotating vertical cylinder. Multidiscip. J. Empir. Res. 2(2):53-63.
Ukpaka CP, Oboho EO (2006). Biokinetics for the production of nitrogen in a natural aquatic ecosystem polluted with crude oil. J. Model. Simul. Control (AMSE) 67(2):39-58.

Sliding mode application in speed control of an induction machine coupled in a hydraulic system

Mahmoud Ellouze[1,2], Riadh Gamoudi[1]* and Abdelkader Mami[1,3]

[1]Laboratory of Analysis and Control Systems (LACS), ENIT, Tunis, C.U Belvedere 1060 Tunis, Tunisia.
[2]Superior Institute of Technology and Data processing, 2080, Ariana, Tunisia.
[3]Department of Physics, Faculty of Sciences of Tunis, Electronic Laboratory, 2092 El Manar Tunis, Tunisia.

In this paper, a nonlinear control of an induction motor (IM) supplied with a photovoltaic generator to assure the level control of two coupled tanks is designed. The global system is decomposed into two separate models thus, coupled are assured by discontinuous command. In the first step, we propose a sliding mode technique to make the speed and the flux control of the IM robust to parameter variations. Then, the aim consists to assure the stability of the system autonomously towards a desired state (water level), by varying the speed of the IM. The use of the nonlinear sliding mode method provides very good performance for motor operation and robustness of the control law despite the external/internal perturbation. Simulation results are given to highlight the performance of the proposed control technique under load disturbances and parameter uncertainties.

Key words: Sliding mode control, photovoltaic energy, water level, coupled reservoirs, induction motor-pump.

INTRODUCTION

The sliding mode controller is designed for a class of non linear dynamic systems to tackle the problems with model uncertainties, parameter fluctuations and external disturbances. By this design, the bounds of the uncertainties are not required to be known in advance.

The variable structure control (VSC) possesses this robustness using the sliding mode control that can offer many good properties such as good performance against unmodelled dynamics, insensitivity to parameter variation, external disturbance rejection and fast dynamic (Utkin, 1977).

These advantages of sliding mode control can be employed in the position and speed control of an alternative current system.

In this paper, we begin with the IM oriented model in view of the vector control, next the rotor flux Φ_r, is estimated. We, then, present the sliding mode theory and design the sliding mode controllers of rotor flux and motor speed. Finally, we give some conclusion remarks on the control proposed of IM using sliding mode.

The induction machine is largely used in industry, mainly due to its reliability and relatively low cost. The control of the induction machine (IM) must take into account machine specificities: the high order of the model, the nonlinear functioning as well as the coupling between the different variables of control.

In the second stage, we interested to the level control in the coupled tanks. Several researchers have investigated the problem of controlling liquid flow of a single or multiple tanks (Khan and Spurgeon, 2006; Pan et al., 2005). The speed variations of the IM carry the level regulation control, the relation between speed and the flow of pumped water is given by (Caro and Bonal, 1997).

PROCESS DESCRIPTION

The increasing demand of water in rural zones and isolated sites made that a growing interest is done to the utilization of photovoltaic (PV) generator as energy source for several motor-pumps. In fact, the realization of autonomous, reliable pumping systems with a good efficiency, gives a practical and economical solution to

*Corresponding author. E-mail : riadh.gamoudi@gmail.com.

Figure 1. Scheme of the global process.

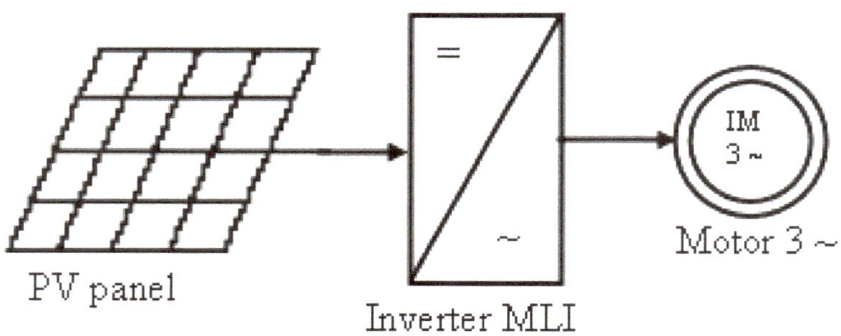

Figure 2. Diagram block of the PV pumping system.

the water lack problem in desert regions (Dhafer et al., 2007). For this reason, we have choice the system done in Figure 1 to study the feasibility in real case.

The studied system is a combination of two subsystems. The PV-powered water pumping system investigated in this paper consists mainly of a PV generator, DC–AC converter, and induction motor-pump. The second subsystem is composed of two coupled tanks.

METHODOLOGY

Dynamic model of induction motor

The first party of the global system investigated in this article consists mainly of a photovoltaic generator (PV), a three phase inverter MLI, an induction motor-pump. The scheme of the studied process is indicated in Figure 2.

We suppose, the inverter behaves as a perfect transferring power organ. In same, the characteristic of the generator PV is supposed ideal; we can assimilate in a classic power source.

The photovoltaic generator is characterized (at an illumination of $1000W/m^2$), by an opened circuit voltage $V_{co} = 471V$, an optimal operation point M (378 V; 2.27 A), and $I_{pH} = 2.47A$ $I_s = 8.143\times10^{-6}A$, $V_T = 37.5V$. Figure 3 show the characteristics of the PV system.

Afterward, we use the equivalent model of a three-phase asynchronous motor in the Park transformation (Olorunfemi, 1991; Mimouni et al., 2004).

So, an induction machine model can be described by the

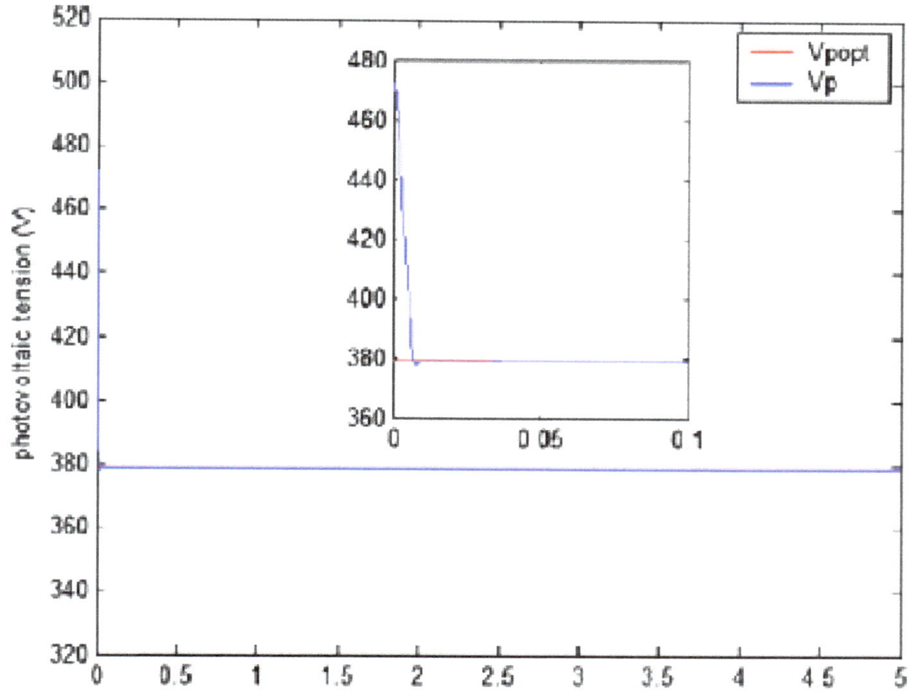

Figure 3. Simulation responses of the PV in closed loop.

following state equations in the synchronous reference frame whose axis d is aligned with the rotor flux vector:

$$\left(\Phi_{rq}=0\right) \quad : \quad \frac{di_{ds}}{dt}=-\gamma i_{ds}+\omega_s i_{qs}+\frac{K}{T_r}\Phi_{rd}+\frac{1}{\delta L_s}u_{ds} \tag{1}$$

$$\frac{di_{qs}}{dt}=-\omega_s i_{ds}-\gamma i_{qs}-p\Omega K\Phi_{rd}+\frac{1}{\delta L_s}u_{qs} \tag{2}$$

$$\frac{d\Phi_{rd}}{dt}=\frac{M_{sr}}{T_r}i_{ds}-\frac{1}{T_r}\Phi_{rd} \tag{3}$$

$$\frac{d\Phi_{rq}}{dt}=\frac{M_{sr}}{T_r}i_{qs}-\left(\omega_s-p\Omega\right)\Phi_{rd} \tag{4}$$

$$\frac{d\Omega}{dt}=\frac{pM_{sr}}{JL_r}\left(\Phi_{rd}i_{qs}\right)-\frac{f}{J}\Omega-\frac{1}{J}C_r \tag{5}$$

with:

$$T_r=\frac{L_r}{R_r},\quad \delta=1-\frac{M_{sr}^2}{L_sL_r},\quad K=\frac{M_{sr}}{\delta L_sL_r},$$

$$\gamma=\frac{R_s}{\delta L_s}+\frac{R_rM_{sr}^2}{\delta L_sL_r^2}$$

Where the electromagnetic torque is given in d-q frame:

$$C_{em}=p\frac{M_{sr}}{L_r}\left(\Phi_{rd}i_{qs}\right) \tag{6}$$

Where Φ_{rd}, Φ_{rq} are rotor flux components, u_{sd}, u_{sq} are stator voltage components, i_{ds}, i_{qs} are stator current components, δ is leakage factor and p is number of pole pairs. R_s and R_r are stator and rotor resistances, L_s and L_r denote stator and rotor inductances, whereas M_{sr} is mutual inductance. C_{em} is the electromagnetic torque, C_r is the load torque, *J* is the moment of inertia of the Induction Motor, Ω is mechanical speed, ω_s is stator pulsation, *f* is damping coefficient, T_r is rotoric time-constant.

Sliding mode control design

Consider a nonlinear system which can be represented by the following state space model in a canonical form (Slotine and Li, 1998):

$$\begin{aligned} x^{(n)}(t)&=f(x(t),t)+g(x(t),t)\,u+d(t)\\ y(t)&=x(t)\end{aligned} \tag{7}$$

Where $x=\left[x(t)\ \dot{x}(t)\cdots x^{(n-1)}(t)\right]^T$ is the state vector, $f(x(t),t)$ and $g(x(t),t)$ are nonlinear functions, *u* is the control input, *d*(*t*) is the external disturbances.

The objective of the control is to determine a control law *u*(*t*) to force the system output *y*(*t*) in (7) to follow a given bounded reference signal $y_d(t)$, that is, the tracking error $e(t) = y_d(t) - y(t)$ and its forward shifted values, defined as:

$$\begin{aligned} e^{(i)}(t) &= y_d^{(i)}(t) - y^{(i)}(t) \\ &= x_d^{(i)}(t) - x^{(i)}(t), \quad (i = 1, \cdots, n-1) \end{aligned} \tag{8}$$

should be small.

The design of SMC involves two tasks. The first one is to select the switching hyperplane to prescribe the desired dynamic characteristics of the controlled system. The second one is to design the discontinuous control such that the system enters the sliding mode $s(x,t) = 0$ and remains in it forever (Slotine and Li, 1998).

In this paper, we use the sliding surface proposed par J.J. Slotine,

$$s(x,t) = \left(\frac{d}{dt} + \lambda \right)^{n-1} e(t) \tag{9}$$

In which $e = x_d(t) - x(t)$, λ is a positive coefficient, and *n* is the system order. It remains to be shown that the control law can be constructed so that the sliding surface will be reached. The surfaces are chosen as functions of the error between the references input signal and the measured signals (Utkin, 1993). Then, a sliding hyperplane can be represented as $s(x,t) = 0$.

Consider a Lyapunov function:

$$V = \frac{1}{2} s^2 \tag{10}$$

From Lyapunov theorem we know that if $\dot{V}$ is negative definite, the system trajectory will be driven and attracted toward the sliding surface and remain sliding on it until the origin is reached asymptotically (Buhler, 1986).

$$\dot{V} = s\,\dot{s} \tag{11}$$

The simplified 1st order problem of keeping the scalar $s(x,t)$ at zero can be achieved by choosing the control law *u*(*t*). A sufficient condition for the stability of the system is:

$$\frac{1}{2}\frac{d}{dt} s^2 \leq -\eta |s| \tag{12}$$

Where η is a positive constant.

The Equation (12) is called reaching condition or sliding condition. *s*(*t*) verifying (12) is referred to as sliding surface, and the system's behaviour once on the surface is called sliding mode.

If the control input is so designed that the inequality (12) is satisfied, together with the properly chosen sliding hyperplan, the state will be driven toward the origin of the state space along the sliding hyperplane from any given initial state. This is the way of the SMC that guarantees asymptotic stability of the systems.

The process of sliding mode control can be divided in two phases, that is, the approaching phase and the sliding phase. The sliding mode control law *u*(*t*) consists of two terms, equivalent term $u^{eq}(t)$, and switching term $u^s(t)$.

In the sliding phase, where $s(x,t) = 0$ and $\dot{s}(x,t) = 0$ the equivalent term $u^{eq}(t)$ is designed to keep the system on the sliding surface. In the approaching phase, where $s(x,t) \neq 0$, the switching term $u^s(t)$ is designed to satisfy the reaching condition (12).

While in sliding phase we have:

$$\dot{s}(x,t) = 0 \tag{13}$$

By solving the above equation formally for the control input, we obtain an expression for *u* called the equivalent control $u^{eq}(t)$, which can be interpreted as the continuous control law that would maintain $\dot{s}(x,t) = 0$ if the dynamics were exactly known.

In order to satisfy sliding conditions (12) and to despite uncertainties on the dynamic of the system, we add a discontinuous term across the surface $s(x,t) = 0$, so the sliding mode control law *u*(*t*) has the following form:

$$\begin{aligned} u &= u^{eq} + u^s \\ u^s &= -K \operatorname{sgn}(s(x,t)) \end{aligned} \tag{14}$$

Where *K* is the control gain.

For a defined function φ:

$$\operatorname{sgn}(\varphi) = \begin{cases} 1, & \text{if } \varphi \succ 0 \\ 0, & \text{if } \varphi = 0 \\ -1, & \text{if } \varphi \prec 0 \end{cases} \tag{15}$$

The controller described by the Equation (14) presents high robustness, insensitive to parameter fluctuations and disturbances (Utkin, 1977; Slotine and Li, 1998; Utkin, 1978; Khalil, 1992), but it will have high-frequency switching (chattering phenomena) near the sliding surface due to sgn function involved. These drastic changes of input can be avoided by introducing a boundary layer with width ε (Slotine and Li, 1998; Khalil, 1992). Thus replacing $\operatorname{sgn}(s(t))$ by $sat(s(t)/\varepsilon)$ in (14), we have:

$$u = u^{eq} - K\, sat(s(x,t)) \tag{16}$$

Where $\varepsilon \succ 0$

$$sat(\varphi) = \begin{cases} \operatorname{sgn}(\varphi) & \text{if } |\varphi| \geq 1 \\ \varphi & \text{if } |\varphi| < 1 \end{cases}$$

After this step, the objective is to determine a control law which drives the state trajectories along the surface.

Speed control

To control the speed of the induction machine, three surfaces are chosen. Variables of control are the rotation speed Ω and the flux Φ_{rd}. The flux will be maintained at its nominal value to have a maximal torque. We take *n=1*, the speed control manifold equations can be obtained as:

$$S(\Omega)=\Omega_{ref}-\Omega$$
$$\dot{S}(\Omega)=\dot{\Omega}_{ref}-\dot{\Omega} \qquad (17a)$$

Substituting the expression of $\dot{\Omega}$ equation *(5)* in equation *(17 a)*, we obtain:

$$\dot{S}(\Omega)=\dot{\Omega}_{ref}-\left[\frac{pM_{sr}}{JL_r}\left(\Phi_{rd}i_{qs}\right)-\frac{f}{J}\Omega-\frac{1}{J}C_r\right] \qquad (17b)$$

we take

$$i_{qs}=i_{qs}{}^{eq}+i_{qs}{}^{s} \qquad (17c)$$

During the sliding mode and in permanent regime, we have:

$$S(\Omega)=0,\quad \dot{S}(\Omega)=0,\quad i_{qs}{}^{s}=0 \qquad (17d)$$

Where the equivalent control is:

$$i_{qs}{}^{eq}=\frac{JL_r}{pM_{sr}\Phi_{rd}}\left(\dot{\Omega}_{ref}+\frac{f}{J}\Omega+\frac{C_r}{J}\right) \qquad (17e)$$

During the convergence mode, the condition $S(\Omega)\dot{S}(\Omega)\leq 0$ must be verified. We obtain:

$$\dot{S}(\Omega)=-\frac{p^2M_{sr}\Phi_{rd}}{JL_r}i_{qs}{}^{s} \qquad (17f)$$

Therefore, the correction factor is given by:

$$i_{qs}{}^{s}=-k_{\Omega}.sat(S(\Omega)) \qquad (17g)$$

To verify the system stability condition, the parameter k_{Ω} must be positive.

Stator current control and limitation

In order to limit all possible overshoot of the current i_{qs}, we add a limiter of current defined by

$$i_{qs}{}^{\lim}=i_{qs}{}^{\max}.sat\left(i_{qs}\right) \qquad (18a)$$

The current control manifold is

$$S\left(i_{qs}\right)=i_{qs}{}^{\lim}-i_{qs}$$
$$\dot{S}\left(i_{qs}\right)=\dot{i}_{qs}{}^{\lim}-\dot{i}_{qs} \qquad (18b)$$

Substituting the expression of $\dot{i}_{qs}$ equation *(2)* in equation *(18 b)*, we obtain:

$$\dot{S}\left(i_{qs}\right)=\dot{i}_{qs}{}^{\lim}-\left(-\omega_s i_{ds}-\gamma i_{qs}-p\Omega K\Phi_{rd}+\frac{1}{\delta L_s}u_{qs}\right) \qquad (18c)$$

The control voltage is

$$u_{qs}{}^{ref}=u_{qs}{}^{eq}+u_{qs}{}^{s} \qquad (18d)$$

$$u_{qs}{}^{eq}=\delta L_s\left(\dot{i}_{qs}{}^{\lim}+\omega_s i_{ds}+\gamma i_{qs}+p\Omega K\Phi_{rd}\right) \qquad (18e)$$

$$u_{qs}{}^{s}=-k_{iqs}.sat\left(S\left(i_{qs}\right)\right) \qquad (18f)$$

To verify the system stability condition, the parameter $k_{i_{qs}}$ must be positive.

Flux control

In the proposed control, we take n=2 to appear control $u_{ds}{}^{ref}$, the manifold equation can be obtained by:

$$S(\Phi_{rd})=\lambda\left(\Phi_{rd}{}^{ref}-\Phi_{rd}\right)+\left(\dot{\Phi}_{rd}{}^{ref}-\dot{\Phi}_{rd}\right)$$
$$\dot{S}(\Phi_{rd})=\lambda\left(\dot{\Phi}_{rd}{}^{ref}-\dot{\Phi}_{rd}\right)+\left(\ddot{\Phi}_{rd}{}^{ref}-\ddot{\Phi}_{rd}\right)$$

(19a)

The control voltage is

$$u_{ds}{}^{ref}=u_{ds}{}^{eq}+u_{ds}{}^{s} \qquad (19b)$$

$$u_{ds}{}^{eq}=\delta L_s\left(\left(\ddot{\Phi}_{rd}{}^{ref}+\lambda\dot{\Phi}_{rd}{}^{ref}+\left(\frac{1}{T_r}-\lambda\right)\dot{\Phi}_{rd}\right)\frac{T_r}{M_{sr}}-\left(-\gamma i_{ds}+\omega_s i_{qs}+\frac{K}{T_r}\Phi_{rd}\right)\right) \qquad (19c)$$

$$u_{ds}{}^{s}=-k_{\Phi rd}.sat\left(S\left(\Phi_{rd}\right)\right) \qquad (19d)$$

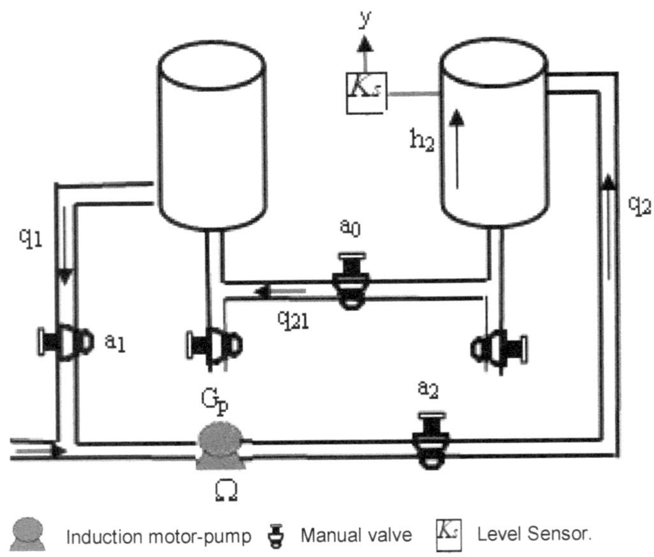

Figure 4. Descriptive scheme of hydraulic system.

To verify the system stability condition, the parameter $k_{\Phi_{rd}}$ must be positive.

The selection of coefficients k_{Ω}, $k_{i_{qs}}$, $k_{\Phi_{rd}}$ and λ must be done in order to satisfy following requirements:

1. Existence condition of the sliding mode, which requires that the state trajectories are directed toward the sliding manifold,
2. Hitting condition, which requires that the system trajectories encounter the manifold sliding irrespective of their starting point in the state space(insure the rapidity of the convergence),
3. Stability of the system trajectories on the sliding manifold,
4. Not saturate the control to allow the application of the control discontinuous.

Dynamic model of hydraulic system

The second party of the global procedure is formed by a coupled tank, considered as a benchmark for the study and the analysis command of hydraulic systems. This device, allowed examining the law order in the liquid level reservoirs, while varying the debit from the variation of the speed of the pump.

This process behaves two vertical tanks coupled by a flow canal, a manual valve used to change the canal section, in consequently, to change the characteristics of flow between the reservoirs (Figure 4). A level sensor is installed in the top of every reservoir. The relation between the speed of the IM Ω and the entry debit of the second reservoir q_2:

$$q_2 = a_1.a_2.G_p.\Omega \tag{20a}$$

The second reservoir can be filled from the first reservoir by the intermediary of the canal 1, while the debit of the second reservoir towards the first is assured by the canal 2. The equilibrated equation of the flow, for the first reservoir:

$$\dot{h}_1 = \frac{1}{S}\left(q_{21} - q_1\right) \tag{20b}$$

For the second reservoir:

$$\dot{h}_2 = \frac{1}{S}\left(q_2 - q_{21}\right) \tag{20c}$$

Such as:

$$q_1 = s_1.\sqrt{2gh_1} \tag{20d}$$

$$q_{21} = s_2.a_0\sqrt{2g\left(h_2 - h_1\right)} \tag{20e}$$

with:
h_1: the level in the first tank
h_2: the level in the second tank,
q_1: the inlet flow rate,
q_2: the flow rate from Tank1 to Tank2,
q_{12}: the flow rate out of Tank2,
g: the gravitational constant,
S: the cross-section area of Tank1 and Tank2,
s_1, s_2 the area of the coupling orifice,
a_1, a_2 the coefficients of the manuals valve1 and valve2,
a_0 the offloads coefficient valve 3,
G_p: the pump gain

Finally, the hydraulic system is governed by the following non linear differential equation:

$$\begin{cases} \dfrac{dh_1}{dt} = \dfrac{1}{S}\left(s_2.a_0\sqrt{2g\left(h_2 - h_1\right)} - s_1.\sqrt{2gh_1}\right) \\ \dfrac{dh_2}{dt} = \dfrac{1}{S}\left(a_1.a_2.G_p.\Omega - s_2.a_0\sqrt{2g\left(h_2 - h_1\right)}\right) \\ y = K_s.h_2 \end{cases} \tag{20f}$$

K_s: the level sensor gain.

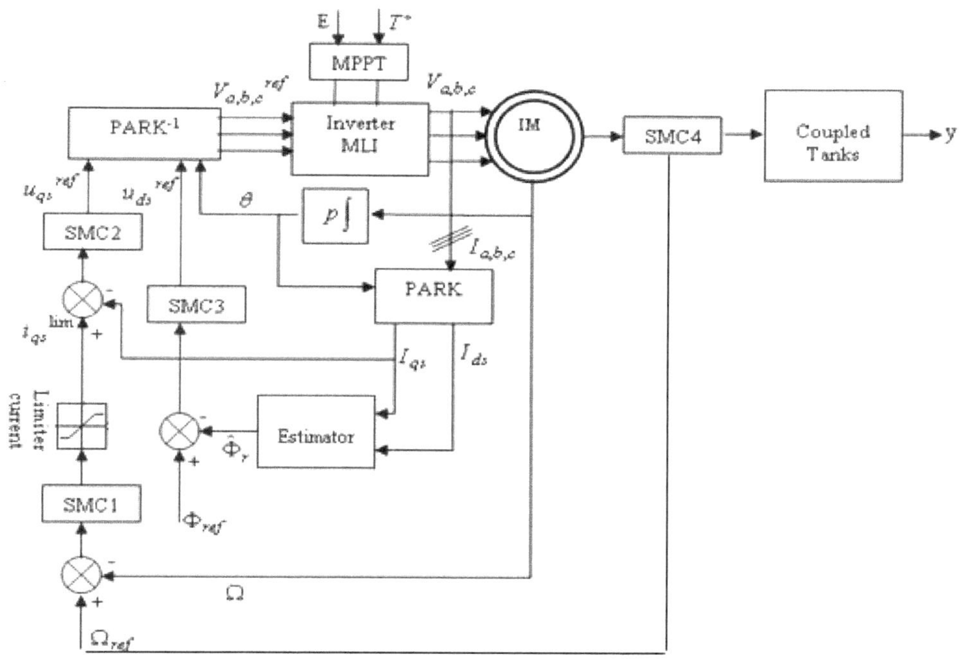

Figure 5. Block diagram of the proposed control scheme of IM coupled in hydraulic system using the sliding mode controllers.

The centrifugal pump is also described by characteristic H(Q) given by (Caro and Bonal, 1997):

$$b_0\Omega_m{}^2 + b_1\Omega_m Q + b_2 Q^2 = H_p + XQ^2 \quad (20g)$$

Where b_0, b_1, b_2 and X are constant parameters. *Hp* and *Q* are respectively the geometric head and the flow of pumped water.
In the model of the Figure 2, Ω_{min} being the minimal speed from which the pump starts to generate a pumped water flow. It is given by the following relation (Caro and Bonal, 1997):

$$\Omega_{\min} = \sqrt{\frac{-4(b_2 - X)H_p}{b_1{}^2 - 4(b_2 - X)b_0}} \quad (20h)$$

Three cases are possible:

1. if $\Omega < \Omega_{min}$ no flow rate are generated :

$$Q=0 \quad (20i)$$

2. if $\Omega = \Omega_{min}$ the pump starting to pumped water:

$$Q = Q_{\min} = \frac{b_1\Omega_{\min}}{2(b_2 - X)} \quad (20j)$$

3. if $\Omega > \Omega_{min}$ the flow pumped water is given by :

$$Q = \frac{-b_1\Omega - \sqrt{(b_1\Omega)^2 - 4(b_2 - X)(b_0\Omega^2 - H_p)}}{2(b_2 - X)} \quad (20k)$$

Description of the system

The bloc diagram of the proposed robust control scheme is presented in Figure 5. The blocks SMC1, SMC2, SMC3, and SMC4 represent the proposed sliding mode controllers. To avoid the appearance of an inadmissible value of current, a saturation bloc is used.

The rotor flux Φ_r is estimated by the "flux rotor observer". The block 'IM' represents the induction motor. We implement the previous sliding mode algorithms using a MATLAB/Simulink simulator.

RESULTS

In order to validate the control strategies as discussed above, we have studied the speed performances with current limitation described in Figure 5. The feedback control algorithms were iterated until best simulation results were obtained. Figure 7 shows the robustness tests in relation to inertia variations. Figure 8 shows the robustness tests to the stator and rotor resistance variations. The speed and flux references are respectively (Figure 6):

$$\Omega_{ref} = 100 rd/s = 960 rpm, \quad \Phi_{r0} = 0.6 wb$$

In the second party of the global system, the bloc SMC4 permits to find out the pursuit of the liquid level in the second tank. Figure 10 shows the robustness tests in relation to inertia variations. Figure 11 shows the

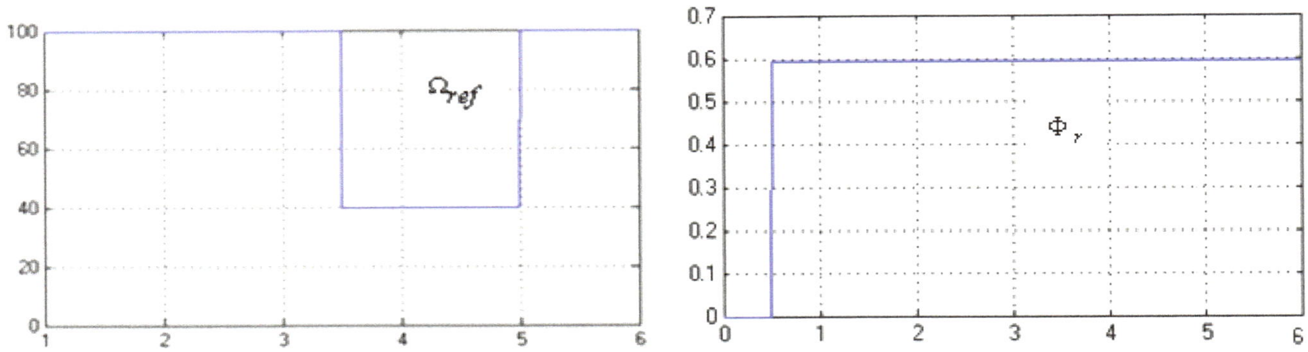

Figure 6. Reference of speed and rotor flux.

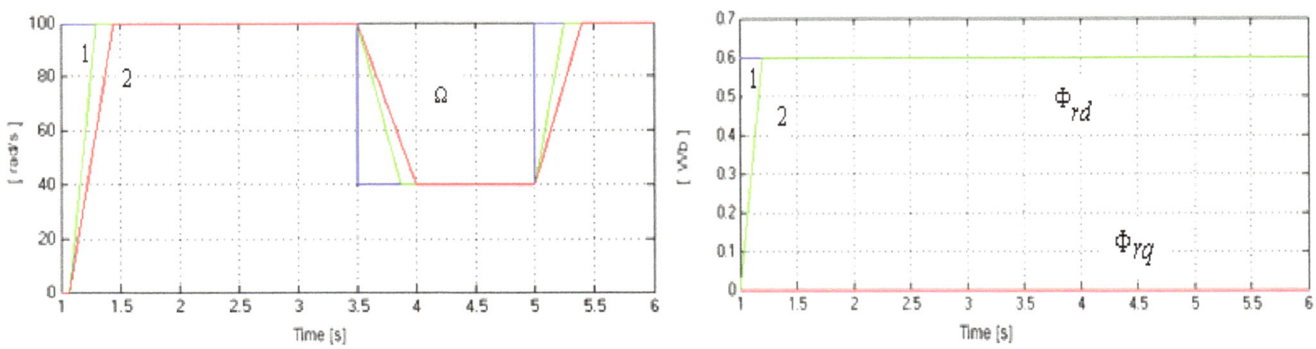

Figure 7. System responses. (Curves **1**: Nominal case; curves **2**: an increase of Inertia 2J).

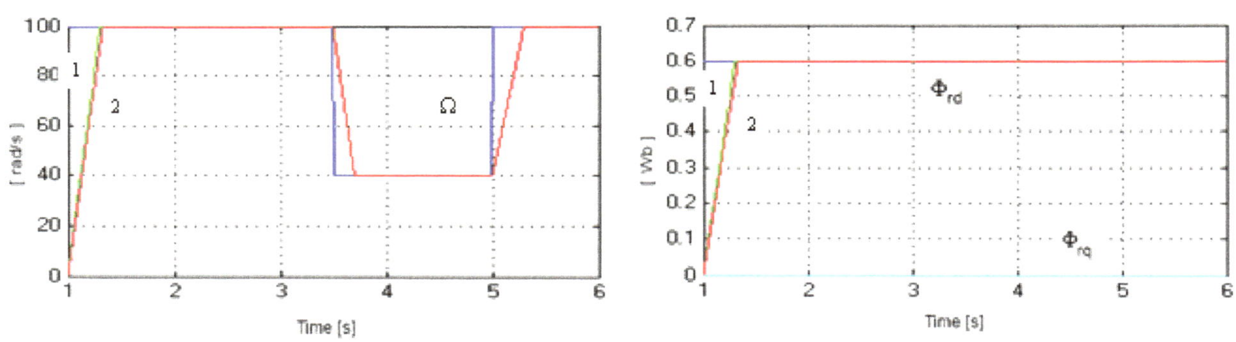

Figure 8. System responses. (Curves **1**: Nominal case; curves **2**: an increase of resistances $1.5R_s$, $1.5R_r$).

robustness tests to the stator and rotor resistance variations.

The SMC4 law (Figure 12) is given by:

$$S(h_2) = h_{2ref} - h_2 \tag{21a}$$

$$u = u^{eq} + u^s \tag{21b}$$

where

$$u^{eq} = \frac{S}{a_1.a_2.G_p}.\dot{h}_{2ref} + \frac{s_2.a_0}{a_1.a_2.G_p}.\sqrt{2.g}.\sqrt{h_2 - h_1}$$

$$u^s = -K^{'}.sat(S(h_2)) \tag{21c}$$

$K^{'}$ a positive constant

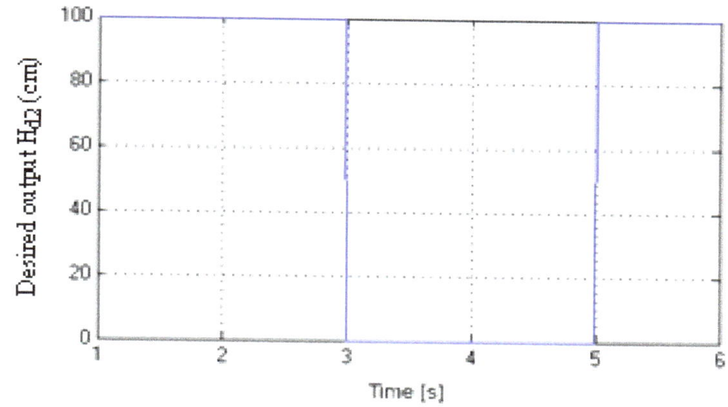

Figure 9. Reference of the level in the second tank.

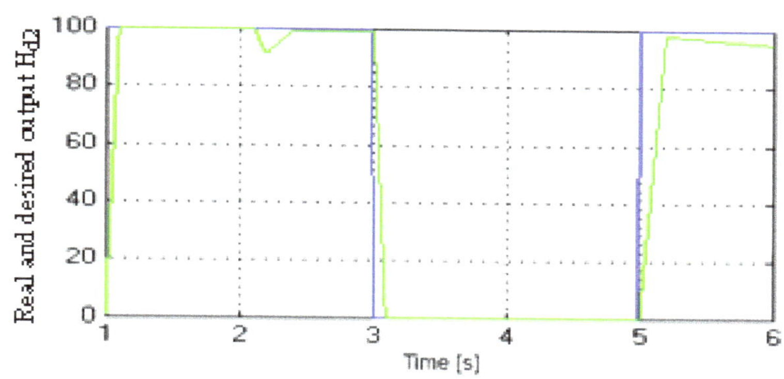

Figure 10. Real and desired output Hd_2, case an increase of Inertia 2J.

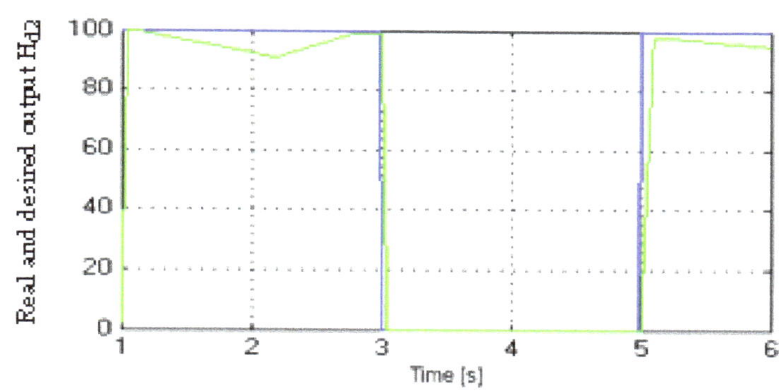

Figure 11. Real and desired output Hd_2, case an increase resistances $1.5R_s$, $1.5R_r$.

Conclusion

A sliding mode control method has been proposed and used for the control of an induction machine. It has been shown the robustness of proposed control. The speed control operates with enough stability and has strong robustness to parameter variations.

The control of two coupled tanks is assured by the speed variation. The performances of the controlled system are studied under variations in system parameters and in the presence of external disturbance.

The simulation results indicate that the proposed

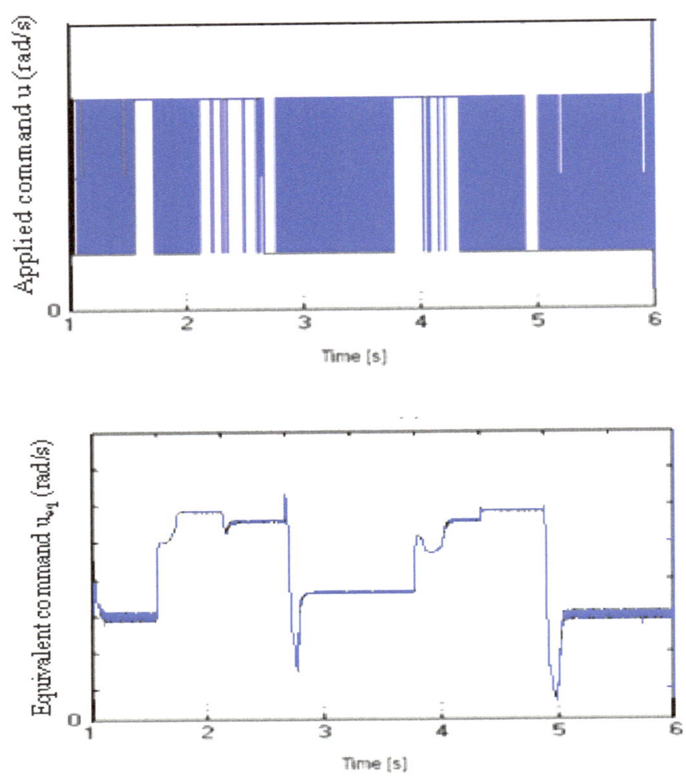

Figure 12. Command applied to the entry of the hydraulic system.

control schemes work very well and are robust to change in the parameters of the system as well as to disturbances acting on the global system.

Furthermore, this regulation presents a simple robust control algorithm that has the advantage to be easily implantable in calculator

REFERENCES

Buhler H (1986). Adjusting sliding mode, EPFL Press romandes, Lausanne.

Caro JM, Bonal J (1997). Electrical Variable Speed Drives. Lavoisier, p. 1.

Dhafer M, Andouls, R, Mami A, Dauphin-Tanguy G (2007). Bond graph modelling of a photovoltaic system feeding an induction motor-pump. Simulation Modelling Practice and Theory, (Elsevier), 15: 1224–1238.

Khan MK, Spurgeon SK (2006). Robust MIMO water level control in interconnected twin-tanks using second order sliding mode control. Control Eng. Prac., (14): 375–86.

Khalil HK (1992). Non linear system, New York: MacMillan.

Mimouni MF, Mansouri MN, Benghanem B, Annabi M (2004). Vectorial command of an asynchronous motor fed by a photovoltaic generator. Renewable Ener., (29): 433-442.

Olorunfemi O (1991). Analysis of current source induction motor drive fed from photovoltaic energy source. IEEE Trans. on Energy Conversion, 6(6): 99-106.

Pan H, Wong H, Kapila V, Queiroz MS (2005). Experimental validation of a nonlinear backstepping liquid level controller for a state coupled two tank system. Control Eng. Prac., (13): 27-40.

Slotine JJE, Li W (1998). Applied nonlinear control, USA: Prentice-Hall.

Utkin VI (1977). Variable structure systems with sliding modes. IEEE Trans. Automat. Contr., 2(22): 212-222.

Utkin VI (1993). Sliding mode control design principles and applications to electric drives. IEEE Trans. Ind. Electronic, (40): 23-36.

Utkin VI (1978). Sliding modes and their application in variable structure system. Moscow: MIR.

APPENDIX

Motor parameters

Nominal power output P = 750W
Nominal voltage: 220V
Nominal current: 1.6A
Number of pole pairs: p=2
Rotoric resistance: Rr = 0.79 Ω
Statoric resistance: Rs = 1.47 Ω
Rotoric cyclic inductance Lr = 0.094 H
Statoric cyclic inductance Ls = 0.105 H
Mutual cyclic inductance Msr = 0.094 H
Moment of inertia J = 0.00256 Kgm^2
Friction coefficient f = 0.0029 Kgm^2s^{-1}

Pump parameters

Geometric head Hp = 0.1 m
b0 = 1.61 10-04 m s^2 rd^{-2}
b1 = 2.584 10-03 m s^2 l^{-1} rd^{-1}
b2 = -0.49 m l^2 s^{-2}
X=0.98388 m s l^{-1}

Coupled reservoirs

The cross-section area of Tanks: S = 1.1310 m^2
The area of the coupling orifices s_1, s_2 0.0079 m^2
The coefficients of the manuals valves a_1, a_2, a_0 ~ 1
The pump gain G_p = 7.5 m^3. s^{-1}.v^{-1}
The level sensor gain K_s = 24.5 v.m^{-1}
The gravitational constant g = 9.8 m.s^{-2}

5

Formulation and characterization of self compacting concrete with silica fume

Kennouche .S.[1] , Zerizer .A.[1], Benmounah .A.[1], Hami .B.[1], Mahdad .M.[2], Benouali .H.[2] and Bedjou .S.[2]

[1]Université de Boumerdès, Unité de recherche Matériaux-Procédé-Environnent, FSI, cite Frantz Fanon, 35000 Boumerdès, Algérie.

[2](C.N.E.R.I.B) Centre National d'Etudes et de Recherches Intégrées du Bâtiment, Cité Nouvelle El Mokrani Souidania – Alger.

Self-compacting concrete (SCC) was elaborated using local materials and silica fume (SF) as admixture in 15% of cement quantity, two different Portland cements (PC) and two different superplasticizer that the chemical nature is polycarboxylate and plynaphtalene, the aggregates used are (AG 3/8 mm, AG 8/15 mm), coarse and fine sand (SC, SF) witch fineness modulus 3.2 and 1 in the order. The dosage of the different superplasticizer used is chosen after experimental spreading tests of each self compacting concrete formulation. Results of fresh concrete tests executed, as L-box and segregation resistance are on concordance whit values recommended by the French association of civil engendering. Also the mechanical characterization was conducted by compressive strength and splitting compression testing procedure, results values are in the range higher than 20 Mpa at the seven day by the compressive test for the all compositions, and the highest value was 40.93 MPa at the 28 day bay compressive test of the fourth's formulation specimens, the values of splitting compressive tests of al formulation specimens at 7, 14 and 28 days, was situated between 2.01 and 4.40 MPa. In order to determine the superplasticizer saturation assay in of cement pasts used in self compacting concrete, the stady was completed by a rheological stady with a variable velocity gradient, so as to estimate the quantity of saturation assay of superplasticizer and the formulation, also the flow models of cement paste.

Key words: Self-compacting concrete (SCC), silica fume, rheology, stress.

INTRODUCTION

Self-compacting concrete (SCC) have high fluidity; their implementation that allows concrete to be placed in complicate forms without any strain vibration and prohibit segregation. Superplasticizers and other additives are used production of SCC; superplasticizers impact high fluidity and prevent segregation, the additions of fillers reduce the quantity of superplasticizers used in SCC mixes, compared to normal concrete (Yahia et al., 1999). Also, the use of self-compacting concrete in construction reduces the achievement time at 40% compared to normal concrete (Perssoiv et al., 1998; Nocher, 2001).

Silica fume (SF) is one of several types of industrial byproducts generated. With increased environmental awareness which is a byproduct of the smelting process in the silicon and ferrosilicon industry. Silica fume is very effective in the design and development of high strength high performance concrete (Rafat, 2011).

The addition of fillers and the replacement of cement in SCC participate to reduce the gas emissions, in cement production Malhotra and Mehta (2002). Miura et al. (1993)

studied the effect of replacing 30% of cement by fly ash in concrete, they concluded at the amelioration of fluidity and workability of concrete, also Bouzoubaa and Lachemi (2001) in their work replaced 50% of the cement used for the formulation of SCC, the result in compressive test was 35 Mpa at 28 days.

Silica fume is used at a dosage corresponding between 5 and 10% of the total weight of cementitious materials (Steven et al., 2004). The silica fume is used in a high level impermeability application required and the high strength concrete. When the concrete is exposed to deicing salts, the maximum amount of silica fume should be at 10% of the mass of cementitious materials, (Steven et al., 2004). The addition of silica fume is very advantageous, addition of 5, 10 and 15% of silica fume cement supply, allows a reduction of the phenomenon of swelling of 121% to 16 day trial. In others papers we found the variation of the dosage of silica fume used in SCC from 0 to 20% (Raharjo et al., 2013).

The rheology of cement pastes was studied bay (Zhang and Han (2000) they conclude that the adding of mineral mixtures in cement pastes decrease the viscosity and the elastic strain limit compared to the pastes without additions. Zhang et al. (2000) and Park et al. (2005) noted that the adding of silica fume in paste cement decrease the viscosity and elastic stress limit. The variation of the dosage on silica fume from 10 inutile 25% in paste cement, with fixed superplasticizers at 3%, produces a small increase of shear stress (Cyr et al., 2000).

The investigation conducted by Salem et al. (2002) on the effect of adding silica fume for 10, 20 and 50%, with a (W / C) ratio = 0.6, concluded at hysteresis effect, which explains the thixotropic behavior of cement pastes.

The objective of this work is the formulation of a self-compacting concrete with local materials with the addition of silica fume, and the study of the rheological behavior of fresh SCC mix.

MATERIALS AND METHODS

Aggregates

In this part of the work, physical and mechanical characterization of local aggregates used was determined, with previous correction of the sand which was levy from the river of Baghlia (Boumerdes, North of Algeria), with fineness modulus Mf = 3.2. It was considered as rude sand and correction done by fine sand in which a fineness modulus is f = 1, resulting from BOUSSAADA (South of Algeria). Sand corrected by the Abrams formula, the fineness of sand corrected is Mf = 2.72.

Characteristics of materials

Table 1 summarizes the characteristics of the aggregates used in this work.

SCC mix proportions

In this work, the calculation of mix proportions for SCC was done using the Japanese method. The specifications of this method are as follows (Philippe, 2004).

(i) The volume of gravel is fixed according to the AG / S; ration (Agregate/Sand) is close to 1.
(ii) The volume of the dough is made between 330 and 350 kg/m^3.
(iii) The weight of the cement varies from 300 to 400 kg/m^3.
(iv) The mass additions up to 200 kg/m^3.
(v) For the superplasticizer dosage close to the saturation dosage is recommended.

Superplasticizers

Superplasticizers used in this work are polycarboxylate (Medaflow 30), which is high range reducing water; it is called the third superplasticizer generation, and polynaphthalene (Medaplast SP 40); a high range water reducing superplasticizer type polynaphthalene sulfone (PNS), to obtain concrete and mortars of high quality. It comes from the Granitex company, a packaging company and distributor of various adjuvants used in construction in Algeria.

Silica fume

Silica fume is an industrial product in the form of very fine powder, with a SiO_2 content of over 90 and 3% in loss on ignition.

Self-compacting concrete formulation

SCC formulation is performed according to the Japanese method. We set the G / S 0.9, the cement is 400 k/m^3, the ratio (W / C + F) is set at 0.42, the mass of the silica fume is 15% of the weight of the cement.

Stock volume is calculated by the following formula:

The cement paste = cement + water + addition (silica fume) + superplasticizer.

Therefore the total volume of the cement paste is the sum of the volumes of these constituents.

(i) Volume of cement: 400/3.1 = 129.03 L;
(ii) Volume of silica fume: 60/2.25 = 26.66 L;
(iii) The volume of water: E / (C + SF) = 0.42 then E = 193.2 L;
(iv) The volume of dry aggregates: 1000 - the volume of the cement paste;
(v) The volume of superplasticizer dosage is determined according to its saturation obtained by averaging the test to Abrams cone.

Table 2 summarizes the amounts of the four volume formulations BAP.

Characterization of self-compacting concrete

Fresh concrete

After making concrete, we went to his characterization fresh. We used to test the box L, a box that we made to the dimensions specified in the theory.

The results obtained during tests on fresh concrete as follows: L box testing, sieve stability and spreading the Abrams cone are shown in the Table 3.

The results obtained during tests in L-box (Figure 1 are higher than 80% (Table 3), this is due to the fluidity of different SCC. And segregation resistance testing sieve, give results lower than 15% (Table 3). Values and spreading the Abrams cone is between (650

Table 1. Physical and mechanical properties of materials.

Aggregates 3/8	**Aggregates 8/15**
Density 1400 kg/m^3	1400 kg/m^3
Density 2620 kg/m^3	2620 kg/m^3
Los Angeles 26%	25%
Micro Deval 17%	15%
Sand Fineness modulus (fM)	
fM sand Baghlia*	2
fM sand BOUSSAADA*	1
fM sand corrected	2.78

*Aggregates quarry in Algeria.

Table 2. Quantity (kg) of the components in each formulation.

Formulation Code	Cement	Water	Silica fume	Aggregate 3/8 (mm)	Aggregate 8/15 (mm)	Coarse sand	Fine sand	Superplasticizer SP
F I	400	193.2	60	533.14	265.04	582.92	243.78	7.2
F II	400	193.2	60	534.27	264	582.41	241.66	8.0
F III	400	193	60	534.08	264.99	584.64	242.55	6
F IV	400	193.1	60	534.27	264	243.69	241.46	8.0

Table 3. Test on fresh concrete.

Formulation Code	L-Box	Segregation resistance (%)	Staggering test (cm)
F I	0.81	3	68
F II	0.84	5	71
F III	0.82	7	67
F IV	0.86	6	67.5

Figure 1. L-Box.

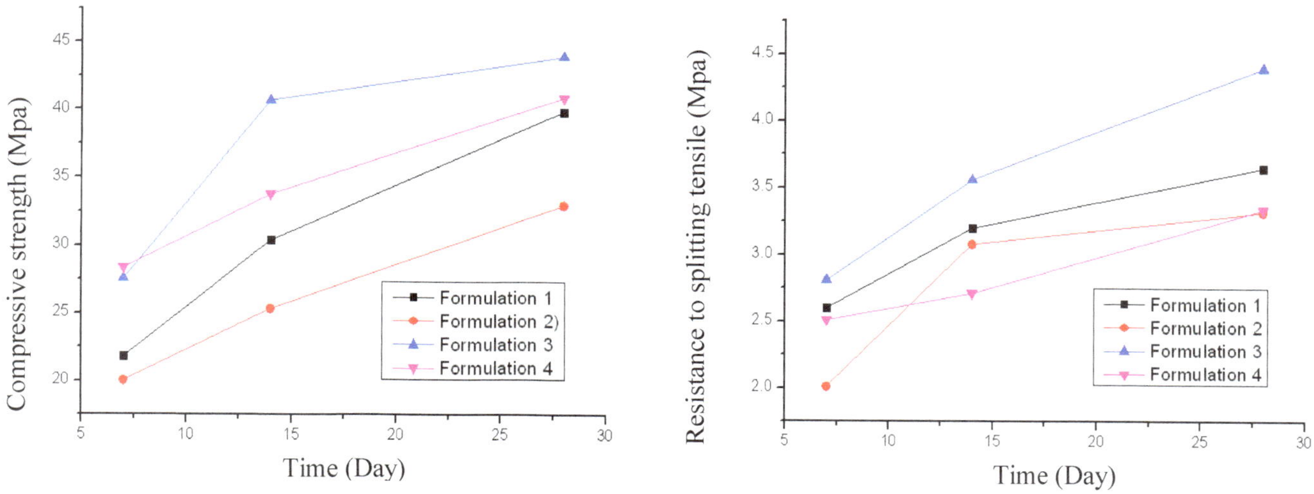

Figure 2. Evolution of mechanical strength in tension and compression.

Figure 3. Concrete specimen.

and 750 mm) (Table 3), in summary, all these values are within the range recommended by the AFGC to obtain a SCC.

Characterization of SCC hardened state

Developed cylindrical specimens were tested in compression and splitting tensile, as shown in histograms (Figure 2).

RESULTS

The results of fresh properties of all SCC mixtures are included in (Table 3), as seen all SCC mixtures exhibited satisfactory slump flows in the range 650 and 750 mm, which is an indication of a good deformability. In addition L-box test is more sensitive to blocking. There is a risk of blocking of the mixture when the L-box blocking ratio is below 0.8. Blocking ratio of SCC produced with all formulations is also given in Table 3. Blocking ratio (h_2/h_1) must be between 0.8 and 1.00. Maximum L-box test and all the mixtures of SCC have remained in target range which is as per EFNARC standards. Also the percentages of milt obtained by segregation resistance testing on sieve can be explained by the absence of segregation, due to the viscosity of the paste that stops separation of aggregates from the past.

The hardened properties that were stated for all the mixtures were the compressive strength and splitting tensile of SCC specimens (Figure 3) of each formulation. It can be seen from Figure 2 that the compressive strength values are in the range of 20 to 26 MPa at 7

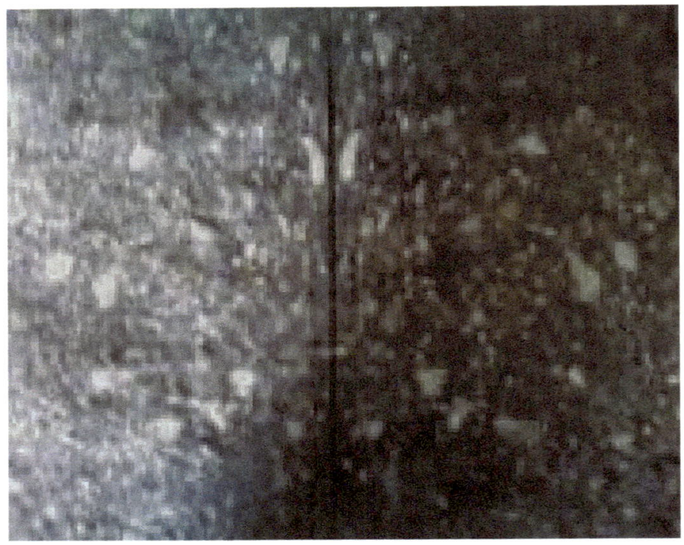

Figure 4. Area of specimen after splitting test.

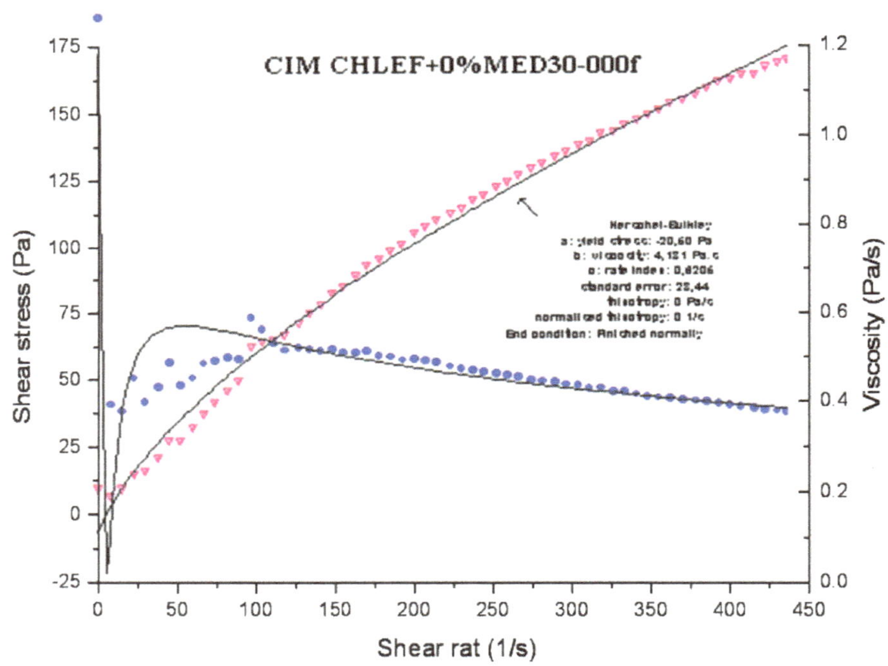

Figure 5. Paste (Cement + 0% Chlef Medaflow 30). First formulation (Cement CHLEF + Medaflow 30).

days, and 35 to 45 Mpa at 25 days (Figures 4 and 5).

Rheological cement pastes

The second part of the work is devoted to the study of rheological cement pastes. The aim of this part of the work is to determine the dosage of superplasticizer saturation which is to be several variations of pulp were analyzed rheometer AR 2000, several dosages in superplasticizers. The rheological curves are shown in Figure 6 to Figure 12 below:

For this purpose, four cement pastes were prepared. The proportions of these paste mixtures were prepared so that they are similar to the pastes in the concrete

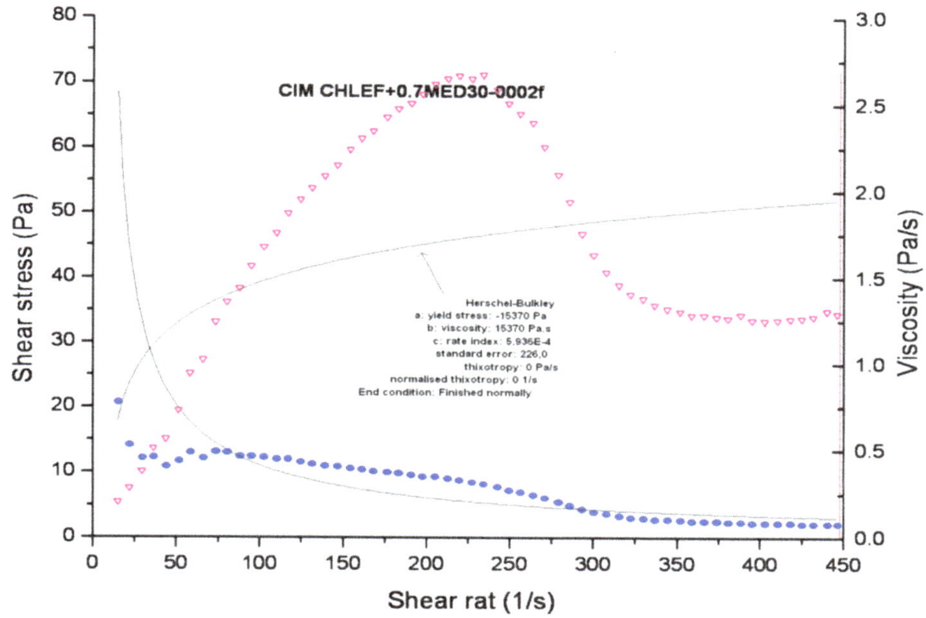

Figure 6. Paste (Cement + 0.75% Chlef Medaflow 30). First formulation (Cement CHLEF + Medaflow 30).

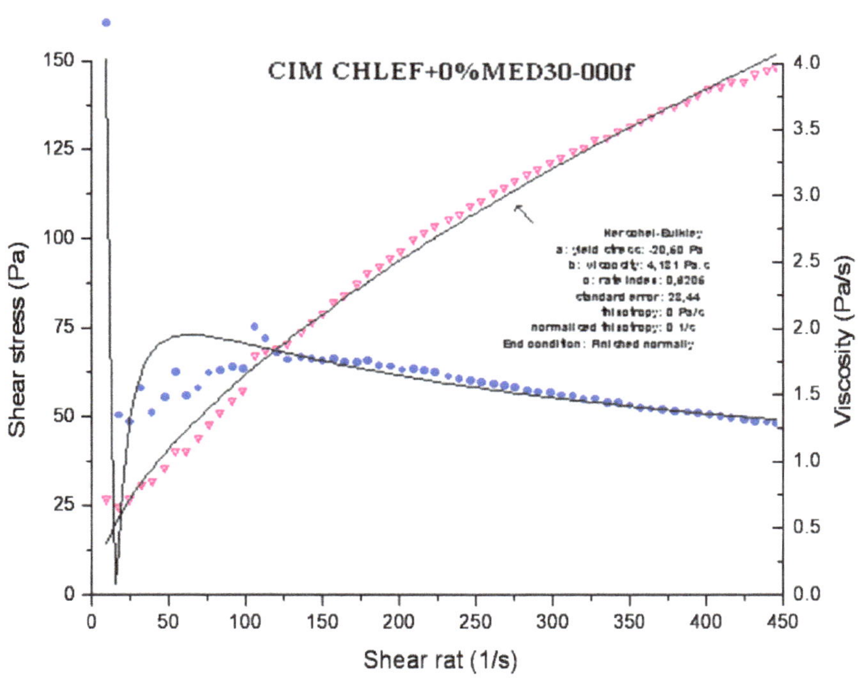

Figure 7. Paste (Cement Chlef +0.25% SP40). Second formulation (Cement Chlef + Medaplast SP 40).

mixtures, so all pastes contained the same quantity of cement, water and silica fume, but the dosage ranging of superplasticizers are varied from 0, 0.25, and 0.75 for Medaflow 30, and from 0, 0.25, and 0.50%. The other hand, the dosage of Medaplast 40 is varied from 0.25 utile 0.5 %, as shown in rheological testes in Figure 7, 8

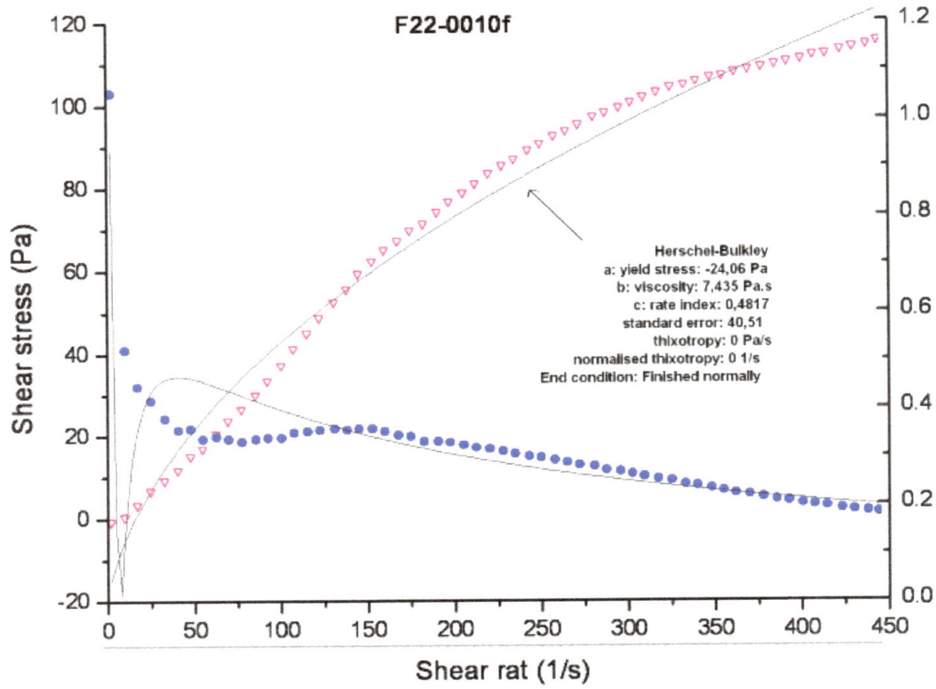

Figure 8. Paste (Cement Chlef +0.75% SP 40). Second formulation (Cement Chlef + Medaplast SP 40).

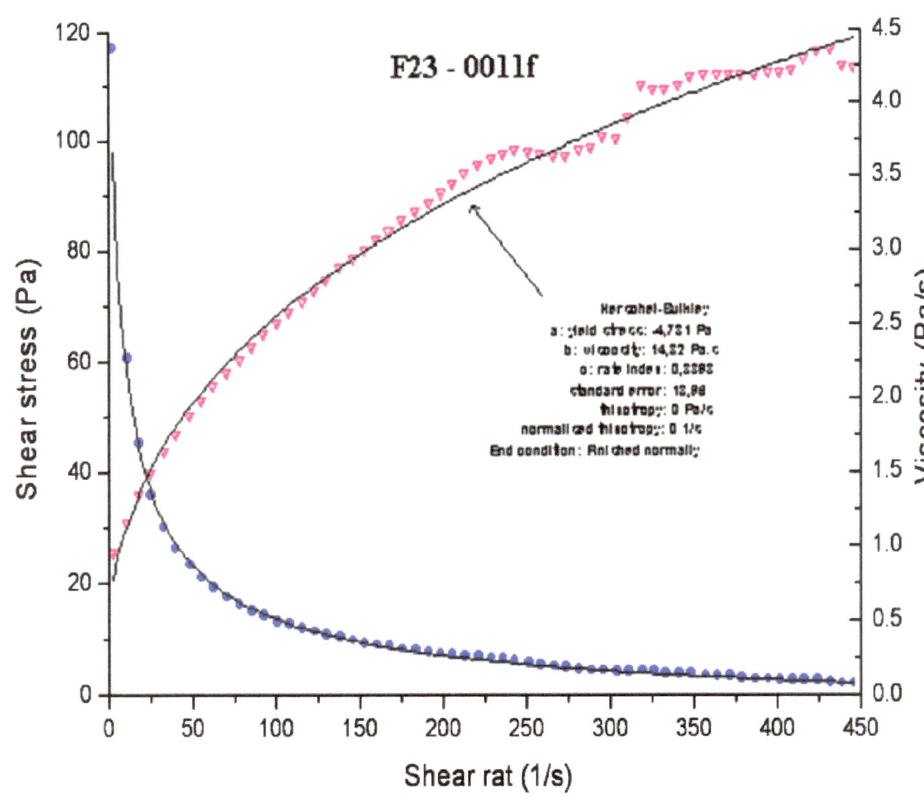

Figure 9. Paste (Cement ACC +0.25% Medaflow 30). Third formulation (Cement ACC + Medaflow 30).

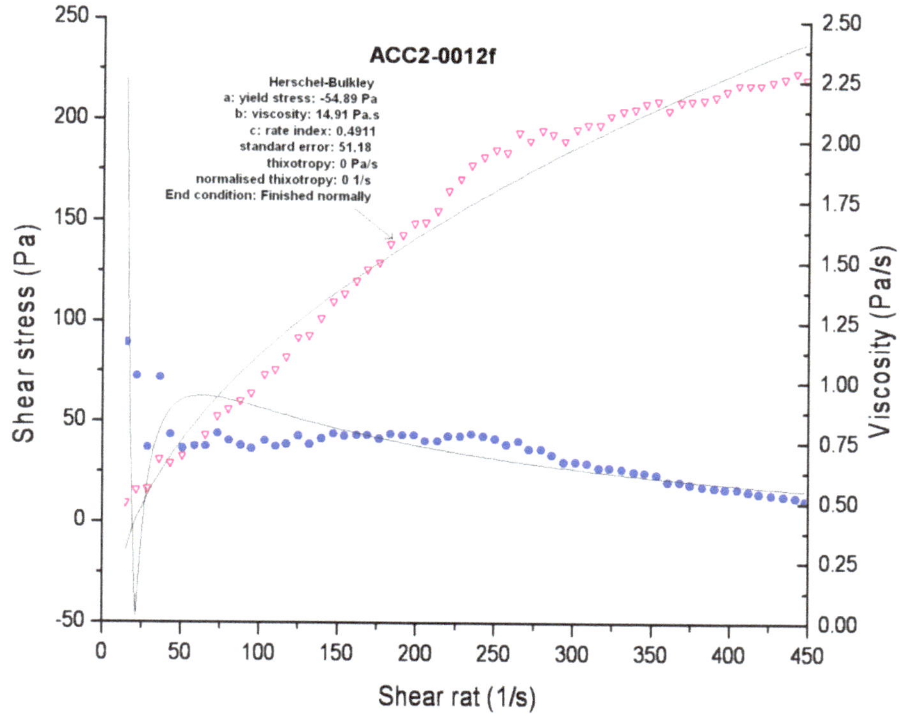

Figure 10. Paste (Cement ACC +0.5% Medaflow 30). Third formulation (Cement ACC + Medaflow 30).

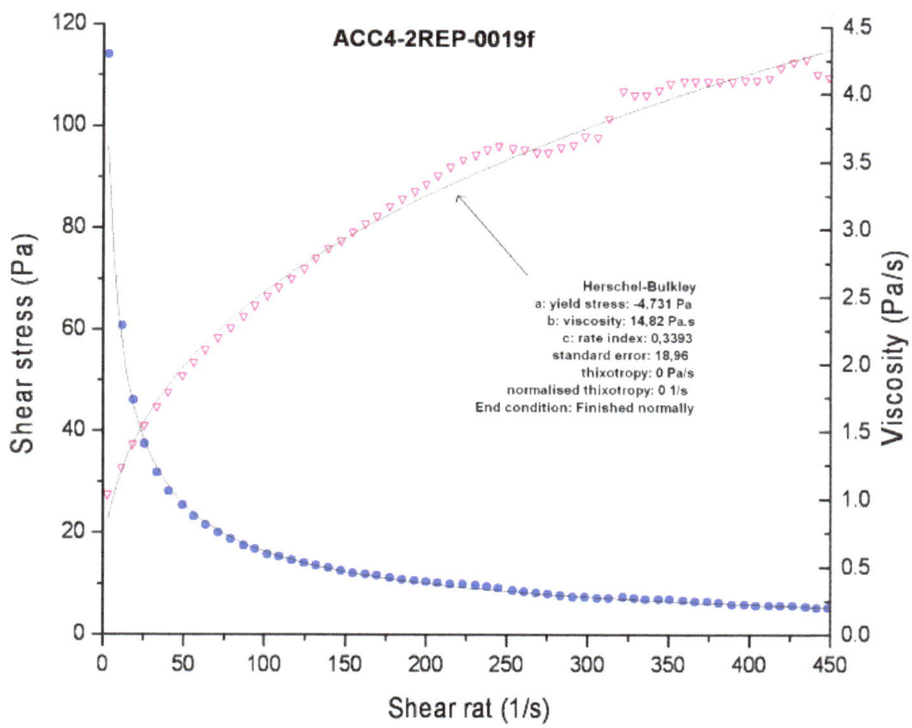

Figure 11. Paste (Cement ACC +0.5% Medaplast SP 40). Fourth formulation (ACC Cement Medaplast + SP 40).

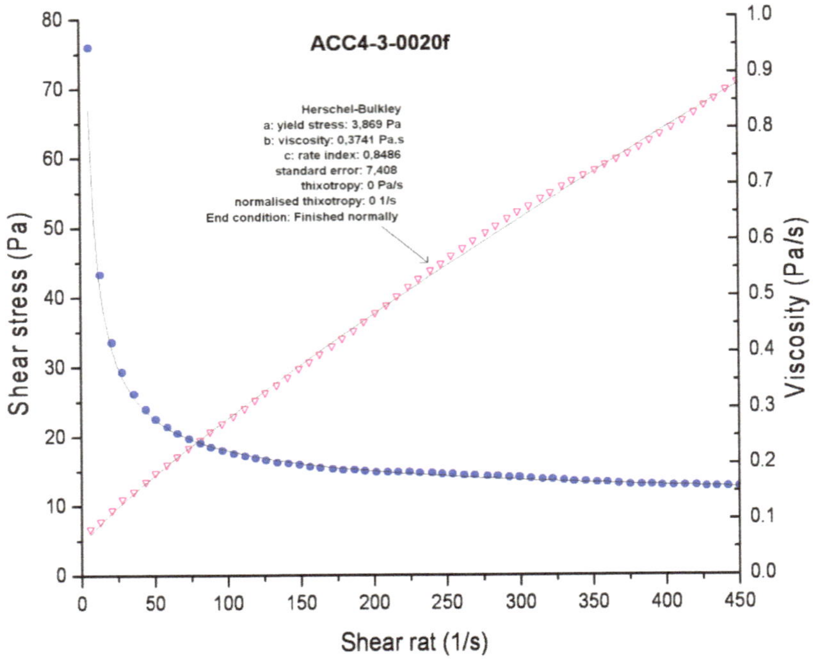

Figure 12. Paste (Cement ACC +0.5% Medaflow 30). Fourth formulation (ACC Cement Medaplast + SP 40).

and 11.

DISCUSSION

Results of rheological test show that the dosage of superplasticizer (Medaflow 30) polycarboxylate is lower than those of type sulfonated polynaphthalene in both of cement paste formulation.

According to mineralogical compositions of the two cements, the percentages of C_3A and C_4AF in cement ACC are higher than those of Chlef cement, as the superplasticizers have a preference to adsorb on the charges is higher than those present on the surfaces of silicate (C_3S and C_2S).

The percentages obtained superplasticizers are high compared to those found in the literature, this is explained by a very high surface area of the silica fume (200 m^2/g), and the architecture and shape of a large superplasticizers influence on the affinity (superplasticizer cement grain), which explains the difference of the percentages of superplasticizers found during the rheological tests and those obtained on SCC fresh.

The rheograms show that cement pastes studied elapse following the Herschel-Bulkley model, which governs the flow threshold materials (flowing with application of low critical stress).

Conclusion

In this paper, SCC elaborated from local materials (Algeria) based on Japanese formulation method. We fixed dosage of silica fume at 15% to cement content, the characterization of fresh concrete, given a good result. The diameter of the cake obtained is between 650 and 750 mm, the percentage of milt in the sieve stability is <15% and the L-box test, confirm that the SCC flows through without blocking in reinforcements with is explained by the high fluidity and absence of segregation aluminates, which the presence on surface of positive given by the incorporation of superplasticizers which suspend the cement.

The mechanical strength results are higher than 25 MPa for compressive strength, which show a very low porosity, this is due to the filler effect of silica fume and the is equal to 0.42 less than the normal concrete.

The rheograms obtained demonstrate the effect of superplasticizers on cement grout flow studied, which are materials flow threshold and follows the rheological model of Herschel-Bulkley. The viscosity of the cement pastes studied is inversely proportional to the percentage of superplasticizer incorporated, up to a saturation assay where the viscosity remains fixe, which corresponds to a minimum threshold stress, and flow problems.

REFERENCES

Bouzoubaa N, Lachemi M (2001). Self-compacting concrete incorporating high volumes of class F fly ash: Preliminary results. J. Cem. Concr. 31:413–20.

Cyr M, Legrand C, Mouret M (2000). Study of the shear thickening effect of superplasticizers on the rheological behaviour of cement

Heba A (2011). Effect of fly ash and silica fume on compressive strength of self-compacting concrete under different curing conditions. Ain Shams University, Department of Structural Engineering, Faculty of Engineering, Zagazig University, Egypt. J. Ain Shams Eng. 2:79–86.

Malhotra VM, Mehta PK (2002). High-performance, high-volume fly ash concrete. Suppl. Cem. Mater. Sustain. Dev. Inc. Canada, 24(7):30-34.

Miura N, Takeda N, Chikamatsu R, Sogo S (1993). Application of super workable concrete to reinforced concrete structures with difficult construction conditions. Special Publication, Am. Concr. Inst. USA 140:163–186.

Nocher M (2001). Flowing Smoothly and Quietly. Advanced Concrete and Masonry Center, Concrete Quarterly, UK. p. 198.

Park CK, Noh MH, Park TH (2005). Rheological properties of cementitious materials containing mineral admixtures. J. Cem. Concr. 35:842–849.

pastes containing or not mineral additives. J. Cem. Concr. 30:1477–1483.

Perssoiv B (1998). Technical Report, Japan Society for the Promotion of Science, Division of Building Materials, Lund Institute of Technology, Japan, 9803:12.

Philippe T (2004). Formulation influence on the withdrawal and cracking of self compacting concrete. Doctoral thesis, Nantes Central School, pp. 20-32.

Rafat S (2011). Utilization of silica fume in concrete: Review of hardened properties, Resources, Conservation and Recycling. pp. 923-932.

Raharjo D, Subakti A , Tavio (2013). Mixed Concrete Optimization Using Fly Ash, Silica Fume and Iron Slag. Procedia Eng. 54:827–839.

Salem TM (2002). Electrical conductivity and rheological properties of ordinary Portland cement-silica fume and calcium hydroxide-silica fume pastes. J. Cem. Concr. 32:1473–1481.

Steven HK, Beatrix K, William CP, Normand FM, Richard JM (2004). Dosage and control of concrete mixtures. Eng. Newslett. Canada 7 (101):57-72.

Yahia A, Tanimura M, Shimabukuro A, Shimoyama Y (1999). Effect of rheological parameters on self compactability of concrete containing various mineral admixtures. Proceedings of the First RILEM International Symposium on Self-compacting Concrete, Stockholm, pp. 523–535.

Zhang X, Han J (2000). The effect of ultra-fine admixture on the rheological property of cement paste. J. Cem. Concr. 30:827–830.

6

Application of nanotechnology in textile engineering: An overview

J. K. PATRA[1] and S. GOUDA[2]

[1]Department of Biotechnology and Medical Engineering, National Institute of Technology, Rourkela, Odisha, India.
[2]Department of Biotechnology and Bioinformatics, AMIT, Bhubaneswar, Odisha, India.

The use of nanotechnology in the textile industry has increased rapidly due to its unique and valuable properties. There is a considerable potential for profitable applications of nanotechnology in cotton and other textile industries. Its application can economically extend the properties and values of textile processing and products. The use of nanotechnology allows textiles to become multifunctional and produce fabrics with special functions, including antibacterial, UV protection, easy-clean, water- and stain repellent and anti-odour. The future success of nanotechnology in textile applications lies in areas where new principles will be combined into durable, multifunctional textile systems without compromising the inherent textile properties, including processability, flexibility etc. The present study highlights the applications of nanotechnology in textile industries, with an emphasis on improving various properties of textiles.

Key words: Antibacterial, antistatic, nanotechnology, wrinkle resistance, water repellence.

INTRODUCTION

The fast development and changes in life style has attracted peoples towards a more comfort and luxurious life. People are moving towards small, safer, cheaper and fast working products which not only reduces the work load but also help them to carry out their works at a much greater pace with minimum efforts. There have been development of gadgets that are much smaller in size like micro-chip, nano capsules, carbon tubes, memory cards, pen drives etc which reduces the problems of transport, and storage and are also much faster and reliable by which we can carry out more of our work in less time. In the formation and development of such products nanotechnology plays a very important and vital role.

The term nanotechnology (sometimes shortened to "nanotech") comes from nanometer – a unit of measure of one billionth of a meter of length. The concept of Nanotechnology was given by Nobel Laureate Physicist Richard Feynman, in 1959. Nanotechnology is defined as the understanding, manipulation, and control of matter at the length scale on nanometer, such that the physical, chemical, and biological properties of materials (individual atoms, molecules and bulk matter) can be engineered, synthesized or altered to develop the next generations of improved materials, devices, structures, and systems (Feynman, 1959; Taniguchi, 1974). Generally, nanotechnology deals with structures that are sized between 1 to 100 nm in at least one dimension and involves developing materials or devices possessing dimension within that size. Nanotechnology creates structure that have excellent properties by controlling atoms and molecules, functional materials, devices and systems on the nanometer scale by involving precise placement of individual atoms (Kathiervelu, 2003). Although nanotechnology is a relatively recent development in scientific research, the development of its central concepts happened over a longer period of time. The emergence of nanotechnology in the 1980s was caused by the convergence of experimental advances. The early 2000s also saw the beginnings of commercial applications of nanotechnology, although these were limited to bulk applications of nanomaterials, such as the silver nano platform for using silver nanoparticles as an antibacterial agent, nanoparticles-based transparent sunscreens, and carbon nanotubes for stain-resistant textiles (Wang et al., 2004; Wang and Chen, 2005).

Throughout history, the textiles sectors have been used worldwide in a wide range of consumer applications. Natural fibers, such as cotton, silk, and wool, along with synthetic fibers, such as polyester and nylon, continue to be the most widely used fibers for apparel manufacturing. Synthetic fibers are mostly suitable for domestic and industrial applications, such as carpets, tents, tires, ropes, belts, cleaning cloths, and medical products. Natural and synthetic fibers generally have different characteristics, which make them ideally suitable mainly for apparel. Depending on the end-use application, some of those characteristics may be good, while the others may not be as good to contribute to the desired performance of the end product. As stated previously, nanotechnology brings the possibility of combining the merits of natural and synthetic fibers, such that advanced fabrics that complement the desirable attributes of each constituent fibre can be produced (Qian and Hinestroza, 2004; Wang and Chen, 2005; Wong et al., 2006).

In the last decade, the advent of nanotechnology has spurred significant developments and innovations in this field of textile technology. By using nanotechnology, there have been developments of several fabric treatments to achieve certain enhanced fabric attributes, such as superior durability, softness, tear strength, abrasion resistance, durable-press and wrinkle-resistance. The (nano)-treated core component of a core-wrap bi-component fabric provides high strength, permanent anti-static behavior, and durability, while the traditionally-treated wrap component of the fabric provides desirable softness, comfort, and aesthetic characteristics.

The use of nanotechnology in the textile industry made it multifunctional which can produce fibres with variable functions and applications such as UV protection, antiodour, antimicrobial etc. In many cases smaller amounts of the additive are required, for the saving on resources. The success of Nanotechnology and its potential applications in textiles lies in various fields where new methods are combined with multifunctional textile systems, durable etc. without affecting the inherent properties of the textiles including softness, flexibility, washbility etc. Keeping the above factors in consideration, the present review highlighted the use of nanotechnology in textile industry and textile engineering, the types and methods of preparation of different nano composites used in textiles.

BASIC NANOTECHNOLOGY

Two main approaches are used in nanotechnology that is, the bottom up approach and the top-down approach. In case of the "bottom-up" approach, the different type of materials and the instruments are made up from different types of molecular components which combine themselves by chemical ways basing on the mechanism of molecular recognition. In case of the "top-down" approaches, various nano-objects are made from various types of components without atomic-level control. Materials reduced to the nanoscale can show different properties compared to what they exhibit on a macro scale, enabling unique applications. The basic premise is that properties can dramatically change when a substance's size is reduced to the nanometer range. For instance, ceramics which are normally brittle can be deformable when their size is reduced, opaque substances become transparent (copper); stable materials turn combustible (aluminum); insoluble materials become soluble (gold) (Sherman, 2003; Burniston et al., 2004).

Nanoparticles can be prepared from a variety of materials such as proteins, polysaccharides and synthetic polymers. The selection of materials in mainly depended on factors like size of the nanoparticles required, inherent properties such as aqueous, solubility and stability, surface characteristics i.e. charge and permeability, degree of biodegradability, biocompatibility and toxicity, release of the desired product, antigenicity of the final product etc. Polymeric nanoparticles have been prepared most frequently be three methods (1) Dispersion of the performed polymers; (2) Polymerization of the monomers and (3) Ionic gelation of the hydrophobic or hydrophilic polymers. However, techniques like supercritical fluid technology and particle repulsion in non- wetting templets (PRINT) have been also used in modern days (Chen, 2002; Kathiervelu, 2003; Anonymous; 2003).

NANOTECHNOLOGY IN TEXTILE INDUSTRY AND TEXTILE ENGINEERING

Of the many applications of nanotechnology, textile industry has been currently added as one of the most benefited sector. Application of nanotechnology in textile industry has tremendously increased the durability of fabrics, increase its comfortness, hygienic properties and have also reduces its production cost. Nanotechnology also offers many advantages as compare to the conventional process in term of economy, energy saving, eco-friendliness, control release of substances, packaging, separating and storing materials on a microscopic scale for later use and release under control condition (David, 2002). The unique and new properties of nanotechnology have attracted scientists and researchers to the textile industry and hence the use of nanotechnology in the textile industry has increased rapidly. This may be due to the reason that textile technology is one of the best areas for development of nanotechnology. The textile fabrics provide best suitable substrates where a large surface area is present for a given weight or a given volume of fabric. The synergy between nanotechnology and textile industry uses this property of large interfacial area and a drastic change in energetic is experienced by various macromolecules or super molecules in the vicinity of a fibre when changing from wet state to a dry state.

The application of nanoparticles to textile materials have been the objective of several studies aimed at producing finished fabrics with different functional performances. Nanoparticles can provide high durability for treated fabrics as they posses large surface area and high surface energy that ensure better affinity for fabrics and led to an increase in durability of the desired textile function. The particle size also plays a primary role in determining their adhesion to the fibers. It is reasonable to except that the largest particle agglomerates will be easily removed from the fibre surface, while the smallest particle will penetrate deeper and adhere strongly into the fabric matrix (Kathiervelu, 2003; Wang and Chen, 2005). Thus, decreasing the size of particles to nano-scale dimensional, fundamentally changes the properties of the material and indeed the entire substance.

A whole variety of novel nanotech textiles are already on the market at this moment. Areas where nanotech enhanced textiles are already seeing some applications include sporting industry, skincare, space technology and clothing as well as materials technology for better protection in extreme environments. The use of nanotechnology allows textiles to become multifunctional and produce fabrics with special functions, including antibacterial, UV-protection, easy-clean, water- and stain repellent and anti-odour (Kathiervelu, 2003).

Types of nanomaterials

Nanocomposite fibers

A composite is a material that combines one or more separate components. Composites are designed to exhibit the best properties of each component. A large variety of systems combining one, two and three dimensional materials with amorphous materials mixed at the nanometer scale (Lee et al., 2003). Nanostructure composite fibers are intensively used in automotive, aerospace and military applications. Nanocomposite fibers are produced by dispersing nanosize fillers into a fibre matrix. Due to their large surface area and high aspect ratio, nanofillers interact with polymer chain movement and thus reduce the chain mobility of the system. Being evenly distributed in polymer matrices, nanoparticles can carry load and increase the toughness and abrasion resistance. Most of the nanocomposite fibers use fillers such as nanosilicates, metal oxide nanoparticles, graphite nanofibers (GNF) as well as single-wall and multi-wall carbon nanotubes (CNT) (Sennett and Welsh, 2003; Dong and Huang, 2002). Some novel CNT reinforced polymer composite materials have been developed, which can be used for developing multifunctional textiles having superior strength, toughness, lightweight, and high electrical conductivity (Lei and Juan, 2003).

Carbon nanofibers and carbon nanoparticles

Carbon nanofibers and carbon black nanoparticles are among the most commonly used nanosize filling materials. Nanofibers can be defined as fibers with a diameter of less than 1 mm or 1000 nm and are characterized as having a high surface area to volume ratio and a small pore size in fabric form (Joshi and Bhattacharyya, 2011; Qian and Hinestroza, 2004). Carbon nanofibers can effectively increase the tensile strength of composite fibers due to its high aspect ratio, while carbon black nanoparticles can improve their abrasion resistance and toughness. Several fibre-forming polymers used as matrices have been investigated including polyester, nylon and polyethylene with the weight of the filler from 5 to 20% (Harholdt, 2003).

There are numerous applications in which nanofibers could be suited (Huang et al., 2003). The high surface area to volume ratio and small pore size allows viruses and spore-forming bacterium such as *anthrax* to be trapped. Filtration devices and wound dressings are just some of the applications in which nanofibers could be utilized. Researchers are investigating textile materials made from nanofibers which can act as a filter for pathogens (bacteria, viruses), toxic gasses, or poisonous or harmful substances in the air. Medical staff, fire fighters, the emergency services or military personnel could all benefit from protective garments made fromnanofibers materials.

Clay nanoparticles

Clay nanoparticles are resistant to heat, chemicals and electricity, and have the ability to block UV light. Incorporating clay nanoparticles into a textile can result in a fabric with improved tensile strength, tensile modulus, flexural strength and flexural modulus. Nanocomposite fibers which utilize clay nanoparticles can be engineered to be flame, UV light resistant and anti-corrosive. Although there have been a number of flame retardant finishes available since the 1970's, the emission of toxic gasses when set ablaze make them somewhat hazardous. Clay nanoparticles have been incorporated into nylon to impart flame retardant characteristics to the textile without the emission of toxic gas. The addition of clay nanoparticles has made polypropylene dyeable. Metal oxide nanoparticles of TiO_2, Al_2O_3, ZnO and MgO exhibit photocatalytic ability, electrical conductivity, UV absorption and photo-oxidizing capacity against chemical and biological species. The main research efforts involving the use of nanoparticles of metal oxides have been focused on antimicrobial, self-decontaminating and UV blocking applications for both military protection gears and civilian health products (Joshi and Bhattacharyya, 2011; Qian and Hinestroza, 2004; Coyle et al., 2007). Nylon fibers filled with ZnO nanoparticles can provide UV shielding function and reduce static electricity on nylon fibers. A composite fibre with nanoparticle of TiO_2 or MgO can provide self-sterilizing function (Harholdt, 2003).

Carbon nanotubes

Carbon Nanotube is a tubular form of carbon with diameter as small as nanometer (nm). A carbon nanotube (CNT) is configurationally equivalent to a two dimensional graphene sheet rolled into a tube. They can be metallic or semiconducting, depending on chirality.CNT are one of the most promising materials due to their high strength and high electrical conductivity. CNT consists of tiny shell(s) of graphite rolled up into a cylinder(s) (Wang et al., 2004; Daoud and Xin, 2004). CNT exhibit 100 times the tensile strength of steel at one-sixth weight, thermal conductivity better than all but the purest diamond, and an electrical conductivity similar to copper, but with the ability to carry much higher currents. The potential applications of CNTs include conductive and high-strength composite fibers, energy storage and energy conversion devices, sensors, and field emission displays. Possible applications include screen displays, sensors, aircraft structures, explosion-proof blankets and electromagnetic shielding. The composite fibers have potential applications in safety harnesses, explosion-proof blankets, and electromagnetic shielding applications. Continuing research activities on CNT fibers involve study of different fibre polymer matrices such as polymethylmethacrylate (PMMA) and polyacrylonitrile (PNA) as well as CNT dispersion and orientation in polymers (Scott and Holly, 2007; Wang et al., 2004).

Nanocellular foam structure

Polymeric materials with nanosize porosity exhibit lightweight, good thermal insulation, as well as high cracking resistance at high temperature without sacrifices in mechanical strength (Daoud and Xin, 2004). By choosing the pretreatment condition to the fibre, the transverse mechanical properties of the composite can be also enhanced through the molecular diffusion across the interface between the fibre and the matrix. The nanocomposites clearly surpass the mechanical properties of most comparable cellulosic materials, their greatest advantage being the fact that they are fully bio-based and biodegradable, but also of relatively high strength. A potential application of cellular structure is to encapsulate functional compounds such as pesticides and drugs inside of the nanosizecells. One of the approaches to fabricate nanocellular fibers is to make use of a thermodynamic instability during supercritical carbon dioxide extrusion and reduce the size of the cellular fibers that can be used as high-performance composite fibers as well as for sporting and aerospace materials (Lei and Juan, 2003).

Properties of nano textile fibres

Water repellence

The water-repellent property of fabric created by nano-whiskers, which are hydrocarbons and 1/1000 of the size of a typical cotton fibre, when added to the fabric create a peach fuzz effect without lowering the strength of cotton. The spaces between the whiskers on the fabric are smaller than the typical drop of water, but still larger than water molecules; water thus, remains on the top of the whiskers and above the surface of the fabric. However, liquid can still pass through the fabric, if pressure is applied to it (Wong et al., 2006; Russell, 2002).

Nanosphere impregnation involving a three-dimensional surface structure with gel forming additives which repel water and prevent dirt particles from attaching themselves are also used. Once water droplets fall onto them, water droplets bead up and, if the surface slopes slightly, will roll off. As a result, the surfaces stay dry even during a heavy shower. Furthermore, the droplets pick up small particles of dirt as they roll, and so the leaves of the lotus plant keep clean even during light rain. By altering the micro and nano-scale surface features on a fabric surface, a more robust control of wetting behavior can be attained. It has been demonstrated that by combining the nanoparticles of hydroxylapatite, TiO_2, ZnO and Fe_7O_3 with other organic and inorganic substances, the audio frequency plasma of fluorocarbon chemical was applied to deposit a nanoparticulate hydrophobic film onto a cotton fabric surface to improve its water repellent property (Wang et al., 2004; Zhang et al, 2003).This sort of surface engineering, which is capable of replicating hydrophobic behavior, can be utilized in developing special chemical finishes for producing water-and/or stain- resistant fabrics while complementing the other desirable fabric attributes, such as breathability, softness and comfort.The surfaces of the textile fabrics can be appreciably modified to achieve considerably greater abrasion resistance, ultraviolet (UV) resistance, electromagnetic and infrared protection properties.

UV-protection

Inorganic UV blockers are more preferable to organic UV blockers as they are non-toxic and chemically stable under exposure to both high temperatures and UV (Yang et al., 2004; El-Molla et al., 2011). Inorganic UV blockers are usually certain semiconductor oxides such as TiO_2, ZnO, SiO_2 and Al_2O_3. Among these semiconductor oxides, titanium dioxide (TiO_2) and zinc oxide (ZnO) are commonly used. It was determined that nano-sized titanium dioxide and zinc oxide are more efficient at absorbing and scattering UV radiation than the conventional size, and are thus better to provide protection against UV rays . This is due to the fact that nano-particles have a larger surface area per unit mass and volume than the conventional materials, leading to the increase of the effectiveness of blocking UV radiation (Kathiervelu, 2003; Yang et al., 2003).

Various researchers have worked on the application of UV blocking treatment to fabric using nanotechnology. UV blocking treatment for cotton fabrics are developed using the sol-gel method. A thin layer of titanium dioxide is formed on the surface of the treated cotton fabric which provides excellent UV protection; the effect can be maintained after 50 home launderings (El-Molla et al., 2011). Apart from titanium dioxide, zinc oxide nano rods of 10 to 50 nm in length are also applied to cotton fabric to provide UV protection. According to the studies on the UV blocking effect, the fabric treated with zinc oxide nanorods were found to have demonstrated an excellent UV protective factor (UPF) rating (Wong et al., 2006). This effect can be further enhanced by using a different procedure for the application of nanoparticles on the fabric surface. When the process of padding is used for applying the nanoparticles on to the fabric, the nanoparticles get applied not only on the surface alone but also penetrates into the interstices of the yarns and the fabric, i.e. some portion of the nanoparticles get penetrate into the fabric structure. Such Nanoparticles which do not stay on the surface may not be very effective in shielding the UV rays. It is worthwhile that only the right (face) side of the fabric gets exposed to the rays and therefore, this surface alone needs to be covered with the nanoparticles for better UV protection. Spraying (using compressed air and spray gun) the fabric surface with the nanoparticles can be an alternate method of applying the nanoparticles.

Antimicrobial

Although many antimicrobial agents are already in used for textile, the major classes of antimicrobial for textile include organo-silicones, organo-metallics, phenols and quaternary ammonium salts. The bis- phenolic compounds exhibits a broad spectrum of antimicrobial activity. For imparting antibacterial properties, nano-sized silver, titanium dioxide,zinc oxide, triclosan and chitosan are used (Burniston et al., 2004). Nano-silver particles have an extremely large relative surface area, thus increasing their contact with bacteria or fungi and vastly improving their bactericidal and fungicidal effectiveness. Nano-silver is very reactive with protein and shows antimicrobial properties at concentrations as low as 0.0003 to 0.0005%. When contacting bacteria and fungi, it will adversely affect cellular metabolism and inhibits cell growth. It also suppresses respiration, the basal metabolism of the electron transfer system, and the transport of the substrate into the microbial cell membrane (Wong et al., 2006). Furthermore, it inhibits the multiplication and growth of those bacteria and fungi which cause infection, odour, itchiness and sores (Yang et al., 2003). Some synthetic antimicrobial nano particles which are used in textiles are as follows. Triclosan, a chlorinated bis- phenol, is a synthetic, non-ionic and broad spectrum antimicrobial agent possessing mostly antibacterial alone with some antifungal and antiviral properties. Chitosan, a natural biopolymer, is effectively used as antibacterial, antifungal, antiviral, non-allergic and biocompatible. ZnO nanoparticles have been widely used for their antibacterial and UV-blocking properties.

Antistatic

An antistatic agent is a compound used for treatment of materials or their surfaces in order to reduce or eliminate buildup of static electricity generally caused by the triboelectric effect. The molecules of an antistatic agent often have both hydrophilic and hydrophobic areas, similar to those of a surfactant; the hydrophobic side interacts with the surface of the material, while the hydrophilic side interacts with the air moisture and binds the water molecules (Dong and Huang, 2002). As synthetic fibers provide poor anti-static properties, research work concerning the improvement of the anti-static properties of textiles by using nanotechnology has been at large. It was determined that nano-sized particles like titanium dioxide, zinc oxide whiskers, nanoantimony-doped tin oxide (ATO) and silanenanosol could impart anti-static properties to synthetic fibers. Such material helps to effectively dissipate the static charge which is accumulated on the fabric (Wong et al., 2006). On the other hand, silanenanosol improves anti-static properties, as the silica gel particles on fibre absorb water and moisture in the air by amino and hydroxyl groups and bound water. Electrically conductive nano-particles are durably anchored in the fibrils of the membrane of teflon, creating an electrically conductive network that prevents the formation of isolated chargeable areas and voltage peaks commonly found in conventional anti-static materials. This method can overcome the limitation of

Table 1. Some finishing based developments through nanotechnology in textile industries.

S. No	Application in textile	Nanomaterial used	References
1.	Electro conductive and antistatic	Carbon black, Carbon nanotubes (CNT), Cu, Polypyrrole, Polyaniline	Sennette and Welsh (2003), Dong and Huang (2002), Anonymous (2003)
2.	Increase durability	Al_2O_3, SiO_2, CNT, ZnO, Polybutylacrylate	Burnistan et al. (2004)
3.	Antibacterial	Ag, Chitosan, SiO_2 (as matrix), TiO_2, ZnO	Burnistan et al. (2004) Wong et al. (2006)
4.	Self-cleaning/ dirt and water repellent	CNT, Fluoroacrylate, SiO_2 (as matrix), TiO_2	Lei and Juan (2003), Zhang et al (2003)
5.	Moisture absorbing	TiO_2	Burnistan et al. (2004)
6.	Improved staining / reduce fade	Carbon black, Nanoporous hydrocarbon on Nitrogen coating, SiO_2 (as matrix)	Wong et al. (2006) Song et al. (2001)
7.	UV protection	TiO_2, ZnO	Burnistan et al. (2004) Wong et al. (2006) Kathiervelu (2003), Yang et al. (2003)
8.	Fire proof	CNT, Boroxosiloxane, Montmorillonite (nano clay), Sb_3O_2	Zhang et al (2003)
9.	Controlled release of active agents, medicinal products or fragrances	Montmorillonite (nano clay), SiO_2 (as matrix)	Harholdt (2003)

conventional methods, which is that the anti-static agent is easily washed off after a few laundry cycles (Anonymous, 2003).

Wrinkle resistance

To impart wrinkle resistance to fabric, resin is commonly used in conventional methods. However, there are limitations to applying resin, including a decrease in the tensile strength of fibre, abrasion resistance, water absorbency and dye-ability, as well as breathability. To overcome the limitations of using resin, some researchers employed nano-titanium dioxide and nano-silica to improve the wrinkle resistance of cotton and silk respectively (Wong et al., 2006). Nano-titanium dioxide employed with carboxylic acid as a catalyst under UV irradiation to catalyses the cross-linking reaction between the cellulose molecule and the acid. On the other hand, nano-silica when applied with maleic anhydride as a catalyst could successfully improve the wrinkle resistance of silk (Song et al., 2001; Zhou et al., 2003). More over the wrinkle recovery of the fabrics can also be improved to a great extent by imparting techniques like padding and exhaustion beside the use of nano-materials to the fabrics. Studies also have suggest that treatment of fabrics with microwaves are more wrinkle resistant as comparable to oven curing, because it generates higher frequency and volumetric heating which minimizes the damage from over drying.

Nanoparticles in textiles finishing

Fabric treated with nanoparticles of TiO_2 and MgO replaces fabrics with active carbon, previously used as chemical and biological protective materials. The photocatalytic activity of TiO_2 and MgO nanoparticles can break harmful and toxic chemicals and biological agents. These nanoparticles can be pre-engineered to adhere to textile substrates via spray coating or electrostatic methods. Textiles with nanoparticles finishing are used to convert fabrics into sensor-based materials which has numerous applications. If nanocrystalline epiezoceramic particles are incorporated intofabrics, the finished fabric can convert exerted mechanical forces into electrical signals enabling the monitoring of bodily functions such as heart rhythm and pulse if they are worn next to skin.

Fabric finishing by using nanotechnology

Finishing of textile fabrics made of natural and synthetic fibers to achieve desirable surface texture, color and other special aesthetic and functional properties, has been a primary focus in textile manufacturing industries. In the last decade, the advent of nanotechnology has spurred significant developments and innovations in this field of textile technology (David, 2002). Fabric finishing has taken new routes and demonstrated a great potential for significant improvements by applications of nanotechnology (Table 1). The developments in the areas of surface engineering and fabric finishing have been highlighted in several researches. There are many ways in which the surface properties of a fabric can be manipulated and enhanced, by implementing appropriate surface finishing, coating, and/or altering techniques, using nanotechnology. A few representative applications of fabric finishing using nanotechnology are schematically displayed in the Figure 1.

Nanotechnology provides plenty of efficient tools and techniques to produce desirable fabric attributes, mainly by engineering modifications of the fabric surface. For example, the prevention of fluid wetting towards the development of water or stain-resistant fabrics has always been of great concerning textile manufacturing.

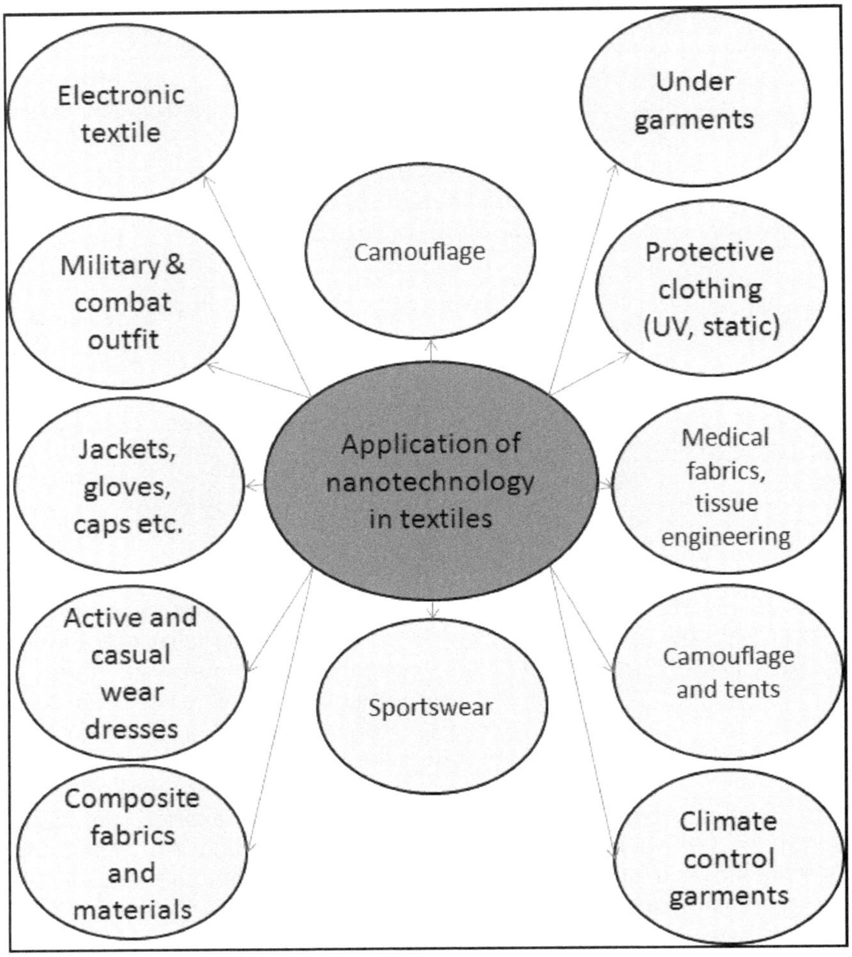

Figure 1. Some representative applications of nanotechnology in textiles.

The basic principles and theoretical background of "fluid-fabric" surface interaction are well described in recent manuscript. It has been demonstrated that by altering the micro and nano-scale surface features on a fabric surface, a more robust control of wetting behavior can be attained. The alteration in the fabric's surface properties enables to exhibit the "Lotus-Effect," which demonstrates the natural hydrophobic behavior of a leaf surface. This sort of surface engineering, which is capable of replicating hydrophobic behaviour, can be utilized in developing special chemical finishes for producing water and/or stain- resistant fabrics (Song et al., 2001; Russell, 2002). In recent years, several attempts have been made by researchers and industries to utilize similar concepts of surface-engineered modifications through nanotechnology to develop high performance textile and smart textile. The concept of surface engineering and nanotextile develops hydrophobic fabric surfaces that are capable of repelling liquids and resisting stains, while complementing the other desirable fabric attributes, such as breathability, softness, and comfort (Anonymous, 2003; Chen, 2002; Song et al., 2001; Russell, 2002). The Table 2 describes about the nano-materials used in textile application and their possible function functions.

CONCLUSION

Nanotechnology is growing by leaps and bounds and it has been introduced in many fields including the textile industries. There is a considerable potential for profitable applications of nanotechnology in cotton and other textile industries. Its application can economically extend the properties and values of textile processing and products. By deploying nanotechnology, ultra strong, durable and specific function oriented fabrics can be efficiently produced for a number of applications including medical, military and industrial apparels etc. As mentioned, nanotechnology overcomes the limitations of applying conventional methods to impart certain properties to textile materials. There is no doubt that in the next few years, nanotechnology will penetrate into every area of textile industry.

Table 2. Overview of the nanomaterials used in textile application and their functions.

S. No	Nano products	Application and properties
1.	Nano-Tex	The fabric shows very good wrinkle-resistance and changes to the fibre do not affect the natural hand feel and breathability of the fabric
2.	Nanosphere	Schoeller developed an Impregnating treatment; Nanosphere that makes fabrics water and soil resistant.
3.	Nano-Pel	Water-and-oil repellent finishing is effective for cotton, linen, wool and silk, polyester, nylon and acrylic
4.	Smart clothes or textile with nano finishing	Silver is used due to its natural anti-bacterial and anti-fungal properties. It inhibits the multiplication and growth of those bacteria and fungi which cause infection, odour, itchiness and sores. Nano silver particles are widely applied in socks to prohibit the growth of bacteria. It is also applied to a range of other healthcare products such as dressings, for burns, scald, skin donor and recipient site.
5.	Magnetic Nanoparticles (mNPs)	Composed of magnetic elements (iron, nickel and cobalt) which are coated with a layer of fatty acids and are used as new means of confirming specific ailments or releasing drugs at exact points within a living system.

REFERENCES

Anonymous (2003). Nanotechnologies keep feet healthy. Advance in Textiles Technology. 3:10-11.

Burniston N, Bygott C, Stratton J (2004). Nano Technology Meets Titanium Dioxide. Surface Coatings International Part A., pp. 179-814.

Chen RQ (2002). Nanometer materials and health-care textiles. Dyestuff Ind. 39:24-28.

Coyle S, Wu Y, Lau KT, Rossi DD, Wallace G, Diamond D (2007). Smart Nanotextiles: A Review of Materials and Applications. MRS Bull. 32:434-442.

Daoud WA, Xin JH (2004). Low temperature sol-gel processed photocatalytictitania coating. J. Sol-Gel Sci. Technol. 29:25-29.

David SS (2002). Nanoparticle-based permanent treatments for textiles. United State Patent, No 6,607: 994.

Dong WG, Huang G (2002). Research on properties of nano polypropylene/TiO_2 composite fiber. J. Textile Res. 23:22-23.

El-Molla MM, El-Khatib EM, El-Gammal MS, Abdel-fatteh SH (2011). Nanotechnology to improve coloration and antimicrobial properties of silk fabrics. Indian J. Fibre Textile Res. 36:266-271.

Feynman R (1959). Lecture at the California Institute of Technology. December 29.

Harholdt K (2003). Carbon Fiber, Past and Future. Ind. Fabric Prod. Rev. 88(4):14-28.

Huang ZM, Zhangb YZ, Kotakic M, Ramakrishnab S (2003). A review on polymer nanofibers by electrospinning and their applications in nanocomposites. Compos. Sci. Technol. 63:2223–2253.

Joshi M and Bhattacharyya A (2011). Nanotechnology – a new route to high performance functional textiles, Textile Progr. 43(3):155-233.

Kathiervelu SS (2003). Applications of nanotechnology in fiber finishing. Synth. Fibres. 32:20-22.

Lee HJ, Yeo SY, Jeong SH (2003). Antibacterial effect of nano sized silver colloidal solution on textile fabrics. J. Mater. Sci. 38:2199-2204.

Lei Q, Juan PH (2003). Application of nanotechnology for high performance textile, pp. 1-8.

Qian L, Hinestroza JP (2004). Application of nanotechnology for high performance textiles. J. Textile Apparel Technol. Manage. 4(1):1-7.

Russell E (2002). Nanotechnologies and the shrinking world of textiles. Textile Horiz. 9/10:7-9.

Scott M, Holly H (2007). Axtell, The Next Generation of Chemical and Biological Protective Material Utilizing Reactive Nanoparticles, Gentex Corporation, Carbondale, PA 18407.

Sennett M, Welsh E (2003). Dispersion and Alignment of Carbon Nanotubes in Polycarbonate. Appl. Phys. A. 76: 111-113.

Sherman J (2003). Nanoparticulate Titanium Dioxide Coatings, and Processes for the Production and Use thereof, Pat. No 736738.

Song XQ, Liu A, Ji CT, Li HT (2001). The effect of nano-particle concentration and heating time in the anti-crinkle treatment of silk. J. Jilin Inst. Technol. 22:24-27.

Taniguchi N (1974). On the basic concept of nano-technology Proceedings of the International Conference on Production Engineering Tokyo Part II Japan Society of Precision Engineering.

Wang CC, Chen CC (2005). Physical properties of cross-linked cellulose catalyzed with nano titanium dioxide. J. Appl. Polym. Sci. 97:2450-2456.

Wang RH, Xin JH, Tao XM, Daoud WA (2004). ZnO nano rods grown on cotton fabrics at low temperature. Chem. Phys. Lett. 398:250-255.

Wong YWH, Yuen CWM, Leung MYS, Ku SKA, Lam HLI (2006). Selected applications of nanotechnology in textiles. AUTEX Res. J. 6(1):1-8.

Yang H, Zhu S, Pan N (2004). Studying the Mechanisms of Titanium Dioxide as Ultraviolet-Blocking Additive for Films and Fabrics by an Improved Scheme. J. Appl. Polym. Sci. 92:3201–3210.

Yang HY, Zhu SK, Pan N (2003). Studying the Mechanisms of Titanium Dioxide as Ultraviolet-Blocking Additive for Films and Fabrics by an Improved Scheme. J. Appl. Polym. Sci. 92:3201-3210.

Zhang J, France P, Radomyselskiy A, Datta S, Zhao J, Ooij WV (2003). Hydrophobic cotton fabric coated by a thin nanoparticulate plasma film. J. Appl. Polym. Sci. 88:1473-1481.

Zhou ZW, Chu LS, Tang WM, Gu LX (2003). Studies on the antistatic mechanism of tetrapod-shaped zinc oxide whisker. J. Electrostat. 57:347-354.

An evolutionary programming embedded Tabu search method for hydro-thermal scheduling with cooling – banking constraints

Nimain Charan Nayak[1] **and C. Christober Asir Rajan**[2*]

[1]Department of Electronics Engineering (EEE), Sathyabama University, Chennai, India.
[2]Department of Electronics Engineering (EEE), Pondicherry Engineering College, Puducherry, India.

This paper presents a new approach for solving the unit commitment problem (UCP) in hydro-thermal power system. The main objective of this paper is to find the generation scheduling by committing the generating units such that the total operating cost can be minimized by satisfying both the forecasted load demand and various operating constraints. It is a Global optimization technique for solving UCP, operates on a system, which is designed to encode each unit's operating schedule with regard to its minimum up/down time. In this method, the unit commitment schedule is coded as a string of symbols. An initial population of parent solutions is generated at random. Here the parents are obtained from a pre-defined set of solutions that is, each and every solution is adjusted to meet the requirements. Then, random recommitment is carried out with respect to the unit's minimum down times. Tabu search (TS) is a powerful optimization procedure that has been successfully applied to a number of combinatorial optimization problems. It avoids entrapment at local optimum by maintaining a short term memory of recently obtained solutions. The memory structure assists in forbidding certain moves that deteriorates the quality of the solution by assigning Tabu status to the forbidden solutions. The Tabu status of a solution can be overruled if certain conditions are satisfied expressed in the form of aspiration level (AL). AL adds flexibility in TS by directing the search towards attractive moves. The best population is selected by evolutionary strategy (ES). Numerical results are shown comparing the cost solutions and computation time obtained by using the proposed hybrid method with conventional methods like Dynamic Programming, and Lagrangian Relaxation etc.

Key words: Evolutionary programming, Tabu search, unit commitment, dynamic programming, lagrangian relaxation.

INTRODUCTION

The electrical power system has daily and weekly cycles. The optimization problem is how to schedule generation to minimize the fuel cost or to maximize the profit over a study period of typically a day, satisfying various constraints. The daily load pattern for a given system may exhibit large differences between minimum and maximum demand. It is not proper and economical to run all the units available all the time. Since the load varies continuously with time, the optimum condition of units may alter during any period.

Therefore, determining the units of a plant that should operate economically for a given load is the problem of unit commitment (UC). For total number of units of higher order, the problems associated with UC have generally been difficult to solve because of uncertainty of particular aspects of the problem. For instance the availability of fuel in precise, load forecast variable costs affected by the loading of generator units and the losses caused by reactive flows are some of the unpredictable issues. In order to reach a feasible solution for UCP, different considerations must be considered.

*Corresponding author. E-mail: asir_70@pec.edu.

A number of numerical optimization techniques have been employed to solve the complicated UCP. The major limitations of the numerical techniques are inability to handle problem dimensions, large computation time, more memory space and complexity in programming.

The two-stage method (Ferrero et al., 1998) has smaller computational requirements than that of the Simulated Annealing algorithm. The optimal generation from hydro and thermal resources is computed simultaneously in the two stage algorithm; there is no need for assuming constant operation of some reservoirs as in the Simulated Annealing method. No discretization of state and control variables is needed in the proposed method. The required storage as well as computing time in the proposed method are reduced as compared to those in the successive-approximations algorithm. The results (Martinez and Soares, 2002) revealed that the partial open-loop feedback control policy provided somewhat higher average and standard deviation for hydroelectric generation in all simulations performed. The higher standard deviation provided, however, not being compensated for by a slightly higher average generation, lead to higher final operating costs. The closed-loop feedback control policy was more efficient in the synthetically simulations. This advantage, however, reduced with the historical simulations, when the different control policies led to almost equivalent performances.

LR-DP method (Benhamida, 2009) is efficiently and effectively implemented to solve the UC problem. LR total production cost over the scheduled time horizon is less than conventional methods especially for the larger number of generating units. The augmented Lagrangian approach (Salem, 2001) presented in this paper accommodates further for pumped-storage units and line flow limitations and concurrently can produce accurate scheduling results. The approach produces feasible schedules and requires no iteration with economic dispatch algorithms. LR approach (Ngundam et al., 2002) to solve the UC Problems was found that it provides faster solution but it will fail to obtain solution feasibility and solution quality problems and becomes complex if the number of units increased. The overall results obtained by the implemented Lagrangian relaxation approach (Alberto et al., 2003) are of very good quality and they are reached within little iteration. The proposed method could be of help for the solution of UC of hydrothermal power generation systems in the uncertain environment of the competitive electricity markets. The results revealed that the proposed method (Ruey-Hsun et al., 2009) is very effective in reaching an optimal generation schedule.

Test results and numerical experiences show that the solution technique (Gary et al., 2001) can give a near-optimal or optimal solution for the mixed integer linear programming (MILP) problem in an acceptable time. The solution of the hydrothermal problem not only provides MW schedules for hydro units and plants, but also indicates the hydro unit commitment status while minimizing the unit startup costs. MILP model allows to accurately represent most of the hydroelectric system characteristics, and turns out to be computationally solvable for a planning horizon of one week, proving the high efficiency of modern MILP software tools, both in terms of solution accuracy and computing time (Alberto et al., 2008). With the proposed (Costas et al., 2009) distributed implementation, even a small-sized generation company can perform overnight large Monte Carlo simulations on the company's personal computers exploiting their idle times under a simple hydrothermal problem communication protocol. The upper and lower bound estimates of the optimal value of the objective function are available with each iteration; a feasible solution to the original problem is available with each iteration; prior experience and feasible existing schedules can be directly incorporated into the computational procedure, introducing additional exclusion rules to improve the efficiency of the restricted integer algorithm (Baptistella and Gerome, 1980). Maximum intensity projection (MIP) methods (Mohan et al., 1992) for solving the unit commitment problems fail when the number of units increases because they require a large memory and suffer from great computational delay. The presented decomposition scheme (Wilfredo and Alberto, 2007) is simple, pure, and robust, even for dominant hydraulic power systems. It is also easy to implement, because it uses well-known, optimized, fast techniques, such as MIP and ac optimal power flow (OPF) algorithms. The chosen decomposition of the problem allows considering the network's entire modeling, with little impact on computing CPU time. Complex hydraulic chains or additional constrains can be easily modeled and/or added. In our studies (Tiago and Andre, 2009), reductions in the overall CPU time to solve the problem depended on the case, with a minimum reduction of two times as compared to the classical Multi Stage Benders Decomposition approach and four times as compared to the single linear program approach. This work (Srinivasa et al., 2009) builds a fuzzy rule base with the use of the area control error and rate of change of the error. The simulation results show that the proposed FLA based controller yields improved control performance than the dual mode controller.

The heuristic search algorithms are efficiently used to commit fuel constrained units, pumped-storage units, and repairing violations due to ramp rate and transmission constraints. The proposed method (Vo and Weerakorn, 2008) obtains less production costs and faster computational time than Augmented Hopfield Neural Network and hybrid Lagrangian relaxation and quadratic programming. SAM (Kirkpatrick et al., 1983; Shokri et al., 1991; Zhuang and Galiana, 1990; Mantawy et al., 1998) is a powerful, general-purpose stochastic optimization technique, which can theoretically converge asymptotically to a global optimum solution with probability one.

But it takes too much time to reach the near-global minimum. It is an iterative improvement procedure that starts with some initial feasible solution and improves it to reach the better solution with computation time. It has the special characteristic of escaping the local optima by employing a flexible memory system. SAM utilizes a short term memory of recent solutions to lead the algorithm to a different direction away from the local optimum region to obtain better solutions that are near to global optimum. TSM (Mantawy et al., 1998; Whei-Min et al., 2002) is a powerful, general-purpose stochastic optimization technique, which can theoretically converge asymptotically to a global optimum solution with probability one. But it will take much time to reach the near-global minimum.

TSD (Xaiomin and Shahidehpur, 1996) has considered the time varying start-up costs as well as the non-linearity in the hydrothermal systems. It can be used as post processor for existing generation scheduling methods or in cases where rescheduling of units is required due to change in the system status. And the application of the modified Benders decomposition method is to solve with constraints that are difficult to formulate. In order to obtain the better results, the experience of the operators in applying some system specific conditions has been included in Tabu Search method. The simulation results (Rudolf and Bayrleithner, 1999) reveal that the features of easy implementation, convergence in an acceptable time, and highly optimal solution in solving the unit commitment problem can be achieved. GA (Yong-Gang and Chun-Ying, 2000) is a general-purpose stochastic and parallel search method based on the mechanics of natural selection and natural genetics. It is a search method to have potential of obtaining near-global minimum. And it has the capability to obtain the accurate results within short time and the constraints are included easily. The proposed GA (Esteban et al., 2003), using new specialized operators, have demonstrated excellent performance in dealing with this kind of problem, obtaining near-optimal solutions in reasonable times and without sacrificing the realism of the electric and economic models. Developed algorithms provide optimal unit commitment and also optimal MW values for energy, spinning reserve and non-spin. Presented algorithm and analysis could be beneficial to GENCO with big number of generators to maximize the profit and bid in competitive electricity market (Mariappane and Thyagarajah, 2009; Lal and Christober, 2011).

With this new approach (Werner and Verstege, 1999), decomposition into sub problems for the hydro and the thermal system is not necessary. Numerical experiments with a hydrothermal test system demonstrate the ability of the proposed method to solve the complex optimization problem with its wealth of constraints. Cau and Kaye (2002) performance compares favorably with constructive DP which is known to be faster than standard LP. It can be used for a rapid approximate optimal scheduling for large scale complex system with multiple cascaded and pumped storage. Results (Chakrabarti and Chattopadhyay, 2003) show that with quadratic thermal cost and without prohibited discharge zones, all EP-based algorithms converge faster during initial stages while Fast Evolutionary Programming and Classical Evolutionary Programming slow down in the latter stages compared to Improved Fast EP. Improved Fast EP performs the best amongst the three in solving this problem in terms of execution time, minimum cost, and mean cost.

From test results, Shyh-Jier (2001), and Venkatesan and Sanavullah (2012) have demonstrated the feasibility of ACM in the hydro-generation scheduling study. This method is also suitable to be implemented under a parallel computer system. The solution speed can be thus further improved. There is no obvious limitation on the size of the problem that must be addressed (Christober, 2010), for its data structure is such that the search space is reduced to a minimum; No relaxation of constraints is required; instead, populations of feasible solutions are produced at each generation and throughout the evolution process; Multiple near optimal solutions to the problem involving multiple constraints and conflicting objectives can be obtained in a reasonable time with the use of heuristics; It works only with feasible solutions generated based on heuristics, thus avoiding the computational burden entailed by the Genetic Algorithm methods which first generate all feasible solutions and then purge the infeasible ones (Nayak and Christober, 2011; Javadi et al., 2011, a).

The flexibility in the demand constraint both in terms of possibility of buying and selling in the market gives better indication of the likely future scenarios so that better bidding strategy can be made (Javadi et al., 2011, b; Chitra et al., 2009; Ganesan and Subramanian, 2012). More improvements could be made to the proposed algorithm in order to increase the speed convergence of the algorithm (Ahmed et al., 2006) and its execution time by improving the gradient method, and by adjusting adequately the penalty weight factor.

Hence, in this paper, an attempt has been made to couple EP with TSM [electromagnetic propagation tool (EPTSM)] with cooling-banking constraints for meeting these requirements of the UCP, which gives the better solution than the individual EP and TS methods with reasonable time. In case of TSM, the temperature and demand are taken as control parameter. Hence the quality of solution is improved. Classical optimisation methods are a direct means for solving this problem. EP seems to be promising and is still evolving. EP has the great advantage of good convergent property and, hence, the computation time is considerably reduced. EP does not suffer from the drawback of handling non-continuous or non-differentiable objective functions as in some plants like Combined Cycle Co-generation plants. Encoding and decoding schemes essential in the GA approach are not needed, considerable computation time is thus saved. The EP combines good solution quality for TSM with rapid convergence for EP. And the selection process is done using ES. By doing so, it can help to find the optimum solution rapidly and efficiently. The validity and effectiveness of the proposed integrated algorithm has been tested with an IEEE test system consisting of 4 hydro generating units and 10 thermal generating units. The results are compared with the other methods.

PROBLEM FORMULATION

The main objective of UCP is to determine the on/off status of the generating units in a power system by meeting the load demand at a minimum operating cost in addition to satisfying the constraints (Allen and Wollenberg, 1984) of the generating units. The problem formulation includes the quadratic cost characteristics, startup cost of thermal power system and operating constraints of thermal and hydro generating units. The power generation cost for thermal power system is given in Equation 1a.

$$F_{s,it}(P_{s,it}) = A_i + B_i P_{s,it} + C_i P_{s,it}^2 \text{ (Rs/h)} \quad (1a)$$

where,

A_i, B_i, C_i - The Cost Function parameters of unit i (Rs/h, Rs/MWh, Rs/MW^{2}h).

$F_{s,it}(P_{s,it})$ - The generation cost of unit i at time t (Rs/h).

$P_{s,it}$ - The output power from unit i at time t (MW).

The overall objective function (Alberto et al, 2008) of UCP that is to be minimized is given in Equation 1b.

$$F_T = \sum_{t=1}^{T} \sum_{i=1}^{N} (F_{it}(P_{it})U_{it} + S_i V_{it}) \text{ (Rs/h)} \quad (1b)$$

where,

U_{it} – Unit i status at hour t

V_{it} – Unit i start up/ shut down status at time t

F_T – Total operating cost over the schedule horizon (Rs/h)

S_{it} – Startup cost of unit i at time t (Rs)

Constraints

Load power balance constraint

The real power generated by thermal and hydro generating units must be sufficient enough to meet the load demand and must satisfy the equation

$$\sum_{i=1}^{N} P_{s,it} + \sum_{j=1}^{M} P_{h,it} = P_{D,i} + P_{L,i} \quad 1 \le t \le T \quad (2)$$

Spinning reserve constraint

Spinning reserve is the total amount of generation available from all units synchronized on the system minus the present load plus the losses being supplied. The reserve is usually expressed as a percentage of forecasted load demand. Spinning reserve is necessary to prevent drop in system frequency and also to meet the loss of most heavily loaded unit in the power system.

$$\sum_{i=1}^{N} P_{\max,i} U_{it} \ge (P_{D,i} + R_t) \quad 1 \le t \le T \quad (3)$$

Thermal constraints

A thermal unit undergoes gradual temperature changes and this increases the time period required to bring the unit online. This time restriction imposes various constraints on generating unit. Some of the constraints are minimum up/down time constraint and crew constraints.

If the units are already running there will be a minimum time before which the units cannot be turned OFF and the constraint is given in Equation 4.

$$T_{on,i} \ge T_{up,i} \quad (4)$$

If the units are already OFF there will be a minimum time before which they cannot be turned ON and the constraint is given in Equation 5.

$$T_{off,i} \ge T_{down,i} \quad (5)$$

Must run units

Some units in the power system are given must run status in order to provide voltage support for the network.

Unit capacity limits

The power generated by the thermal unit must lie within the maximum and minimum power capacity of the unit.

$$P_{s,i}^{\min} \le P_{s,i} \le P_{s,i}^{\max} \quad (6)$$

Hydro constraints

Hydro plant generation limits

The power generated by the hydro units must be within the maximum and minimum power capacity of the unit (Ferrero, 1998)

$$P_{h,i}^{\min} \leq P_{h,i} \leq P_{h,i}^{\max} \quad (7)$$

Hydraulic network constraints

Physical limitations on reservoir storage volumes and discharge rates.

$$V_{h,i}^{\min} \leq V_{h,i} \leq V_{h,i}^{\max} \quad (8)$$

$$Q_{h,i}^{\min} \leq Q_{h,i} \leq Q_{h,i}^{\max} \quad (9)$$

The initial volume and the final volume that is to be retained at the end of scheduling period.

$$V_{h,it}^{t=0} = V_{h,i}^{begin} \quad (10)$$

$$V_{h,it}^{t=T} = V_{h,i}^{end} \quad (11)$$

The Continuity equation for hydro reservoir network is given in Equation 12.

$$V_h(i,t) = V_h(i,t-1) + I_h(i,t) - S_h(i,t) - Q_h(i,t) - \sum_{m=1}^{Ru}[Q_h(m,t-\Gamma(i,m)) + S_h(m,t-\Gamma(i,m))] \quad (12)$$

Hydro plant unit power generation characteristics

The hydro power generated is related to the reservoir characteristics as well as water discharge rates. Hydro power output is a function of the volume of the reservoir and discharge rate. The equation representing the hydro power generation characteristics is given in Equation 13.

$$P_h(i,t) = C_{1,i}V_h(i,t)^2 + C_{2,i}Q_h(i,t)^2 + C_{3,i}[V_h(i,t)Q_h(i,t)]C_{4,i}V_h(i,t) + C_{5,i}Q_h(i,t) + C_{6,i} \quad (13)$$

ELECTROMAGNETIC PROPAGATION *TOOL (EPTSM) FOR HYDRO THERMAL* UNIT COMMITMENT PROBLEM (*UCP)*

The TSM is integrated with evolutionary programming algorithm to escape local optima. Since Tabu search improves the solution without entrapment in local minima the population obtained in EP is refined using TSM.

Implementation of *electromagnetic propagation* tool (EPTSM)

In the proposed algorithm, an initial set of parent vectors are formed at random. The objective function values of all the parents are evaluated and their startup cost is added to get the operating cost of each parent vector. Mutation is performed to all the parents and off springs is formed. The objective function values of off springs are evaluated as in the case of parents. Then the parents and off springs are combined to get the total population. Each individual from the population is refined using Tabu search algorithm. Then a tournament competition and selection process is performed to refined population from Tabu Search to obtain better half population. They are given as parents to the next generation. The above described process is repeated until maximum number of iterations is reached.

Improvements performed in the algorithm

1. Each parent is mutated twice to get two off springs. Therefore the total population will be the sum of Np parents and 2Np off springs that is, totally 3Np population. This allows in exploring more areas of the possible solution space and increases the probability of obtaining global optimum solution.
2. In the mutation process, if a parent yields good offspring then the particular mutation value is given a score. If at each time the particular mutation value gives good offspring then its score is increased and other mutation values scores are reduced. Hence during the next iteration the mutation values with more scores are applied to get offspring with less objective function values and thereby increasing the probability of obtaining global optimum solution.

Electromagnetic propagation *tool (EPTSM) algorithm*

The proposed integrated algorithm combines EP and TS techniques to solve the UCP problem. The EP technique, hold the main responsibility of finding the optimal point and TS assists EP to converge towards the optimum point quickly. The search is basically done with EP, but additionally the TS is used to escape the search path from local optimum point. The algorithm for the proposed method is as follows:

1. Commit all the M hydro units and considering discharge rates ($Q_h(i,t)$) between the limits, calculate the volumes ($V_h(i,t)$) of the reservoirs from 1 to M.

$$V_h(i,t) = V_h(i,t-1) + I_h(i,t) - Q_h(i,t) - S_h(i,t) + \sum_{m=1}^{Ru}[Q_h(m,t-\tau(i,m) + S_h(m,t-\varsigma(i,m))] \quad (14)$$

2. Calculate the power produced by each hydro unit ($P_h(i,t)$) from the values of discharge rates and volumes.

$$P_h(i,t) = C_{1,i}V_h(i,t)^2 + C_{2,i}Q_h(i,t)^2 + C_{3,i}(V_h(i,t)*Q_h(i,t)) + C_{4,i}V_h(i,t) + C_{5,i}Q_h(i,t) + C_{6,i} \quad (15)$$

3. Sum up all the hydro powers for each period and subtract the total hydro power from the power demand for each period.
4. Find the remaining load demand to be met with thermal power such that

$$\sum_{i \in Rs} Ps(i,t) + \sum_{i=Rh} Ph(i,t) = PD(t) + PL(t) \quad (16)$$

5. Obtain the power (P_{dt}) to be produced by thermal unit,

$$PD_t = PD - PD_h \quad (17)$$

and for the thermal system Unit Commitment is performed as below.
6. An initial population of "parent" solutions S_k, k=1,2,3…,M (where M is the number of parents), is generated at random.
7. The objective function value associated with each solution S_k is calculated by economically dispatching the hourly load to the operating units and by computing the total fuel and start-up/shut-down costs, that is,

$$TC(S_k) = TFC(S_k) + TSUC(S_k) + TSDC(S_k) \quad (18)$$

8. An offspring S_k' is created from each parent by adding a Gaussian random variable $N(0, \sigma^2_k)$ to the elements a_{ijk} of parent S_k:

$$a'_{ijk} = a_{ijk} + N(0, \sigma^2{}_k) \quad (19)$$

$$\sigma_k = \beta_i * \frac{TC(S_k) * p_i}{TC_{\min}} \quad (20)$$

Here, the value of β_i is chosen in such a manner that product β_i x p_i should guarantee a minimum variance. Normally constant scaling factor is used in conventional EP. In this non-linear scaling factor is used for better convergence. For the first 40% of the total number of generations (N1) the decrement in scaling factor 'g1' is calculated as

$$g1 = \frac{(\beta_{\max} - \beta_{mid})}{N1} \quad (21)$$

For the remaining 60% of the total number of generations (N2) the decrement in β is calculated as 'g2' as

$$g2 = \frac{(\beta_{mid} - \beta_{\min})}{N2} \quad (22)$$

9. Each feasible offspring S_k' is evaluated according to 7.
10. For each feasible candidate, parent or offspring, a value W_k is assigned.

$$W_k = \sum_{\zeta=1}^{c} W_\zeta \quad (23)$$

$$W_\zeta = \begin{cases} 1, if TC(S_k) \langle TC(S_r) \\ 0, otherwise; \end{cases} \quad (24)$$

where $r = [2Mu + 1]$, *r not equal to k*, *[x]* denotes the greatest integer less than or equal to *x*, *c* is the number of competitions, and *u* is a uniform random number ranging over *[0, 1]*. Here, *c* is set at *1/10* of the population.

11. The feasible competitors are ranked in descending order of W_k. The first *M* solutions survive and are transcribed along with their elements to form the basis for TS Algorithm.
12. In TS Algorithm the temperature variable (C_p) is initially assigned to be relatively higher value.
13. The number of iteration 'n' to be performed for refining each individual solution is obtained and the process is done to every individual independently.
14. The initial solution is assigned as the current best solution 'Ui', the function to be checked is assumed to be minimum, in our case it is the cost 'Fi'.
15. Random perturbation is done to the current solution and the neighbouring solution 'Uj' is obtained whose feasibility is examined by checking to see if there is an uptime or downtime constraint.
16. Check if the cost Fj $\leq$ Fi, if less replace Uj and Fj as current solutions for Ui and Fi, if greater check if exp [(Fi-Fj)/Ck] $\geq$ U (0, 1), if satisfied, set Ui =Uj.
17. The iteration count 'n' is decremented and another neighbouring solution is generated. When the iteration count **'n'** reaches zero, the temperature variable C_p is lowered to a new value.
18. The entire process terminates when sufficient iterations have occurred at the specified lowest temperature and this process is repeated to all the individual solution till all the Np solutions are refined.
19. The refined Np number of population is passed on to

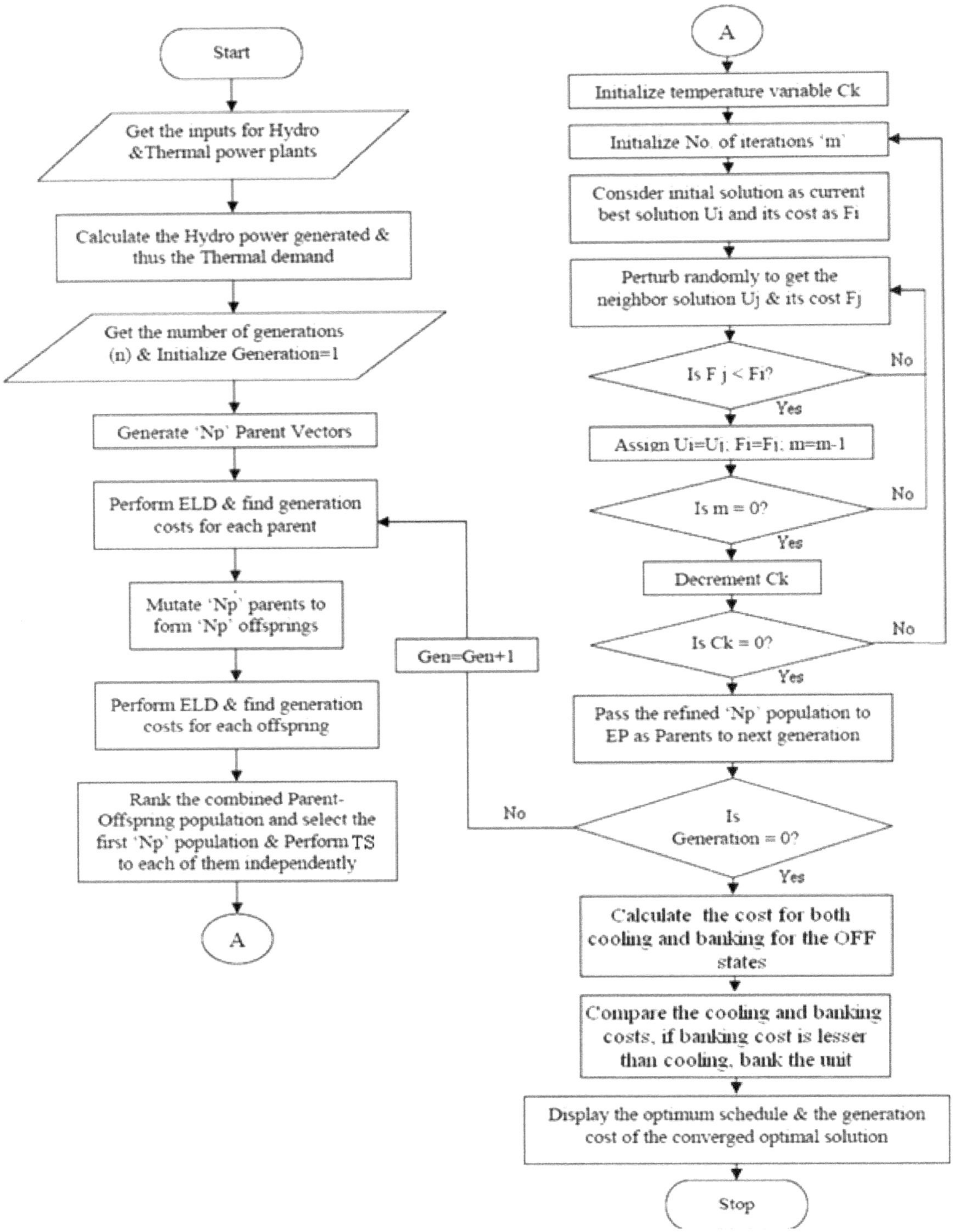

Figure 1. EPTSM flowchart for Hydro-Thermal UCP.

the EP part as the parents for next generation. And this process is repeated till the convergence in production cost is reached along with the optimum schedule having satisfied the constraints

20. For the units, which are in the off states, calculate the cost for both cooling and banking.

21. Compare the cooling and banking costs, if banking cost is lesser than cooling, bank the unit.

22. Print the optimum schedule.

The diagrammatic description of the proposed hybrid EPTSM algorithm is shown in Figure 1.

Table 1. IEEE thermal test system.

Unit	Pmax (MW)	Pmin (MW)	A ($ / H)	B ($/ MWH)	C ($/ W^2H)	Min Up (H)	Min Down (H)	Hot Start Cost ($)	Cold Start Cost ($)	Cold Start Hours (H)	Initial Status (H)
1	455	150	1000	16.19	0.00048	8	8	4500	9000	5	8
2	455	150	970	17.26	0.00031	8	8	5000	1000	5	8
3	130	20	700	16.60	0.002	5	5	550	1100	4	-5
4	130	20	680	16.50	0.00211	5	5	560	1120	4	-5
5	162	25	450	19.70	0.00398	6	6	900	1800	4	-6
6	80	20	370	22.26	0.00712	3	3	170	340	2	-3
7	85	25	480	27.74	0.00079	3	3	260	520	2	-3
8	55	10	660	25.92	0.00413	1	1	30	60	0	-1
9	55	10	665	27.27	0.00222	1	1	30	60	0	-1
10	55	10	670	27.79	0.00173	1	1	30	60	0	-1

Table 2. Hydrodischarge coefficient.

Plant No.	C1	C2	C3	C4	C5	C6
1	-0.0042	-0.42	0.03	0.9	10	-50
2	-0.004	-0.3	0.015	1.14	9.5	-70
3	-0.0016	-0.3	0.014	0.55	5.5	-40
4	-0.003	-0.31	0.027	1.44	14	-90

Termination criterion of the algorithm

The algorithm can be terminated at any time if it satisfies certain conditions. There may be several possible conditions for termination of the algorithm. But the best conditions are selected by the quality of the solution obtained after termination. In this algorithm two possible conditions for termination have been applied. The algorithm will be terminated if the following conditions are satisfied:

1. Given maximum number of iterations have been performed (or)
2. The best operating cost obtained repeats successively for certain number of iterations.

CASE STUDY

An IEEE test system consisting of 4 hydro generating units and 10 thermal generating units has been considered as a case study (Mohan et al., 1992). A time period of 24 h is considered and the unit commitment problem is solved for these 10 units power system. The required inputs for solving the UCP are tabulated below. The IEEE thermal test system is shown in Table 1, hydro discharge coefficients, reservoir volumes and discharge limits and inflows to the reservoir are shown in Tables 2, 3 and 4. The daily load pattern considered is shown in Table 5.

The cost convergence graph of EPTSM and hydro and thermal generations are shown in Figures 2 and 3. The operating cost comparison of EPTSM with EP, TSM, LR and DP is shown in Table 6. By analyzing the graphs between the cost and iterations, as iterations increased the cost will be reduced with the slight increase of computation time. From the results obtained, we observed that EPTSM with cooling-banking constraints approaches to near optimal solution.

Conclusion

This paper gives an efficient, fast and robust

Table 3. Reservoir volume and discharge limits (× 10^{-4} M^3).

Plant No.	Vmin	Vmax	Vini	Vend	Qmin	Qmax	Ph min (MW)	Ph max (MW)
1	80	150	100	120	5	15	0	500
2	60	120	80	70	6	15	0	500
3	100	240	170	170	10	30	0	500
4	70	160	120	140	13	25	0	500

Table 4. Inflows to the reservoir (× 10^{-4} M^3).

Unit	1	2	3	4	5	6	7	8
1	12	12	11.8	11.7	11.7	11.6	11.5	11.4
2	6	6	6.5	6.7	6.8	6.9	7	7.2
3	3	4	5	6	7	8	8.8	8.9
4	3	2	2	0	0	0	0	0

Table 5. Load pattern for 24 h.

Hour	Demand (MW)	Hour	Demand (MW)
1	1358	13	1567
2	1357	14	1539
3	1187	15	1374
4	1321	16	1356
5	1500	17	1555
6	1501	18	1372
7	1468	19	1380
8	1298	20	1390
9	1292	21	1469
10	1176	22	1391
11	1521	23	1276
12	1399	24	1553

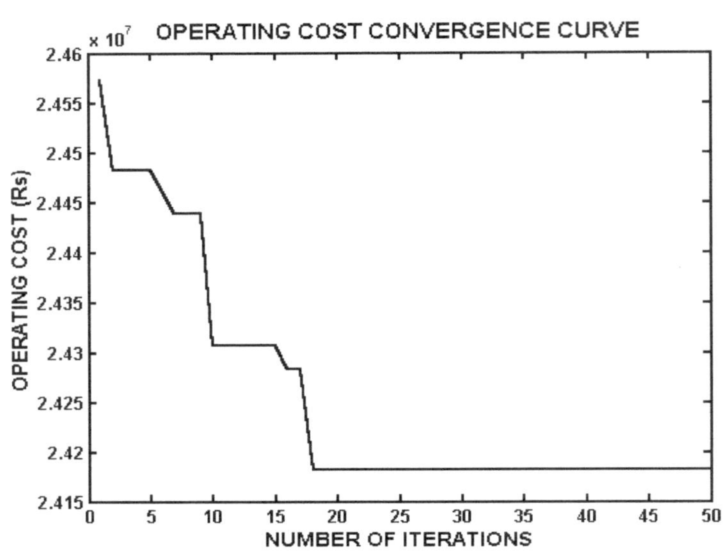

Figure 2. Cost convergence characteristics for 50 Iterations.

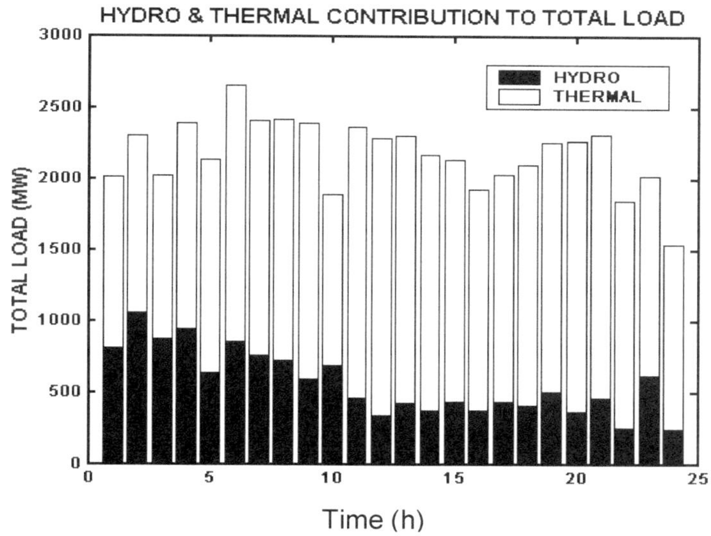

Figuer 3. Hydro and thermal generations.

Table 6. Production for different technique.

Techniques used	Iterations	Production cost (Rs.)	Computation time (s)
DP	-	32774013	83.90
LR	-	30656586	73.65
TSM	25	24363089	39.07
EP	25	24240650	26.91
EPTSM	25	24181789	24.45
EPTSM (C & B)	25	24164478	24.10

UCP through EPTSM with cooling-banking constraints. EP is characterized by its good convergent property and significant speedup over traditional GA's. An initial parent vector is formed at random. Then mutation is applied to obtain offspring. Best solutions are selected from the combined population of both parents and off springs for next generation. TSM is characterized by its ability to escape from the local optima by employing short term memory structure. Also it has a strategy called AL which directs the search towards attractive moves leading to a better solution. The effectiveness of the algorithm is proved by considering IEEE thermal and IEEE hydro test systems. The good convergent property of EP and TSM ability to avoid entrapment in local optima are integrated to form new hybrid algorithm for solving UCP which showed better results. Further improvements in the application of mutation operator to find better off springs will help in searching more areas in solution space resulting in global optimum solution. On comparing the results obtained from the different techniques, EPTSM with cooling-banking constraints obviously displays a satisfactory performance. Thus, the solution obtained from EPTSM has better quality in terms of economy and computation time.

Nomenclature

E_c : Energy of the current configuration
E_{Config}: Energy of a given configuration
E_t : Energy of the trail configuration
$F_{it}(P_{it})$: Production cost of unit i at a time t (Rs/h.)
F_T: Total operating cost over the scheduled horizon (Rs/H)
K : Constant
N: Number of available generating units
P_{Config} : Probability of a given configuration
PD_t : System peak demand at hour t (MW)
P_{it} : Output power from unit i at time t (MW)
P_{maxi} : Maximum generation limit of unit i (MW)
P_{mini} : Unit i minimum generation limit (MW)
R_t : Spinning reserve at time t (MW).
S_{it} : Start up cost of unit i at hour t (Rs).
S_{oi} : Unit i cold start – up cost (Rs).
T : Scheduled time horizon (24 h)
$Tdown_i$: Unit i minimum down time (H)
$Toff_i$: Duration for which unit i is continuously OFF (H)
Ton_i : Duration for which unit i is continuously ON (H)
$Tshut_i$: Instant of shut down of a unit i (H)
$Tstart_i$: Instant of start up of a unit i (H)

Tup_i : Unit i minimum up time (H)
U (0,1) : Uniform distribution with parameters 0 and 1
U_{it} : Unit i status at hour t = 1 (if unit is ON) = 0 (if unit is OFF)
UD(a,b) : Discrete uniform distribution with parameters a and b.
V_{it} : Unit i start up /shut down status at hour t = 1 if the unit is started at hour t and 0 otherwise.
F: Composite cost function
F_i : Fuel cost of ith thermal unit in Rs/h
$P_s(i,t)$: Generation of ith thermal unit at time t in MW
$P_h(i,t)$: Generation of ith hydro unit a time t in MW
$V_h(i,t)$: Storage volume of ith reservoir at time t in m^3
$Q_h(i,t)$: Water discharge rate of ith reservoir at time t in m^3
$P_D(t)$: Power demand at time t in MW
$P_L(t)$: Total Transmission line losses at time t in MW
$S_h(i,t)$: Spillage of ith reservoir at time t in m^3
$I_h(i,t)$: Inflow rate of ith reservoir at time t in m^3
$H_i(t)$: Net head of ith reservoir at time t in m^3
α,β,γ : Thermal generation cost coefficients
$C_{i,1}$ to $C_{i,6}$: Hydro power generation coefficients
$\tau_{i,m}$: Water transport delay from reservoir σ to i
R_u : Set of upstream units directly above i^{th} hydro unit
R_h / R_s : Set of Hydro/Thermal plants in the system
i,m : Reservoir index, index of reservoir upstream of the ith reservoir
t,T : Time index, scheduling period
V,i^{begin} : Initial storage volume of ith reservoir in m^3
V,i^{end} : Final storage volume of ith reservoir in m^3
P_i : Output generation for unit i in MW
P_L : Total current system load in MW
PT_L : Total system transmission losses in MW
OBJ : Objective cost function
***F_i* :** *Cost function for unit i.*

REFERENCES

Ahmed B, Abdelmalik B, Salah EZ (2006). Optimal Hydropower Generation Management. J. Electr. Eng. 6(2):21-26.

Alberto B, Antonio F, Fabrizio L, Carlo AN (2003). Lagrangian Heuristics Based on Disaggregated Bundle Methods for Hydrothermal Unit Commitment. IEEE Trans. Power Syst. 18(1):313-323.

Alberto B, Claudia D, Andrea L, Silvano M (2008). An MILP Approach for Short-Term Hydro Scheduling and Unit Commitment with Head-Dependent Reservoir. IEEE Trans. Power Syst. 23(3):1115-1124.

Allen JW, Wollenberg BF (1984). Power Generation, Operation and Control. John Wiley & Sons p. 444.

Baptistella LFB, Gerome JC (1980). Decomposition approach to problem of unit commitment schedule for hydrothermal systems. IEE Proc. Gener. Transm. Distrib. 127(6):250-258.

Benhamida F (2009). A Hybrid Lagrangian Relaxation – Dynamic Programming Solution to the Unit Commitment Problem. J. Electr. Eng. 9(4):31-40.

Cau TDH, Kaye RJ (2002). Evolutionary Optimization Method for multistorage hydrothermal scheduling. IEE Proc. – Gener. Transm. Distrib. 149(2):152-156.

Chitra SS, Kumudini DRP, Christober ARC (2009). A Hybrid Approaches for the Profit Based Unit Commitment Problem in the Deregulated Markets. J. Electr. Eng. 9(3):35-41.

Christober ARC (2010). Hybridizing Evolutionary Programming, Simulated Annealing, Tabu Search Method to Solve the Unit Commitment Problem. J. Electr. Eng. 10(1):34-41.

Costas GB, Stylianos EP, Anastasios GB (2009). Simulation of Optimal Medium-Term Hydro-Thermal System Operation by Grid Computing. IEEE Trans. Power Syst. 24(3):1208-1217.

Esteban G, Julian B, Hugh R (2003). Short-Term Hydrothermal Generation Scheduling Model Using a Genetic Algorithm. IEEE Trans. Power Syst. 18(4):1256-1264.

Ferrero RW, Rivera JF, Shahidehpour SM (1998). A Dynamic Programming Two-Stage Algorithm For Long-Term Hydrothermal Scheduling Of Multi reservoir Systems. IEEE Trans. Power Syst. 13(4):1534-1540.

Gary WC, Mohamed A, James GW, José M, Tony B, Steve R, Christoforidis M (2001). Experiences With Mixed Integer Linear Programming Based Approaches on Short-Term Hydro Scheduling. IEEE Trans. Power Syst. 16(4):743-749.

Kirkpatrick S, Gelatt CD, Vecehi MP (1983). Optimisation by Simulated Annealing. Science 220(1):4598.

Javadi MS, Meskarbashi A, Azami R, Hematipour G, Javadinasab A (2011). Security Constrained Unit Commitment in Iran's Electricity Market. Int. Rev. Model. Simul. 4(6):3104-3112.

Javadi MS, Meskarbashi A, Azami R, Hematipour G, Javadinasab A (2011). Emission Controlled Security Constrained Unit Commitment Considering Hydro-Thermal Generation Units. Int. Rev. Model. Simul. 4(6):3243-3250.

Mantawy AH, Youssef L, Abdel-Magid, Shokri ZS (1998). A Simulated Annealing Algorithm for Unit Commitment. IEEE Trans. Power Syst. 13(1):197-204.

Mantawy AH, Youssef LA, Shokri ZS (1998). A Unit Commitment By Tabu Search. IEE Proc. Gener. Transm. Distrib. 145(1):56-64.

Mariappane E, Thyagarajah K (2009). A Genetic Algorithm Approach to Price-Based Unit Commitment. J. Electr. Eng. 9(4):41-46.

Martinez L, Soares S (2002). Comparison between Closed-Loop and Partial Open-Loop Feedback Control Policies in Long Term Hydrothermal Scheduling. IEEE Trans. Power Syst. 17(2):330-336.

Mohan MR, Kuppusamy K, Abdullah KM (1992). Optimal Short-term Hydro-Thermal Scheduling using Decomposition Approach and Linear Programming Method, Int. J. Electr. Power Energy Syst. 14(1):39-44.

Ngundam JM, Kenfack F, Tatietse TT (2002). Optimal Scheduling of Large–Scale Hydrothermal Power Systems Using the Lagrangian Relaxation Technique. Int. J. Electr. Power Energy Syst. 22(1):237-245.

Nidul S, Chakrabarti R, Chattopadhyay PK (2003). Fast Evolutionary Programming Techniques for Short-Term Hydrothermal Scheduling. IEEE Trans. Power Syst. 18(1):214-220.

Nimain CN, Christober CAR (2011). Thermal Unit Commitment Scheduling Problem in Utility System by Simulated Annealing Embedded Evolutionary Programming Method. Int. Rev. Model. Simul. 4(6):3188-3194.

Lal RRS, Christober CAR (2011). A Hybrid Particle Swarm Optimization Employing Genetic Algorithm for Unit Commitment Problem. Int. Rev. Electr. Eng. 6(7):3211-3217.

Rudolf A, Bayrleithner R (1999). A Genetic Algorithm for Solving the Unit Commitment Problem of a Hydro Thermal Power System. IEEE Trans. Power Syst. 14(4):1460-1468.

Ruey-Hsun L, Ming-Huei K, Yie-Tone C (2009). Co evolutionary Algorithm Based on Lagrangian Method for Hydrothermal Generation Scheduling. IEEE Trans. Power Syst. 24(2):499-507.

Ganesan S, Subramanian S (2012). Reliability Constrained Thermal Unit Commitment Solution Using a Hybrid Approach. Int. Rev. Model. Simul. 5(1):434-441.

Salem A (2001). Hydrothermal Scheduling by Augmented Lagrangian: Consideration of Transmission Constraints and Pumped-Storage Units. IEEE Trans. Power Syst. 16(4):750-756.

Shokri ZS, Alsultan K (1991). A Simulated Annealing Algorithm for the Clustering Problem. Patt. Recognit. 24(10):1003-1008.

Shyh-Jier H (2001). Enhancement of Hydroelectric Generation Scheduling Using Ant Colony System Based Optimization Approaches. IEEE Trans. Energy Convers. 16(3):296-301.

Srinivasa RC, Siva NS, Sangameswara RP (2009). Automatic generation control of TCPS based hydrothermal system under open market scenario: A fuzzy logic approach. Int. J. Electr. Power Energy Syst. 31(1):315–322.

Venkatesan T, Sanavullah MY (2012). Implementation of Modified SFLA for the Thermal Unit Commitment Problem. Int. Rev. Model Simul. 5(1):450-457.

Tiago Norbiato dos Santos, Andre LD (2009). A New Multi period Stage Definition for the Multistage Benders Decomposition Approach Applied to Hydrothermal Scheduling. IEEE Trans. Power Syst. 24(3):1383-1392.

Vo ND, Weerakorn O (2008). Enhanced merit order and augmented Lagrange Hopfield network for hydrothermal scheduling. Int. J. Electr. Power Energy Syst. 30(1):93–101.

Werner TG, Verstege F (1999), An Evolution Strategy for Short-Term Operation Planning of Hydrothermal Power Systems. IEEE Trans. Power Syst. 14(4):1362-1368.

Whei-Min L, Fu-Sheng C, Ming-Tong T (2002). An Improved Tabu Search for Economic Dispatch With Multiple Minima, IEEE Trans. Power Syst. 17(1):108-112.

Wilfredo SS, Alberto V (2007). Hydrothermal Scheduling Using Benders Decomposition: Accelerating Techniques. IEEE Trans. Power Syst. 22(3):1351-1359.

Xaiomin B, Shahidehpur SM (1996). Hydro-Thermal Scheduling by Tabu Search and Decomposition. IEEE Trans. Power Syst. 11(2):968-974.

Yong-Gang W, Chun-Ying H (2000). A Diploid Genetic Approach to Short-Term Scheduling of Hydro-Thermal System. IEEE Trans. Power Syst. 15(4):1268-1274.

Zhuang F, Galiana FD (1990). Unit Commitment by Simulated Annealing. IEEE Trans. Power Syst. 5(1):311-318.

Application of chemical injection on cooling treatment technology control of corrosion and fouling in petrochemical plant: Case study of Indorama Plc, Akpajo-Eleme

C. P. Ukpaka

Department of Chemical/Petrochemical Engineering, Rivers State University of Science and Technology, Nkpolu, P. M. B. 5080, Port Harcourt, Nigeria. E-mail: chukwuemeka24@yahoo.com.

Corrosion and fouling has been a major menace to the operation, integrity and life expectancy of process equipment in the production line of refinery and petrochemical equipment. This research focuses on effective cooling water treatment technology by inhibitive and bio-remediative chemical injection to control fouling and minimizes corrosion in process equipment such as boiler, heat exchangers and heaters which form the livewire of a petrochemical plant. The research has been streamlined to consider Indorama Plc, formerly Eleme Petrochemical Plant network as a case study. Models for corrosion rate prediction and treatment level recommendation adopted uses a combination of laboratory, field and experiential data. The models provide tools for evaluating a cooling system under varying conditions to determine the impact of operating parameter changes (for example, acid feed, concentration ratio, pH) upon water corrosivity and corrosion inhibitor requirements. The study reviews the effectiveness and suitability of the existing chemical treatment program in place and measures it with performance indicators such as models, theories and empirical values such as Langelier saturation index (LSI), Ryznar stability index (RSI) to predict the tendency of corrossivity scaling of water, especially cooling water. Relevant data from daily analysis, routine checks, inspection and monitoring over a period of 1 year form the basis of the presented result and deductions. Corrosion coupons and relevant test instrument were employed for the research. Result obtained shows a good inhibitive treatment and consistent bio-activity control from effective chemical dosage given corrosion rate as low as 2 mpy on the average. The cycle of concentration that the cooling tower maintained had consistently been between 4 and 5. Calculated value of LSI and RSI on the average agrees with literature value for good treatment that enhances better performance. LSI and RSI values was obtained respectively to be, -0.1 to +0.1, and 0.4 giving credence to low corrosive and scaling tendencies that breeds fouling.

Key words: Cooling, treatment, control, corrosion, fouling.

INTRODUCTION

The petrochemical plant unit carrying a cooling agent may suffer from internal corrosion, if there is water present in the system. In the cooling treatment unit the corrosivity will vary independence of factors such as the temperature, zinc concentration, ortho-phosphate concentration, calcium hardness ($CaCO_3$), pH, turbidity, total iron, silica, cycles of concentration (based on silica) and Langeller saturation index, aerobic bacteria counts, bioscan counts, flow conditions, use of inhibiting chemicals etc. The corrosion allowance is based on assumption of the corrosivity during the production period. In order to maintain integrity, many cooling treatment plants are subject to intelligent cathodic protection/pig inspection at certain time intervals.

Corrosion monitoring at fixed locations is sometimes used to verify the efficiency of inhibiting chemicals (Singley, 1981; Pisigan and Singley, 1984, 1985; Mc Neill and Edwards, 2002; Imran, 2003).

Corrosion may be defined as the destruction of a material by action of the surrounding environment. Material resistance to corrosion depends on many variables such as materials properties itself, environment characteristics and others.

Various scientists (Pisigan and Singley, 1987; Rossum and Merrill, 1983; Frigeribaum et al., 1988; Ferguson, 1992) have reported that controlling a cooling treatment plant unit from further corrosion by the application of chemical injection operation involves: identification of flow characteristic of cooling and inhibiting fluid, respectively, and determination of the concentration of the physico-chemical parameters of interest.

Research conducted by various groups revealed that when the water phase is dominated by the reservoir water (brine), there is normally no significant variation of the ion concentrations along the pipeline. In wet gas systems the situation is different; the water phase will be dominated by the water condensing from the gas as the temperature drops along the pipeline. This water has no salts dissolved, and the ions accumulating in the water phase have to come from the dissolution of gases (CO_2 and H_2S) and the corrosion process. Dissolution of CO_2 will cause formation of ions like bicarbonate (HCO_3). During corrosion, Fe^{2+} and HCO_3^- ion are released as a result of the electrochemical reaction. For wet gas systems, therefore, it is necessary to adjust the Fe^{2+} and HCO_3^- concentrations along the pipeline due to the corrosion process. This may have a significant impact on the pH profile along the pipeline (Rossum et al., 1990; NNC, 1992; NNPC, 1994; Davernport, 1997; Goody, 2008).

When the chemical is injected into the cooling treatment plant unit, it acts as an inhibitor in the process plant, hereby changing the chemical composition of the cooling water by reducing corrosion rate. The injection of chemical is useful in inhibiting the functional parameters that facilitate the rapid rate of corrosion of cooling treatment plant unit (Cullen, 2002; Flakeysar et al., 1997; Montgomery, 1985; Davis, 1993).

MATERIALS AND METHODS

Research Instrument

This comprises field test instruments and instruments used in the laboratory for analysis.

Field test instruments

This includes, chemical dosing tank network, dosing pumps, infra red, fluke thermometer, coupons, bio-film box, data monitoring panel and colourimeter DR/890.

Laboratory test instruments

This includes, reagents pillows, chemical reagent, conical flasks, beakers, burettes, pH meters, indicators, buffer solutions, refrigerator, oven, weighing machine, sample bottles/cans, spectrophotometer DR/2400, spectrophotometer DR/890 and sample cells.

Research method and procedures

Field tests

Basically this is categorized into daily routine tests, based on daily inspection and monitoring. On the other hand, special attention was given to observable changes and noticeable side effect as direct fallout from field operations and material handling.

Cooling tower operation data

This includes, system designation - unit 23, number of systems – 1 × 10 cell tower, system volume – 1100 m^3, total NEL + blow down – 100 m^3, Make up rate – 508 m^3/h, max. cycle of concentration (COC) – 6, Operating duration – 24 h/day/365 day/yr, recirculation rate total – 27,000 m^3/h, system evaporation rate – 408 m^3/h, supply temperature – 32°C and return temperature – 44°C. The above data forms the basis of the operation of the cooling tower process.

Cooling tower treatment

Corrosion is an electrochemical process by which metal returns to its natural state. For example mild steel will oxidise to form $Fe(OH)_2$. On further exposure to the environment it is further attacked by water and oxygen forming iron (III) oxide, Fe^{3+} which precipitate out of the water as reddish brown colour that causes rust, deterioration, loss of metal properties, fouling of water, deposition and eventual corrosion. Corrosion effect, as a result of inadequate treatment manifest on a short or long term on galvanic metals due to factors such as: corrosive environment, pitting causing perforation on metals, localization, intergranulation along grain boundaries of metal or alloy, erosion, crevice formation, cavitations and stress corrosion cracking accelerated by internal or external stress or a form of impingement attack from turbulent flow regime around bends or inlet into a unit.

Deposition: The constituent of deposits formed on the internal heat surface of a process unit for example, boiler, heat exchanger, as a result of inefficient chemical treatment programme generally comes from water. More of the constituent belongs to one or more of the following groups: iron, oxides, silica, silicates, metallic copper, carbonate, phosphates, calcium and magnesium salts, etc. Poor flow regime- low velocities, can also allow carryover and build up of these substances by settling to form scales, under deposit corrosion, causing clogging of tubes in heat exchangers thus reducing the rate of heat transfer which if not controlled could lead to possible failure of the equipment.

Cooling water corrosion treatment chemicals

GE treatment facilities control corrosion and deposition by a continuous dose of 70 to 90 L of Dianodic DN2108 daily and 10 to 20 L of Flogard MS6208 daily. These are patented chemicals from GE Infrastructures – Water and Process Technologies. The Dianodic DN108 supplies the phosphate active treatment regime formation while the Flogard MS6208, provide the zinc phosphate active treatment regime formation. This was found adequate so the concentration of each chemical can be controlled separately. Dianodic DN2108 is a dispersant blend.

Cooling water micro-biological treatment

Cooling tower can provide an environment suitable for growth of bacteria, fungi, yeast, and algae. These microbes cause fouling, corrosion, formation of gases, wood attack and can protect one another from biocide attacks. Recently, of great concern is a bacterium namely *legionella pneumophila*, which causes 'Legionnaires disease' in human. The metabolic activity of these micro-organism can either directly or indirectly cause deterioration of metal by corrosion process. This activity can: cause a corrosion environment, create electrolytic concentration cells on metal surface, alter the resistance of surface film and can influence the rate of anodic or cathodic reaction, alter the environmental composition. Disinfection to arrest or reduce to the barest minimum the invasion and activities of these microbes is therefore imperative. Disinfection is a process of removing bacteria, algae, and other micro-organism from water either by one or a combination of the following method: chemical injection for example, chlorine gas, calcium hypochlorite (HTH), sodium hypochlorite etc, ozonation, ultraviolet light, boiling and microfilteration. A GE operation employs chemical injection method to combat the activities of microbes in cooling water treatment.

Micro-biological treatment chemical: GE controls the activities of primary biocides through a timely shot dose of oxidising and non-oxidising biocide chemicals.

Oxidising biocides: 135 kg calcium hypochlorite daily dose supplies the cooling system calcium, and needed 110 kg of Spectrus OX909 and liquid bromine dosed alternately on weekly basis. The above chemicals uses halogen release agent, to meet the microbiological treatment the system requires.

Non-oxidising biocides: 800 kg of Spectrus NX1101 is dosed fortnightly and 800 kg of spectrus NX1164 dosed alternately fortnightly. A secondary treatment employed the use of non-oxidising biocides dosed fortnightly in an alternate manner to cover the month thereby complementing the dose of oxidising biocide to combat the activities of microbes in the system. This is so done to avoid the micro-organism developing a resistance to a specific chemical dose. The chemical dosed has been found effective to destroy the cyst of the microbes, and washing them off the cell walls of the tower where they are entrapped in the strainer for scooping off the water system. Bio-Dispersant: 100 kg of spectrus BD1500 dosed once in the week on Saturday and another chemical dosed is spectrus BD1500, a bio-enhancer, surface active agent and bio-dispersant that provides a good interface for the effectiveness of calcium hypochlorite, spectrus OX909 (Liquid bromine) and enhances the effectiveness of other biocide treatment chemical dosed by keeping the suspension in the water until blow down is considered necessary.

Field test/Monitoring

Monitoring: This involves daily assessment of the client facility; the tower assembly, the cells (1 × 10), flow in the cells, cooling water outlook and quality. Also included in daily monitoring of GE treatment facilities; the panel assembly, taking of field readings/parameters on TDS, pH, corrator, wet bulb, tower supply and return temperature from control room. On daily basis, sampling is done from cooling water return line sampling point for laboratory analysis to ascertain if the major control parameters are within control limit established.

Field activity: On daily basis, we carry out residual free chlorine test to ascertain if there is sufficient disinfection/micro-biological treatment for the cooling process water. Treatment target before now had been between 0.3 to 0.6 mg/L, temperatures of the cells, that makes up the tower is taken respectively and reported on daily basis, approach temperature is also monitored for the cells with reference to the supply temperature.

Coupon installation and analysis

Coupon analysis was also undertaken by exposing prototype sample coupon of distinct material as available in the plant make up such as mild steel, admiralty brass and copper into the return line of the cooling water. The coupons which may be treated or untreated are a replica of process unit material as the heat exchangers, furnaces, boilers, column etc where cooling water passes. These coupons are exposed differently for 30 and 90 days respectively in the cooling water test rack. A corrater dipped in the water in one of the compartment measures the rate of corrosion while the sample coupons on removal are sent outside the country for analysis for onward result to the client. The outlook on removal depicts and reveals what is happening in the entire system with the same process unit material as available in the plant. Result of previous exposures shows values averaging 1 mpy/year.

Emergency operation

Outside the daily routine operation we are always at the beckon and call of our client to attend to emergency situation that may arise in the plant regarding our operation that requires our attention. Some of these in recent past include: sampling of opened exchangers, analysing sample and reporting observation, troubleshooting for source of leakages into cooling tower, rouble-shooting serious deviation of certain parameters from normal to arrest the situation and bring it under control for example, silica increase in furnaces or boiler feed water, need for passivation of a single exchanger and passivation from the cooling water source. The above are some of the major challenges encountered in the cooling water monitoring, few of which will be discussed in detail later.

Passivation

Passivation is a pre-treatment programme applied to the cooling water system in order to ensure all new and/or recently cleaned carbon steel surfaces receive the best possible protection from corrosion. Pre-treatment promotes the rapid formation of a uniform passive film that stifles the corrosion reaction. Once the film has been established, it can be maintained through continuous, low treatment levels to deter the accumulation of corrosion products. Pre-treatment of equipment in cooling water systems is recommended following initial installation, chemical cleaning, and prior to start-up after turnaround or inspection. Pre-treatment, followed by ongoing treatment programs, minimizes corrosion for improved heat transfer, longer service life, and reduced plant maintenance. It is expected that the treatment will be applied twice in the first year of operation as below: Application could be general and central from the cooling tower sump, with circulation done via the tower to the entire equipment network in the plant. It is necessary to drain the tower sump, clean the basin out and refill the system. The system will then be put in partial operation (that is, not all plant on line). It is proposed to carry out the Pre-clean and Passivation process after the tower is initially re-filled and the treatment will also be applied once any new heat exchangers are

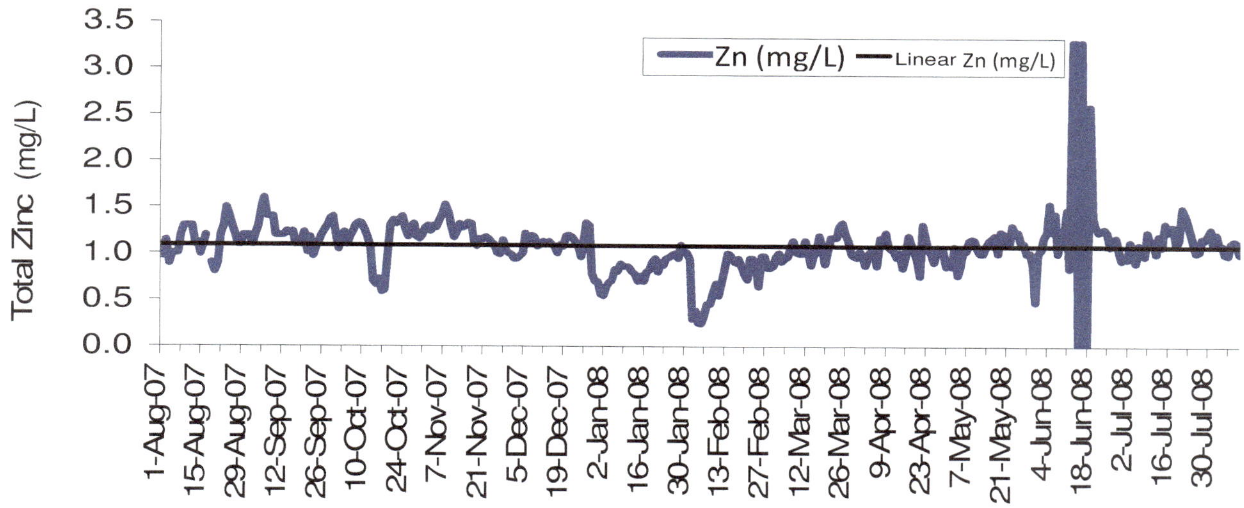

Figure 1. Zinc concentration versus time.

brought into service or when plant not previously pre-treated is brought online.

Treatment programme: The specific products recommended for the Pre-clean/Passivation treatment programme are: GE Betz Dianodic DN2300: a surfactant dispersant containing GE Betz HPS co-polymer, GE Betz Dianodic DN2200: a dispersant blend containing ortho-phosphate, poly phosphate, HEDP and TTA. In the last one year, the general passivation below is the passivation process undertaken- single unit passivation of exchanger 1-E-301A at Polyethylene with dianodic DN2108 in March '08 due to unavailability of DN2300 and DN2200. Recirculation period was 12 h, and general passivation from cooling water sump with Nalco Chemical equivalent of GE Betz chemical. Due to delay in clearing and delivery of GE Betz passivation the chemicals and single unit passivation of exchanger 1-E-301B at polyethylene with DN2300 in September 5th/6th 2008 circulated for 24 h.

Note: Necessary calculation was carried out with available data to ascertain the chemical requirement for the treatment and for general passivation blow-down is not meant to be effected until considered necessary. Test analysis was carried out and phosphate, pH, and Turbidity level were been monitored on daily basis until an acceptable level comfortable to cooling tower standard treatment range was obtained before blow-down was initiated and pre-cleaning is necessary before commencing full treatment for effective passivation.

Laboratory analysis of parameters

In cooling water treatment GE Water and Process Technologies analyses and monitors, consistently on daily basis, up to 14 parameters called 'analytes' to ensure each is in order with the range of design and operational data given to our client. This informs the proactive action taken whenever there is any deviation from normal in any of the parameter from the results of analysis. The analytes are zinc, filtered and unfiltered phosphate, calcium hardness, magnesium hardness, methyl alkalinity, total iron, turbidity, silica, bioscan count, aerobic bacteria count, and free residual chlorine. Also, we check the azole level in the cooling water system from the continuous dose of dianodic chemical when the need arises but not on daily basis. Tolytriazole is a blend constituent chemical in dianodic that function to provide protection for material unit made of copper alloy forming a protective film on the surface to inhibit corrosion. Each of these parameters has positive or negative effect on the treatment program as the case may be.

Routine operations

Chemical stock

Monitors in-house/ laboratory analysis, chemical reagent usage and depletion, reporting same for the purpose of order placement and replenishing, stocks of corrosion control and micro-biological treatment chemical is monitored and reported on daily and monthly basis and monthly report of chemical consumption and stock balance is sent to the client for record purpose and ordering procedure decision when the need arises.

RESULTS AND DISCUSSION

The results were presented in figures. Results include: monitoring and analyzing parameters from cooling water sample, monitoring of control parameters and performance of process equipments with special attention on boiler in olefins unit of the plant, and calculated values of models and performance indices used in predicting the efficiency of treatment program.

Cooling tower monitoring and analysis result

Daily laboratory analysis of cooling water sample and monitoring for an estimated period of 1 year as could be seen in the result of cooling water treatment monitoring of parameters between August, 2007 and July, 2008 for 1 year is shown in Figures 1 to 11 for different parameters.

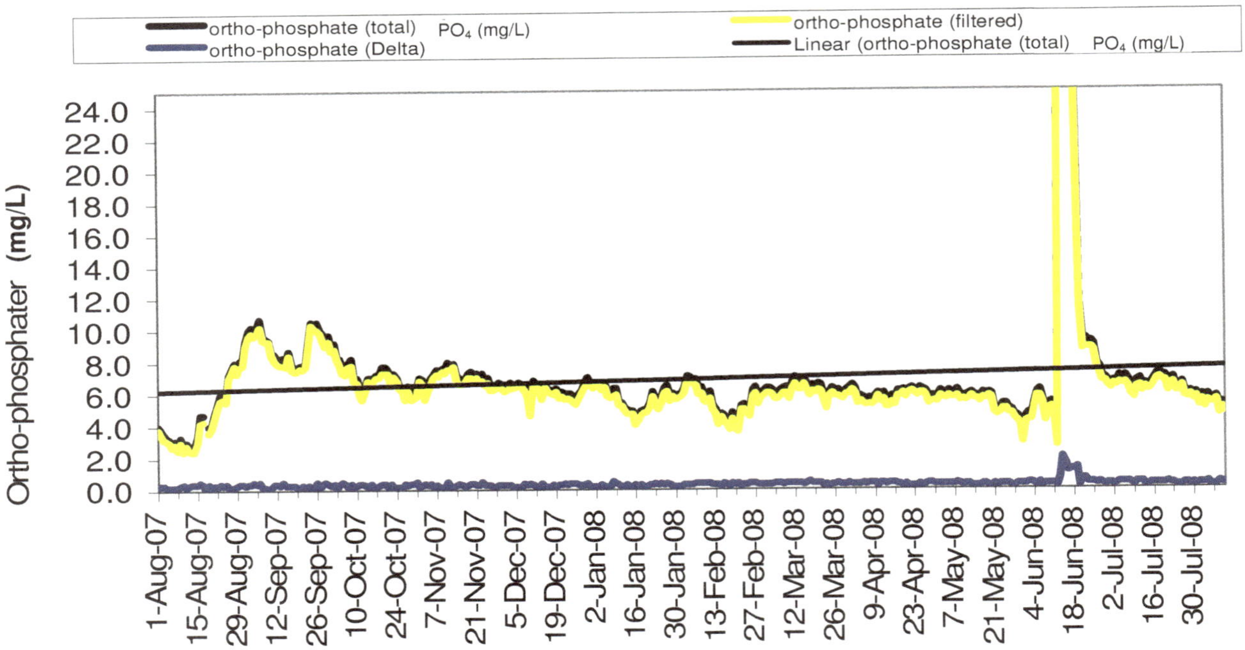

Figure 2. Ortho-phosphate concentration versus time for filtered component.

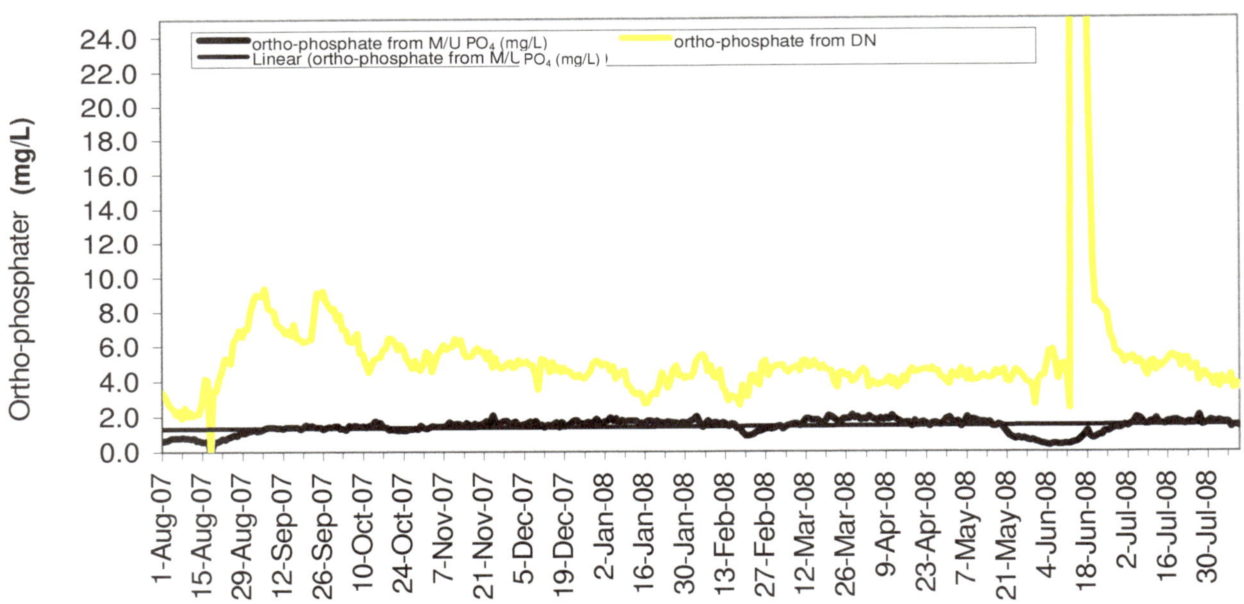

Figure 3. Ortho-phosphate concentration versus time for DN component.

The plot of zinc concentration versus time shows a great degree of compliance with standard treatment regime of 0.8 to 1.2 mg/L, zinc been a major harmful element to the environment, hence discharge must be minimal to the environment. Although certain deviation from standard value as high as >3.0 occurs and control were often regained on timely manner through effective treatment (Figure 1).

Orthophosphate concentration within the period was effectively managed within the specified control limit of 4 to 6 mg/L, however certain deviations occurs within June 2008 after which control was regained. Offset went as far

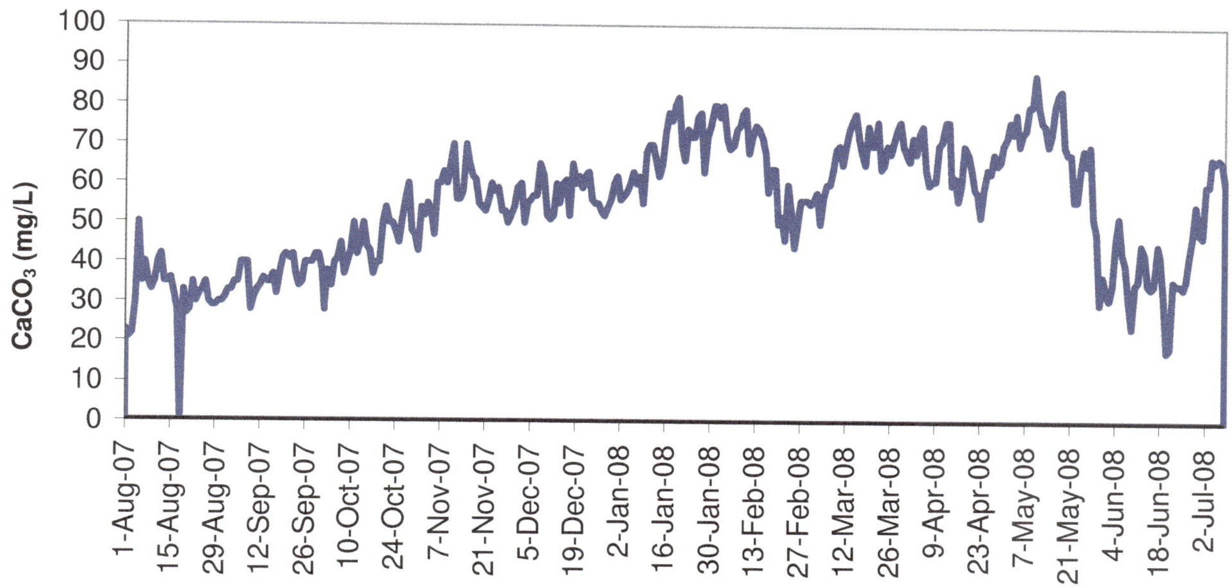

Figure 4. Calcium hardness $CaCO_3$ versus Time.

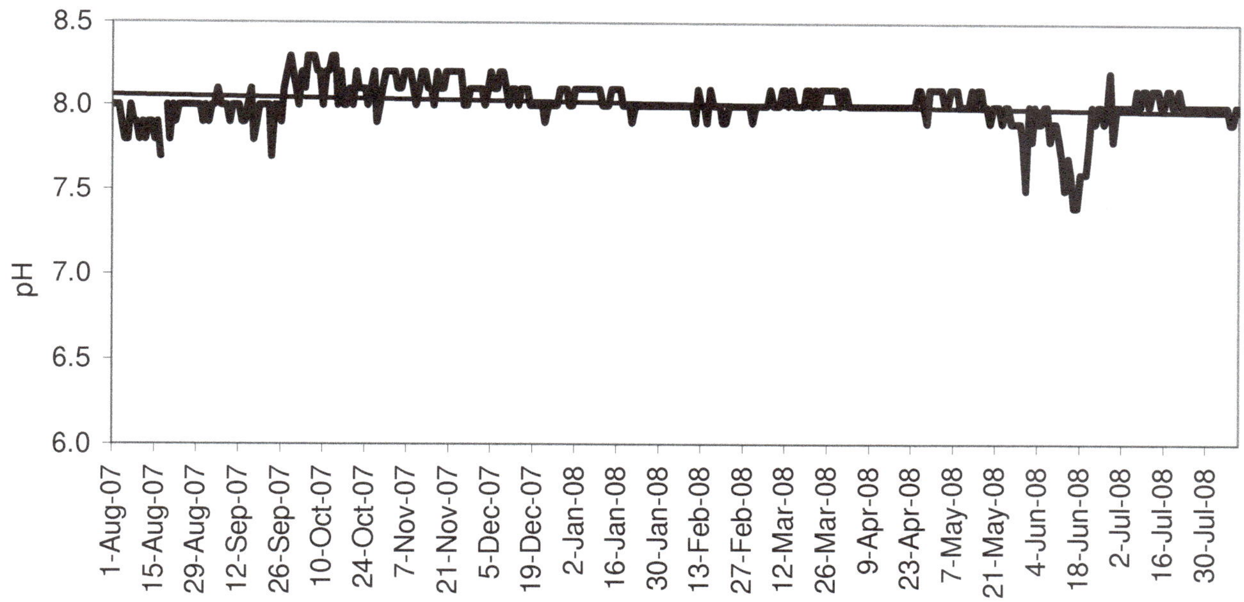

Figure 5. pH versus Time.

as >20 mg/L (Figure 2).

The Figure 3 illustrates the characteristics of the ortho-phosphate concentration upon the influence of time on the DN component as well as the linear ortho-phosphate concentration in the system.

Calcium hardness concentration was excellent during the period of treatment showing the effectiveness of calcium supplement from 135 kg of calcium hypochlorite daily dose, 6 days within the week. Control was kept with the treatment range specified as 40 to 100 mg/L (Figure 4).

The pH concentration plot against time shows alkaline behavior within the given control treatment of 8.0 to 8.5 mg/L, although on few occasion a departure to acidity scale was experienced which is not good for the system. At such period, neutralization is effected by dosing 4 to 5

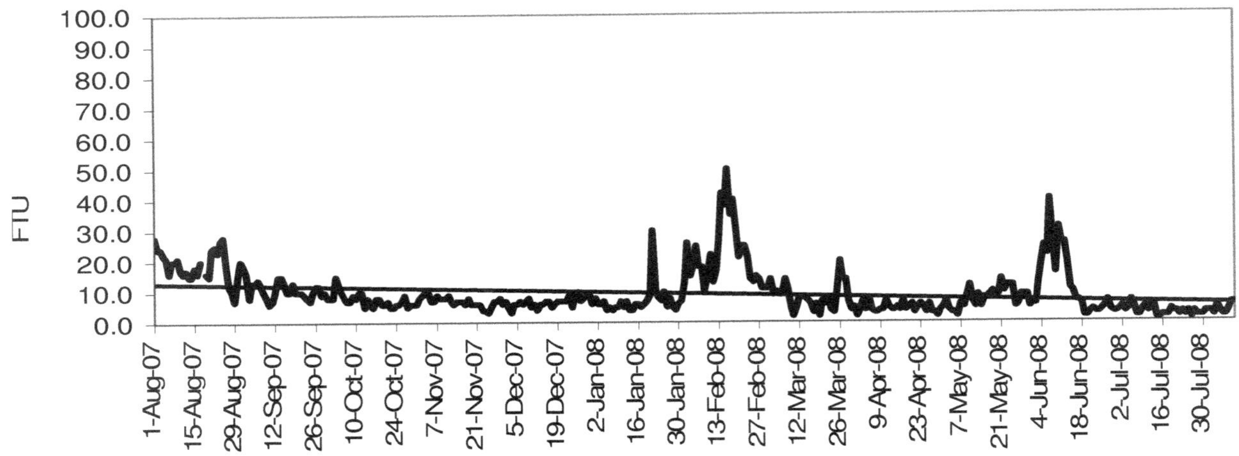

Figure 6. Turbidity versus time.

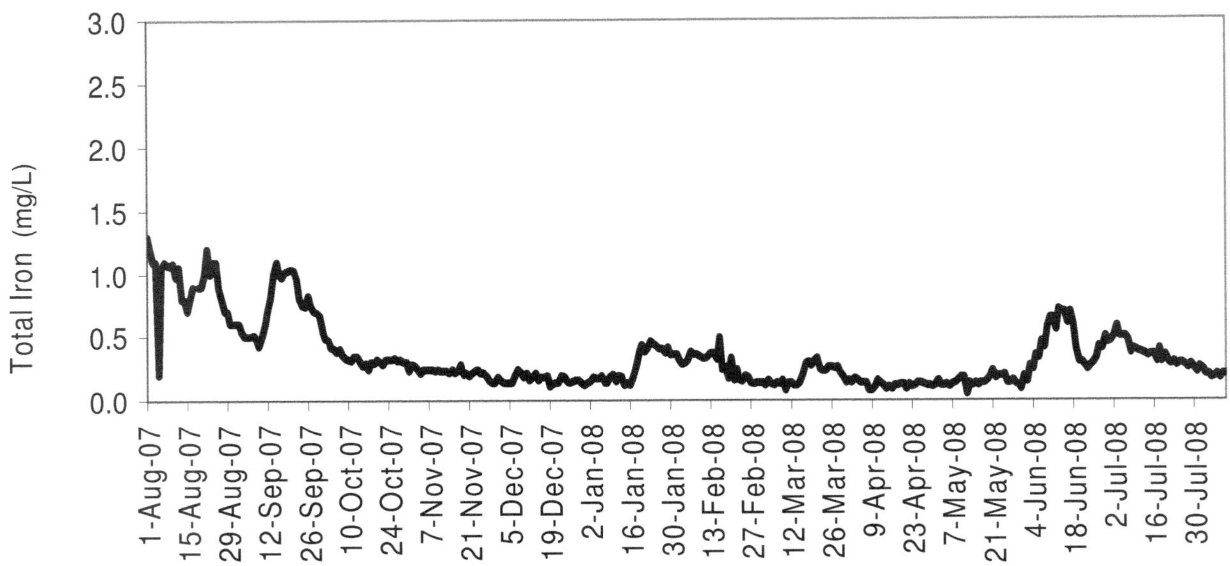

Figure 7. Total Iron versus time.

bags of sodium hydroxide (NaOH) to regain back the alkaline condition (Figure 5).

Turbidity of the water specified control treatment was given as 0 to 25 NTU (nephelometric turbidity units). Treatment result shows conformity to a great extent, also some deviations from standard were experienced within February, 2008 and June, 2008. Increased zinc, Fe and leakages into the system are possible causes of a corresponding increase in turbidity Figure 6.

Iron concentration over the period of treatment was excellently kept with the standard control limit given as 0 to 1.5 mg/L, hence not exposing the system to major corrosion (Figure 7).

Silica content in the treatment was effectively managed also within the control limit for recirculation cooling water of 0 to 150 mg/L. This is a major determinant for a good cycle of concentration (Figure 8).

Cycle of concentration over the period of treatment was majorly 4> COC>5, a value advised for good treatment. This is what informs the time lapse for blow-down and needed for replenishing of the system. The maximum operating COC for the tower is 6 (Figure 9). Langelier saturation index, a major indices for the prediction of corrosion tendency shows positive values <1.0 which shows good treatment with low corrosion tendency for the plant.

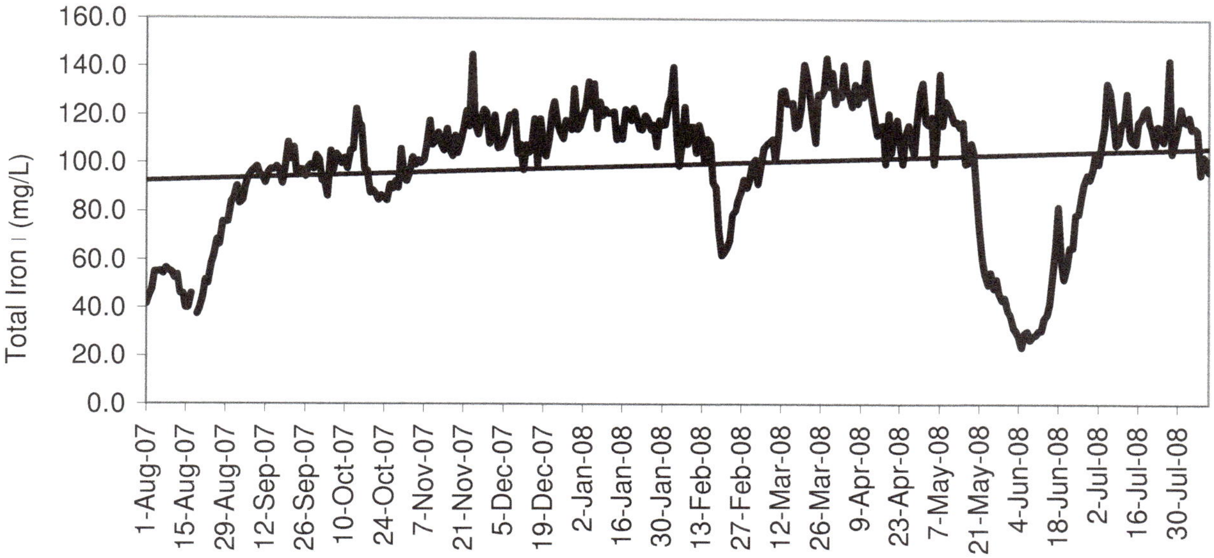

Figure 8. Plot of silica versus time.

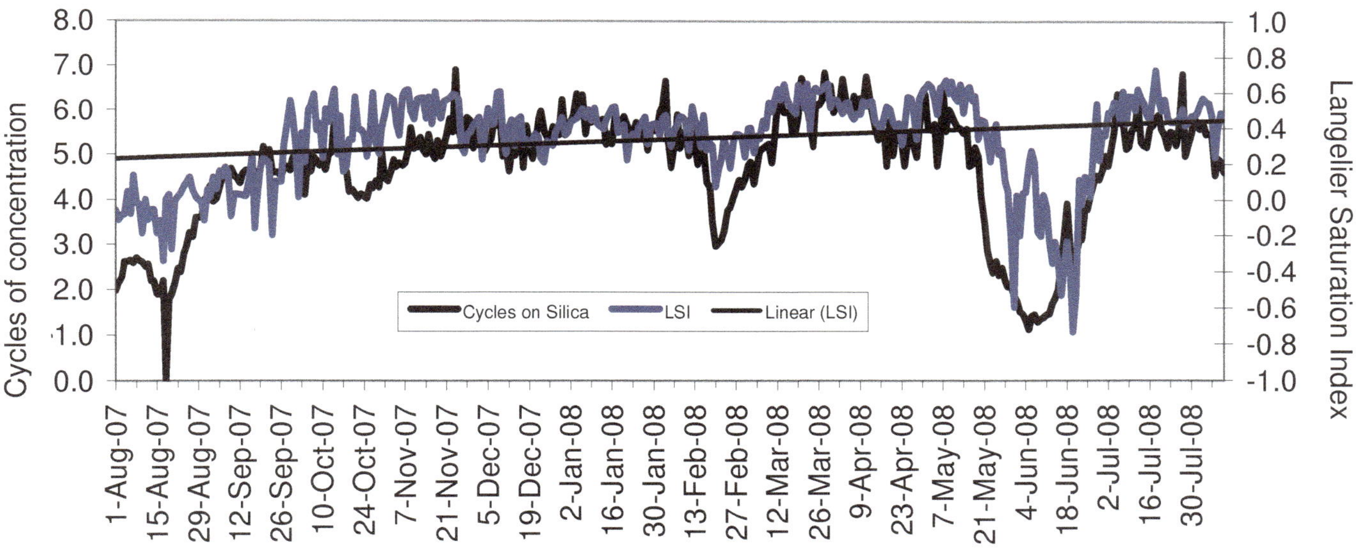

Figure 9. Cycles of concentration (based on silica) and Langelier saturation index versus time.

Aerobic bacteria count in the water from result shows low bacteria activity and invasion with values as low as 10<ABC<100 over the treatment period (Figure 10).

Bioscan Count shows the count of active and non-active biocide in the water which reveals a good treatment been kept with the specified treatment limit of 0 to 10,000 (Figure 11).

In all of the above, the major deviations experienced were attributed to process leaks, from existing plant units by backflow to the tower (for example, olefins, polyethylene, polypropylene), critical exchangers leaks, hydrocarbon and oil leaks from seal of equipment etc. Control is however regained by trouble shooting, isolation of area of disturbance, creating necessary by-passes and treatment with continuous dose of appropriate chemicals.

Boiler result from olefins unit monitoring

Daily laboratory analysis result of boiler drums and steam condensates boiler feed water (BFW) and Quench water while monitoring the operation of the equipment itself and other ancillary units for example, dosing pumps. Due to the boiler being a steam generating unit, emphasis will be

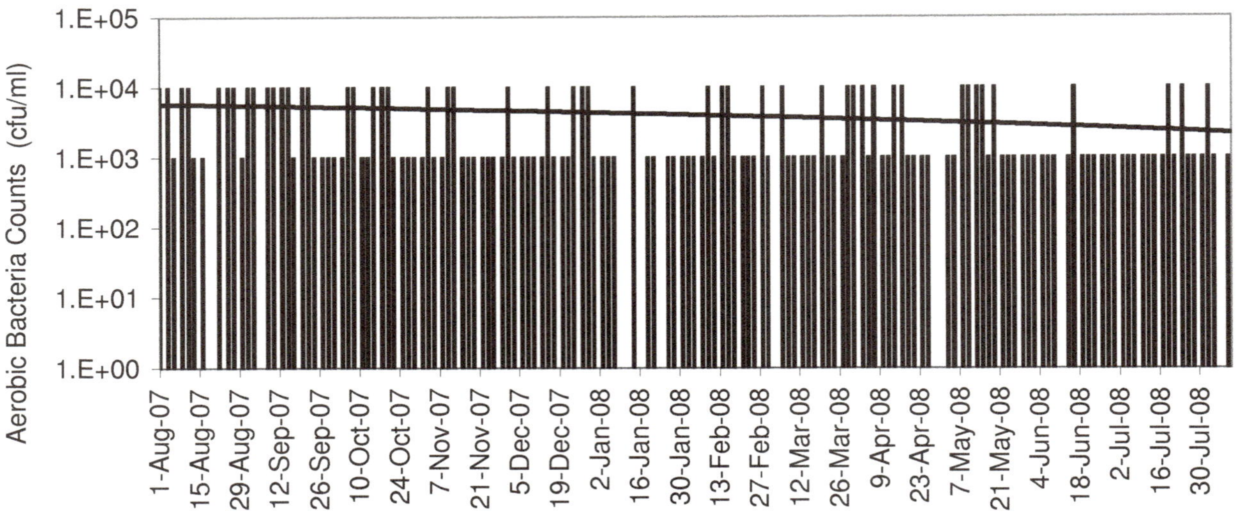

Figure 10. Aerobic bacteria counts versus time.

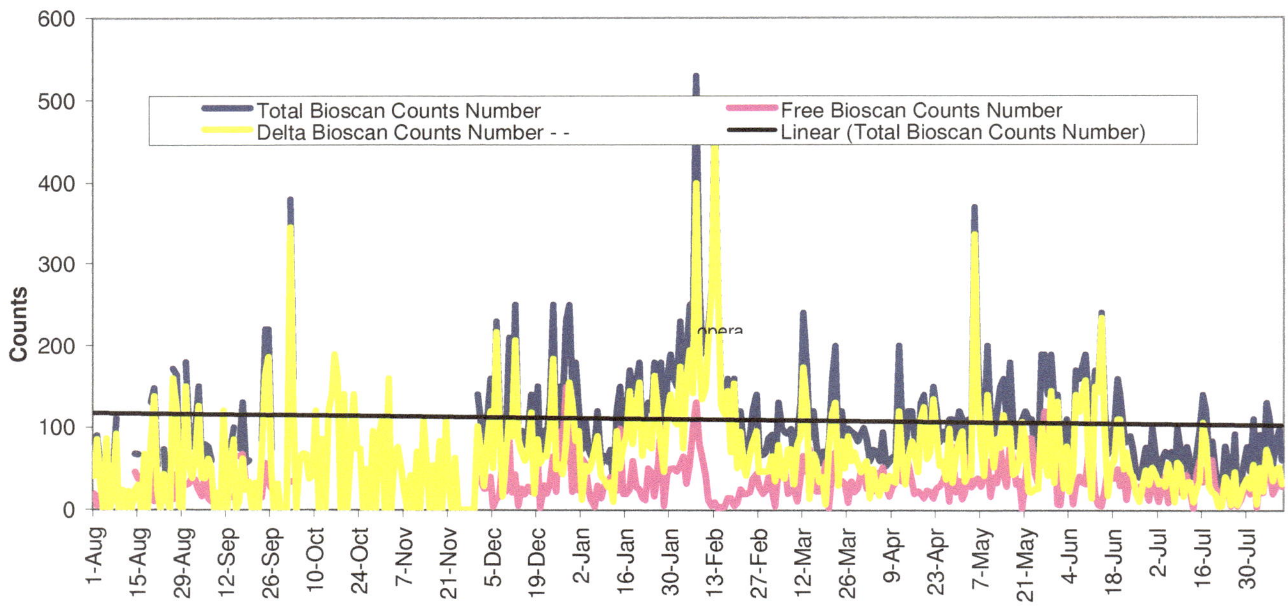

Figure 11. Bioscan counts versus time.

based majorly on boiler A steam. Boiler A pH within the period of monitoring and treatment conform to the treatment standard 8.0 to 8.5 (Figure 12).

Silica content in the boiler was kept as low as possible, standard control limit being 0.01 to 0.02 mg/L. Silica is not good for the boiler tubes and system; its carryover causes hotspot, reduced heat transfer surfaces, rupture, tube leakages and poor performance (Figure 13).

Ortho-phosphate level in the boiler was effectively managed within the control limit of 0 to 0.05 mg/L (Figure 14). The above result altogether shows a good treatment and handling of the boiler.

Exchanger's inspection, process unit passivation and general passivation from the cooling tower sump results are given in Tables 1 and 2.

Application of relevant model

This comprises result from: deduction and calculation of

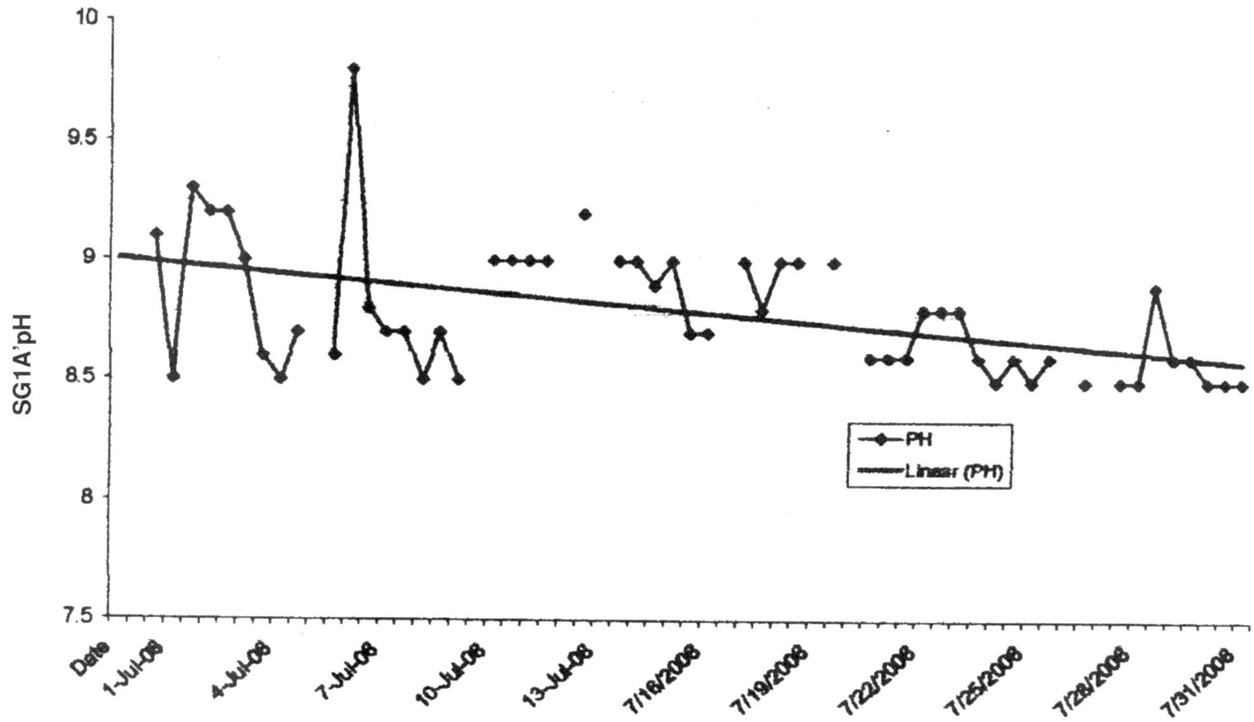

Figure 12. Boiler A steam pH versus time.

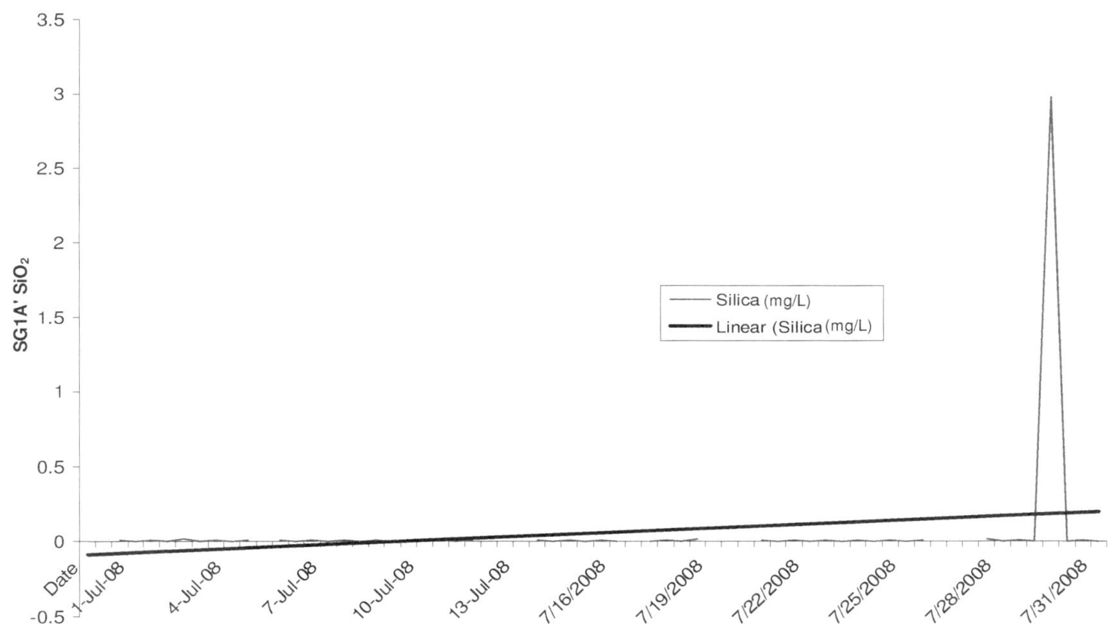

Figure 13. Boiler A steam A silica versus time.

Langelier saturation index (LSI), deduction and calculation of Reynar stability index (RSI), relationship between LSI and RSI, and relevance of the above and other models in predicting effectiveness of chemical inhibitors versus estimated corrosion rate.

Calculations

Langelier Saturation Index (LSI): The LSI index is calculated at two temperatures: 25°C (room temperature) and 82°C (cage wash cycle). The colder incoming water

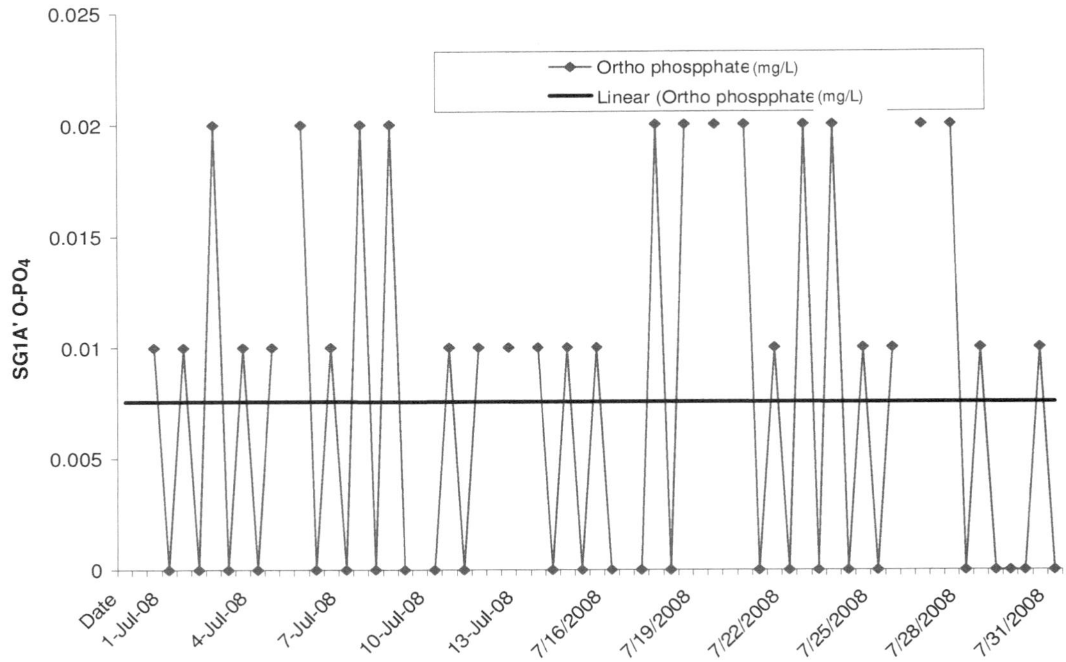

Figure 14. Boiler A steam A ortho-phosphate versus time.

will warm to room temperature in the manifolds. Residual water in the rack manifold can be heated to 82°C when the rack is in the cage washer. At this process conditions, the following data were obtained.

Water analysis: pH = 7.5, TDS = 320 mg/L, Calcium = 150 mg/L (or ppm) as $CaCO_3$ and Alkalinity = 34 mg/L (or ppm) as $CaCO_3$.

LSI formula: LSI = pH - pH_s, pH_s = (9.3 + A + B) - (C + D) where: A = (Log_{10}[TDS] - 1)/10 = 0.15, B = -13.12 × Log_{10}(°C + 273) + 34.55 = 2.09 at 25°C and 1.09 at 82°C, C = $Log_{10}[Ca^{2+}$ as $CaCO_3]$ - 0.4 = 1.78, D = Log_{10}[alkalinity as $CaCO_3$] = 1.53.

Calculation at 25°C: pH_s = (9.3 + 0.15 + 2.09) - (1.78 + 1.53) = 8.2, LSI = 7.5 - 8.2 = - 0.7, hence no tendency to scale calculation at 82°C, pH_s = (9.3 + 0.15 + 1.09) - (1.78 + 1.53) = 7.2, LSI = 7.5 - 7.2 = + 0.3, hence slight tendency to scale, also, for chemical treatment program at Indorama process operation, practically LSI from empirical data is calculated automatically by direct system unit computations and reported on daily basis as follows:

LSI=[PH-(9.4-LOG(Ca)-0.94*LOG(M.Alk)+LOG(Conduct)/10.7+3.24*EXP(-(65*1.8)+32)/191)]

The controlling parameters were some of the major parameters upon which researcher "Pisigan" based his corrosion rate prediction model equation.

Reynar stability index (RSI)

The Reynar stability index (RSI) attempts to correlate an empirical database of scale thickness observed in municipal water systems to the water chemistry. Like the LSI, the RSI has its basis in the concept of saturation level. Reynar attempted to quantify the relationship between calcium carbonate saturation state and scale formation. Applying the formula below, the Reynar index takes the form:

RSI = 2(pH_s) - pH

At the same condition as that applicable to LSI, RSI can also be determined as follows:

At 25°C, pHs = 8.2, RSI = 2(8.2) – pH, At 82°C, pHs = 7.2, RSI = 2(7.2) – pH

Hence, RSI can be determined, taking direct measurement of the water PH=7.5, RSI = (2(8.2) -7.5) – 2(7.2) – 7.5), RSI = 8.9 – 6.9 = 2.0. Where pH is the measured water pH, and pHs the pH at saturation in calcite or calcium carbonate, thus applying the above values to Indorama Plc cooling water treatment process gives an operating data temperature of; supply temperature = 32°C, return temperature = 44°C. Average LSI and RSI by applying the above formula, all conditions been normal it was obtained that: LSI = pH - pH_s, pH_s = (9.3 + A + B) - (C + D) where: A = (Log_{10}[TDS] - 1)/10 = 0.15, B = -13.12 x Log_{10}(°C + 273) + 34.55 = 1.96 at 32°C and 1.74 at 44°C, C = $Log_{10}[Ca^{2+}$ as $CaCO_3]$ - 0.4 = 1.78,

Table 1. Chemical composition of cooling water treatment chemical.

OPO4 (Ortho-phosphate)	Carbon steel corrosion inhibitor	Anodic and cathodic
HPS (Co-polymer)	Dispersant for iron scale formation	-Control scale formation. -Stabilizes salt of zinc phosphate and calcium phosphate from precipitating.
HEDP(Organic phosphonates)		-Calcium carbonate stabilizer. -Control salts formation to avoid scaling.
TTA (Tolytriazole)	Copper alloy corrosion inhibitor	-Forms Cu-azole passive film on copper alloys.
Flogard MS6208 on the other hand contains:		
Zn (Zinc)	Carbon steel corrosion inhibitor	Cathodic

Table 2. Effect of analyses parameters in cooling water and controls.

S/N	Analytes	High effect	Low effect	Control action
1	Zinc (Zn)	Environmental	Cathodic corrosion from deposit	Optimize dosage
2	Phosphate (PO_4)	Precipitation to form salt deposit	Ineffective corrosion inhibition	Optimize dosage
3	Calcium and Magnesium hardness	Scaling	Corrosion	High- initiate blowdown, low- optimised dosage to maintain natural inhibition from film
4	PH	Fouling/Scaling, removal of surface protective film	Corrosion	Low- dose suitable quantity of caustic soda
5	Conductivity	Better performance	Poor performance	Monitor cycle of concentration
6	Total Iron	Deposition/Scaling/Corrosion	Under control	Maintain pH and auto-phosphate treatment range
7	Turbidity	Possibility of leakage or interference	Better	High-check for leakages and contain it by dosing dispersant; Initiate blowdown
8	Silica	Good, gives high cycle of concentration (COC)	Not good	Minimise blowdown rate and check for leakages and possible loss of water
9	Chloride	Could form $CaCl_2$ salt with calcium and weak acid with hydrogen causing pH depression and possible corrosion		Optimise the dose of dispersant to stabilise salt formation and inhibit possible corrosion
10	Bioscan	High biocide activity	Low biocide activity	Ensure sufficient oxidising and non-oxidising treatment
11	Aerobic bacteria count	Excess bio-activity	Low bio-activity	Ensure sufficient oxidising and non-oxidising treatment
12	Free residual Chlorine (Cl_2)	Form weak acid with loose hydrogen/possible corrosion	Insufficient dose of Calcium hypochlorite/increased bio-activity	Maintain sufficient dose of calcium hypochlorite or other supplement for example, liquid bromine

D = Log_{10}[alkalinity as $CaCO_3$] = 1.53.

Note: Cooling water design treatment data for operation of Indorama Plc Cooling Tower is given by the range 8.0 to 8.2. Hence for Indorama treatment parameter:

Calculation at 32°C: pH_s = (9.3 + 0.15 + 1.96) - (1.78 + 1.53) = 8.1, LSI = 8.0 - 8.1 = - 0.1; hence no tendency to scale.

Calculation at 82°C: pH_s = (9.3 + 0.15 + 1.74) - (1.78 + 1.53) = 7.7, LSI = 8.0 - 7.7 = + 0.1; hence slight tendency to scale.

Also Average Reynar stability index (RSI) is calculated as

follow: At 32°C, pHs = 8.1, RSI = 2(8.1) – pH; At 44°C, pHs = 7.9, RSI = 2(7.9) – pH. Hence, RSI can be obtained, taking direct measurement of the water pH=8.0, RSI = (2(8.1) -8.0)–2(7.9) – 8.0). RSI = 8.2-7.8 = 0.4.

Conclusion

A combination of models (for example, corrosion inhibitor and scale inhibitor models) can be used to optimize the ratio of ingredients in a blended formulation. Corrosion rate and inhibitor 2 dosage models can provide guidelines to assist an experienced cooling system professional in optimizing treatment rates and controllable operating parameters. As with any predictive tool, the models are a supplement to experience - not a substitute for it.

Corrosion and Fouling in Indorama petrochemical plant has been effectively managed and kept at minimal level from continuous chemical dosage treatment, of inhibitors, oxidising and non- oxidising biocide given to cooling water that runs through the entire plant. Installed corrosion rate from coupon result has been kept as low as 1.5 mpy.

Process equipment that forms the network of the plant as boilers, exchangers, heaters etc receives adequate protection due to general passivity initiated from the cooling tower forming resistance passive film to corrosion and microbial invasion against scale tendency and eventual fouling. Chemical consumption in treatment of cooling water and olefins operation has been excellent. Necessary modifications still further need to be implemented to improve performance. Daily challenges faced that has to do with control of parameters and arresting of emergency situations were adequately managed through surveillance inspection, troubleshooting and reporting of incidences. Leakages were traced, pumps were shut, and optimizations in chemical dosage were carried out as occasion demands.

The fouling observed is as a result of a number of factors pertaining to both the treatment of the cooling water, and the design and engineering of the plant and exchangers. This further highlights the need for a well-managed and maintained treatment programme, for additional monitoring as detailed above and for ongoing review of results with a view to implementing continuous improvement. A great stride has been achieved as a testimonial of efficiency in cooling water monitoring while improvement plans are still necessary to fully be on top of olefins monitoring programme. Predictive models, theories, Index and empirical relationship applied in the research agrees with the treatment regime approved for the chemical treatment program in place of Indorama Plc. Average circle of concentration (COC) during treatment was between 4 to 5. Average LSI and RSI values have been between -0.1 to 0.3 and 0.3 respectively. This shows low corrosive tendency and reduction in heavy scale tendency for possible fouling.

CONTRIBUTION TO KNOWLEDGE

This research work has been able to ascertain the suitability of cooling water treatment as a medium for control of corrosion, scaling and bio-fouling in Petrochemical plant. If this is maintained consistently, increased efficiency and improved performance will be the result. This will further extend the useful life on Petrochemical plant. For future consideration the standard treatment must be sustained while some improvement procedure and necessary modification may be useful. One of such modification needed for Indorama Plc Petrochemical plant is 'true sense' but for the cost which would be a major limitation, new set of chemicals that goes with true sense will be introduced and dosage opted for example, SPECTRUS BD15001E, CONTINUM AEC3192, INHIBITOR AZ101, GENGARD GN7004 and SPECTRUS NX 1422. Some parameters treatment range shall be reviewed upward and dosage possibly increased for example, free residual chlorine formerly 0.3 - 0.6 mg/L to 0.6 - 1.0 mg/L, Orthophosphate formerly 4 - 6 mg/L to 8 - 10 ppm, Zinc formerly 0.8 - 1.2 mg/L to 0.7 - 0.8 mg/L. Dose of Spectrus OX909 will be reduced to optimize cost and dosing of calcium hypochlorite will be increased to span 4 to 5 h for good retention in the system against Bio-activity.

Operation of cooling tower and monitoring of olefins operations has been quite successful and also challenging. This informs certain modifications to further improve the operation and service delivery as applicable to both cooling tower and olefin monitoring. The supposed modifications some of which has commenced implementation are presented below: increase dosage of calcium hypochlorite to increase the killing effect of biocides, reducing the rate of blow-down so that oxidising biocides and bio-dispersant can stay longer in the system for effective treatment and maintainance of an optimum cycle of concentration, change and modification in the entire treatment programme. Dosage of non-oxidising chemical on a weekly spread basis, as against fortnight step-up of free residual chlorine treatment ranges from 0.3 - 0.6 to 0.6 - 1.0. Possibly warehouse is hired for chemical storage for stock control purpose. Similarly, for Olefins: installation of new dosing pumps, modification of some dosing point and possible changes occur. Introduction of support chemical petroflo 21Y621 to complement petroflo 22y630 for adequate control of pH in the system and monitoring increased to cover 24 h for adequate service delivery.

REFERENCES

Cullen CJ (2002). Equilibration of pilot-scale distribution systems. MS thesis. University of Central Florida. Orlando pp. 12-20.

Davenport HR (1997). ETF Series Cooling Tower. Davenport Co. Ltd Bradford pp. 1- 3.

Davenport NC (1997). Cooling Water Treatment. Davenport Co. Ltd Bradford, pp. 1-2.

Davis RV (1993). Investigation of Factors Influencing Mild Steel Corrosion Using Experimental Design Paper (280).

Ferguson RJ (1992). Developing Scale Inhibitor Dosage Models. Proceedings of WaterTech '92. Houston, TX.

Frigenbaum C, Gal-Or L, Yahalom J (1988). Scale protection criteria in natural waters. Corrosion 34(4):133–137.

GE Water and Process Technologies (2006). Proposal for Cooling Water Chemical Treatment, pp. 1-20 GE Water Purification Handbook Chapter 23 – 30 from http//. www.gewater.com (accessed December 12th 2007).

Googy D (2008). Technical Training Lecture for Indorama Plc. Boiler / Cooling Water Optimal. Efficient treatment and calcium carbonate saturation indexes. J. Am. Water Works Assoc. (827):71–77.

Googy D (2008). Technical Training Lecture for Indorama Plc. Boiler / Cooling Water Optimal. Efficient treatment and calcium carbonate saturation indexes copper. J. Am. Water Works Assoc. 79(2):62–70.

Imran SA (2003). The effect of water quality on red water release from iron distribution systems. PhD dissertation, University of Central Florida, Orlando.

FlaKeysar S, Hasson D, Semiat R, Bramson D, (1997). Corrosion protection of mild steel by a calcite layer. Ind. Eng. Chem. Res. 36(8):2903–2909.

McNeill LS, Edwards M (2002). The importance of temperature in assessing iron pipe corrosion in water distribution systems. Environ. Monit. Assess. 77(3):229–242.

Montgomery JM (1985). Water treatment principles and design Wiley New York, pp. 271-282.

Nigerian National Petroleum Corporation (NNPC) Port Harcourt Nigeria (1992). Ethylene / Propylene Plant Operation Instruction Manual, p. 31.

Nigerian National Petroleum Corporation (NNPC) (1994). Eleme Petrochemical Complex Operation Manual for Cooling Water System Unit 23. Vol. 1 Operation Instruction Book 1 of 1, HIYODA Corp. Yokohama Japan pp. 135-150

Pisigan RA, Singley JE (1984). Evaluation of water's corrosivity using Langelier index and relation to corrosion rate. J. Am. Water Work Assoc. 77(2):81.

Pisigan RA, Singley JE (1985). Effects of water quality parameters on the corrosion of galvanized steel. J. Am. Water 77(11):76-82.

Pisigan RA, Singley JE (1987). Influence of buffer capacity chlorine residual and flow rate on corrosion of mild steel. J. Am. Water Work Assoc. 82(7):300.

Rossum JR, Merrill DT (1983). An evaluation of the calcium carbonate saturation indexes. J. Am. Water Works Assoc. 75(13):172-180.

Rossum JR, Scanlon LP, Singley JE, Whitney JR (1990). Suggested methods for calculating and interpreting. Wiley, New York pp. 73-150.

Singley JE (1981). The search for a corrosion index. J. Am. Water Works Assoc. 73(11):578–582.

Review on fly ash-based geopolymer concrete without Portland Cement

Mohd Mustafa Al Bakri[1]*, H. Mohammed[2], H. Kamarudin[1], I. Khairul Niza[3] and Y. Zarina[1]

[1]School of Material Engineering, Universiti Malaysia Perlis (UniMAP), P. O. Box 77, d/a Pejabat, Pos Besar, 01007 Kangar, Perlis, Malaysia.
[2]King Abdul Aziz City Science and Technology (KACST), P. O. Box 6086, Riyadh 11442, Kingdom of Saudi Arabia.
[3]School of Environmental Engineering, Universiti Malaysia Perlis (UniMAP), P. O. Box 77, d/a Pejabat, Pos Besar, 01007 Kangar, Perlis, Malaysia.

The consumption of Ordinary Portland Cement (OPC) caused pollution to the environment due to the emission of CO_2. As such, alternative material had been introduced to replace OPC in the concrete. Fly ash is a by-product from the coal industry, which is widely available in the world. Moreover, the use of fly ash is more environmental friendly and save cost compared to OPC. Fly ash is rich in silicate and alumina, hence it reacts with alkaline solution to produce aluminosilicate gel that binds the aggregate to produce a good concrete. The compressive strength increases with the increasing of fly ash fineness and thus the reduction in porosity can be obtained. Fly ash based geopolymer also provided better resistance against aggressive environment and elevated temperature compared to normal concrete. As a conclusion, the properties of fly ash-based geopolymer are enhanced with few factors that influence its performance.

Key words: Fly ash-based geopolymer, alkaline solution, compressive strength.

INTRODUCTION

Ordinary Portland Cement (OPC) becomes an important material in the production of concrete which act as its binder to bind all the aggregate together. However, the utilization of cement causes pollution to the environment and reduction of raw material (limestone). The manufacturing of OPC requires the burning of large quantities of fuel and decomposition of limestone, resulting in significant emissions of carbon dioxide (Kong and Sanjayan, 2008). As such, geopolymer concrete had been introduced to reduce the above problem. Geopolymer concrete also showed good properties such as high compressive strength, low creep, good acid resistance and low shrinkage (Lodeiro et al., 2007). The role of binder in geopolymer concrete is replaced by fly ash which also possess pozzolanic properties as OPC and rich with alumina and silicate. Fly ash is residue from the combustion of coal which is widely available worldwide and lead to waste management proposal. Hence, fly ash-based geopolymer concrete is a good alternative to overcome the abundant of fly ash. In fly ash-based geopolymer concrete, the silica and the alumina present in the source materials are first induced by alkaline activators to form a gel known as aluminosilicate.

This gel binds the loose aggregates and other un-reacted materials in the mixture to form the geopolymer concrete (Wallah, 2009). Besides that, the reaction also depends on a few parameters such as size of aggregates, chemical composition of fly ash, amount of vitreous phase in fly ash, nature, concentration and pH of activators. The curing process of geopolymer concrete play shows a great influence on the development of microstructure, and subsequently on the mechanical characteristics of geopolymer (Komljenovi´c et al., 2010).

This paper summarizes the properties of fly ash-based geopolymer which make it better compared to normal concrete.

*Corresponding author. E-mail: mustafa_albakri@unimap.edu.my.

ALKALINE SOLUTIONS

The common materials used as alkaline solution in producing fly ash-based geopolymer are sodium silicate and potassium hydroxide. Usually either of this material was mixed with sodium hydroxide to produce the alkaline solution and the molarity (M) of alkaline solution is 7 to 10 M. The alkaline solution was prepared a day before it is mixed with fly ash. Then, the materials are mixed together with fine aggregate and coarse aggregate to form concrete and curing process been done.

Chindaprasirt et al. (2007) found that, to produce higher strength geopolymer, the optimum sodium silicate to sodium hydroxide ratio was in range of 0.67 to 1.00. Meanwhile, the concentration of NaOH between 10 and 20 M give small effect on the strength.

CURING PROCESS

Setting time of geopolymer depend on many factors such as composition of alkaline solution and ratio of alkaline liquid to fly ash by mass. However, the curing temperature is the most important factor for geopolymer. As the curing temperature increases, the setting time of concrete decreases (Chanh et al., 2008). During curing process, the geopolymer concrete experience polymerization process. Due to the increasing of temperature, polymerization become more rapid and the concrete can gain 70% of it strength within 3 to 4 h of curing (Kong and Sanjayan, 2008).

PROPERTIES OF GEOPOLYMER

Workability of fresh geopolymer

Water also play important role in geopolymer concrete as much as normal concrete. The used of water in geopolymer is to improve the workability, but it will increase the porosity in concrete due to the evaporation of water during curing process at elevated temperature (Sathia et al., 2008).

Chindaprasirt et al. (2007) discovered an increase in sodium hydroxide and sodium silicate concentration will reduce the flow of mortar. The workable flow of geopolymer mortar was in the range of 110 ± 5 to 135 ± 5%.

To improve the workability of mortar, superplasticiser or extra water can be added. However, the use of superplasticiser had an adverse effect on the strength of geopolymer. As such, extra water gives higher strength than addition of superplasticiser.

Compressive strength

Compressive strength is an essential property for all concrete where it also depends on curing time and curing temperature. When the curing time and temperature increase, the compressive strength also increases. With curing temperature in range of 60 to 90ºC, within time in 24 to 72 h, the compressive strength of concrete can be obtained about 400 to 500 kg/cm^2 (Chanh et al., 2008). In addition, the compressive strength of geopolymers also mainly depended on the content of fly ash fine particles (smaller than 43 μm).

The compressive strength was increase when the finest of fly ash increase. Hence the nature and the concentration of the activators were dominant factors in the reaction of alkali activation. The highest compressive strength was obtained using a solution of sodium silicate as an activator (n = 1.5; 10% Na_2O). Sodium silicate is the most suitable as alkaline activator because it contains dissolved and partially polymerized silicon which reacts easily, incorporates into the reaction products and significantly contributes to improving the mortar characteristics (Komljenovi´c et al., 2010).

Resistance against aggressive environment

Fly ash-based geopolymer had been proved by many studies to provide better resistance against aggressive environment. As such, this advantage can be used to construct structure that exposed to marine environment (Chanh et al., 2008).

The exposure of geopolymer in acid solution shows that the weight loss due to the exposure is only 0.5% compared to normal concrete when immersed in 3% acid sulfuric (Sathia et al., 2008).

According to Bakharev (2005) in acidic exposure, high-performance geopolymer materials deteriorate with the formation of fissures in amorphous polymer matrix, while low-performance geopolymers deteriorate through crystallisation of zeolites and formation of fragile grainy structures. The formation of aluminosilicate gel is important to determine the stability of geopolymer. More crystalline geopolymer material prepared with sodium hydroxide was more stable in the aggressive environment of sulfuric and acetic acid solutions than amorphous geopolymers prepared with the sodium silicate activator.

Thokchom et al. (2009) expose the geopolymer mortar in 10% sulfuric acid and found specimens still intact and did not show any recognizable change in colour after 18 weeks. When observed under optical microscope, the expose surface reveled a corroded structure and it progress with time of exposure. In addition, the weight loss results obtained in this study showed better performance than OPC and the specimens with higher alkali content were observed to lose more weight than specimens with lower alkali content. At 18 weeks, the specimens were fully dealkalized by the sulfuric acid as in Figure 1, but it still had substantial residual compressive strength to prove the higher resistance against acid exposure (Thokchom et al., 2009). Thokchom et al. (2009) investigated the effect of sodium oxide content on durability of geopolymer in sulphuric acid. Specimens with

(a) Before immersion (b) After 6 weeks (c) After 12 weeks (d) After 18 weeks

Figure 1. Residual alkalinity in 10% acid sulfuric solution.

Figure 2. Fly ash-based geopolymer specimens with white deposit after 24 weeks exposure.

varying alkali content showed varying degree of deterioration when exposed to sulphuric acid. Beside that, the specimens showed no visible sign of structural disintegration but under optical microscope observation, the surface deterioration was clearly visible and it appeared more severe in specimen with less alkali content. The specimen with maximum alkali content experienced more reduction in weight compared to minimum alkali content. One the other hand, specimen with minimum alkali content loss more strength. Hence, specimen with higher alkali content performed much better than those with lower alkali content in terms of residual compressive strength.

When the fly ash-based geopolymer was exposed to magnesium sulphate environment for 24 weeks, the white deposit were observed on the surface on the specimens as shown in Figure 2. Nevertheless, the observations under optical microscope shown, only fine cracks were seen on a few specimens but there are no visible cracks observed on the specimen. The gain of specimen weight when immersed in magnesium sulphate depends of the sodium oxide content produced from alkaline activator. The maximum weight gain was produced by specimen with minimum content of sodium oxide. Even so, all the specimens only recorded a small gain weight compared to normal concrete (Thokchom et al., 2010). Lodeiro et al. (2007) discovered that geopolymer are less prone to formation of alkali silica gel.

BEHAVIOUR OF GEOPOLYMER AT ELEVATED TEMPERATURE

Fire resistance of concrete was other properties that are always considered due to the safety of user. Kong and Sanjayan (2008) said that the ratio of fly ash to alkaline solution influence the general strength and fire resistance of geopolymer. It was found that the fly ash-based geopolymer displayed increase in strength after temperature exposure.

Kong and Sanjayan (2010) observed the behavior of geopolymer concrete under elevated temperature affected by the size of aggregates. The aggregate with smaller sized (<10 mm) could lead to spalling and also extensive cracking of geopolymer but the larger aggregate (>10 mm) were more stable. In addition, the thermal incompatibility between the geopolymer matrix and its aggregate components was the most likely cause of strength loss in geopolymer concrete specimens at elevated temperatures. It can be proved by comparison between geopolymer concretes made two different aggregates with distinctly different thermal expansion characteristics. The geopolymer concrete with greater incompatibility led to higher strength loss during elevated temperature. Thus, the expansion of aggregate with respect to temperature was a factor that controls the performance of geopolymer.

Bakharev (2006) observed the thermal stability of the geopolymer materials prepared with sodium containing activators was rather low and significant changes in the microstructure occurred. At 800℃, the strength of the concrete was reduced due to the increase in the average pore size where amorphous structures were replaced by the crystalline Na-feldspars. The reverse situation was observed when potassium silicate was used as activators because it can remain mostly amorphous up to 1200℃. After firing these materials, it reduced average pore size and improved compressive strength of geopolymer. Fly ash based geopolymer prepared using class F fly ash with sodium and potassium silicate show high shrinkage as well as large changes in compressive strength with increasing fired temperature in the range 800 to 1200℃ (Bakharev, 2006).

CONCLUSION

Fly ash-based geopolymer is better than normal concrete in many aspects such as compressive strength, exposure to aggressive environment, workability and exposure to high temperature.

REFERENCES

Bakharev T (2006). Thermal behaviour of geopolymers prepared using class F fly ash and elevated temperature curing. Cement Concrete Res., 36: 1134-1147.

Bakharev T (2005). Resistance of geopolymer materials to acid attack. Cement Concrete Res.. 35: 658-670.

Chindaprasirt P, Chareerat T, Siricicatnanon V (2007). Workability and strength of coarse high calcium fly ash geopolymer. Cement Concrete Compos., 29: 224-229.

Chanh NV, Trung BD, Tuan DV (2008). Recent research geopolymer concrete. The 3rd ACF International Coference –ACF/VCA.

Kong DLY, Sanjayan JG (2008). Damage behavior of geopolymer composites exposed to elevated temperature. Cement Concrete Compos., 30: 986-991.

Kong DLY, Sanjayan JG (2010). Effect of elevated temperatures geopolymer paste, mortar and concrete. Cement Concrete Res., 40; 334-339.

Komljenovi´c M, Bascarevic Z, Bradic V (2010). Mechanical and microstructural properties of alkali-activated fly ash geopolymer. J. Hazardous Mater., 181: 35-42.

Lodeiro G, Palomo A, Jiménez F (2007). Alkali-aggregate reaction in activated fly ash system. Cement Concrete Res., 37: 175-183.

Sathia R, Babu KG, Santhanam M (2008). Durability study of low calcium fly ash geopolymer concrete. The 3rd ACF International Coference –ACF/VCA.

Thokchom S, Ghosh P, Ghosh S (2010). Performance of Fly ash Based Geopolymer Mortars in Sulphate Solution. J. Eng. Tech. Rev., 3(1): 36-40.

Wallah SE (2009). Drying Shrinkage of Heat-Cured Fly Ash-Based Geopolymer Concrete. Modern Appl. Sci., 3: 12.

Flow characteristics of fluid and its effectiveness on orifice plate using pneumatic proportional control

C. P. Ukpaka[1] and V. M. Ndor[2]

[1]Department of Chemical/Petrochemical Engineering, Rivers State University of Science and Technology, Nkpolu, P.M.B. 5080, Port Harcourt, Nigeria.
[2]Department of Mechanical Engineering, Rivers State University of Science and Technology, Nkpolu, P.M.B. 5080, Port Harcourt, Nigeria.

A number of studies on effect of temperature on the flow characteristics of various fluids have been carried out. The aim of this work was to examine the flow characteristics of hexane upon the influence of temperature as well as the effectiveness response of orifice plate using pneumatic proportional control. Knowledge of the past on temperature impact as one aspect of the information set held by various scientific and a well established finding in engineering is that changes in composition can change the effect of temperature on the flow characteristics of hexane in a flow line system. The particular aspects of temperature impact on the research work highlighted are density, viscosity and pressure characteristics. The effect of temperature on flow characteristics of hexane was examined within the temperature range of 283 to 323 K and Bernoulli's equation knowledge of the past was used in developing the model for this paper. The mathematical models developed were simulated using the numerical concept of polynomial expression of the best fit. The effects of temperature on the functional parameters were examined and proportional gain at various error range of E = 0.1 to 0.5 were considered during the investigation. These effects on the variation of the results were attributed to the temperature, flow characteristics of the hexane and its effectiveness on orifice plate using pneumatic proportional control.

Key words: Effect, temperature, hexane, effectiveness, flow characteristics, orifice plate.

INTRODUCTION

The current worldwide complain against pneumatic proportional control has attracted the attention of engineers. In process of investigating the fluid flow characteristics and its influence on the orifices plate using pneumatic proportional control, may result to analysis of some functional parameters, such as the effect of temperature pressure and viscosity, on the flow rate of the substance (hexane) under investigation.

The investigation has to be carried out in such a way as to protect and preserve the failure of the orifices plate while still achieving desire economic benefiting of the research work. In Nigeria today, most of the industrial equipment in operation uses pneumatic system of both upstream and downstream sectors as their operations have rapidly increased. The essence of the studies is to examine the pneumatic proportional control principles on the orifices plate upon the influence of flow characteristics as well as the possible error range effect on the system. The research is conducted with the aim to compare and contrast the variations experienced due to constant area, which its influence was examined and reported to be depended of pressure drop and temperature (Perry and Green, 1984; Byron et al., 1976; Williams, 1989; HS, 1978; Bergen et al., 1999 Eckman, 1958; Gerard, 1998). Investigation conducted revealed that variation in temperature and pressure may either increase or decrease the density and viscosity of the substance as reported (Winkler, 1997; LENTECH, 2005; Chrysikopoulos et al., 2003; Glaso, 1980; Al-Marhoun, 2003). As the fluid flow through the orifices plate it pressure decreases and the fluid then is forced to converge to pass through the small hole; the point of maximum convergence actually occurs shortly downstream of the physical orifice, at the so-called vena contracta point (Donald, 1950). As it does so, the velocity and the pressure changes. Beyond the vena contracta, the fluid expands and the velocity and pressure change once again. By measuring the difference in fluid pressure between the normal pipe section and at the vena contracta, the volumetric and mass flow rates can be obtained from Bernoulli's equation, as reported by various researchers (Labedi, 1990; Uyigue and Umoh, 2007; Bradford et al., 2006; Ukpaka, 2012).

Similarly, research conducted revealed that various works have been done in flow characteristics; attempt has been made to monitor and predict the pressure drop at constant area in a horizontal orifice plate using different fluids as presented (Cunningham, 1951; Fried, 2000; Winkler, 1997; JICA, 2000; Zhn et al., 2001). Only a small number of rivers are appropriate for the use of this technology, since the plate must remain completely immersed. That is, the approach pipe must be full, and the river must be substantially free of debris. In the natural environment large orifice plates are used to control onward flow in flood relief dams. In these structures a low dam is placed across a river and in normal operation the water flow through the orifice plate unimpeded as the orifice is substantially larger than normal flow cross section as presented by (Jiang et al., 2002, Consider, 1957; Eckman, 1958). However, in floods, the flow rate rises and floods out the orifice plate which can then only pass a flow determined by the physical dimensions of the orifice. Flow is then held back behind the low dam in a temporary reservoir which is slowly discharged through the orifice when the flood subsides.

MATERIALS AND METHODS

Materials used for the investigation includes: hexane, 2 inche pvc pipe, pump, thermometer, heater, batch reactor, speed sound density meter, cylindrical viscometer.

Sample collection

The hexane sample was collected from the Chemical/Petrochemical Engineering Laboratory in Rivers State University of Science and Technology Nkpolu, Port Harcourt. The sample was used in setting up the main experimental work for the investigation.

The flow diagram of the experimental set-up

The flow diagram illustrating the experimental set up for the investigation conducted upon, the effect of temperature on the flow characteristics.

Experimental procedure

The hexane sample was collected from the Department of Chemical/Petrochemical Engineering Laboratory in Rivers State University of Science and Technology, Nkpolu, Port Harcourt using glass bottles of 10 L each. All the samples collected were introduced into the batch reactor set-up to achieve 50 L of hexane as raw material for the investigation. Thermometer was immersed into the glass reactor where a heater was inserted and glass reactor was filled with water as shown in Figure 1, above. The electric heater was connected to the electrical source, which heated up the water component in glass batch reactor. The temperature of the heater was controlled at various operating conditions of 283 to 323K. The glass bottle containing the fluid experienced the process of conduction, conversion and radiation. Similarly, another thermometer was immersed into the glass bottle containing the hexane, this was necessary to enable us know the actual operating temperature of hexane before subjecting into flow characteristics. A pump of 0.5 Hp was inserted and regulated within the speed sound density meter inserted along the flow line as shown in Figure 1. The density of the fluid was measured using speed sound density meter and viscosity capillary viscosity respectively at different operating temperature. The pressure drop on the flow line was examined based on the effectiveness the orifice plate. The output temperature of the hexane was measured, as well as the density and viscosity. The system was design in a manner in which the proportional gain was estimated using mathematical tools and operational principle. The Bernoulli's equation knowledge of the past was applied in resolving the problem of orifice plate effect on the effectiveness of flow in pipe system using the principles that governs flow characteristics as shown in Figure 2.

The model

In developing the mathematical model the following assumptions were considered:

1. The research was conducted at various operating temperature
2. The process of conduction, convention and radiation of heat from the water zone into the hexane zone was experienced as thermometer is immense before allowing the flow of the raw material under investigation to ascertain the required temperature in the reactor.
3. The investigation was conducted using 2 inches pvc pipe connected to achieve a total length of 21 m with the displacement velocity of 1m/s in the system.

By assuming steady-state, incompressible (constant fluid density) laminar flow in a horizontal pipes no change in elevation with negligible frictional losses. Bernoulli's equation reduces to an equation relation the conservation of energy between two points on the same stream line is given as

$$P_2 + \frac{1}{2}\rho V_2^2 = P_1 \quad (1)$$

Where P_1 is output pressure and P_2 is final pressure (N/m^2), ρ is

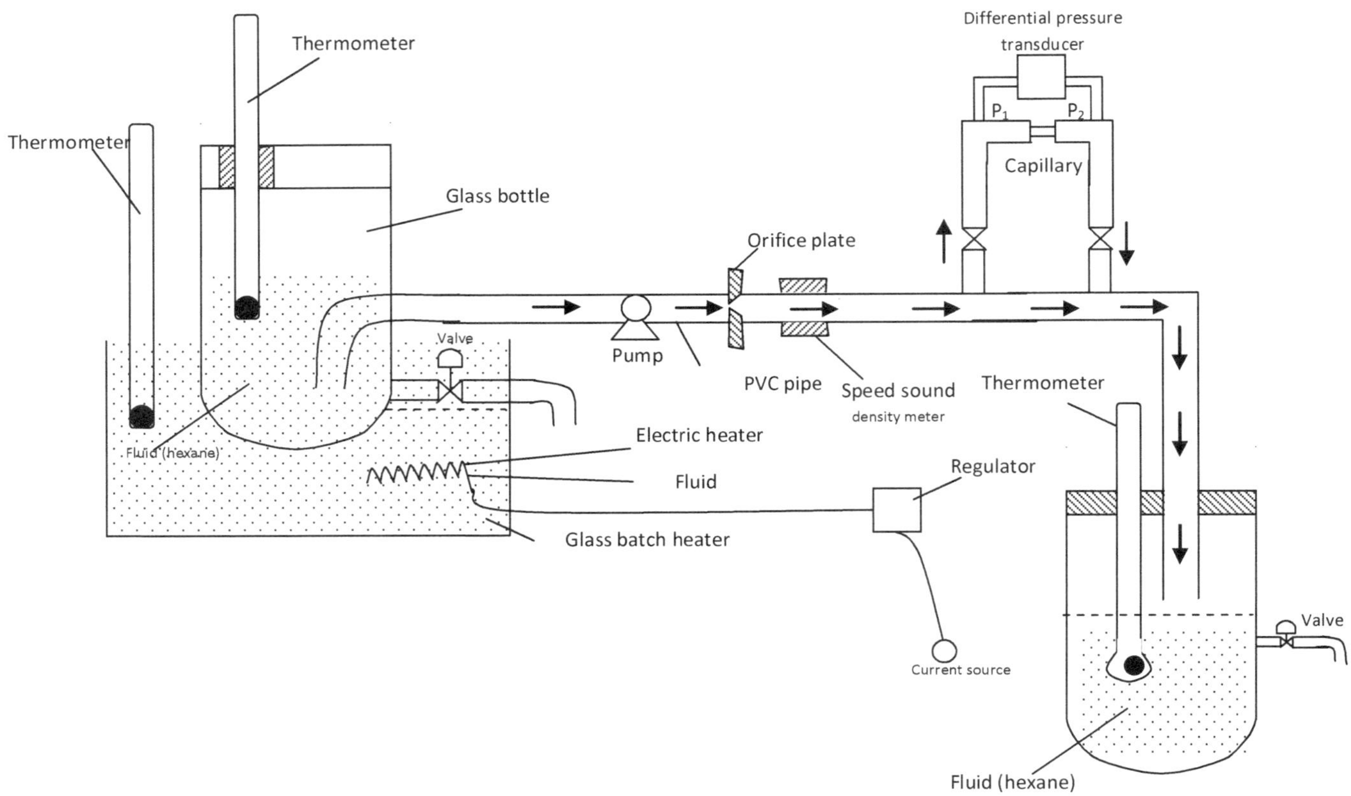

Figure 1. Experimental set-up to investigate the effect of temperature on flow characteristics of fluid.

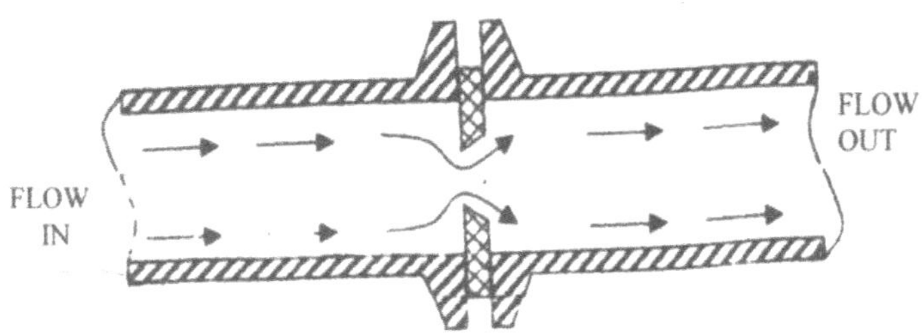

Figure 2. Simple diagram of fluid flowing through a pipe.

density (kg/m^3), V_2 is final velocity (m/s). Recalling the process known as continuity equation, the following expression can be obtained as

$$A_1V_1 = A_2V_2 \tag{2}$$

Where A_1 is the cross-sectional area of the pvc pipe (mm), A_2 is the cross-sectional area of the orifices (mm), ϕ is the flow (m^3/s)

$$\text{but } V_1 = \frac{\phi}{A_1} \text{ and } V_2 = \frac{\phi}{A_2} \tag{3}$$

substituting Equation (3) into (1) we have

$$P_1 - P_2 = \frac{1}{2}\rho\left(\frac{\phi}{A_2}\right) - \rho\left(\frac{\phi}{A_1}\right)^2 \tag{4}$$

Where P_1 is input pressure or initial pressure (N/m^2), P_2 is output or final pressure N/m^2

Solving for ϕ from Equation (4) we have

$$\phi = A_2\sqrt{\frac{2(P_1 - P_2)/\rho}{1 - (A_2/A_1)^2}} \tag{5}$$

Expressing Equation (5) in term of diameter relationship of

$$A_2/A_1 = \frac{(2\not{\pi} r_2)^2}{(2\not{\pi} r_1)^2} = \left[\frac{(d_2^2)}{(d_1^2)}\right]$$

Since r = d/2 and $r_2^2 \neq r_1^2$ therefore equation (5) can be written as

$$\phi = A_2\sqrt{\frac{1}{1 - (d_2/d_1)^4}}\sqrt{\frac{2(P - P_{21})}{\rho}} \tag{6}$$

The above expression for Equation (6) gives the theoretical volume flow rate. Introducing the beta factor into Equation (6)

Factor $\beta = d_2/d_1$ as well as the coefficient of discharge Cd, therefore Equation (6) becomes

$$\phi = C_d A_2 \sqrt{\frac{1}{1-\beta^4}} \sqrt{\frac{2(P_1 - P)_2}{\rho}} \quad (7)$$

Finally, introducing the meter coefficient C which is defined as $C = \frac{Cd}{\sqrt{1-\beta^4}}$ to obtain the final equation of the volumetric flow of the fluid through the orifice, we have

$$\phi = CA_2 \sqrt{\frac{2(P_1 - P_2)}{\rho}} \quad (8)$$

Equation (9) can be obtained by multiplying the density of the fluid with the mass flow rate at any section in the pipe is given as

$$\dot{M} = \rho\phi = CA_2 \sqrt{2\rho(P_1 - P_2)} \quad (9)$$

A fluid passing through the orifice construction above will experience a drop in pressure across the orifice. This charge can be used to measure the flow rate of the fluid. Hence, to calculate the flow rate (ϕ) of the fluid passing through an orifice plates enters the parameters above. Hence, we have

$$\phi = CA_2 \sqrt{\frac{2(P_1 - P_2)}{\rho}} \quad (10)$$

Considering material balance of the orifice and the fluid passing through the orifice is at steady state condition. Therefore, the rate of accumulation and consumption is equal to zero. Therefore Equation (8) can be expressed as:

$$\frac{2(P_1 - P_2)}{\rho} = \frac{\phi^2}{C^2 A_2^2}$$

$$\Delta P = P_1 - P_2 = \frac{\phi^2 \rho}{2C^2 A_2^2}$$

Equation (10) can also be written as

$$\Delta P = \frac{\phi^2 \rho}{2\, C^2 A_2^2} \quad (11)$$

Orifice plate and pneumatic proportional control model

Considering the mathematical expression as presented in Equation (12)

$$\Delta P = P_1 - P_2 = \frac{\phi^2 \rho}{2C^2 A_2^2} \quad (12)$$

Where Δ p = change in pressure (N/m^2)

Recalling the pneumatic proportional control model where the output pressure can be expressed as;

P = Po + Kc E

Considering the concept that $P = P_1 - P_2 = \Delta P$, where P = output signal, P_o is output signal when there is no error, K_c is proportional gain, E is error

Therefore Equation (10) and (12) combing yields

$$\frac{\phi^2 \rho}{2C^2 A_2^2} = P_0 + K_c E \quad (13)$$

Therefore rearranging Equation (13) we have

$$\phi^2 \rho = 2C^2 A_2^2 \left(P_o + K_c E\right) \quad (14)$$

$$\phi^2 = \frac{2C^2 A_2^2 \,(P_o + K_c\, E)}{\rho} \quad (15)$$

If $2C^2 = \gamma$, therefore substituting the expression into Equation (15) we have

$$\phi = \sqrt{\frac{\gamma(P_o + K_c E)}{\rho}\, A_2^2} \quad (16)$$

From Equation (16) making density (ρ) the subject of the formula, we have

$$\rho = \frac{2C^2\, A_2^2\, (P_o + K_c\, E)}{\phi^2} \quad (17)$$

From Equation (16) making E the subject of the formula we have

$$K_c\, E = \left[\frac{\phi^2 \rho}{2\, C^2 A_2^2} - P_o\right] \quad (18)$$

$$E = \frac{\left[\frac{\phi^2 \rho}{2\, C^2 A_2^2} - P_o\right]}{K_c} \quad (19)$$

Similarly, from Equation (16) making K_c the subject formula, we have

Table 1. Evaluation of functional parameters of hexane.

Component	Temperature (K)	Densiy (kg/m^3)	Pressure (pa)	Viscosity (Ns/m^2)	Flow rate φ (m^3/s)
Hezane	283	678	10	0.62	0.095837251
Hezane	293	668	20	0.57	0.136545051
Hezane	303	658	30	0.51	0.168498825
Hezane	313	649	40	0.45	0.195910108
Hezane	323	639	50	0.4	0.220741388

$$K_c\ E + P_o = \frac{\phi^2 \rho}{2C^2 A_2^2} \qquad (20)$$

where P_o is output signal when there is no error

$$K_c = \left[\frac{\phi^2 \rho}{2C^2 A_2^2} - P_o\right] /E \qquad (21)$$

$$K_c = \frac{\left[\frac{\phi^2 \rho}{2\ C^2 A_2^2} - P_o\right]}{E} \qquad (22)$$

Computational procedure

The following parameters were arbitrarily chosen, orifice flow coefficient (dimensionless) (C) = 0.62; cross sectional area of the orifice A_2 = 0.9 m^2, temperature within the range of 283 to 323K and including all experimental data obtained from the investigation as presented in this paper.

RESULT AND DISCUSSION

The results obtained from the research work are presented in Figures. The relationship among, density, temperature, viscosity, pressure, flow rate, proportional gain is presented in Figures and Tables 1, 2 and 3.

The graph presented in Figure 3 illustrates the relationship between density and temperature. It is seen that increase in temperature resulted to decrease in density. The equation of the best fit was established using polynomial and it is given as P = - 0.97T + 687.5 and the square root is given as R^2 = 0.9997 which is the best fit of the root.

From Figure 4, the variation in pressure was studied with the variation in temperature. The results obtained in Figure 4 illustrate increase in pressure with increase in temperature. The equation of the best fit was P = T with R^2 = 1.

Figure 5 illustrates the relationship between viscosity and temperature. It is seen that from Figure 5, viscosity decreases with increase in temperature. The polynomial expression is given as V = - 0.0056T +0.678 and R^2 = 0.9987.

The relationship between flow rate and temperature is illustrated in Figure 6. The results obtained showed that increase in temperature yielded increase in flow rate. The polynomial expression of the curve is given as ϕ = 0.0031 T + 0.0708 with R^2 = 0.9899

The result presented in Figure 7 illustrates the relationship between proportional gain and temperature. Increase in proportional gain is experience with the temperature of 283 to 290K and suddenly decrease within the temperature range of 290 to 300K and finally increased within the temperature range of 300 to 323K at different values of KcE = 0.1, 0.2, 0.3, 0.4 and 0.5. The polynomial expression of the graph as shown in Figure 7 is given as KcE(0.1) = - $0.0008T^4$ + $0.1081T^3$ – $47147T^2$ + 79.788T – 408.56 with R^2 = 1, KcE (0.2) = $-0.004T^4$ + $0.054T^3$ – $2.3574T^2$ + 39.894T – 204.28 with R^2 = 1, KcE (0.3) = - $0.0003T^4$ + $0.0361T^3$ – $1.5716T^2$ 26.596T – 136.19 with R^2 = 1 and $K_cE(0.4)$ = - $0.002T^4$ + $0.027T^3$ – $1.1787T^2$ + 19.947T – 102.14 with R^2 = 1, $K_cE(0.5)$ = - $0.0002T^4$ + $0.0216T^3$ – $0.9429T^2$ + 15.958T – 81.712 with R^2 = 1.

The result presented in Figure 8 illustrates the relationship between the proportional gain and density. It is that a decrease in proportional gain for the various values of KcE = 0.1, 0.2, 0.3, 0.4 and 0.5 was observed with increase in density at 650 kg/m^3 and suddenly increased in proportional gain within the density range of 650 to 660 kg/m^3 and finally a decrease in proportional gain was observed with increase in density as presented in Figure 8. The variation in density can be attributed to the variation in proportional gain as well as the functional parameters. The polynomial expression of the graph as shown in Figure 8 is given as KcE(0.1) = - $0.001\rho^4$ +$2.6419\rho^3$ – $2598.2\rho^2$+ IE + 06ρ – 2E+08 and R^2 = 1, $K_cE(0.2)$ = - $0.0005\ \rho^4$ + $1.3221\ \rho^3$ – $1299.1\ \rho^2$ + $567648\ \rho$ - 9E + 07 with R^2 = 1, $K_cE(0.3)$ = - $0.0003\rho^4$ + $0.8806\rho^3$ -$866.06\rho^2$ + 378432ρ - 6E + 07 with R^2 = 1, $K_cE(0.4)$ = - $0.00032\rho^4$ + $0.66052\rho^3$ – $649.542\rho^2$ + 2838242ρ - 5E + 07 with R^2 = 1, $K_cE(0.5)$ = - $0.0002\rho^2$ + $0.5284\rho^3$ – $519.63\rho^2$ + 227059ρ - 4E+07 and R^2 = 1.

The relationship between proportional gain and

Table 2. Mathematical computation of the functional parameters at various operating temperature, density, pressure and viscosity as well at various errors (E).

Component	Temperature K	Density (kg/m^3)	Pressure (N/m^2)	Viscosity (NS/m)	Flow rate φ(m^3/s)	Proportional gain (Kc)	Coefficient of proportional gain (Kc *10e14)
				E = 0.1			
Hezane	283	678	10	0.62	0.09584	1.77636E-14	17.7636
Hezane	293	668	20	0.57	0.13655	3.55271E-14	35.5271
Hezane	303	658	30	0.51	0.16850	0	0.0000
Hezane	313	649	40	0.45	0.19591	7.10543E-14	71.0543
Hezane	323	639	50	0.4	0.22074	2.13163E-13	213.1628
				E = 0.2			
Hezane	283	678	10	0.62	0.09584	8.88178E-15	8.881784197
Hezane	293	668	20	0.57	0.13655	1.77636E-14	17.76356839
Hezane	303	658	30	0.51	0.16850	0	0
Hezane	313	649	40	0.45	0.19591	3.55271E-14	35.52713679
Hezane	323	639	50	0.4	0.22074	1.06581E-13	106.5814104
				E = 0.3			
Hezane	283	678	10	0.62	0.09584	5.92119E-15	5.921189465
Hezane	293	668	20	0.57	0.13655	1.18424E-14	11.84237893
Hezane	303	658	30	0.51	0.16850	0	0
Hezane	313	649	40	0.45	0.19591	2.36848E-14	23.68475786
Hezane	323	639	50	0.4	0.22074	7.10543E-14	71.05427358
				E = 0.4			
Hezane	283	678	10	0.62	0.09584	4.44089E-15	4.440892099
Hezane	293	668	20	0.57	0.13655	8.88178E-15	8.881784197
Hezane	303	658	30	0.51	0.16850	0	0
Hezane	313	649	40	0.45	0.19591	1.77636E-14	17.76356839
Hezane	323	639	50	0.4	0.22074	5.32907E-14	53.29070518
				E = 0.5			
Hezane	283	678	10	0.62	0.09584	3.55271E-15	3.552713679
Hezane	293	668	20	0.57	0.13655	7.10543E-15	7.105427358
Hezane	303	658	30	0.51	0.16850	0	0
Hezane	313	649	40	0.45	0.19591	1.42109E-14	14.21085472
Hezane	323	639	50	0.4	0.22074	4.26326E-14	42.63256415

Table 3. Evaluation of functional parameters at constant temperature, density, viscosity, pressure.

E	Flow rate (φ)	Proportional gain (Kc)	Coefficient of proportional gain (Kc*10e14)
		At temperature = 283 K	
0.1	0.095837251	1.77636E-14	17.76356839
0.2	0.095837251	8.88178E-15	8.881784197
0.3	0.095837251	5.92119E-15	5.921189465
0.4	0.095837251	4.44089E-15	4.440892099
0.5	0.095837251	3.55271E-15	3.552713679

Table 3. Contd.

		At temperature = 293 K	
0.1	0.136545051	3.55271E-14	35.52713679
0.2	0.136545051	1.77636E-14	17.76356839
0.3	0.136545051	1.18424E-14	11.84237893
0.4	0.136545051	8.88178E-15	8.881784197
0.5	0.136545051	7.10543E-15	7.105427358
		At temperature = 303	
0.1	0.168498825	0	0
0.2	0.168498825	0	0
0.3	0.168498825	0	0
0.4	0.168498825	0	0
0.5	0.168498825	0	0
		At temperature = 313K	
0.1	0.195910108	7.10543E-14	71.05427358
0.2	0.195910108	3.55271E-14	35.52713679
0.3	0.195910108	2.36848E-14	23.68475786
0.4	0.195910108	1.77636E-14	17.76356839
0.5	0.195910108	1.42109E-14	14.21085472
		At temperature = 323K	
0.1	0.220741388	2.13163E-13	213.1628207
0.2	0.220741388	1.06581E-13	106.5814104
0.3	0.220741388	7.10543E-14	71.05427358
0.4	0.220741388	5.32907E-14	53.29070518
0.5	0.220741388	4.26326E-14	42.63256415

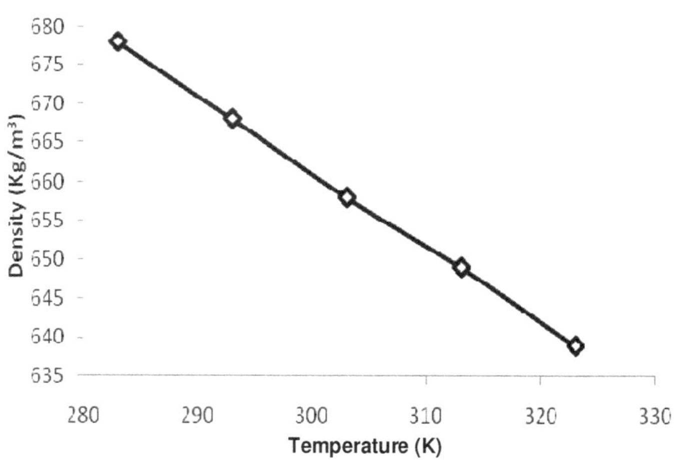

Figure 3. Graph of density against temperature for hexane.

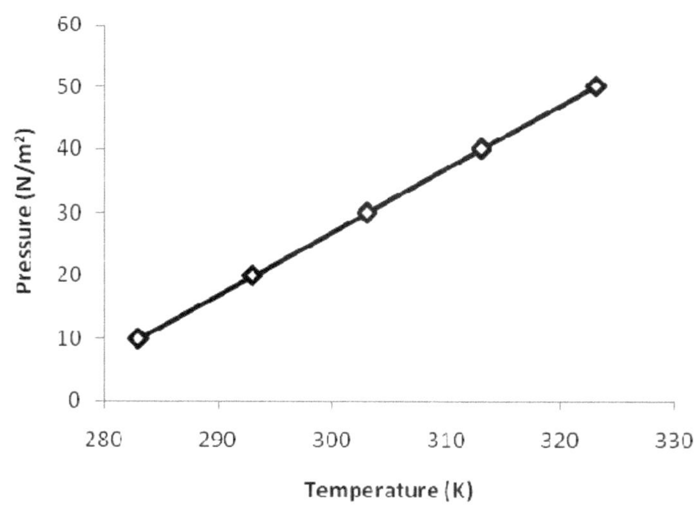

Figure 4. Graph of Pressure against temperature for hexane.

pressure is shown in Figure 9. Increase in proportional gain is experienced with increase in pressure within the range of 20 to 30 N/m^2 and sudden decrease was observed in proportional gain within the range of 30 to 40 N/m^2. Above 40 N/m^2 an increase in pressure for the value of KcE = 0.1, 0.2, 0.3, 0.4 and 0.5. The variation in the proportional gain can be attributed to the variation in

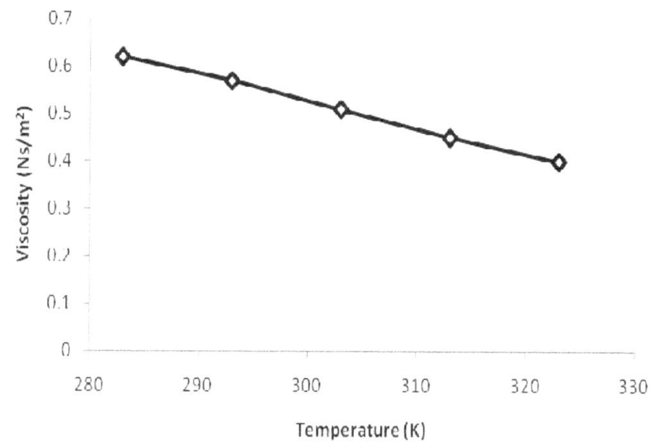

Figure 5. Graph of viscosity against temperature for hexane.

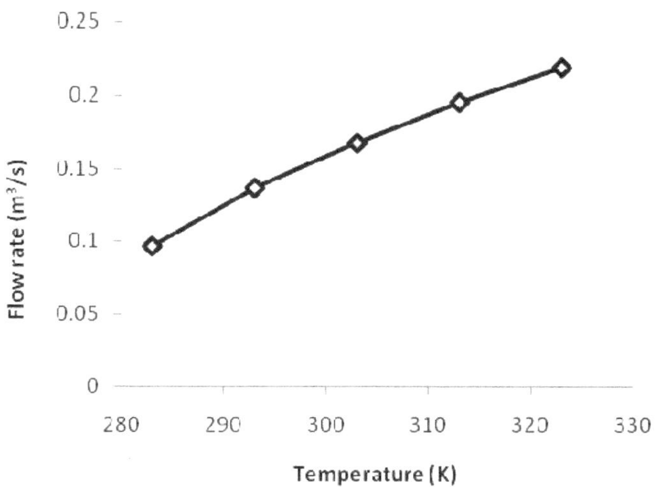

Figure 6. Graph of flow rate against temperature for hexane.

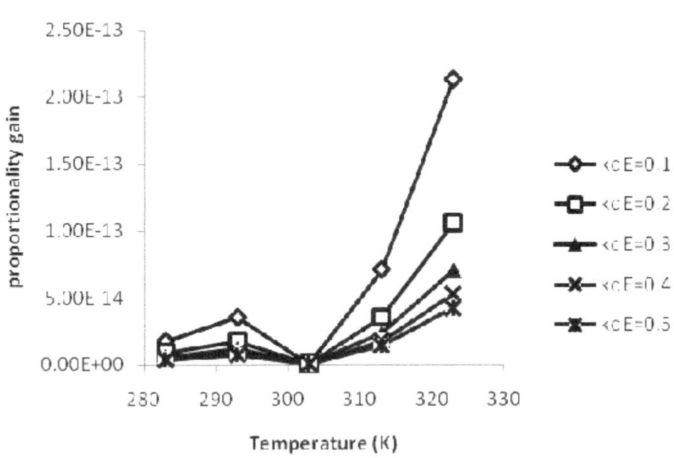

Figure 7. Graph of proportional gain verses temperature for various component of E.

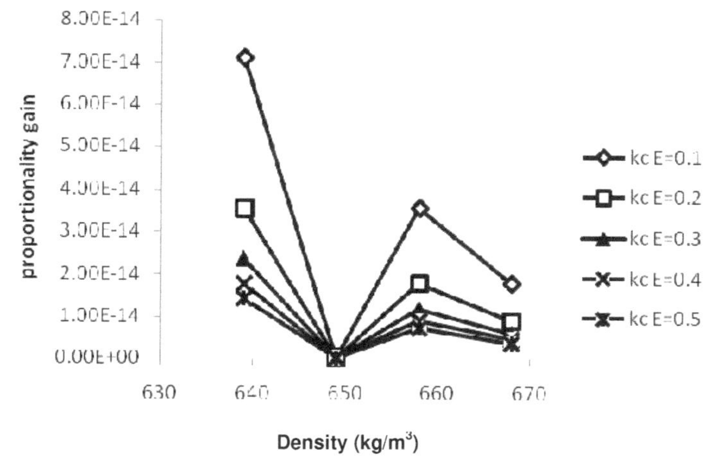

Figure 8. Graph of proportional gain verses density for various component of E.

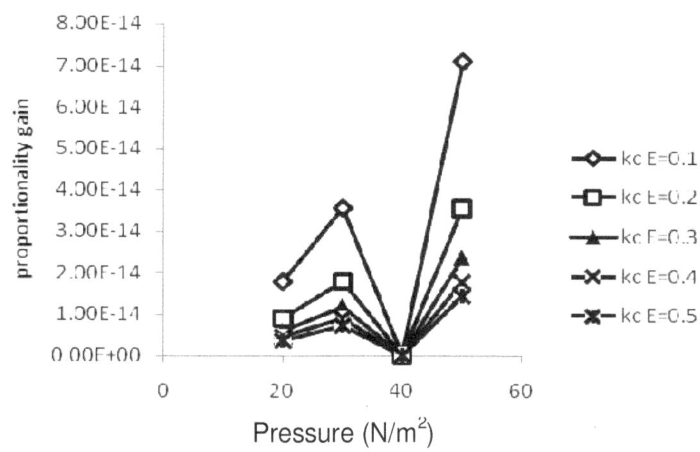

Figure 9. Graph of proportional gain verses pressure for various component of E.

pressure as well as effect of temperature in the system. The polynomial expression of the graph as shown in Figure 9 is given as $KcE(0.1) = -0.1081P^3 - 4.7147P^2 + 79.788P - 408.56$ with $R^2 = 1$, $K_cE(0.2) = -0.0004P^4 + 0.054P^3 - 2.3574P^2 + 39.894P - 204.28$ with $R^2 = 1$, $K_cE(0.3) = -0.002\rho^4 + 0.027\rho^3 - 1.1787\rho^2 + 19.947\rho - 102.14$ with $R^2 = 1$, $K_cE(0.4) = -0.0002P^4 + 0.0216P^3 - 0.929P^2 + 15.958P + 15.958P - 81.712$ with $R^2 = 1$, $K_cE(0.5) = -0.0002P^4 + 0.0216P^3 - 0.929P^2 + 15.958P + 15.958P - 81.712$ with $R^2 = 1$.

Figure 10 illustrates the variation in the proportional gain flow rate. Increase in proportional gain was observe with increase in flow rate within the range of 0.15 to 0.17 m^3/s and suddenly the proportional gain decreased with increase in flow rate with the range of 0.175 to 0.2 m^3/s. An increase in proportional gain was finally experience

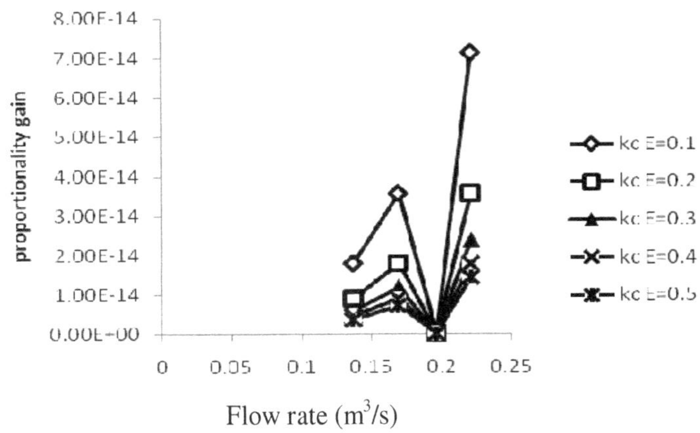

Figure 10. Graph of proportional gain verses flow rate for various component of E.

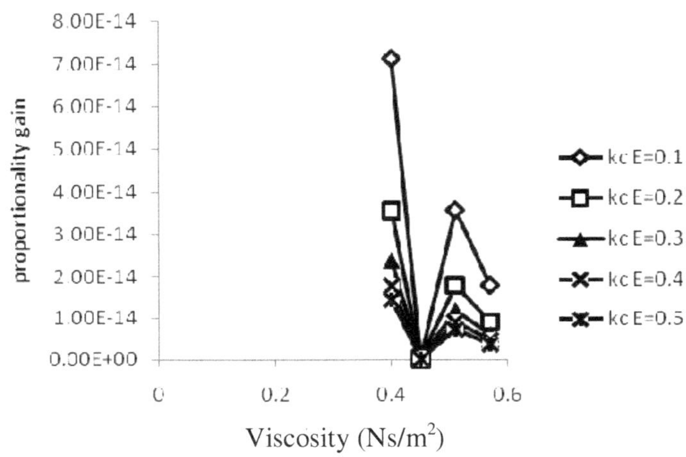

Figure 11. Graph of proportional gain verses viscosity for various component of E.

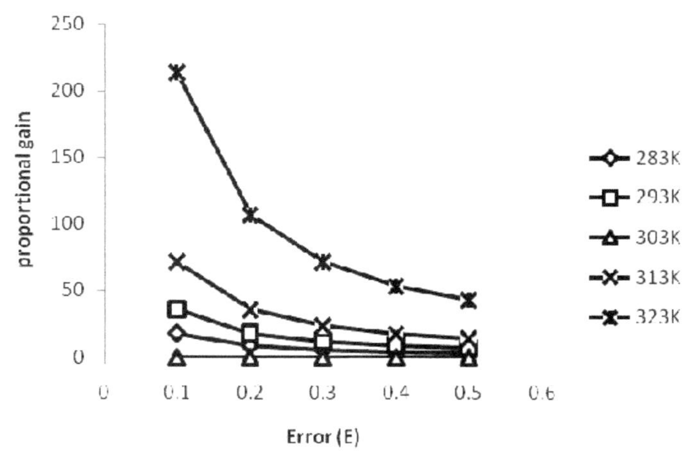

Figure 12. Graph of proportional gain verses Error at different operating temperature.

with increase flow rate for the various values of KcE = 0.1, 0.2, 0.3 0.4 and 0.5. The polynomial expression of the graph as shown in Figure 10 is given as KcE(0.1) = $518185\phi^3 - 219220\ \phi^2 + 30008\ \phi$ - 1299.7 with R^2 = 0.991, $K_cE(0.2)$ = 3E + $06\phi^4$ + 2E + 06 $\phi^3 - 631494\ \phi^2 + 67649\ \phi$ - 2570.5 with R^2 = 1, $K_cE(0.3)$ = - 2E + $06\phi^4$ + 2E + 06 ϕ^4 + 2E + $06\phi^3 - 420996\ \phi^2 + 451000\phi$ - 1713.7with R^2 =1, $K_cE(0.4)$ = - IE + $06\phi^4 + 994512\ \phi^3 - 252597\phi^2 + 27060\phi$ - 1028.2 with R^2 = 1, K_cE (0.5) = - 1E + $06\phi^4 + 994512\phi^3 - 252597\phi^2 + 27060\phi$ - 1028.2 with R^2 = 1.

The relationship between the proportional gain and viscosity is presented in Figure 11. And results obtained indicate decrease in proportional gain with increase in viscosity within the range of 0.4 to 0.49 Ns/m². And increase was observed in proportional gain within an increase in density for a range of 0.49 to 0.57 Ns/m² and suddenly decreased with increase in viscosity for the various components of KcE = 0.1, 0.2, 0.3, 0.4 and 0.5.The variation in the proportional gain can be attributed to the variation in viscosity. The polynomial expression of the graph as shown in Figure 11 is given as KcE(0.1) = $8707.3V^2 - 9633.8V + 2661.3$ with R^2 = 0.9218, $K_cE(0.2) = -206297V^4 + 397625V^3 - 280742V^2 + 85537V - 9391.9$ with R^2 = 1, $K_cE(0.3) = -123778V^4 + 23857V^3 - 168445V^2 + 51322V - 5635.1$ with R^2 = 1

Figure 12 illustrates the relationship between the proportional gain and error at different operating temperature. The result obtained indicates decrease in proportional gain with increase in error at different operating temperature conditions. The variation in the proportional gain can be attributed and other functional parameter and coefficient. The polynomial expression of the graph as shown in Figure 12 is given as KcE(0.1) = $1.7764E^{-1}$ and R^2 = 1, $K_cE(0.2) = 3.5527E^{-1}$ and R^2 = 1, $K_cE(0.3)$ = 0, R^2 = N/A, $K_cE(0.4) = 7.1054E^{-1}$ and R^2 = 1, $K_cE(0.5) = 21.316E^{-1}$ and R^2 = 1.

Conclusion

In this present study baseline experiments were performed to determine the concentration of the density viscosity of hexane. To this end the hexane substance was investigated for the flow characteristic and temperature potential impact on the effectiveness of the orifice plate in term of response to action. The Bernoulli's equation was used as a guideline in the establishment of the developed mathematical model which was correlated to the pneumatic proportional control. The correlation was useful in the evaluation of the proportional gain and its effect upon the functional parameters considered on this study.

This study has also demonstrated that there is a significant difference in flow characteristics due to the application of different temperature which leads to different polynomial expression as presented in Figure. Some caution is needed in extrapolating from these findings, given that this is a particular hexane substance with relatively intensive in industrial activities. A reduced temperature system, for examples, may result in a more favorable outcome for petrochemical process, depending on the process condition needed.

REFERENCES

Al-Marhoun MA (2003). The Coefficient of Isothermal Compressibility. SPE paper 81432. Paper Prepared for Presentation at the SPE 13th Middle East Oil Show and Conference to be Held in Bahrain 5-8 April: pp. 207-210.

Bergen A, Kiko H, Weisenborn T (1999). Research Note on Phase Behaviour Guidelines, EPT-Ro, SIEP RTS Rijswijk, 7:37. Dec.

Bradford SA, Tadassa YF, Jin Y (2006). Transport of Coliphage in the Presence and Absence of Manure Suspension. J. Environ. Qual. 35:1692-1701.

Byron RB, Warren ES, Edwin NL (1976). Transport Phenomena. Quinn-wood Brine-Inc Publisher , New York London, pp. 51-184.

Chrysikopoulos CV, Hsuan PY, Fyrillas MM, Lee KY (2003). Mass Transfer Coefficient and Concentration Boundary Layer Thickness for a Dissolving (Non-aqueous phase liquid) NAPL pool in Porous Media. J. Hazard. Mater. 7:245-205.

Consider DM (1957). Process Instruments and Controls, Handbook New York McGraw-Hill, pp. 3-10.

Cunningham RG (1951). "Orifice Meters with supercritical compressible Flow." Trans. ASME 73:625-638.

Donald QK (1950). Process Heat Transfer, International Student Edition, McGraw-Hill Book Co. Publisher New York, London, pp. 102-251.

Eckman DP (1958). Automatic Process Control. New York John Wiley, pp. 127-300.

Fried JR (2000). Polymer Science and Technology. Second Edition, Prentice-Hall Incorporated, USA, pp. 201-300.

Gerard K (1998). Environmental Engineering, International Editions, Inwin/McGraw-Hill Book Co. Publisher New York London, pp. 56-300.

Glaso O (1980). Generalized Pressure-Volume Temperature Correlations. JPT 4(2):785-795.

Halliburton Services HS (1978). Instruction Manual No. 800.68515. High Pressure Consistometer for Instruments 800.685 and 800.686, Duncan, Ok. USA, p. 72.

Jiang JQ, Graham N, Andre C, Kelsalt GH, Brandon N (2002). Laboratory Study of Electro-Coagulation Flotation for Water Treatment. Water Res. 36:4064-4078.

JICA (2000). The Study on Water supply System for Siem Reap Region in Cambodia Final Report, Prepared by Japan International Cooperation Agency (JICA) for the Ministry of Industry, Mines and Energy. The Royal Government of Cambodia. pp. 10-13.

Labedi R (1990). Use of Production Data to Estimate Volume factor Density and Compressibility of Reservior Fluids. J. Pet. Sci. Eng. 4:350-377.

LENTECH (2005). Lentech Water Treatment and Air Purification Holding B.V Rotterdamseweg, pp. 7-42.

Perry H, Green W (1984). Perry's Chemical Engineers' Handbook (Sixth Edition) McGraw Hil. pp. 1021-1452.

Uyigue L, Umoh JR (2007). Modelling the Effect of Extruder Screw Speed on the Mechanical Properties of High Density Polyethylene Blow Film. AMSE 22(1):75-85; 2:75-83.

William LL (1989). Process Modelling, Simulation and Control for Chemical Engineering, Second Edition, McGraw-Hill, Book Co. New York, San Francisco, California. pp. 285-400.

Winkler G (1997). Extrusion of Blown Films in Plastic Extrusion Technology. Second Edition Hanser, New York, pp. 72-270.

Zhn J, Ndegwa PM, Luo A (2001). Effect of Solid-Liquid Separation on BOD and VFA in Swine Manure. Environ, Technol. 22:1237-1243.

Dielectric relaxation, electrical conductivity and impedance response of Barium titanate (BT) and Strontium titanate (ST) doped $Ba(Fe_{0.5}Nb_{0.5})O_3$ ceramics

N. K. Singh[1]*, Pritam Kumar[1], A. Kumar[1] and S. Sharma[2]

[1]Department of Physics, Veer Kunwar Singh (V. K. S) University, Arrah, Bhojpur, 802 301 Bihar, India.
[2]Department of Physics, A. N. College, Patna, Bihar, India.

Lead-free polycrystalline ceramics of $Ba(Fe_{0.5}Nb_{0.5})O_3$ and its solid solutions $0.91Ba(Fe_{0.5}Nb_{0.5})O_3$-$0.09BaTiO_3$ (BFN-BT) and $0.91Ba(Fe_{0.5}Nb_{0.5})O_3$-$0.09SrTiO_3$ (BFN-ST) were fabricated by a solid-state reaction process, and their electrical properties were characterized in a broad frequency range (100 Hz to 1 MHz) at a temperature range from 30 to 385°C. The prepared ceramics are single phased with monoclinic structure as confirmed by X-ray diffraction. The microstructure analysis was done by scanning electron microscope. Dielectric response show low temperature dielectric dispersion at low frequency showing space charge phenomena arising due to defects. Existence of sharp transition around (T_c~255°C) and broad transitions around (T_c~205°C) of dielectric constant (ε′) versus temperature is demonstrated in the complex perovskite in BFN-BT and BFN-ST ceramics respectively at different frequencies (10.82, 20.2 and 32.2 kHz). The relaxor property was analyzed by the broadening of the maximum dielectric permittivity as well as its shifting to high temperatures with the variation of frequency measurements. In the pure BFN ceramic, dielectric studies confirmed that the compound do not have dielectric anomaly at different frequencies (10.82, 20.2 and 32.2 kHz) and temperature ranges (30 to 385°C). The frequency-dependent electrical data are also analyzed in the framework of conductivity and impedance formalisms.

Key words: Perovskite oxides, dielectric constant, electrical properties, scanning electron micrographs.

INTRODUCTION

The dielectric properties of material are intrinsic properties expressed by the relative complex permittivity $\varepsilon^* = \varepsilon' - j\varepsilon''$, where ε' is the dielectric constant and represents the ability of a material to store electrical energy and ε'', is the loss factor and represents the loss of electric energy in the material. Amount of loss is described by a parameter loss tangent (tanδ). The dielectric parameters are generally dependent on frequency, temperature, density and other factors such as material structure and composition (Bansal et al., 2001; Nelson, 1992, 1993). The electrical properties are often represented in terms of some complex parameters like complex impedance (Z^*), complex dielectric modulus (M^*), and loss tangent (tanδ). They are related to each other as follows: $Z^* = Z' - j Z'' = R_s - \frac{j}{\omega C_s}$, $M^* = M' + j M'' = j\omega\varepsilon_0 Z^*$ and loss tangent, $\tan\delta = \frac{M''}{M'} = -\frac{Z'}{Z''}$; where j = is

*Corresponding author. E-mail: singh_nk_phy27@yahoo.com.

the imaginary factor, R_s and C_s are the resistance and capacitance in series (s), ω (=2πf) is the angular frequency and ε_0 is the permittivity in free space (8.854 × 10^{-12} F/m). The above three expressions offer a wide scope for graphical representation. Complex impedance spectroscopy (CIS) technique will be used to study the impedance properties/parameters of the compounds using a computer controlled LCR meter/impedance analyzer at an ac signal over a wide range of frequency and temperature. Impedance analysis basically involves the display of the impedance data in different formalism and provides us the maximum possible information about the materials. The display of impedance data in the complex plane plot appears in the form of a succession of semicircles attributed to relaxation phenomena with different time constants due to the contribution of grain (bulk), grain boundary and interface/polarization in a polycrystalline material. Hence, the contributions to the overall electrical property by various components in the material are separated out easily.

Complex perovskite oxides with high dielectric constants play an important role in various technological applications such as wireless communication systems, microelectronics and global positioning systems. High dielectric constant allows one to fabricate smaller capacitive components, thus offering the opportunity to decrease the size of the electronic devices (Agrawal et al., 2009). Various relaxation processes seem to coexist in complex perovskite ceramics, which contain a number of different energy barriers due to point defects appearing during their fabrication. Therefore, the departure of the response from an ideal Debye model in ceramic samples, resulting from the interaction between dipoles, cannot be disregarded (Bharti et al., 2009).

Simple perovskite compounds have the chemical formula ABO_3. In the cubic phase, the oxygen atoms form a cubic lattice of corner-sharing octahedra with the B cations at their centers, while the A cations form a second interpenetrating cubic sublattice located at the 12-fold coordinated sites between octahedra. Interestingly, most of the perovskite compounds that are of greatest technological interests are not simple systems, but rather complex oxides with two different kinds of B atoms of the form $A(B'B'')O_3$. Ferroelectric perovskites have been the subject of extensive studies due to their promising electrical characteristic which has a potential usefulness in fundamental research and technological applications. Investigation of the electrical properties of these materials is desirable to predict their suitability for electronic applications.

Being a lead free perovskite material, $Ba(Fe_{0.5}Nb_{0.5})O_3$ is an environmental friendly material, thus making it a good substitute for Pb containing compounds for various applications. Both A and B–sites dopants have been used to modify the electrical and dielectric properties of $Ba(Fe_{0.5}Nb_{0.5})O_3$. Not much work has been reported on B- site doping and there is hardly any extensive report on the strontium titanate and barium titanate doping in barium iron niobate. In this article, we report dielectric relaxation, electrical conductivity and impedance response of BT and ST doped $Ba(Fe_{0.5}Nb_{0.5})O_3$ ceramics.

EXPERIMENTAL PROCEDURE

Sample preparation

In analogy to our previous work (Singh et al., 2011a, b; Singh and Kumar, 2011a) the conventional solid state reaction technique is used for the synthesis of BFN, BFN-BT and BFN-ST ceramics. The starting materials are reagent grade $BaCO_3$, $SrCO_3$, Fe_2O_3, Nb_2O_5 (M/S-Loba Chemie Pvt. Ltd.) and TiO_2 (M/S-S.D. fine), taken in stoichiometric ratio. The materials are mixed in acetone medium for 6 h and the mixture is calcined at 1200°C for 8 h and brought to room temperature by cooling at the rate of 100°C /h. The calcined sample is palletized into a disk using polyvinyl alcohol (PVA) as binder. Finally, the disk is sintered at 1250°C for 6 h and cooled down to room temperature at the cooling rate of 1°C /min.

Electrical measurement

For electrical measurement, the sintered disk (of thickness of the range of 1.28 to 1.56 mm and diameter of the range of 11.79 to 12.23 mm) is polished, electroded with fine silver paint (Alfa Aesar). Impedance parameters (Z), phase angle (θ), capacitance (Cp) and loss tangent (tanδ) were measured as a function of frequency (100 Hz to 1 MHz) at different temperatures (30 to 385°C), using a PSM (Comm.1735) in conjunction with a laboratory-made sample holder and heating arrangement with an AC signal (1.2 V).

RESULTS AND DISCUSSION

Structural and microstructural analysis

The X-ray diffraction pattern of BFN (inset), BFN-BT and BFN-ST measured at room temperature is shown in Figure 1. The pattern is characteristic of a perovskite structure. All the reflection peaks of the X-ray profile are indexed and lattice parameters are determined using a least-squares method with the help of a standard computer programme (POWDMULT) (Wul and PowdMult, 1999). Good agreement between the observed and calculated d-values (Table 1) suggests that the BFN, BFN-BT and BFN-ST ceramics crystallize in monoclinic phase at room temperature. The cell parameters with cell volume are shown in Table 2. X-ray diffraction confirms that the specimen is single phase.

The scanning electron micrograph of the samples was recorded by FEI Quanta 200 equipment to check proper compactness of all the samples is shown in Figure 2. The nature of the micrographs exhibits the polycrystalline texture of the material having highly distinctive and compact rectangular/cubical grain distributions. The grain size of the pellet sample was found to be in the range of 2 to 3 μm.

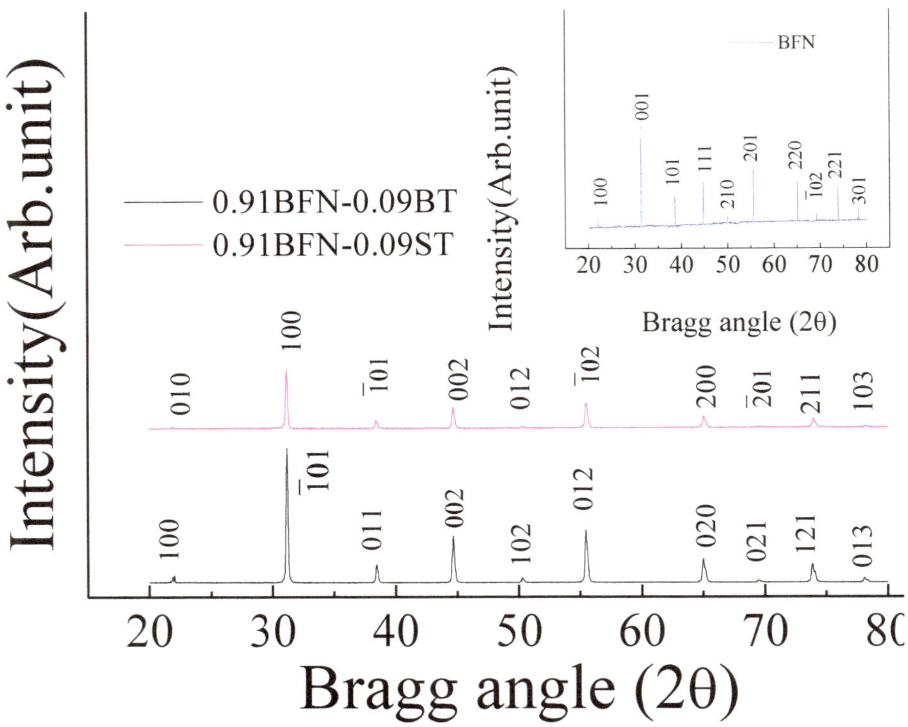

Figure 1. XRD patterns of BFN, 0.91BFN-0.09BT and 0.91BFN-0.09ST ceramics at room temperature.

Dielectric studies

The variation of relative dielectric constant (ε′) and tangent loss (tanδ) of BFN, BFN-BT and BFN-ST ceramics with frequency at different temperature is shown in Figures 3 and 4 respectively. For all the samples initially, both ε′and tanδ decreases with increase in frequency, which is a general feature of polar dielectric materials irrespective of composition of the specimens. The fall in ε′ arises from the fact that polarization does not occur instantaneously with the application of the electric field, which is further due to the inertia of the dipoles and the delay in response towards the impressed alternating electric field leads to dielectric loss and decline in ε′. At low frequencies, all types of polarization contributes and as the frequency is increased, polarizations with large relaxation times cease to respond and hence the decrease in ε′ (Chopra et al., 2004). At lower frequencies ε′ is maximum because the contributions from the space charge polarization is large. At higher frequencies, contributions from the polarizations having high relaxation time cease resulting in the decrease in ε′ (Kittel, 1995). The same type of frequency-dependent dielectric behavior is found in many other perovskite ceramic systems (Singh et al., 2010a, b, 2008; Kumar et al., 2011a, b; Singh and Kumar 2011b). In BFN-BT and BFN-ST ceramics, the tanδ peaks were observed which shift towards the higher frequency region (with rise in temperature). This type of feature suggests the presence of dielectric relaxation in the compound (Shantha and Varma, 1997; Saha and Sinha, 2006). The range of variation of ε′ in BFN-BT is higher than in BFN-ST and BFN ceramics, the range of ε′ is 153 to 7370 for BFN-BT, 207 to 1013 for BFN-ST and 150 to 1999 for BFN ceramics at 30 to120°C.

$BaTiO_3$ (BT) with a perovskite structure is a strong dielectric material, which has far reaching applications in the electronics industry for transducers, actuators and high-k dielectrics, and is mostly used to make multilayer ceramic capacitor (MLCC) materials (Mao et al., 2003; Cui et al., 2006; Pengfei et al., 2007). On the other hand, dielectric properties of $SrTiO_3$ (ST) based materials have been widely studied (Muller and Burkhard, 1979; Bianchi et al., 1995; Ang and Yu, 1992). Fe doped $SrTiO_3$ (ST) materials, combining the required stability and interesting transport properties at relatively high temperatures, have been considered for application as electrochemical electrodes and possibly also for resistive oxygen sensors. For these applications, special attention has been paid to the electrical transport properties at high temperatures. However, work on both the dielectric and conductive properties, especially at low frequencies and low

Table 1. Comparison of some observed (obs) and calculated (cal) d-values (in Å) of some reflections of $Ba(Fe_{0.5}Nb_{0.5})O_3$, $0.91Ba(Fe_{0.5}Nb_{0.5})O_3$-$0.09BaTiO_3$, and $0.91Ba(Fe_{0.5}Nb_{0.5})O_3$-$0.09SrTiO_3$ compounds at room temperature with intensity ratio I/I_0. The estimated error in d is ± 0.0001 Å.

(hkl)	BFN d (Å)	(hkl)	$0.91Ba(Fe_{0.5}Nb_{0.5})O_3$-$0.09BaTiO_3$ d (Å)	(hkl)	$0.91Ba(Fe_{0.5}Nb_{0.5})O_3$-$0.09SrTiO_3$ d (Å)
(100)	[o] 4.0337 (17) [c] 4.0337	(100)	[o] 4.0350 (20) [c] 4.0350	(010)	[o] 4.0623 (5) [c] 4.0623
(001)	[o] 2.8553 (100) [c] 2.8553	$(\bar{1}01)$	[o] 2.8643 (100) [c] 2.8643	(100)	[o] 2.8643 (100) [c] 2.8643
(101)	[o] 2.3410 (43) [c] 2.3410	(011)	[o] 2.3375 (27) [c] 2.3375	$(\bar{1}01)$	[o] 2.3410 (17) [c] 2.3410
(111)	[o] 2.0213 (55) [c] 2.0213	(002)	[o] 2.0264 (45) [c] 2.0264	(002)	[o] 2.0264 (37) [c] 2.0264
(210)	[o] 1.8181 (26) [c] 1.8226	(102)	[o] 1.8124 (19) [c] 1.8090	(012)	[o] 1.8124 (4) [c] 1.8133
(201)	[o] 1.6581 (66) [c] 1.6537	(012)	[o] 1.6554 (49) [c] 1.6552	$(\bar{1}02)$	[o] 1.6548 (43) [c] 1.6556
(220)	[o] 1.4331 (56) [c] 1.4331	(020)	[o] 1.4336 (31) [c] 1.4337	(200)	[o] 1.4324 (21) [c] 1.4321
$(\bar{1}02)$	[o] 1.3578 (30) [c] 1.3573	(021)	[o] 1.3520 (18) [c] 1.3517	$(\bar{2}01)$	[o] 1.3513 (4) [c] 1.3510
(221)	[o] 1.2816 (52) [c] 1.2800	(121)	[o] 1.2823 (28) [c] 1.2813	(211)	[o] 1.2808 (17) [c] 1.2807
(301)	[o] 1.2175(29) [c] 1.2154	(013)	[o] 1.2226 (19) [c] 1.2224	(103)	[o] 1.2211 (5) [c] 1.2211

Table 2. Structural data for $Ba(Fe_{0.5}Nb_{0.5})O_3$, $0.91Ba(Fe_{0.5}Nb_{0.5})O_3$-$0.09BaTiO_3$, and $0.91Ba(Fe_{0.5}Nb_{0.5})O_3$-$0.09SrTiO_3$ compounds.

Sample	Structure	a	b	c	β (in degree)	V
BFN	Monoclinic	4.0337 Å	4.0734 Å	2.8553 Å	90.12	46.92
$0.91Ba(Fe_{0.5}Nb_{0.5})O_3$-$0.09BaTiO_3$	Monoclinic	4.0350Å	4.0539Å	2.8674Å	90.17	46.90
$0.91Ba(Fe_{0.5}Nb_{0.5})O_3$-$0.09SrTiO_3$	Monoclinic	4.0623Å	4.0529Å	2.8643Å	90.09	47.16

temperatures, has been rarely done and needs to be accomplished.

The temperature-dependent relative dielectric constant (ε') and tangent loss (tanδ) at 10.82, 20.20 and 32.30 kHz of BFN, BFN-BT and BFN-ST ceramics is shown in Figures 5 and 6 respectively. The variation in ε′ with temperature shows that it is strongly frequency and temperature dependent. In BFN ceramics, the relative dielectric constant increases linearly with rise in temperature at different frequencies (10.82, 20.2 and 32.2 kHz). However, no phase transition was observed in BFN in the experimental temperature range. A dielectric anomaly was observed in BFN-BT and BFN-ST ceramics. It is observed that peaks of the curves that is, the maximum value of dielectric constant decreases with increasing frequency. In BFN-BT and BFN-ST ceramics, the dielectric constant increases gradually with an increase in temperature up to transition temperature (Tc)

Figure 2. SEM micrographs of BFN, 0.91BFN-0.09BT and 0.91BFN-0.09ST, ceramics at room temperature.

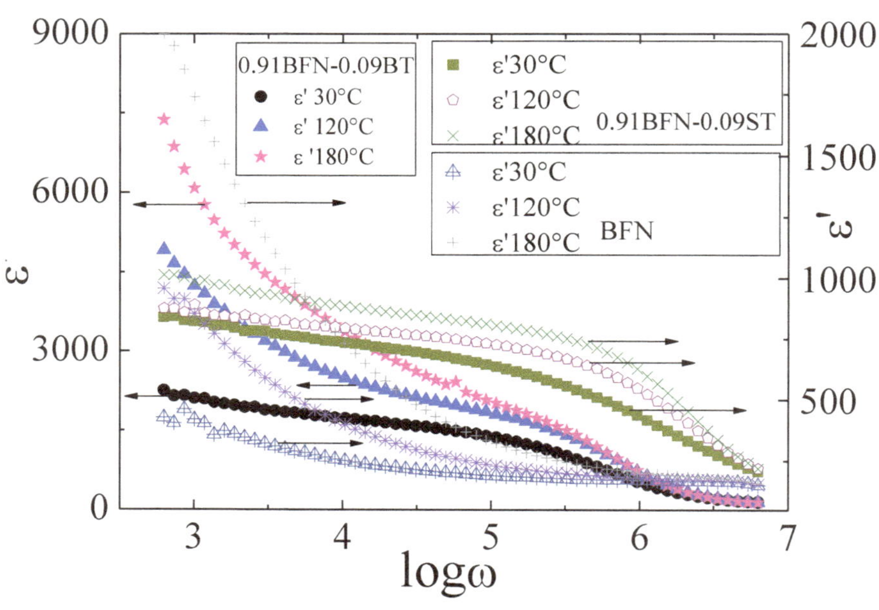

Figure 3. Logarithmic angular frequency dependence of dielectric constant (ε′) of BFN, 0.91BFN-0.09BT and 0.91BFN-0.09ST, ceramics, at various temperatures.

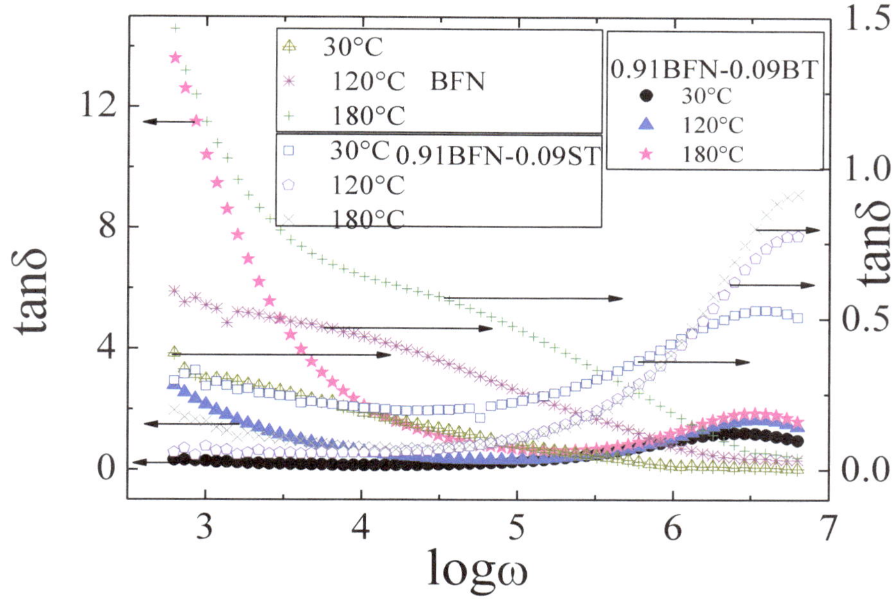

Figure 4. Tangent loss (tanδ) of BFN, 0.91BFN-0.09BT and 0.91BFN-0.09ST, as a function of logarithmic angular frequency at various temperatures.

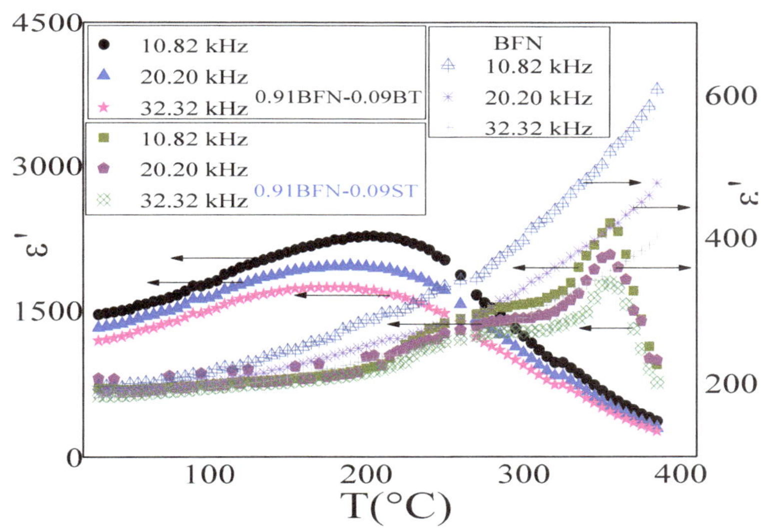

Figure 5. Dielectric constant (ε′) of BFN, 0.91BFN-0.09BT and 0.91BFN-0.09ST, ceramics as a function of temperature at selected frequencies.

and then decreases. The region around the dielectric peak is apparently broadened. The broadening or diffuseness of peak occurs mainly due to compositional fluctuation and/or substitution disordering in the arrangement of cations in one or more crystallographic sites of the BFN-BT and BFN-ST structure. It is observed from Figure 6, as the temperature increases tangent loss (tanδ) is found to increase. At low temperatures, the molecules can not orient themselves in polar dielectrics. When the temperature rises, the orientation of dipoles is facilitated and this increases tangent loss (tanδ). At high temperatures, the dielectric losses caused by the dipole mechanism reach their maximum value and the degree of dipole orientation increases. Apart from dipole losses, electrical conduction also increases with increase temperature. These factors would cause the increase in

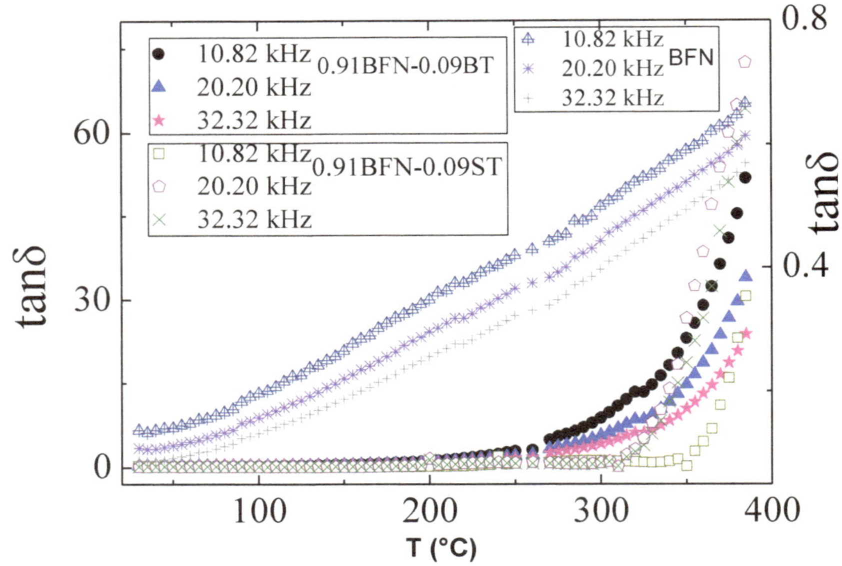

Figure 6. Tangent loss (tanδ) of BFN, 0.91BFN-0.09BT and 0.91BFN-0.09ST, ceramics as a function of temperature at selected frequencies.

tangent loss of BFN ceramics with increase in temperature (Taeev, 1975). In BFN-BT and BFN-ST, it is observed that tanδ remains almost constant up to ~305°C for BFN-ST and ~210°C for BFN-BT ceramics; afterwards, it increases rapidly.

Impedance formalism

In order to understand the dynamics of the mobile ions in BFN, BFN-BT and BFN-ST ceramics, we have plotted the angular frequency dependence of real (Z') and imaginary (Z'') parts of complex electrical impedance of above ceramics in Figures 7 and 8 respectively at various temperatures. Typical curves are observed in figures. The temperature affects strongly the magnitude of resistance. At lower temperatures, Z' and Z'' decreases monotonically with increase in frequency upto certain frequency and then becomes frequency independent. The higher values of Z' and Z'' at low frequencies and low temperatures means the polarization is larger. The temperatures where this change occurs vary in the material with frequencies. This also means that the resistive grain boundaries become conductive at these temperatures. This also shows that the grain boundaries are not relaxing even at very high frequencies even at higher temperatures (Singh et al., 2011b).

The plot of the imaginary (Z'') versus real (Z') parts of the complex impedance (Z*) (Cole–Cole plot) at temperature 180°C for BFN (inset), BFN-BT and BFN-ST ceramics is shown in Figure 9. A single arc of the impedance spectrum indicates that the electrical process in the material arises due to its bulk resistance only (Macdonald, 1987). The intercept of the semicircular arc on x-axis is an estimate of the bulk resistance (R_b) of the material and indicates the departure from the ideal Debye behavior. Infact, this behavior exhibits the non-Debye type of relaxation phenomenon in the materials.

Conductivity formalism

It is to be noted that the high values of dielectric constant (ε') and tangent loss (tanδ) in lower frequency region do not generally correspond to bulk effect. The high values of ε' interestingly observed only at very high temperature and very low frequencies may be attributed to the fact that the free charges buildup at interfaces within the bulk of the sample (interfacial Maxwell–Wagner (MW) polarization) (Kyritsis et al., 1995) and at the interface between the sample and the electrodes space-charge polarization (Jonscher, 1983). In order to elucidate this point, the frequency dependant ac conductivity (σ_{ac}) at various temperatures of BFN, BFN-BT and BFN-ST ceramics is plotted in Figure 10. The real part of ac conductivity and dielectric constant (ε') and conductivity (σ) is as follows

$$\varepsilon' = \frac{C}{C_0} \tag{1}$$

$$\sigma = \omega\varepsilon_0\varepsilon'' \tag{2}$$

Where ε_0 is the dielectric permittivity in air, C/C0 the

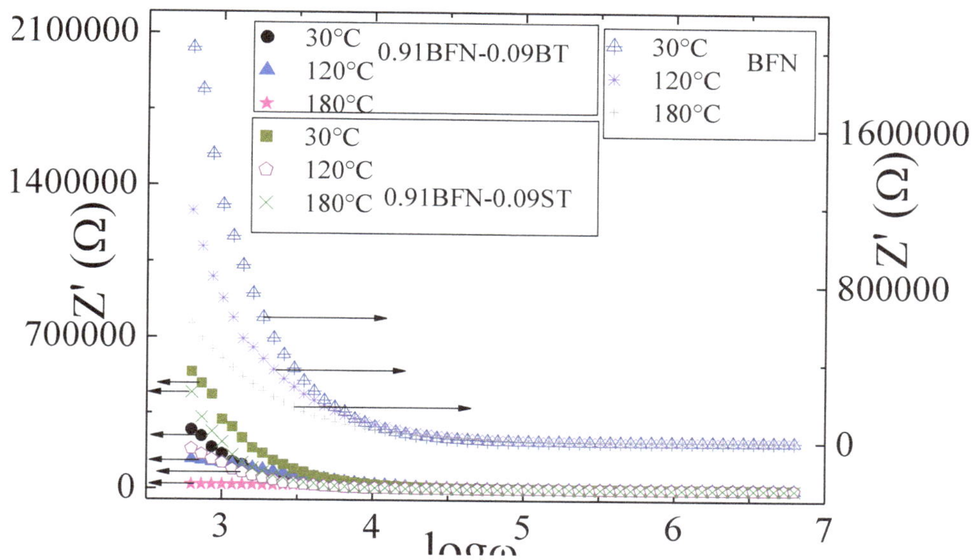

Figure 7. Z' of BFN, 0.91BFN-0.09BT and 0.91BFN-0.09ST, ceramics as a function of logarithmic angular frequency at various temperatures.

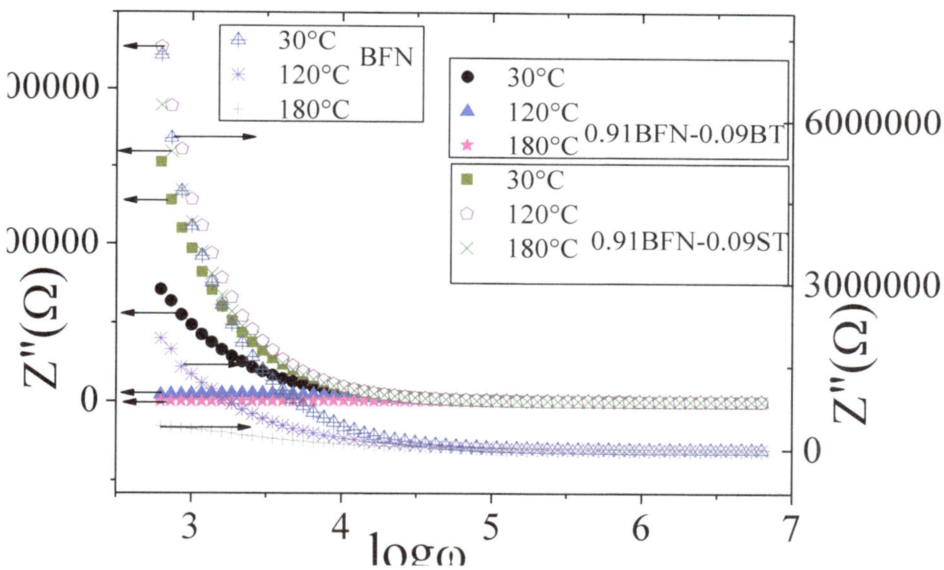

Figure 8. Z″ of BFN, 0.91BFN-0.09BT and 0.91BFN-0.09ST, ceramics as a function of logarithmic angular frequency at various temperatures.

ratio of capacitance measured with dielectric and without dielectric, ω the angular frequency and $\varepsilon'' = \tan\delta\ \varepsilon'$.

The conductivity increases with increasing frequency and increasing temperatures. For all the samples the conductivity shows dispersion which shifts to higher frequency side with the increase of temperature Notice that at low frequencies, random diffusion of charge carriers via hopping gives rise to a frequency independent conductivity. The increasing trend of $\sigma(\omega)$ in the low-frequency range may be due to the disordering cations between the neighboring sites and the presence of space charges that vanishes at higher temperatures and frequencies. The high-frequency conductivity dispersion may be attributed to ac conductivity whereas the frequency independent region of the conductivity pattern corresponds to the dc conductivity (σ_{dc}) of the

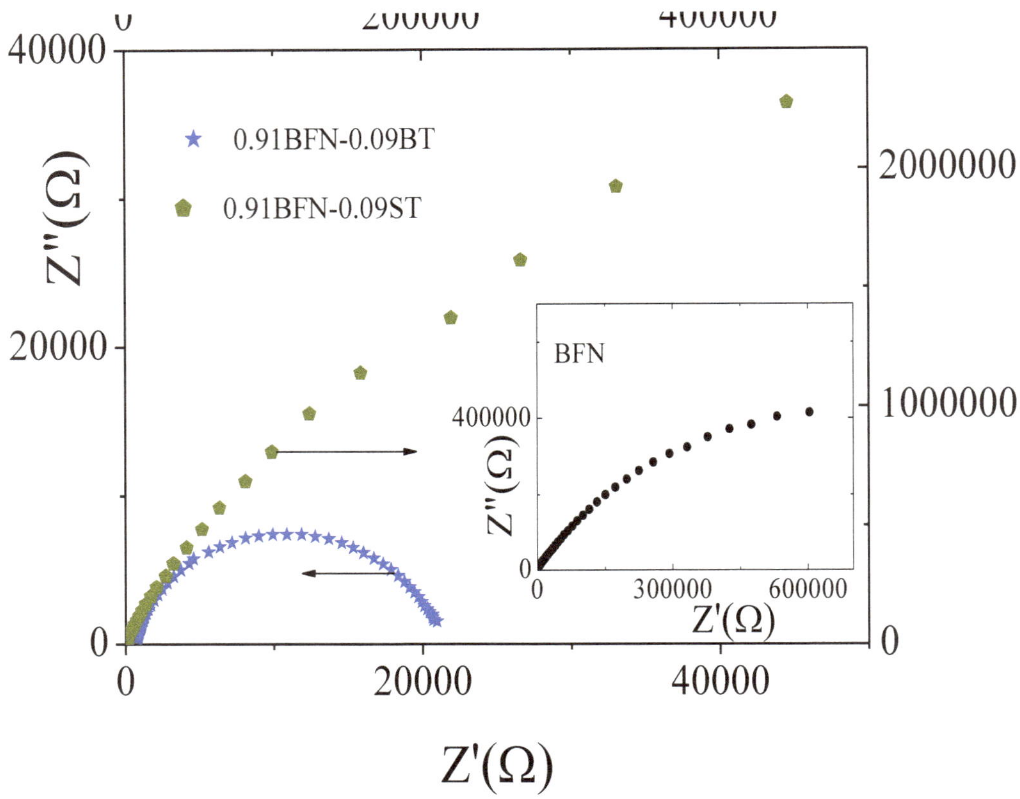

Figure 9. Complex plane impedance plot of BFN, 0.91BFN-0.09BT and 0.91BFN-0.09ST, ceramics, at 180°C.

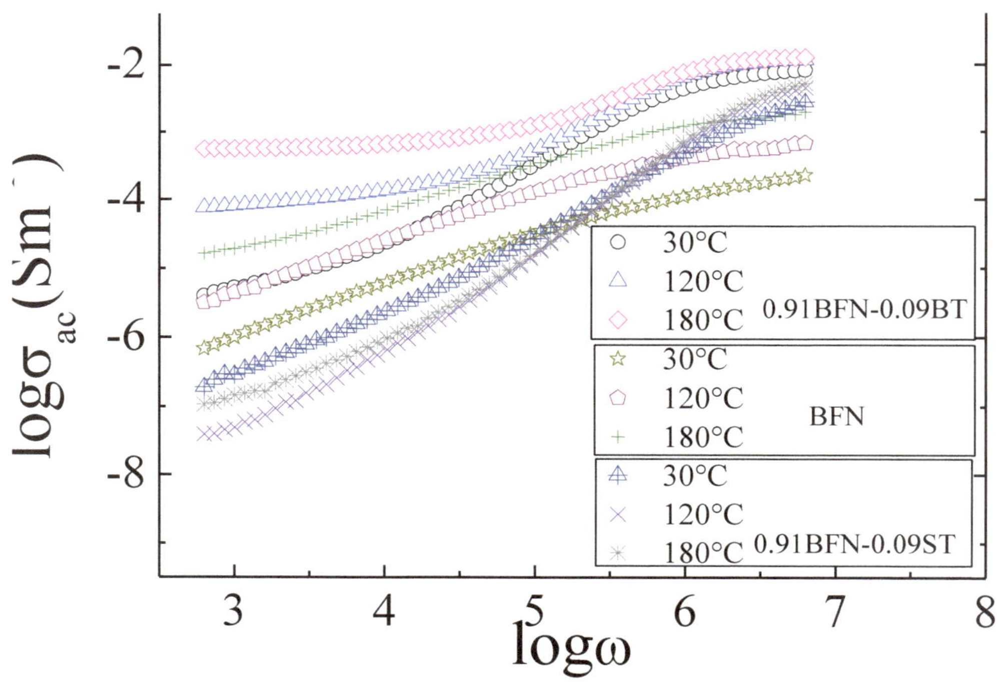

Figure 10. Logarithmic angular frequency dependence of ac conductivity (σ_{ac}) for BFN, 0.91BFN-0.09BT and 0.91BFN-0.09ST ceramics at various temperatures.

material.

Conclusion

The frequency-dependent dielectric dispersion of $Ba(Fe_{0.5}Nb_{0.5})O_3$ (BFN), $0.91Ba(Fe_{0.5}Nb_{0.5})O_3$-$0.09BaTiO_3$ (BFN-BT) and $0.91Ba(Fe_{0.5}Nb_{0.5})O_3$-$0.09SrTiO_3$ (BFN-ST) ceramics were prepared by a high temperature solid-state reaction technique and is investigated in the temperature range at 30 to 385°C and in the frequency range at 100 Hz to 1 MHz. The crystal structures of BFN, BFN-BT and BFN-ST ceramics determined by powder X-ray diffraction shows monoclinic phase at room temperature. The microstructure of the ceramics was examined by the scanning electron microscopy (SEM), and shows the polycrystalline nature of the samples with different grain sizes, which are inhomogeneously distributed through the sample surface. The variation of relative dielectric permittivity (ε') and tangent loss ($\tan\delta$) may be attributed to hopping of trapped charge carriers, which resulted in an extra dielectric response in addition to the dipole response.

ACKNOWLEDGEMENT

The authors are grateful to DRDO, for providing financial assistance for this research grant.

REFERENCES

Agrawal L, Singh BP, Sinha TP (2009). Dielectric Relaxation in Complex Perovskite Oxides $In(Ni_{1/2}Zr_{1/2})O_3$. Mater. Res. Bull. 44:1858–1862.

Ang C, Yu Z (1992). Dielectric properties and complex defect in $(Sr_{1-x}Bi_{2/3x})TiO_3$., J. Appl. Phys. 71:4451.

Bansal AK, Singh PJ, Sharma KS (2001). Microwave dielectric relaxation study of binary mixtures of cyclochexylamine with acetophenone. Indian J. Pure Appl. Phys. 39:799-803.

Bharti C, Dutta A, Shannigrahi S, Choudhary SN, Thapa RK, Sinha TP (2009). Impedance Spectroscopy, electronic structure and X-ray Photoelectron spectroscopy studies of $Pb(Fe_{1/2}Nb_{1/2})O_3$. J. Elect. Spectrosc. Relat. Phenomena 169:80–85.

Bianchi U, Dec J, Kleemann W, Bednorz JG (1995). Cluster and Domain-State Dynamics of ferroelectric $Sr_{1-X}Ca_XTiO_3$(X=0.007). Phys. Rev. B. 51:8737.

Chopra S, Sharma S, Goel TC, Mendiratta RG (2004). Effect of Annealing temperature on the Microstructure of chemically deposited Ca modified Lead Titanate Effect thin films. Appl. Sur. Sci. pp. 230-207.

Cui B, Yu PF, Tian J, Chang ZG (2006). Preparation and characterization of Co-doped $BaTiO_3$ nanosized powders and ceramics. Mater. Sci. Eng. B 133:205.

Jonscher AK (1983). Dielectric Relaxation in Solids. Chelsea Dielectric Press, London.

Kittel C (1995). Introduction to Solid State Physics. Wiley, New York.

Kumar P, Singh BP, Sinha TP, Singh NK (2011a). Ac Conductivity and Dielectric relaxation in $Ba(Sm_{1/2}Nb_{1/2})O_3$ ceramics, Physica B 406:139–143.

Kumar P, Singh BP, Sinha TP, Singh NK (2011b). X-ray and electrical properties of $Ba(Gd_{1/2}Nb_{1/2})O_3$ ceramic. Adv. Mater. Lett. 2(1):76.

Kyritsis A, Pissis P, Grammatikakis J (1995). Dielectric relaxation spectroscopy in poly(hydroxyethyl acrylates)/water hydrogels. J. Polym. Sci. Polym. Phys. 33:1737.

Macdonald JR (1987). Impedance spectroscopy emphasizing solid materials and system. New York: Wiley

Mao YB, Banerjee S, Wong SS (2003). Large Scale Synthesis of Single Crystalline Perovskite Nanostructures. J. Am. Chem. Soc. 125:15718.

Muller KA, Burkhard H (1979). $SrTiO_3$: An Intrinsic quantum paraelectric below 4K, Phys. Rev. B 19:3593.

Nelson SO (1993). Insect control studies with microwaves and other radio frequency. Bull. Entomol. Soc. Am. 19(3):157-163.

Nelson SO (1992). Measurement and application of dielectric properties of agricultural products. IEEE Trans. Instrum. Meas. 41(1):116-122.

Pengfei Y, Xue W, Bin C (2007). The Sol-gel process using organic monoacid as surfacatant. Scripta Mater. 57:623.

Saha S, Sinha TP (2006). Dielectric relaxation in $SrFe_{1/2}Nb_{1/2}O_3$. J. Phys. Chem. Sol. 67:1484–1491.

Shantha K, Varma KBR (1997). Frequency dependence of the dielectric properties of ferroelectric $Bi_2VO_{5.5}$(BiV) Ceramics. Sol. State Ion. 99:225–231.

Singh NK, Choudhary RNP, Banarji B (2008). Dielectric and Electrical Characteristics of $Nd_2(Ba_{0.5}R_{0.5})_2O_7$,(R=W, M_O). Physica B 403:1673.

Singh NK, Kumar P (2011a). Structural and Impedance Spectroscopy Study of (1-X)$BaFe_{0.5}Nb_{0.5})O_3$+X$BaTiO_3$. ISST J. Appl. Phys. 2(1):29-33.

Singh NK, Kumar P (2011b). Studies of Structural and Electrical behavior of Samarium Barium Tungstate ceramics. J. Adv. Dielectr. 1(4):465-470.

Singh NK, Kumar P, Kumar H, Rai R (2010a). Structural and Dielectric properties of $Dy_2(Ba_{0.5}R_{0.5})_2O_7$(R=W,Mo). Adv. Mater. Lett. 1:79-82.

Singh NK, Kumar P, Roy OP, Rai R (2010b). Structural and Dielectric properties of $Eu_2(B'_{0.5}B''_{0.5})_2O_7$(B'=Ba,B''=W,Mo). J. Alloys Compd. 507:542-546.

Singh NK, Kumar P, Rai R (2011a). Studies of Structural, Dielectric and Electrical behavior of (1-X)$BaFe_{0.5}Nb_{0.5})O_3$-X$SrTiO_3$. J. Alloys Compd. 509:2957–2963.

Singh NK, Pritam K, Sharma AK, Choudhary RNP (2011b). Structural and Impedance Spectroscopy Study of $BaFe_{0.5}Nb_{0.5})O_3$-$SrTiO_3$ Ceramics System. Mater. Sci. Appl. 2:1593-1600.

Taeev B (1975). "Physics of dielectric materials." Mir Publishers, Moscow.

Wul E, PowdMult (1999). PowdMult: An interactive powder diffraction data interpretation and index program, version 2.1, School of Physical Science, Flinders University of South Australia, Bedford Park, S.A. 5042, Australia.

12

Determination of oscillator strength of caffeine and caffeine in tea leaves by the integrated absorption coefficient technique at two different temperatures

Tadelech Atomssa Keyeta* and A. V. Gholap

Faculty of Physical and Chemical Sciences, Physics Department, Addis Ababa University, P.O.Box 1176, Addis Ababa, Ethiopia.

In this paper, oscillator strength of caffeine in dichloromethane, water, chloroform, and ethyl acetate were calculated in the wave number range of 32, 000 to 41,000cm^{-1}. It was found that oscillator strength of caffeine in dichloromethane, water, chloroform, and ethyl acetate were 0.230 ± 0.001, 0.170 ± 0.0003, 0.114 ± 0.001, 0.150 ± 0.001 respectively. Caffeine content of real tea leaves using integrated absorption coefficient method were also calculated for different samples at two different temperatures. The result agrees with that obtained using Beer's law with a little less in the case of integrated absorption coefficient technique with higher value at the boiling temperature and smaller extraction time.

Key words: Tea leaves, integrated absorption coefficient, caffeine, oscillator strength, UV-visible spectroscopy.

INTRODUCTION

Processed tea, which is one of the most popular beverages, is manufactured from the young tender leaves of the plant camellia sinensis (Cabrera et al, 2003). Tea is not only used as fresh drink but also as traditional herb which has many benefits for human health. Recently, several studies led to the acceptance of tea as a useful cancer preventive agent (Jankun et al., 1997; Borrelli et al., 2004; Scalbert et al., 2005; Yang et al., 2004) which increased public awareness of the health protective characteristics of tea. There are two major kinds of tea; black tea and green tea, and they both contain caffeine (1 to 5 % of its dry weight) (Amra et al, 2006). Tea composition varies with climate, season, tea variety, and age of the leaf.

Due to the physiological and psychological effects of caffeine, decaffeination is currently a popular method to minimize the caffeine content in tea leaves. Decaffeination is done by the use of solvent, which can extract caffeine. For this purpose, the currently available solvents are chloroform, dichloromethane, ethyl acetate, super critical carbon dioxide, etc. Dichloromethane is used to decaffeinate a high proportion of conventional teas. It is also a relatively non-toxic solvent and is often used in place of chloroform. Ethyl acetate also removes caffeine from tea leaves effectively; it can also extract other chemical components as well (Mumin et al., 2006). Studies on green tea decaffeinated with ethyl acetate have shown the potential for up to 30% of Epigallocatechin Gallate (EGCG, considered to be the beneficial component in green tea) and other beneficial antioxidant compounds to be extracted along with caffeine (Kirmer, 1998; Johnson, 1988; Jefferson, 1998).

Due to the widespread consumption of caffeine, it is important to collect precise information on their content in foods. Most research activities have been focused on chromatographic methods; however, spectrophotometric determination is preferred because of its rapidity, high accuracy and reproducibility (Komes et al., 2009; Guzin et al., 2002). Furthermore, UV-visible spectrophotometer is cheap and it is found in many laboratories. But caffeine content in tea leaves cannot be determined directly using UV-visible spectrometer due to the matrix effect of UV absorbing substances (Guzin, 2002; Zhang et al., 2005).

Caffeine content in tea leaves can be obtained using UV-visible spectrometer by Beer-Lambert's law. The results obtained using Beer-Lambert's law are satisfactory and reproducible at room temperature; however, when there are temperature and pressure variations or the concentration of solute in the solution is high or a reaction field exists due to the solute and the solvent, line

*Corresponding author. E-mail: tadelechad@yahoo.com.

broadening and a shift will happen, and thus, reliable results may not be obtained with this technique. However, the integrated absorption coefficient is independent of line function which varies due to pressure, temperature, and solute solvent interaction. In addition, measuring the intensity of absorption by the integrated method provides additional information about the nature of the absorbing molecules and establishes accurate evaluation of UV-visible absorption band intensity. Therefore, in this paper we have used UV-visible spectrometer for the determination of caffeine content in tea leaves using integrated absorption coefficient technique at 30°C and boiling temperature of 94°C. The results obtained using integrated absorption coefficient technique and Beer-Lambert's law was also compared. Finally, oscillator strengths of caffeine in different solvents were also calculated.

MATERIALS AND METHODS

Samples

Four different Ethiopian commercial tea samples named Addis, black lion, cinnamon black teas and Ethiopian green tea. All the tea leaves were purchased from supermarkets in Addis Ababa. All the samples were from tea bags except black lion. Cinnamon tea is black tea leaves blended with cinnamon.

Instrumentation

The absorption spectra were recorded on a Perkin-Elmer Lambda 19 spectrometer with wavelength range of 170 to 3200 nm and double monochromator which offers the advantage of low levels of stray light that is significant to measure a high value of absorbance using 1 cm cuvette. The investigated compound was measured in the UV-visible spectrum range of 200 to 500 nm with a spectral band width of 2 nm and scan speed of 480 nm min^{-1}.

Lab apparatus

Microbalance (Mettler At 250, 0.1/0.01 mg) for measuring mass of caffeine, balance (BP 210S, d = 0.1 mg, Max 210 g) for measuring mass of tea leaves, measuring cylinder, magnetic stirrer, glass filter (2 'Pyrex', England), beakers, thermometer, 1 cm quartz cuvette, separating funnel, funnel, ice bath, and distilled water was used.

Chemicals

Dichloromethane (assay: 99.6%, Aldrich Germany), chloroform (assay: 99.8%, European Union), ethyl acetate (assay: 99%, Indian), caffeine (M.W. 194.19 g/mol, Aldrich Germany).

Methods for measuring oscillator strength of caffeine in water, dichloromethane, chloroform, and ethyl acetate

Caffeine concentration of (14.419 to 2.670)10^{-5}M, (9.475 to 1.1560)10^{-5}M, (14.240 to 4.747)10^{-5}M, (10.299 to 2.943)10^{-5}M were prepared in distilled water, dichloromethane, chloroform, and ethyl acetate respectively. Absorbance (optical density); which is a dimensionless quantity versus wavelength of the solution was measured using UV-visible spectrometer. Then, Integrated absorption coefficient and oscillator strength were calculated from the data of each spectra using origin 6.1 soft ware along with Equations 1 and 2.

Methods for determination of caffeine content in tea leaves

50 mg of tea leaves was first dissolved in distilled water at two different temperatures (30°C and boiling temperature of 94°C) and then, caffeine was extracted using dichloromethane. Caffeine was repeatedly extracted four times from tea infusion. To ensure the reliability of the experiment, three independent measurements (n=3) were taken for each sample. For 94°C, we used smaller extraction time than 30°C. Then, the absorption spectrum of the solution were taken, from which, contents of caffeine for four different samples of tea leaves were calculated using integrated absorption coefficient technique. The absorption coefficients of caffeine were calculated in the wave number regions of 32, 000 to 41,000 cm^{-1}.

RESULTS AND DISCUSSION

In this section, the results of the experiments were presented based on the experimental procedures of the previous section. Each data in these experimental activities was found from computerized recorder interfaced with the spectrometer. It was then analyzed using origin 6.1soft ware.

Validation of the method

From the UV-visible absorption spectrum of caffeine in dichloromethane, water, chloroform, and ethyl acetate, the calibration graph correlating the integrated absorption coefficient versus concentration was constructed. With this, the linear regression coefficients of 0.99978, 0.99998, 0.99986, 0.9994 and standard deviations of 31.18222, 9.16031, 17.58995, and 39.14415 were obtained for caffeine in dichloromethane, water, chloroform, and ethyl acetate respectively. Thus, for the analysis of the calibration, a linear data set was obtained. The absorption spectra of pure caffeine and caffeine extracted by dichloromethane exactly overlaps (Figure 1). This indicates that dichloromethane was a good solvent used to extract caffeine from tea leaves.

Integrated absorption coefficient and oscillator strength of caffeine in different solvents

From UV-visible absorption spectra, the integrated absorption coefficient and oscillator strength of caffeine were calculated in dichloromethane, water, chloroform, and ethyl acetate. The absorption coefficient determines how far into a material, light of a particular wavelength can penetrate before it is absorbed. The integrated absorption coefficient a_t which is the sum of absorption

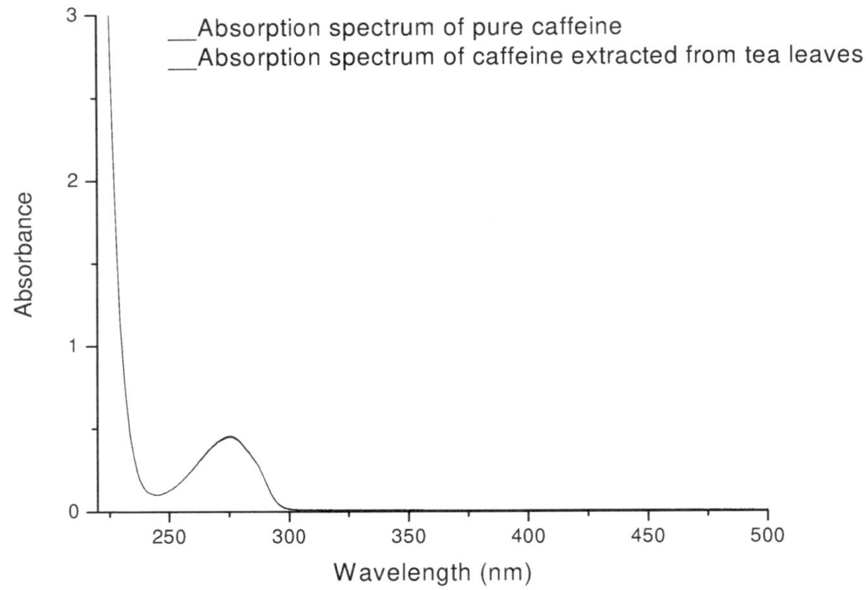

Figure 1. Absorption spectra of pure caffeine and caffeine extracted from tea leaves.

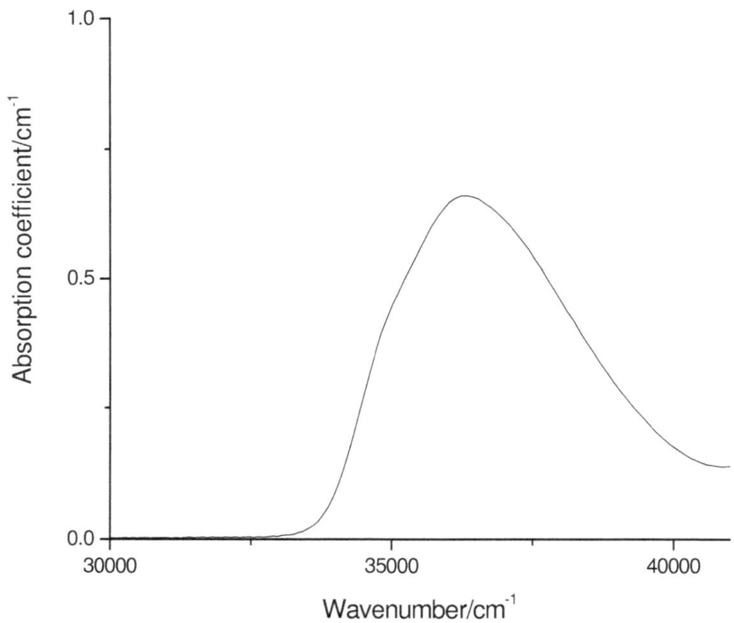

Figure 2. a_ν versus ν of caffeine in dichloromethane.

coefficient a_λ for all frequencies in the band is expressed as

$$a_t = \int a_\lambda d\nu \quad , \tag{1}$$

where ν is the frequency. Figure 2 shows the absorption coefficient versus wave number of caffeine in dichloromethane. The integrated absorption coefficient is independent of line function, which may vary with parameters like pressure, temperature, and solute solvent interaction. It thus provides a measure of the inherent absorbing strength of atoms.

In this experimental activity, the integrated absorption coefficient was found by recalculating the absorbance

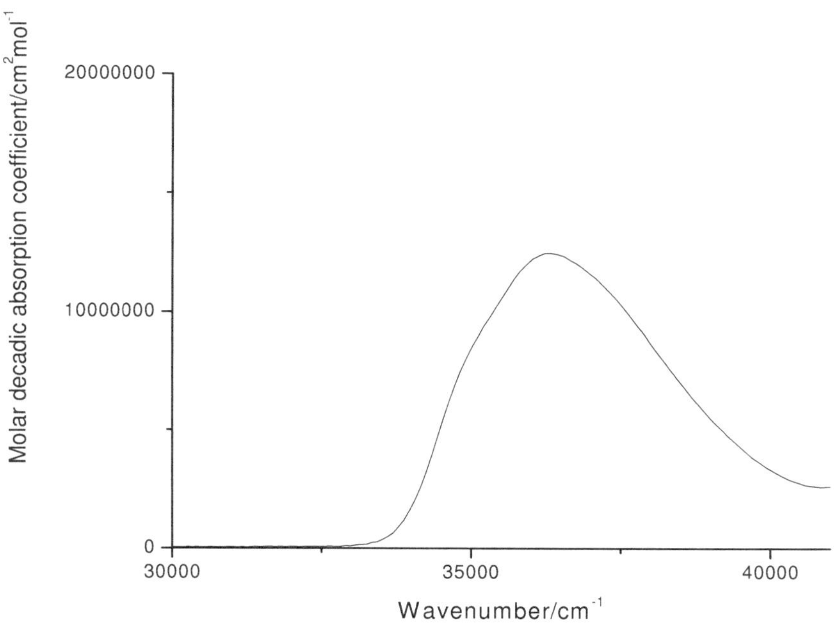

Figure 3. $\varepsilon(\nu)$ versus ν for caffeine in dichloromethane.

versus wave length into absorption coefficient versus wave number in the wave number region of 32, 000 to 41,000cm^{-1}. The other important parameter, which provides the relative strength of electron transition, is the oscillator strength (f). Oscillator strength relates to molar decadic absorption coefficient (ε) by the following equation (Georgakopoulous et al., 2004; Radwan, 2007).

$$f = 4.32 \times 10^{-9} \frac{mol cm^2}{L} \int \varepsilon(\nu) d\nu \quad (2)$$

The molar decadic absorption coefficient measures the intensity of optical absorption at a given wavelength.

Thus in this experimental activity, oscillator strength of caffeine in different solvents was found by recalculating absorbance versus wavelength into molar decadic absorption coefficient versus wave number along with Equation 2 using origin 6.1 software. Figure 3 shows the molar decadic absorption coefficient versus wave number of caffeine in dichloromethane. Furthermore, absorption coefficient is related to the number density and oscillator strength by Equation 3 (Throne, 1988).

$$\int a_\lambda d\nu = 2.65 \times 10^{-6} Nf \quad ...(3)$$

Where N is number density in molecules m^{-3}, f oscillator strength of the transition molecule a_λ absorption coefficient in m^{-1}, and ν is the frequency in Hz. From N, concentration (c) (N = cN_a, where N_a is Avogadro's number) of caffeine was determined, which in turn is used to calculate mass (m) of caffeine (c = m/MV, where M molecular mass of caffeine, and V volume of solution). Finally, percentage of caffeine was calculated using this result.

The integrated absorption coefficient, integrated molar decadic absorption coefficient, and oscillator strength of caffeine in dichloromethane, water, chloroform, and ethyl acetate are indicated in Table 1. As indicated in Table 1, the oscillator strength of caffeine in dichloromethane is greater than in the other solvents.

Caffeine content in tea leaves using integrated absorption coefficient technique

From Equation (3), number density of caffeine in tea leaves was calculated for caffeine in real tea leaves where f is the oscillator strength of pure caffeine in dichloromethane (Table 1). From N, concentration, c, of caffeine was determined, which was in turn used to calculate mass of caffeine. Finally, percentage of caffeine was calculated. Using this method, caffeine content in four Ethiopian tea leaves were calculated at 30°C and boiling temperature. This result was compared with the result obtained using Beer's law (Table 2). From Table 2, for both temperatures, the percentage of caffeine, using both methods are similar except a little smaller in the case of integrated absorption coefficient methods. This small difference will be due to the fact that, in the case of Beer's law it was calculated only at maximum peak of the absorption spectrum. When we compare the content of caffeine at 30°C and boiling temperature, the percentage

Table 1. Integrated absorption coefficient and oscillator strength of caffeine in dichloromethane, water, chloromethane, and ethyl acetate.

Solvent	Integrated absorption coefficient (cm^{-2})	Integrated molar decadic absorption coefficient ($L\ mol^{-1}cm^{-2}$)	Oscillator strength
Dichloromethane	1626.50 ± 36.16	$(5.22\pm0.03)10^7$	0.230±0.001
Water	2519.43±33.66	$(3.92\pm0.01)10^7$	0.170±0.0003
Chloroform	2162.92±29.30	$(2.64\pm0.01)10^7$	0.114±0.001
Ethyl acetate	1890.59±31.62	$(3.39\pm0.02)10^7$	0.150±0.001

Table 2. Comparison of caffeine content in tea leaves using absorption coefficient technique and Beer's law at 30℃ and boiling temperature ,94℃ (n = 3).

	30℃			94℃		
Sample	Concentration of caffeine using integrated absorption coefficient(mol/m^{-3})	Percentage of caffeine using integrated absorption coefficient	Percentage of caffeine using Beer's law	Concentration of caffeine using integrated absorption coefficient(mol/m^{-3})	Percentage of caffeine using integrated absorption coefficient	Percentage of caffeine using Beer's law
Addis black tea	$(3.55\pm0.11)10^{-5}$	1.86±0.05	2.01±0.04	$(3.77\pm0.06)10^{-5}$	2.00±0.02	2.11±0.04
Ethiopian green tea	$(3.51\pm0.06)10^{-5}$	1.88±0.02	2.08±0.01	$(3.78\pm0.12)10^{-5}$	2.01±0.07	2.15±0.12
Black lion tea	$(2.76\pm0.02)10^{-5}$	1.47±0.004	1.50±0.00	$(2.95\pm0.03)10^{-5}$	1.57±0.02	1.60±0.01
Cinnamon black tea	$(3.49\pm0.22)10^{-5}$	1.82±0.11	1.98±0.09	$(3.41\pm0.15)10^{-5}$	1.82±0.09	1.87±0.08

of caffeine at the boiling temperature is greater with smaller extraction time. At both temperatures, Ethiopian green tea leaves has greater caffeine content than the other tea leaves studied.

Caffeine content in tea leaves reported in this research is in the range of values 1 to 5% reported by Amra et al. (2006). As it can be seen in this research, the maximum content of caffeine is almost similar to the results obtained for the caffeine contents of marketed tea by Misra et al. (2009) using HPLC in which the maximum concentration of caffeine in tea samples of marketed tea was 2.145%. Mumin et al. (2006) also reported using HPLC that, caffeine content for black tea was 3.34% and green tea was 2.44%. Furthermore, Guzin et al. (2002) reported that caffeine content in tea, using derivative spectrometric and HPLC methods were 1.53% and 1.51% respectively. Precision of the method was determined by three replications of each sample. The precision (%RSD) of the replications was found to be less than two which is indicative of a precise method.

Conclusion

A new technique for measuring the quantity of caffeine in tea leaves has been described using integrated absorption coefficient technique. Using this technique, contents of caffeine in four different tea samples were calculated at 30℃ and boiling temperature (94℃). The result agrees with the range of values in literature using other analytical methods. The results at the two temperatures reveal that, caffeine is better extracted at the boiling temperature than at 30℃ and more extraction time. Therefore, it can be said that, by taking more extraction time and at higher temperatures, more caffeine can be extracted than at boiling temperature and less extraction time.

This technique is more advantageous for evaluating the UV-Visible absorption intensity since the effect of line broadening due to temperature, pressure, and other variation do not affect the intensity of the absorbing molecule.

REFERENCES

Mumin MD, Akhter KF, Abedin MZ, Hossain MZ (2006). Determination and characterization of caffeine in tea, coffee, and soft drinks by Solid phase extraction and high performance liquid chromatography (SPE – HPLC). Malaysian J. Chem., 8(1): 045-051.

Amra P, Mojca S, Zeljko K, Bernd W, Frank O, Sabine G (2006). Extraction of active ingredients from Green tea (camellia sinensis): Extraction efficiency of major catechins and caffeine. Food Chem., 96: 597 – 605.

Borrelli F, Capasso R, Russo A, Ernst E (2004). Systematic review: green tea and gastrointestinal cancer risk. Alimentary Pharmacol. Therapeutics, 19: 497-510.

Cabrera C, Gimenez R, Lopez CM (2003). Determination of tea components with antioxidant activity. J. Agric. Food Chem., 51: 4427 – 4435.

Georgakopoulous S, Grondelle RV, Zwan GVD (2004). Circular dichroism of carotenoides in bacterial light-harvesting complexes: experimental and modeling. Biophysical J., 87: 3010- 3022.

Guzin A (2002). Derivative spectrophotometer determination of caffeine in some Beverages. Turk J. Chem., 26: 295 – 302.

Jankun J, Selman SH, Swiercz R (1997). Why drinking green tea could prevent cancer. Nature, 387(5): 561- 563.

Jefferson JW (1998). Lithium tremor and caffeine intake: two cases of drinking less and shaking More. J. Clin. Psychiatry., 49: 72.

Johnson GD, Fatis M, Sonnek D (1988). A survey of caffeine use and associated side effects in a college population. J. Drug Educ., 18(3): 211.

Kirmer DA (1988). Caffeine use and abuse in psychiatric clients. J. Psychosoc Nurs. Ment Health Service. 26:20.

Komes D, Horzic D, Belscak A, Kovacevic GK, Boljak A (2009). Determination of caffeine contents in tea and mate tea by using different methods. Czech J. Food Sci., pp. 27

Misra H, Mehta D, Mehta BK, Soni M, Jain DC (2009). Study of extraction and HPTLC – UV method for estimation of caffeine in marketed tea granules. International J. Green Pharm., 3(1): 47– 51.

Radwan RM (2007). Electron induced modifications in the optical properties of polypropylene. J. Phys D: ppl. Phys., 40: 374-379.

Scalbert M, Manach C, Morand C, Romesy C, Jimenez L (2005). Dietary polyphenols and the prevention of disease. Critical Reviews in Food Science and Nutrition. 45:287-308.

Throne AP (1988). Spectrophysics. Chapman and Hall Ltd, London, pp. 318-320.

Yang YC, Lu FH, Wu JS, Wu CH, Chang CJ (2004). The protective effect of habitual tea consumption on hypertension. Archive Int. Med., 164(14): 1534-1540.

Zhang QL, Lian HZ, Wang WH, Chen HY (2005). Separation of caffeine and theophylline in poly (dimethylsiloxane) micro channel electrophoresis with electrochemical detection. J. Chromatography, 1098(A): 172-176.

Microstructure of different NaOH molarity of fly ash-based green polymeric cement

A. M Mustafa Al Bakri[1]*, H. Kamarudin[1], M. Bnhussain[3], I. Khairul Nizar[3], A. R Rafiza[1] and Y. Zarina[1]

[1]Green Concrete@UniMAP, School of Material Engineering, Universiti Malaysia Perlis (UniMAP), P. O. Box 77, d/a Pejabat, Pos Besar, 01007 Kangar, Perlis Malaysia.
[2]School of Environment Engineering ,Universiti Malaysia Perlis (UniMAP), P. O Box 77, d/a Pejabat, Pos Besar, 01007 Kangar, Perlis Malaysia.
[3]King Abdul Aziz City Science and Technology (KACST), P.O. Box 6086, Riyadh 11442, Kingdom of Saudi Arabia.

Every 1 ton of concrete leads to CO_2 emissions which vary between 0.05 to 0.13 tons. About 95% of all CO_2 emissions from a cubic yard of concrete are from cement manufacturing. It is important to reduce CO_2 emissions through the greater use of substitute to ordinary Portland cement (OPC) such as fly ash, clay and others geo-based material. This paper, report on the study of the processing of geopolymer using fly ash and alkaline activator with geopolymerization process. The factors that influence the early age compressive strength such as molarity of sodium hydroxide (NaOH) have been studied. Sodium hydroxide and sodium silicate solution were used as an alkaline activator. The geopolymer paste samples were cured at 70°C for 1 day and keep in room temperature until the testing days. The compressive strength was done at 1, 2, 3 and 7 days. The result showed that the geopolymer paste with NaOH concentration of 12 M produced maximum compressive strength.

Key words: Green polymeric concrete, fly ash, molarity, compressive strength.

INTRODUCTION

The term geopolymer was first applied by Davidovits (1994) to alkali aluminosilicate binders formed by the alkali silicate activation of aluminosilicate materials. Geopolymer (green polymeric concrete) are amorphous to semi-crystalline equivalent of certain zeolitic materials with excellent properties, such as high fire and erosion resistances, and high strength materials. Recent works have shown that the addition of moderate amount of minerals to a geopolymer can give significant improvements on the geopolymer structure and properties (Hu Mingyu et al., 2009).

The alkaline liquid could be used to react with the silicon (Si) and the aluminum (Al) in a source material of natural minerals or in by-product materials such as fly ash and rice-husk ash to produce binders (Davidovits, 1994). The alkaline activation of materials can be defined as a chemical process that provides a rapid change of some specific structures, partial or totally amorphous, into compact cemented frameworks (Fernandez and Palomo, 2003). Alkali activation of fly ash is a process that differs widely from Portland cement hydration, and very similar to the chemistry involved in the synthesis of a large groups of zeolites (Criado et al., 2005). Some researchers described the alkali activation of fly ash (AAFA) as a physical–chemical process in which this powdery solid is mixed with a concentrated alkali solution (in a suitable proportion to produce a workable and mouldable paste) and stored at mild temperatures (< 100°C) for a short period of time to produce a material with good binding properties (Fernandez et al., 2006; Fernandez and Palomo, 2005). At the end of this process, an amorphous alkaline aluminosilicate gel is formed as the main reaction product. In addition, Na-Herschelite-type zeolites and hydroxysodalite are formed as secondary reaction products (Criado et al., 2005;

*Corresponding author. E-mail: mustafa_albakri@unimap.edu.my.

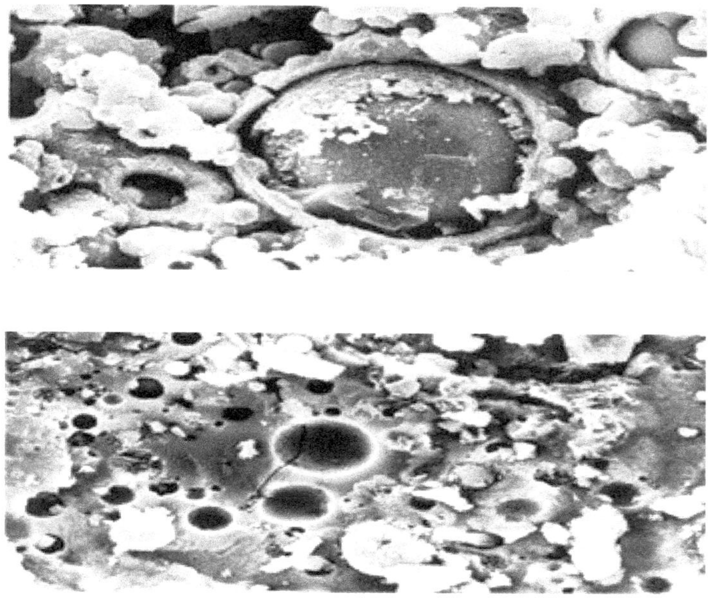

Figure 1. Microstructure of the fly ash-based geopolymer (Palomo et al., 1999a).

Fernandez et al., 2006; Fernandez and Palomo, 2005). The most used alkaline activators are a mixture of sodium or potassium hydroxide (NaOH, KOH) with sodium waterglass ($nSiO_2Na_2O$) or potassium waterglass ($nSiO_2K_2O$) (Davidovits, 1994; Palomo et al., 1999a; Vijaya Rangan, 2008; Fernando et al., 2008). The concentrations of sodium hydroxide solution that can be used are in the range of 8-16 M (Vijaya Rangan, 2008). The studies on the effect of different molarity of NaOH to the geopolymer have been done by some researchers. Puertas et al. (2000) stated that at 28 days of reaction, the mixture of 50% fly ash/ 50% slag activated with 10 M NaOH and cured at 25°C, developed compressive mechanical strengths at about 50 MPa. Rattanasak and Chindaprasirt (2009) concluded that when the mixture was formulated with NaOH molarity of 10 M and ratio of sodium silicate to NaOH of 1.0, the strength of geopolymer mortar up to 70 Mpa was obtained. Palomo et al. (1999a) reported that an activator with a 12 M of NaOH concentration leads to better results than a 18 M of NaOH concentration.

In this study, the effect of various molarity of NaOH to fly ash geopolymer paste was studied. There were six different molarity of NaOH (6, 8, 10, 12, 14, and 16 M) have been utilised in this study. The properties of the geopolymer such as compressive strength, water absorption, porosity, and density will plays a role as an indicator to prove that the geopolymer have similar properties to OPC (ordinary Portland cement).

Microstructure of geopolymers

Hos et al. (2002) systematically analyzed the microstructure of aluminosilicate inorganic polymer. By using scanning electron microscopy (SEM), they observed the nonporous microstructure of their materials. According to Hos et al. (2002), the nonporous microstructure is a result of extensive dissolution of aluminosilicate species that occurs before polycondensation commences and consolidates the shapes of the specimen through a chaotic three-dimensional network of polysodium aluminosilicate.

Moreover, fly ash-based geopolymer has also exhibits a high density of pores or air bubbles. Palomo et al. (1999a) reported that the fly ash-based geopolymer material is very porous and the microspheres (originating from fly ash grains) are surrounded by a curst of reaction product. The curst adhere to the sphere is not very strong and the bond between grains is produced through the necks of reaction product as indicated in (Figure 1A). Crack development is evident in the middle of the matrix and is likely been initiated from the pore (Figure 1B).

EXPERIMENTAL PROCEDURES

Materials

In the research work, low calcium, Class F (American Society for Testing and Materials, 2001) dry fly ash (Hos et al., 2002) obtained from Sultan Abdul Aziz power station in Kapar, Selangor, Malaysia was used as base material to make the geopolymers.

Sodium silicate (Na_2SiO_3) mixed with sodium hydroxide (NaOH) as an alkaline activator has been used in this study. NaOH in pellet form with 97% purity (Vijaya Rangan, 2008; Buchwald and Schulz, 2005; Steenie, 2009) and Na_2SiO_3 consists of Na_2O = 9.4%, SiO_2 = 30.1% and H_2O = 60.5%, with weight ratio SiO_2/Na_2O = 3.20-3.30),

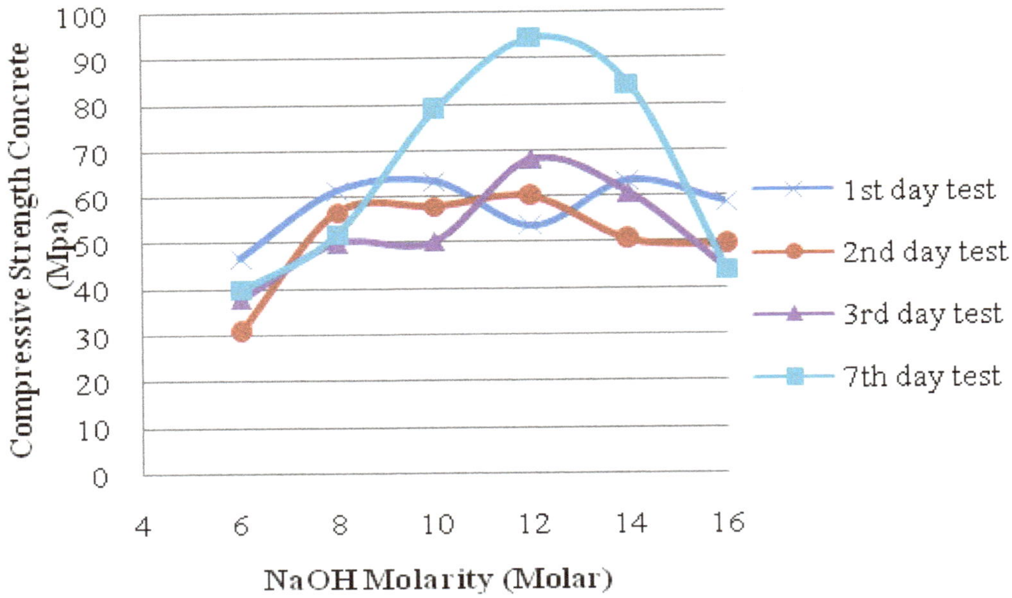

Figure 2. Compressive strength of various molarity of NaOH.

Table 1. Mixture proportions of geopolymer paste.

Mixture of geopolymer	Mass ratios (g)
Ratio fly ash/alkaline activator	2.50
Ratio Na_2SiO_3/NaOH	2.50
Mass of fly ash (g)	334.82
Mass of NaOH (g)	38.27
Mass of Na_2SiO_3 (g)	95.66

and specific gravity at 20ºC = 1.4 gm/cc were used in this study.

Method of mixing

In preparation of NaOH solution, NaOH pellets was dissolved in one liter of distilled water in a volumetric flask for six different concentration of NaOH (6, 8, 10, 12, 14 and 16 M). Alkaline activator with the combination of NaOH and Na_2SiO_3 was prepared just before mixing with fly ash. The addition of sodium silicate is to enhance the process of geopolymerization process (Xu and Deventer, 2000).

The ratio of fly ash/alkaline activator and Na_2SiO_3/NaOH is 2.5 and fixed for all mixture. The used of this ratio is due to Hardjito et al. (2008) which stated that ratio of fly ash/alkaline activator of 2.5 produced the highest compressive strength at 28th day of testing. Hardjito et al. (2004) used ratio of Na_2SiO_3/NaOH 2.5 and 0.4 in the preparation for geopolymer concrete.

The fly ash and alkaline activator were mixed together in the mixer until homogeneous paste was obtained. This mixing process can be handled within 5 min for each mixture with different molarity of NaOH. The total mass of each material used is kept constant for geopolymer paste for all the six different types of samples (6, 8, 10, 12, 14, and 16 M) with different molarity as shown in Table 1.

Casting and curing

The geopolymer paste was placed in 50 x 50 x 50mm cube mould and cured in the oven for 1, 2, 3 and 7 days at constant temperature of 70°C.

Compressive strength test and microstructure

The compressive strength test was performed on geopolymer paste samples accordance to BS 1881-116:1983 by using mechanical testing with Automatic Max (Instron, 5569 USA) in order to obtain the ultimate strength of geopolymer. The samples were loaded with 50 KN and the utilized with speed rate of 5 mm/min. The compressive strength values are determined by the average of 3 samples testing for each different types of mixture.

Scanning electron microscopy (SEM) was performed by using SEM JSM-6460 LA Jeol Japan in School of Materials Engineering, University of Malaysia Perlis (UniMAP) to study the microstructure of the base material (fly ash) and the resulting geopolymer samples. The test was carried out using secondary and backscattered electron detectors. The SEM samples must be prepared in powder form. The prepared samples were cut into 0.5 mm thick slices and then grinded into powder form for the test

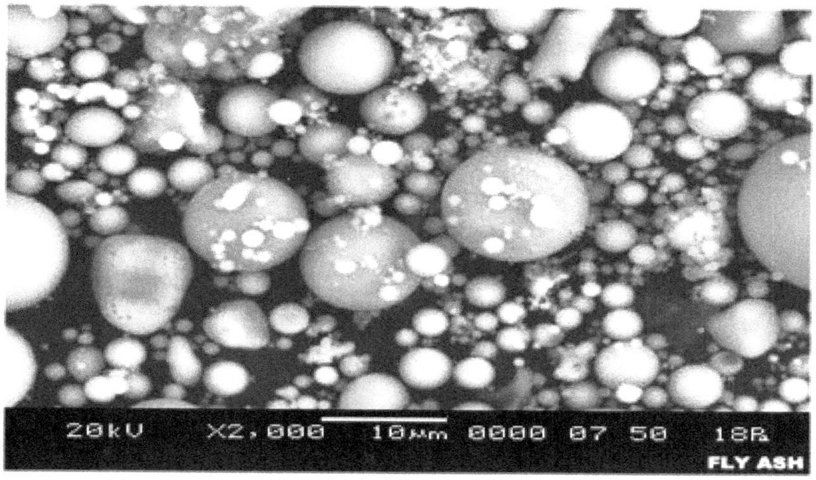

Figure 3. SEM micrograph of fly ash.

requirement.

RESULT AND DISCUSSION

Compressive strength due to different molarity of NaOH solution

The compressive strength result for 1st, 2nd, 3rd and 7th days of testing are shown in Figure 2. For 1st day of testing, 10 M NaOH solutions showed the highest compressive strength of 63.41 Mpa. However, for the 2nd, 3rd and 7th days of testing, samples with 12 M NaOH solution produced highest compressive strength of 60.15, 68.1 and 94.59 MPa respectively.

Study carried by Alonso and Palomo (2001) stated that when activator concentration increased above 10 M, a lower rate of polymer formation was produced resulting in the decrease of strength. This might be due to differences in the type of source materials which their study use metakaolin but in this work, the source material use is fly ash.

The perfect curve has been shown for 7th day testing. After 12 M of NaOH solution, the compressive strength has decreased. Palomo et al. (1999b) also found that alkaline activator with 12 M NaOH solution produced better result compared to 18 M NaOH. However, Hardjito et al. (2008) stated that the increase of NaOH molarity will increase the compressive strength of samples. The strength increased with the increased of NaOH concentration mainly through the leaching out of silica and alumina (Chindaprasirt et al., 2009).

Microstructure of fly ash

The microstructure of fly ash was analyzed using scanning electron microscope (SEM). From Figure 3, the microstructure of fly ash appeared to be a glassy, hollow, spherical particle which is cenospheres (thin walled hollow spheres). The microstructure appearance of the original fly ash is well agreed as reported by Davidovits (1994).

Although fly ash particles are essentially the same, variations do occur in shape (rounded to angular) with some crystals of mullite and iron (Fernandez and Palomo, 2003). Furthermore, surface texture appears to be smooth and dense to highly porous.

SEM analysis for geopolymer paste

The microstructure of geopolymer with different molarity of NaOH are shown in Figures 4(a and b). The formation of pores heterogonous matrix which does not exist in the original fly ash was observed in Figures 4 a-f. The formation of matrix was due to geopolymerization reaction occurred after mixing the original fly ash with the alkaline activator liquid and for further preparation processes. The gel formation was in colloidal form with the sizes <1 µm up to 20 µm. As the molarity of NaOH increased from 6 to 16 M, the microstructure of the resulted geopolymer appeared to contain relatively less unreacted fly ash microspheres proportions. Based on Figure 4 (a) and (b), a large proportion of fly ash still did not completely dissolve. Figure 4 (d) shows most less unreacted fly ash with alkaline activator which give higher compressive strength of 94 MPa for 7th days test. Pores as indicated on the figures by arrow sign, and cracks which could limit the strength of geopolymer also found in the matrix [Figure 4 (a), (b), (c), (e)] that leads to lower compressive strength. The results obtained suggested that the composition of aluminosilicate gel formed by the reaction between fly ash and alkaline activator is variable

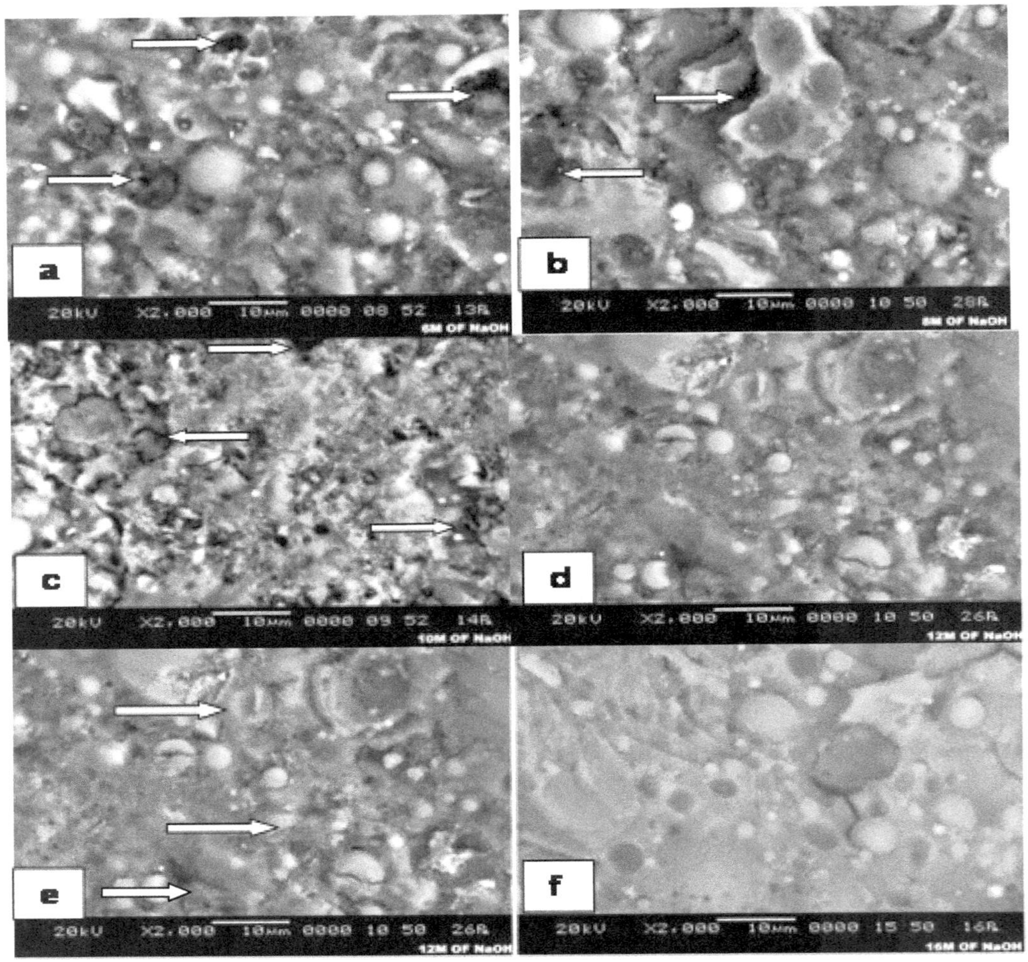

Figure 4. (a) SEM of geopolymer with 6 M of NaOH solution (b) SEM of geopolymer with 8 M of NaOH solution (c) SEM of geopolymer with 10 M of NaOH solution (d) SEM of geopolymer with 12 M of NaOH solution. (e) SEM of geopolymer with 14 M of NaOH solution (f) SEM of geopolymer with 16 M of NaOH solution.

and depends on the reactivity, type and concentration of activators (Duchesne et al., 2010)

Conclusion

Fly ash from Sultan Abdul Aziz power station combined with sodium hydroxide and sodium silicate as geopolymer can be enhanced to the use of OPC which often used in construction due to high compressive strength produced. Fly ash-based geopolymer with 12 M NaOH concentration shows excellent result with high compressive strength (94.59 MPa) for 7th days of testing. The microstructure of the optimum strength geopolymer appears to be homogenous and contains minimum proportion of unreacted fly ash microspheres, with continuous matrices of aluminosilicate and microcracks.

ACKNOWLEDGMENT

Grant from King Abdul Aziz City Science and Technology (KACST) for this research project is gratefully acknowledged.

REFERENCES

Alonso S, Palomo A (2001). Alkaline Activation of Metakaolin and Calcium Hydroxide Mixtures: Influence of Temperature,Activator Concentration and Solid Ratio. Mater. Lett., 47(1-2): 55-62.

Buchwald A, Schulz M (2005). Alkali-activated binders by use of industrial by-products. Cem. Concrete Res., 35: 968-973.

Chindaprasirt P, Chareerat T, Sirivivatnanon V (2007). Workability and strength of coarse high calcium fly ash geopolymer. Cem. Concrete Composit., 29: 224-229.

Chindaprasirt P, Jaturapitakkul C, Chalee W, Rattanasak U (2009). Comparative Study on the Characteristic of Fly Ash and Bottom Ash Geopolymer. Waste Manage., pp. 539-543.

Criado M, Palomo A, Fernandez-Jimenez A (2005). Alkali activation of fly ashes. Part 1: Effect of curing conditions on the carbonation of the reaction products. Fuel, 84: 2048-2054.

Davidovits J (1994). High-alkali cements for 21st century concretes. In Concrete Technology, Past, Present and Future. Metha, P.K, American Concrete Institute, Farmington Hills, MI. Ed., pp. 383–397.

Duchesne J, Duong L, Bostrom T, Frost R (2010). Microstructure Study of Early *in situ* Reaction of Fly Ash Geopolymer Observed by Environmental Scanning Electron Microscopy (ESEM). Waste Biomass Valor. 1: 367-377.

Fernandez-Jimenez A, Palomo A (2003). Characterisation of fly ashes. Potential reactivity as alkaline cements. Fuel, 82: 2259-2265.

Fernandez-Jimenez A, Palomo A (2005). Composition and Microstructure of Alkali Activated Fly Ash Binder: Effect of the Activator. Cem. Concrete Res., 35: 1984-1992.

Fernandez-Jimenez A, Torre AG, Palomo A, Lopez-Olmo G, Alonso MM, Aranda MAG (2006). Quantitative determination of phases in the alkaline activation of fly ash. Part II: Degree of reaction. Fuel, 85: 1960-1969.

Fernando P-T, Joao C-G, Said J (2008). Alkali-activated binders: A review. Part 2. About materials and binders manufacture. J. Constr. Build. Mater., 22: 1315-1322.

Hardjito D, Cheak CC, Lee ICH (2008). Strength and Setting Time of Low Calcium Fly Ash-based Geopolymer Mortar. Mod. Appl. Sci., 2(4): 3-11.

Hardjito D, Wallah SE, Sumajouw DMJ, Rangan BV (2004). On the Development of Fly Ash-Based Geopolymer Concrete. ACI Mat. J., 101(M52): 467-472.

Hos J, McCormick P, Byrne L (2002). Investigation Of The Synthetic Aluminosilicate Inorganic Polymer. J. Mater. Sci., 37: 2311-2316.

Hu M, Zhu X, Long F (2009). Alkali-activated fly ash-based geopolymers with zeolite or bentonite as additives. J. Cem. Concrete Composit., 31: 762-768.

Palomo A, Grutzek MW, Blanco MT (1999a). Alkali-activated fly ashes. A cement for the future. Cem. Concr. Res., 29(8): 1323-1329.

Palomo A, Blanco M, Granizo M, Puertas F, Vazquez T, Grutzeck M (1999b). Chemical Stability of Cementitious Materials Based on Metakaolin. Cem. Concrete Res., 29(7): 997-1004.

Puertas F, MartoAnez-RamoArez S, Alonso S, Va Azquez T (2000). Alkali-activated fly ash/slag cement, Strength behaviour and hydration products. Cem. Concrete Res., 30: 1625-1632.

Rattanasak U, Chindaprasirt P (2009). Influence of NaOH solution on the synthesis of fly ash geopolymer. Min. Eng., 22: 1073-1078.

Steenie EW (2009). Drying Shrinkage of Heat-Cured Fly Ash-Based Geopolymer Concrete. Mod. Appl. Sci., 3(12): 12-21.

Vijaya R (2008). Low-Calcium Fly-Ash-Based Geopolymer Concrete. Faculty of Engineering, Curtin University of Technology, Perth, Australia. Chapter 26, pp. 1-19.

Xu H, Deventer J (2000). The Geopolymerisation of Alumino-silicate Minerals. Int. J. Min. Process., 59(3): 247-266.

The influence of transient strain rate deformation on the microstructure of AA2024 aluminum alloy in the low temperature range

G. R. Ebrahimi[1] and H. Arabshahi[2]*

[1]Metallurgy and Materials Engineering Department, Sabzevar Tarbiat Moallem University, Sabzevar, Iran.
[2]Physics Department, Ferdowsi University of Mashhad, Mashhad, Iran.

Constant and varying strain rate deformation conditions have been applied to AA2024 aluminum alloy using hot compression testing in the temperature range of 200 - 300°C. The initial strain rate of 0.1 s^{-1} and the second pass strain rate of 0.0001, 0.001 and 0.01 s^{-1} have been applied and tested. This research shows that in low temperature range with decreasing strain rate in the second pass, the straining hardening rate is increased. Increasing the strain rate and deformation leads to restoration processes and dynamic coarsening. Decreasing strain rate during the second pass straining increases the mobile dislocation density.

Key words: Strain rate, hardening, deformation, dislocation density.

INTRODUCTION

Approximately 15% of the world's primary aluminium production is used each year in the manufacture of cans for the beverage industry. With increasing demands from can manufacturers to lower production costs there is a growing need for an ever more detailed understanding and control of the micro-structural changes which occur during each stage of can production, particularly during thermo-mechanical processing (Davenport et al., 2006). It is widely recognized that having soundly based physical models of micro-structural evolution during thermo-mechanical processing is highly desirable if truly predictive capabilities in relation to the effect of process variables on subsequent processing and product properties are to be achieved. The benefits of having such predictive capabilities in the aluminum industry are clear (Castro-Fernandez et al., 2004).

During industrial hot working operations of forging, rolling or extrusion the strain rate varies during the deformation process and heat transfer leads to the development of temperature gradients and changes in the mean temperature of work piece. In the case of rolling and forging, sequential passes may also take place at different temperatures and strain rates. By contrast, laboratory tests are usually carried out under constant deformation conditions to obtain relationships between flow stress and the strain (Furu et al., 2006). From such tests, constitutive relationships have been proposed and have been used to derive working forces for the hot working operations by considering the mean strain rate and temperature (Gil et al., 2002). Present models, however, have no predictive power outside their experimentally established range of application. They can not be applied to the complex temperature and strain rate transients encountered during thermo-mechanical processing. In order to progress further and improve knowledge of the link between deformation structures and re-crystallization kinetics a better understanding of the evolution of microstructure during transient deformation conditions is required (Shahani, 2005).

The key variables that describe a deformed structure are the subgrain size, the misorientation between neighbouring subgrains and the internal dislocation density of the subgrains. These influence flow stress and stored energy. The microstructural evolution during cold deformation of metals and alloys has been the subject of many studies over the last three decades (Ricks et al., 1994). Fewer studies have investigated the micro-structural evolution in aluminum alloys at high

*Corresponding author. E-mail: arabshahi@um.ac.ir.

Table 1. Chemical composition of the AA2024 alloy.

Element	Cu	Mg	Mn	Fe	Si	Ti	Zn	Al
Wt%	3.9	1.5	0.2	0.5	0.6	0.1	0.9	Base

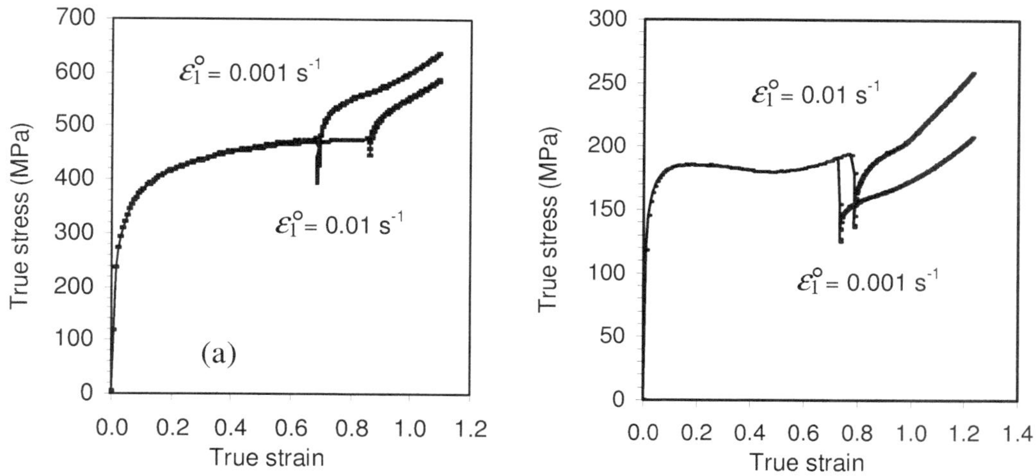

Figure 1. True stress-true strain curves for super saturated specimens deformed at, (a) 200ºC and (b) 300ºC to a strain of 1, at strain rate of 0.1 s^{-1}.

deformation temperatures (Nes et al., 2001).

During thermo-mechanical processing, the re-crystalization behaviour of a material is influenced by the chemical composition, strain rate and temperature of deformation. The latter parameters can be controlled during processing but rarely, if ever, are constant deformation conditions produced. Current models rely on data which have been obtained from laboratory experiments carried out under nominally constant deformation conditions. The tests involves simple deformation paths that reproduce the main characteristics of hot rolling and can be performed at both constant and varying strain rates.

The alloy that is chosen for these investigations is a AA2024 aluminum alloy. The present study is aimed to examine the grain refinement in an extrude AA2024 aluminum alloy using hot compression testing in the temperature range of 200 - 300°C.

MATERIAL AND EXPERIMENTAL PROCEDURE

The material used in the present work is a AA2024 aluminum alloy. The composition is given in Table 1. Cylindrical compression specimens which is 8 mm in diameter and 12 mm length were machined parallel to the extrude axis. The samples were homogenized at 300ºC for 12 h (T_1). The initial microstructure is composed of dendrite lamellas lying parallel to the billet axis. The samples were deformed in compression at constant crosshead speed using a modified servo-hydraulic system with an INSTRON-8500 controller equipped with a heating chamber. Specimens were strained to ε = 0.2 with the constant strain rate of 0.1 s^{-1}, where steady-state flow was achieved and rapidly decreasing strain rate from 0.1 s^{-1} to 0.01, 0.001 and 0.0001 s^{-1}, respectively. The time between the two tests was less than 1 s. The temperature was measured by an internal thermocouple placed very close to the specimens. Specimens were heated in a furnace to the deformation temperature and held for 300 s prior to the deformation.

EXPERIMENTAL RESULTS AND DISCUSSION

The true stress-true strain curves of specimens subjected to a strain rate drop are shown in Figures 1 and 2 for specimens of super saturated and annealed, respectively. In the second test at 300ºC, the flow stress decreases with further straining and, after reaching its maximum value, it is reduced (Figure 1b). The work softening is due to solute depletion, particle coalescence and enhancement of dynamic recovery. It may be seen that in most cases there is a transient period during which the flow stress drops and eventually reaches a stable constant.

At low second strain rate (0.0001 s^{-1}), the strain value reaches to a steady-state of ~ 0.1. In this condition, dynamic softening occurs but at higher second strain rate (0.01 s^{-1}) before reaching to a new steady-state work, hardening is produced. It is clearly shown that three typical features exist during multistage hot deformation of aluminum alloys, namely flow stress which increases directly with increasing strain (AA2024 alloy at 300°C), keeps constant with increasing strain and exhibits a peak

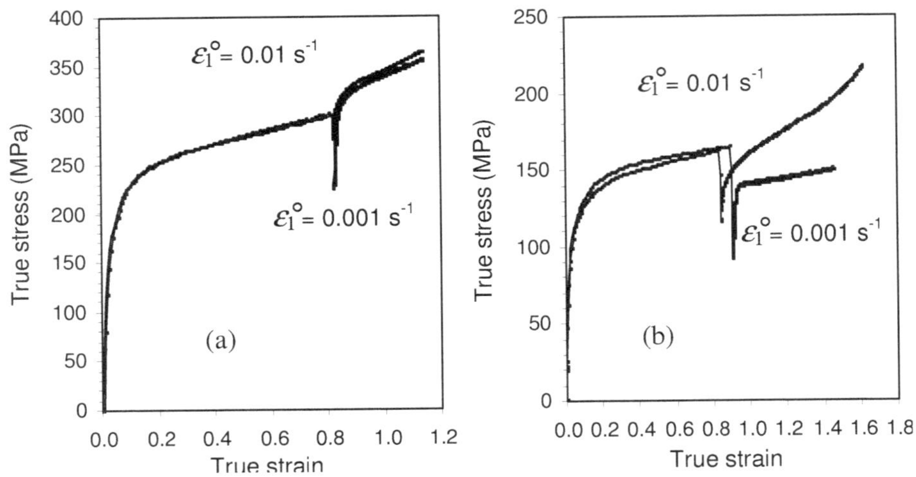

Figure 2. True stress-true strain curves for annealed specimen deformed at, (a) 200ºC and (b) 300ºC to a strain of 1, at strain rate of 0.1 s^{-1}.

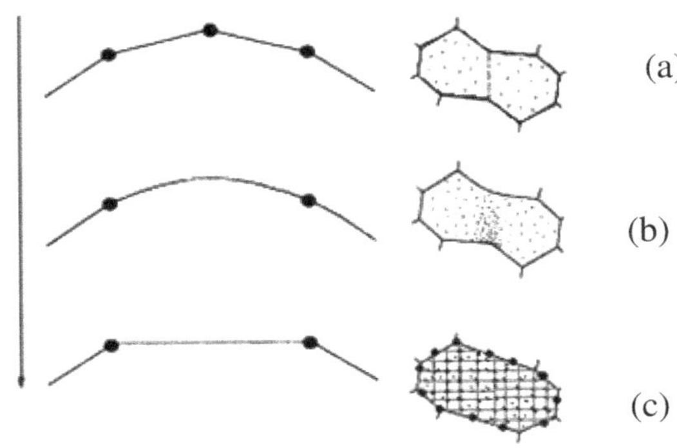

Figure 3. Schematic of the dissociation of a subgrain (microband) boundary to give increased subgrain size.

with a gradual decline with increasing. The stress–strain curves show considerable dynamic softening after reaching the peak at each stage.

Part of this softening, especially at higher deformation temperature, could be due to the dynamic recovery and re-crystallization during deformation. In order to quantify this dynamic softening, Sellars et al. (1994) have considered the relative difference between the value of the peak stress and the value of the stress taken at a strain of 0.25 beyond the peak. Dynamic recovery (DRV) and re-crystallization (DRX) are softening mechanisms that have distinctive effects on shaping the flow curve and retarding fissure formation to raise the hot ductility.

It is concluded that the flow softening of Al–5Mg coincides with the transformation of the dislocation wall structure into subgrains and is not related to dynamic recrystallization. It is found that grain boundary sliding takes place frequently during deformation and the flow softening in the regions of localized deformation results from the apparent grain refinement based on the folding of pancake-shaped grain structures and grain boundary sliding. In additional, the flow stress value at the second deformation seems to be a little smaller than the first deformation value due to the effect of the friction and inhomogeneous deformation at strain > 0.6, and a little dependence on the delay times for the same alloy at the same deformation temperature and strain rate during multistage hot deformation.

The consequences of a subgrain boundary dissociating are illustrated schematically in Figure 3. The initial structure is shown in Figure 3a, with dissociation of the central boundary occurring in Figure 3b and resulting in ejection into the adjacent subgrains of the geometrically necessary dislocations to retain the curvature. This gives an increase in subgrain size and a high internal dislocation density. These dislocations are of the same burgers vector, so they can not annihilate either mutually or to a significant extent with dislocations in the random

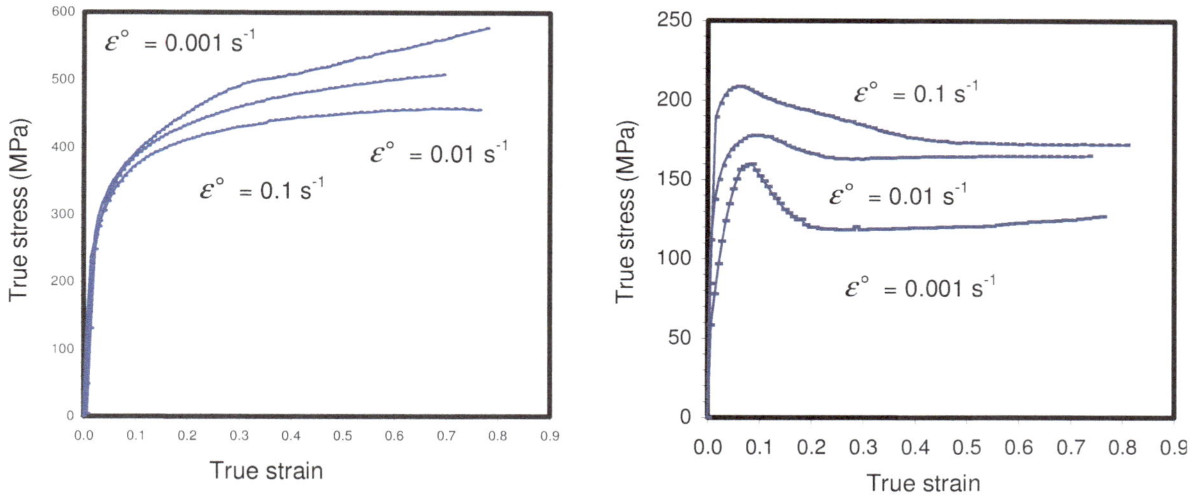

Figure 4. True stress-true strain curves for super saturated aluminum alloy AA2024 deformed at, (a) 200ºC and (b) 300ºC with a constant strain rate.

three-dimensional internal network. With increasing strain, these geometrically necessary dislocations can join the adjacent subgrain boundaries to increase their misorientation and so retain the overall local lattice curvature.

This schematic change in internal dislocation density is clearly consistent with the experimental observations after a decrease in strain rate. This important role of microbands is consistent with observations on the effects of reversing the direction of straining.

Figure 4 shows the behavior of hot deformation samples in 200 and 300°C at strain rate of 0.5 s^{-1}. It is clearly shown that in strain less than 0.8, the behavior of hot deformed samples are different. As it can be seen with decreasing strain rate, there is some factors in the structure which cause dynamical behavior changes. The flow stress is increased with further straining at 300ºC and after reaching the maximum value, it is reduced. The work softening is due to solute depletion, particle coalescence and enhancement of dynamic recovery. At low second strain rate dynamic softening occurs.

Conclusion

The interaction between the microstructural variables subgrain size, subgrain misorientation and dislocation density within the subgrains have been examined. For deformation at constant strain rate, the subgrain size decrease with strain until a steady-state size is reached. It is also shown that the misorientation between microbands and subgrains increases with strain to a steady state value. The dislocation density within the grains also increases with strain until a steady state value is reached. Under transient conditions of reducing strain rate to 1, the subgrain size and misorientation are smaller than in steady-state. During deformation at constant strain rate after the transient, the subgrains grow by a process of dissociation of subgrain boundaries which produces a larger number of dislocations in the subgrains until the values characteristic of steady-state are eventually achieved.

REFERENCES

Castro-Fernandez FR, Sellars CM (2004). Deformation behavior in metals at high temperature range. Phil. Mag. A, 60: 487-493.

Davenport SB, Sparks CN, Sellars CM (2006). Proc. Conf. Thermo-Mechanical Processing in Theory, Modeling and Practice (TMP), Stockholm, Sweden, pp. 249-255.

Furu T, Schercliff HR, Ashby MF (2006). The interaction between the microstructural variables subgrain size and metals. Mater. Sci. Forum., 43: 217-222.

Gil J, Van Houtte P, Kuhlmann-Wilsdorf D (2002). Nano particle simulation at high temperature. Prog. Mater. Sci., 25: 69-77.

Nes E, Vatne HE, Daaland O (2001). Proc. 4th Int. Conf. on Aluminium Alloy (ICAA4), Atlanta, USA, pp. 18-25.

Ricks R, Bolngbroke RK, Orsund R, Raynaud GM (1994). Proc. 5th Riso Int. Symp. on Materials, pp. 99-105.

Sellars CM, Humphreys FJ, Nes E, Jenson J (1994). Proc. 5th Riso Int. Symp. On Materials, pp. 109-116.

Shahani R (2005). PSC reproducibility. Brite Euram report BREU-CT91-0399-0405.

15

Bulk strain estimation on gneisses in Central Nigeria: A preliminary assessment

Goki, N. G.[1], Amadi, A. N.[2]*, Amuneni, O.[1], Agbo, P.[1], Okoye, N. O[2] and Unuevho, C. I.[2]

[1]Department of Geology and Mining, Nasarawa State University, P. M. B. 1022, Keffi, Nigeria.
[2]Department of Geology, Federal University of Technology, P. M. B. 65, Minna, Nigeria.

The challenges of regional metamorphism and the need to investigate the petro-structural strain and tectonic evolution of the Nigerian basement using central Nigeria contact zone is useful in mineral exploration. Meso-structural data were plotted using GEOrient to obtain general trend as well as the axial folds of the Pre-Pan African deformation. Strain estimation was done on photomicrographs of the gneisses cut normal to the planes of foliation across the lithologies in order to establish the homogeneity and amount of strain. The centres of the quartz grains were digitized using the Fry method from which the axial ratios of the strain ellipsoids were used to estimate and compare the amount and orientation of strain across the different rock types in the study area. Petro-structural bulk strain estimation on the gneisses around Kaduna, central Nigeria show a generally moderate but inhomogeneous strain with axial ratios of the finite ellipsoids ranging between 1.6 and 3.8 with strain geometries that are consistent with the NW- SE metamorphic foliation. The amount of strain across the rock types correspond to grain size variation signifying probably that quartz responds differently to strain with decrease in grain size. In response to the generally NW-SE tectono-metamorphic ductile deformation that produced the steeply dipping gneisses of the area, there is heterogeneity in the bulk strain. The orientation is essentially concordant to the metamorphic deformation and probably unconnected with the latter but brittle Pan-African and predominantly NE-SW fracture systems.

Key words: Petro-structural, bulk strain, gneisses, central Nigeria.

INTRODUCTION

Macro and micro-structural strain measurement was carried out on rocks of the areas covering Rijana, Sabon Kasarami and Wuya situated 43 km south of Kaduna along Kaduna–Abuja Road (Figure 1). The study was within Latitude 7°00'00" N and 10°08'40 N and Longitude 7°13'40"E and 7°19'20"E. Strain measurement involves the use of selected relevant strain markers that are consistently distributed within the studied rock body and which can be easily used. The choice of which method to apply is also depended on a good exposure of the long and short axes of the strain markers as well as availability of the processing package (Ogezi, 1977; Ajibade and Wright, 1988; Annor and Olasehinde, 1996). The gneisses around Rijana do not show adequate exposure of macroscopic porphyroblasts that can be easily measured on the field. This posed a major challenge to field estimation of strain hence the Fry (1979) method was adopted. Deformation patterns on the basement rocks of Nigeria are so complex (Rahaman, 1988; Ogezi, 1977; Ajibade and Wright, 1988; Mc Curry, 1976) and hence require a detailed study to understand their strain history. Additionally, the homogeneity of deformation on the rocks which is dependent on the degree of response to the prevailing stress fields is important in elucidating the tectono-metamorphic conditions that affected the rocks. These, in addition to other geological and geochemical parameters can give clues to the geodynamic evolution of this Pan-African mobile terrain (Amadi and Olasehinde, 2010).

*Corresponding author. E-mail: akoamadi@yahoo.com.

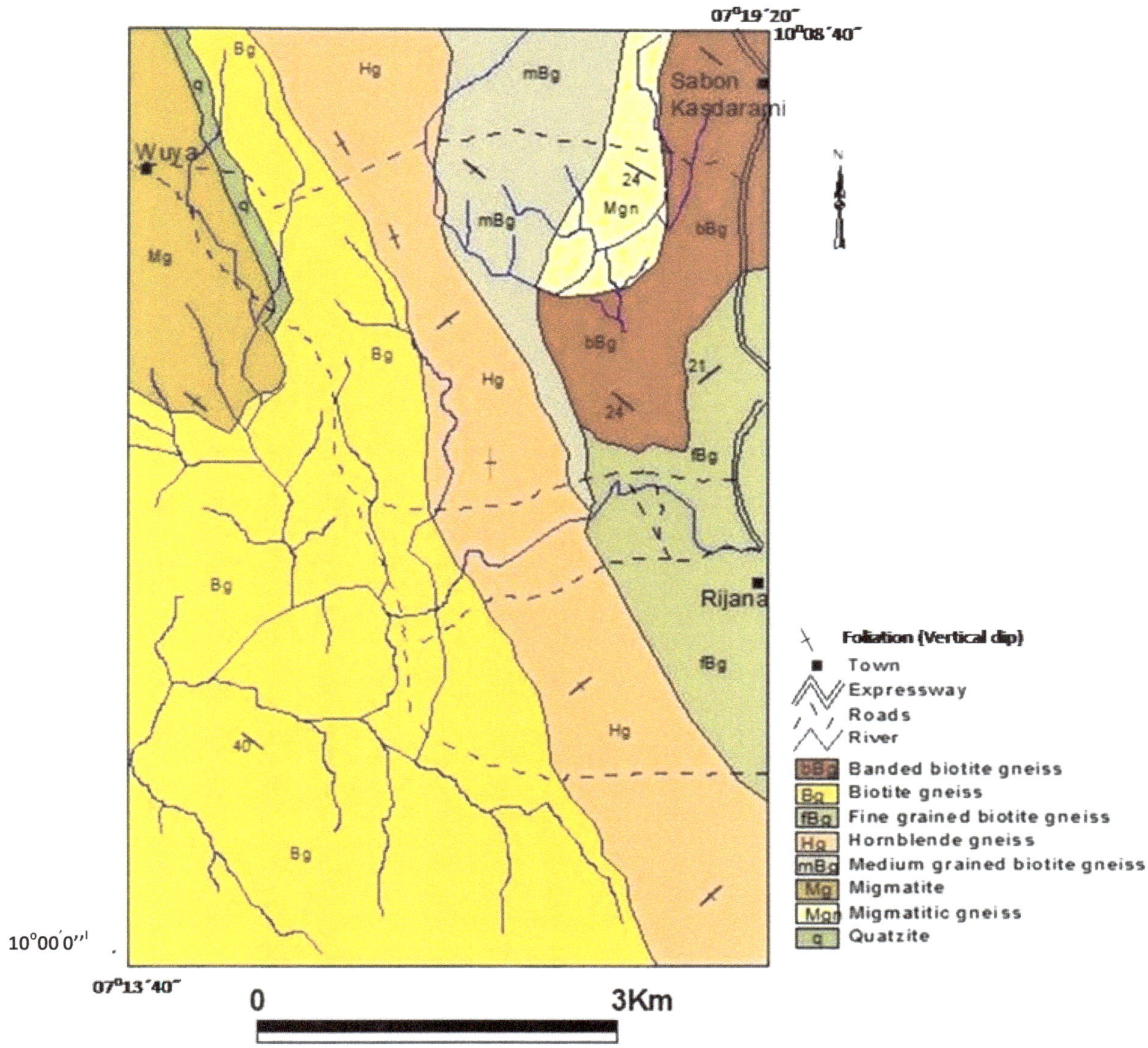

Figure 1. Geological map of the study area (Southern Kaduna, Nigeria).

METHODS OF STUDY

Geology

The area forms part of the basement complex of Nigeria (Figure 1) which consists of migmatites, gneisses, schist, metasedimentary rocks and granitoids (Rahaman, 1988; Dada, 1989; Oyawoye, 1972). The rocks of this area consist mainly of vertically dipping hornblende and biotite gneisses of variable grain sizes and migmatites (Figure 2). Unlike the biotite and migmatitic gneisses, the hornblende gneisses form a generally NW – SE dissecting ridges which are litho-stratigraphically conformal and parallel to quartzitic ridges. Compositionally, the gneisses range from hornblende to biotite banded to foliated gneisses. Petrographically, the gneisses are dominated by undulous and strained quartz, which is obliquely sub-parallel to micas. The biotite display alteration rims and elongate sphene occur in accessory quantity. Plagioclase and microcline coexist in most of the samples studied but in many samples, myrmikites are formed at the expense of alkali feldspar (Amadi et al., 2010).

Locally, the gneisses are foliated and banded with a sub-vertical to vertical dips and a generally complex but general NW-SE structural grain except for some few exposures of alkali feldspar dominated gneisses around Rijana along the River Sarkin Pawa where NE-SW foliation surfaces were observed. Evidence of shearing is displayed by cracks with similar trends in the porphyroblasts in the gneissose bands.

RESULTS AND DISCUSSION

Meso-structural data and interpretation

The structural relationships and their interpretations were investigated and analyzed in detail with the aim of

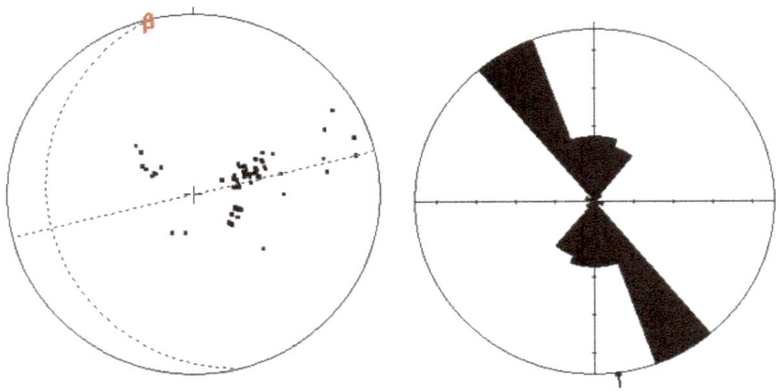

Figure 2. Lamberts equal area projection of foliation surfaces from the area.

unravelling the tectonic evolution of the basement. Measurements and tabulation of structural data as well as their plots are restricted to the study area only. The main planar structures encountered are foliations and fractures. The predominant foliations are mineralogical and textural foliations are vertical to oblique with variable trend patterns but comparatively constant over a large area and often parallel to the gneissose banding.

A total of 240 measurements of dips, strikes of foliation and fracture surfaces were taken during the fieldwork, with 80 measurements for each. The data were plotted on frequency-azimuth rose diagrams and Lambert equal area plots and show that most of the foliation planes have a dominant NE-SW and NW-SE trends with minor NNE- SSW and NNW-SSE and marginal N-S trends. The Lambert equal area plots revealed more than one axial plane for folds with one to the east of the poles while the other west of the poles in which more values plot, and the fold axes averagely plunging inclined. The fold axes trend NNW-SSE to NNE-SSW. The foliation surfaces have low (19 - 40°) to high (60 - 90°) dips predominantly to the NW, while the fracture planes are all near vertical.

Comparison of these data and the plotted diagrams show variation in trend. Generally, the strikes of foliation show a dominant NE-SW and NW-SE trends and the minor NNE-SSW and NNW-SSE trends which could be twin conjugates of the major probably Pan-African NE-SW and NW-SE trends whose effects may have also given rise to the marginal N-S trend. Careful observation and analysis of the plots of fracture surfaces show a great similarity. The similarity in the trends of the fracture surfaces show that they are resulting from the same stresses but with local variations.

Micro-structural strain estimation and interpretation

Strain estimates were carried out to determine bulk strain and to interpret correctly the degree and variability of metamorphism as well as the extent of deformational effects on the different rock types, resulting from the different tectono-metamorphic cycles. Several methods of estimating two dimensional strains from the deformed rocks have been proposed, according to Hari (2009) most of which require the measurement of long and short axes of deformed objects called strain markers. The Fry method, according to Fry (1979) is based on the assumption that initially uniform anti-clustered distribution of points will change after deformation into a non-uniform distribution. In this method, centre of the entire grains screen digitized on the selected micrographs of the gneisses using GeoFry package which creates a vacancy called Fry hallow of ellipsoidal shape giving directly the long and short axes (Figure 3). The orientation of the ellipse also gives the orientation of the strain ellipse. Consequently, the finite strain ellipsoid that is almost circular shows low degree of grain deformation, while the flattened type shows a greater degree of grain deformation.

The ellipsoidal shapes of the grains generated using the Fry method on the photomicrograph of the thin sections of rocks from the study area has revealed heterogenous and varying grain deformation as well as varying orientations corresponding also to the marginal N-S, NE-SW, NW-SE and NNE-SSW and NNW-SSE (Figure 4). The shapes range from near circular type (axial ratio 1.68) on the migmatite, biotite gneiss, migmatitic gneiss and the medium grained biotite gneiss on the western and central portion of the study area, which is indicative of low degree of deformation, hence little grain elongation in the directions of NE-SW, NW-SE, NNE-SSW and NNW-SSE, to the flattening type on the hornblende gneiss, banded biotite gneiss and fine grained biotite gneiss which show the highest grain elongation of axial ratio 3.8 signifying a flattening type (Genier and Epard, 2004) in the directions of the marginal N-S, NE-SW and NW-SE. The result is the strain and structural map shown in Figure (5).

Critical observation of the ellipsoidal shapes of the different rock types across the study area shows

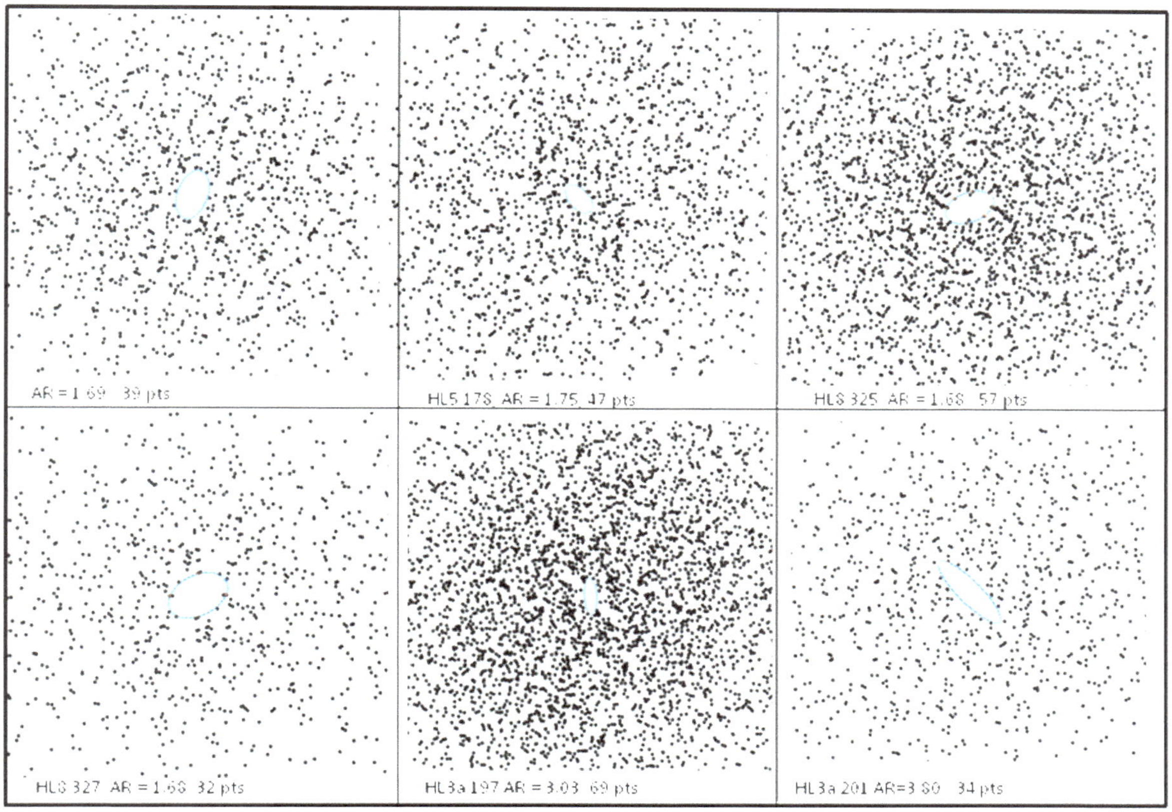

Figure 3. Fry haloes from selected samples generated from micrographs of the study area.

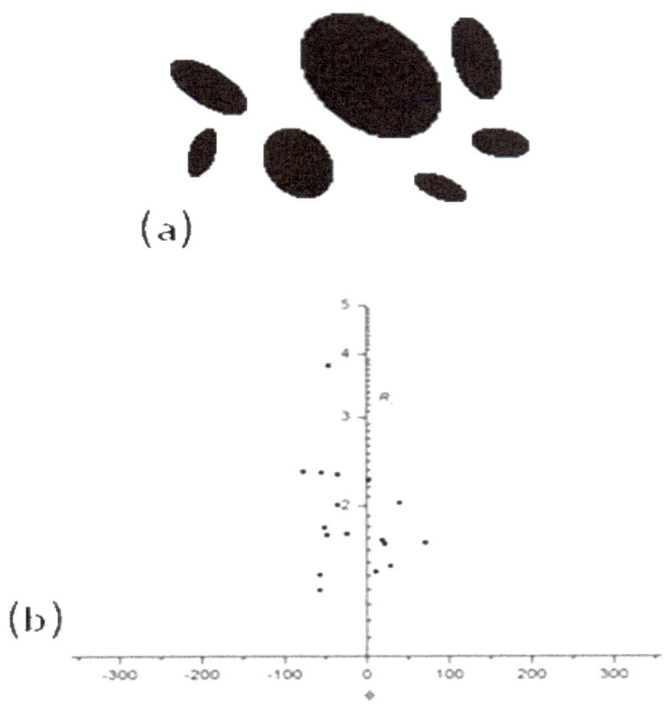

Figure 4. Micro-structural strain data from deformed quartz grains from the rocks. (a) Strain ellipsoids, (b) R_f and φ values for ellipses in (a) plotted on log R_f /linear φ scale.

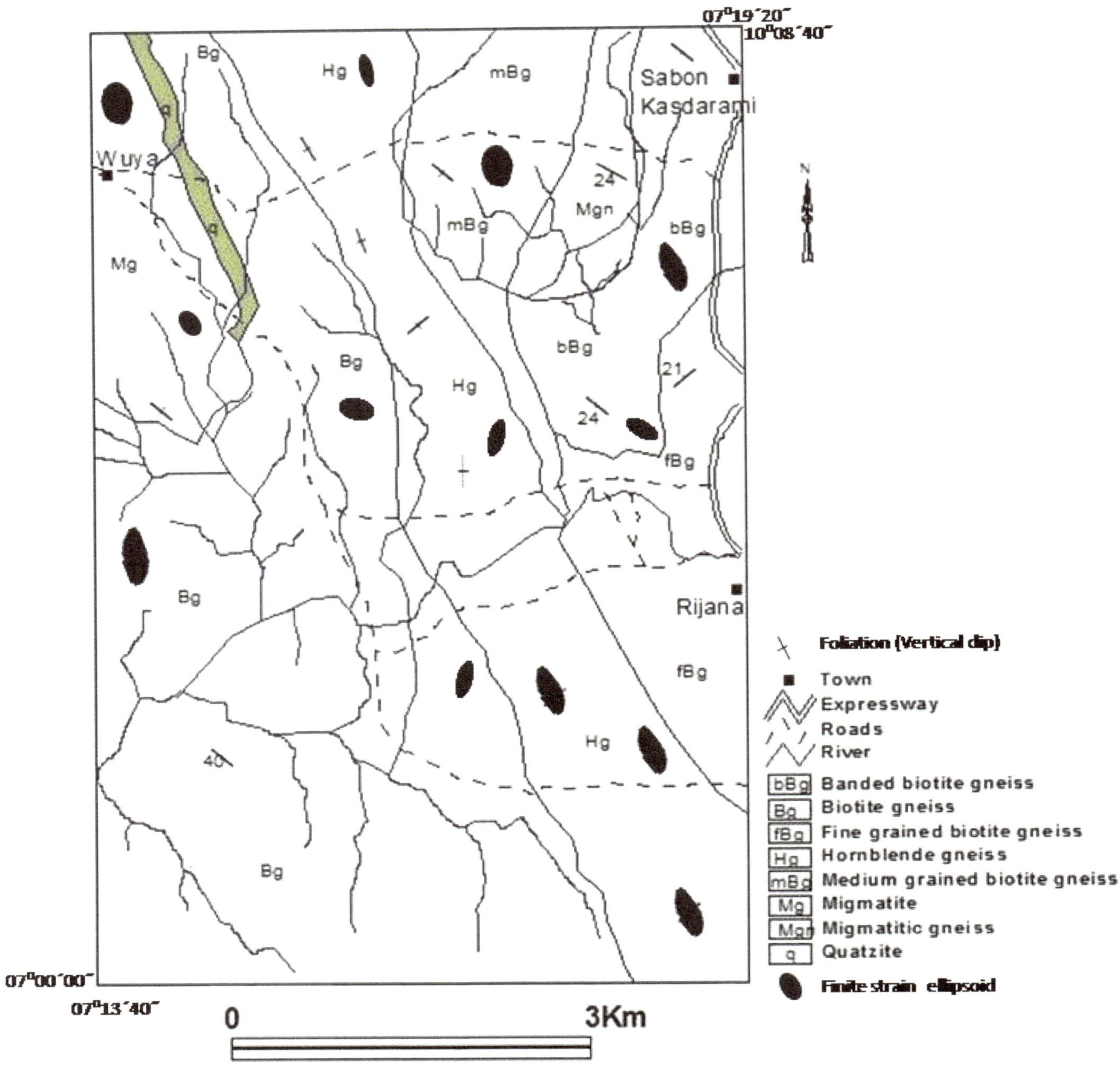

Figure 5. Strain ellipsoids and orientation distribution map of the study area.

progressive grain deformation from the migmatite and the biotite gneiss down to the banded biotite gneiss and fine-grained biotite gneiss corresponding to west-east strain variation. Using grain size distribution, the progressive increase in grain deformation is corresponding to the decrease in grain sizes across the study area. The heterogeneous and varying grain deformation could be attributed to variation in mineral assemblage, heterogeneity of rock composition and probably variation in temperature and pressure conditions of metamorphism across the study area or the entire region.

Conclusions

It concluded from this research that in response to the generally NW-SE tectono-metamorphic ductile deformation that produced the steeply dipping gneisses of the area, there is heterogeneity in the bulk strain. The orientation is essentially concordant to the metamorphic deformation and probably unconnected with the latter but brittle Pan-African and predominantly NE-SW fracture systems.

REFERENCES

Ajibade AC, Wright JB (1988). Structural Relationship in the Schist Belts of North Western Nigeria. In P.O Oluyide et al (Eds).Precambrian geology of Nigeria. A publication of Geological Survey, pp. 103-109.

Amadi AN, Olasehinde PI (2010). Application of remote sensing techniques in hydrogeological mapping of parts of Bosso Area, Minna, North-Central Nigeria. Int. J. Phys. Sci. 5(9): 1465 – 1474.

Amadi AN, Olasehinde PI, Okunlola IA, Okoye NO, Waziri S (2010). A multidisciplinary approach to subsurface characterization in Northwest of Minna, Niger State, Nigeria. Bayero. J. Phys. Math. Sci., 3(1): 74–83.

Annor AE, Olasehinde PI (1996). Vegetational Niche as a remote sensor for subsurface Aquifer: A Geological –Geophysical study of Jere area, Central Nigeria. Water Resour, J. NAH. 7(1 & 2): 26-30.

Dada SS (1989). Geochemistry and petrogenesis of the reworked Archaean gneiss complex of North Central Nigeria: major and trace element studies on Kaduna amphibolites and migmatitic gneisses. Global J. Pure Appl. Sci. (5&4): 535-543.

Fry N (1979). Density distribution techniques and strained length methods for determination of finite strains. J. Struct. Geol., 1(3): 221-244.

Genier F, Epard JL (2004). The Fry method applied to an augen orthogneiss; problems and results. Paper Presented at 2nd Swiss Symposium on Geosciences' Conference, Lausanne Swizerland.

Hari BS (2009). Strain estimation from the fabrics of deformed rocks. Curr. Sci., 97(6): 946-949.

Mc Curry P (1976). The Geology of the Precambrian to Lower Paleozoic Rocks of Northern Nigeria. A review, in Geology of Nigeria, Edited by C.A. Kogbe. Published by Elizabethan Co. Lagos. pp. 15-39.

Ogezi OE (1977). Geochemistry and geochronology of basement rocks from Northwestern Nigeria. Unpublished Ph.D Thesis, University of Leeds.

Oyawoye MO (1972). The Basement Complex of Nigeria. In: T. F. J. Dessauvagie and A. J. Whitement (eds). African Geology, Ibadan, pp.66-102.

Rahaman MA (1988). Recent advances in the study of the basement complex of Nigeria, Precambrian Geology of Nigeria, Geol. Surv. Niger. 34: 11-43.

16

An optimal parametric design to improve pool boiling heat transfer of sintered surfaces

Mao-Yu Wen[1]*, Ching-Yen Ho[2] and kuen-Jang Jang[1]

[1]Department of Mechanical Engineering, Cheng Shiu University, Kaohsiung 833, ROC, Taiwan.
[2]Department of Mechanical Engineering, Hwa Hsia Institute of Technology, Taipei 235, ROC, Taiwan.

This study investigated the effect of design parameters on pool boiling heat transfer on the sintered surfaces of a tube. The pool boiling experiments were conducted in saturated, deionized and degassed water. Data were taken at an atmospherical pressure and a fixed heat flux of 41,000 W/m^2.In the experimentation, the effects of the sintering pressure, sintering time, sintering temperature, heating rate, and particle size on the boiling heat-transfer coefficient of the sintered surface were investigated using the Taguchi method, and an $L_{15}(3^5)$ orthogonal array table was selected as an experimental plan for some parameters earlier mentioned. Based on the results of SN (signal/noise) ratio and ANOVA (analysis of variance), the optimal conditions of specifications of parameters will be provided. It was found that all the chosen sintering factors have significant effects on the pool boiling heat transfer coefficient. Flow visualization is also used to investigate the behavior of the pool boiling phenomena of the present sintered surfaces. Optimum pool boiling heat transfer coefficient of 5.29 kW/m^2 K was achieved with a sintering pressure of 2 atmospheres, a sintering time of 2 h, a sintering temperature of 900℃, a heating rate of 5℃/min and a particle size of 0.35 mm in a nitrogen container.

Key words: Pool boiling, heat transfer coefficient, sintered surface, Taguchi method.

INTRODUCTION

A great amount of applications in a variety of industries such as electronic chip cooling, electric power generation, chemical thermal processes, and refrigeration and heat pump systems, rely on nucleate boiling to remove high heat fluxes from heated surfaces. Several typical approaches have been considered to enhance pool boiling and critical heat flux, including: (a) oxidation or selective fouling of a heater surface to increase the wettability of the liquid; (b) vibration of heaters to promote the departure of bubbles from a heater surface; (c) coating or extending of heater surface to increase the heat transfer area; (d) heater rotation to promote bubble departure from and liquid deposition onto the heater surface; (e) fluid vibration to promote bubble departure and liquid supply; and (f) the application of electric fields to promote bubble departure from the surface by dielectrophoretic force to increase liquid renewal (Chang and Baek, 2003). In addition to these methods, heater surfaces with thin porous layers of particles have been used in the attempt to improve pool boiling heat transfer. A number of methods (Scurlock, 1995; Webb, 1981, 1994; Bergles and Chyu, 1982; Chyu and Bergles, 1982) have been used for producing nucleation sites artificially, either by mechanically treating the existing surface by scratching, scribing, knurling, etc., or by chemical treatments such as etching, coating, or by sintering the surface with a porous layer containing ready-made nucleation sites.

The most significant advances in enhanced heat transfer technology have been made with special surface geometries that promote high-performance nucleate boiling. From 1955 to 1965, fundamental advances were made in understanding the character of nucleation sites and the shape necessary to form stable vapor traps. In

*Corresponding author. E-mail: wmy@csu.edu.tw.

addition, Bankoff (1958) developed a dynamic model to predict the ability of a cavity to serve as a stable vapor trap. The analysis considered triangular and constant width circular cavities and grooves. Bankoff's suggestion that grooves may function as vapor traps provided motivation for later research on cavities with reentrant groove shapes. Moreover, Griffith and Wallis (1960) showed that cavity geometry is important in two ways: the mouth diameter determines the superheat needed to initiate boiling and its shape determines its stability once boiling has begun. Furthermore, Bergles and Webb (1980) also defined 12 enhancement techniques applicable to pool boiling and presented 258 references for work on enhanced pool boiling.

There is no detailed information about the optimum design parameters for pool boiling heat transfer on sintered surfaces available in previous studies. This is because vast numbers of experiments are required to obtain such information. The need for so many experiments increases the experimental cost and time period. This systematically analyzed the influence of various design parameters on the heat transfer characteristics of pool boiling on sintered surfaces. The Taguchi method (Zeng et al., 2010; Li et al., 2010; Taguchi et al., 1989; Tseng et al., 2007; Bilen et al., 2001), known to be a highly reasonable tool in parametric studies, is also applied in the present study. Tests were run by boiling saturated, deionized and degassed water at atmospheric pressure. The results provide main optimum factors which significantly influence the thermal hydraulic performance of pool boiling heat transfer on sintered surfaces in the design process. One of the advantages of the Taguchi method over conventional experimental methods is that in addition to reducing the experimental cost at the minimum level, it also minimizes the variability around the target when bringing the performance value to the target value. Its other advantage is that optimal working conditions, determined from the laboratory study, can also be reproduced in real applications. This study presents an investigation of the effect of design parameters on pool boiling heat transfer and the phenomena on the sintered surfaces of the tube. The effects of five design parameters for the sintered surfaces on the boiling heat-transfer characteristics were determined using the Taguchi method. Contribution ratios for each parameter on the heat transfer were also determined.

EXPERIMENTAL APPARATUS AND PROCEDURE

Instrumentation

Figure 1 is a schematic drawing of the test apparatus, identifying all major components. The tank has a glass view port (100X100 mm) which permits the viewing of the tubes and photographing. A horizontal test section with the sintered surface is shown in Figure 2. The sintered surface is on a copper (k=111 $\mathrm{W/m\ K}$) tube (135.8 X 18.9 X 16.6 mm) comprising the augmented surface. Heating was provided by conducting direct current through electric wire packed into the copper tube. A two-walled structure (MgO + quartz) was sandwiched between the copper tube and the electric wire as an electrical shield. Quartz was used as electrical insulation. MgO distributes the heat energy uniformly. The boiling surface of the test tube was sintered with multiple copper particles. Sintering was performed in various conditions, as shown in Table 1. Table 1 lists the sintering pressure, sintering time, sintering temperature, heating rate and particle size for the sintered surfaces investigated in this study. Tests were run by boiling deionized water at atmospheric pressure. The boiler vessel consisted of a thick-walled stainless steel container with a Plexiglas cover fitted with a rubber O-ring. The top cover had penetrations for four auxiliary cartridge heaters to allow the test liquid to be preboiled at saturation temperature throughout the run. The vapor was condensed and returned to the boiler by gravity from a Pyrex glass condenser using tap water for cooling. An additional thermocouple was placed above the boiling fluid to determine fluid vapor temperature, and two thermocouples were immersed in the fluid pool to monitor fluid bulk temperature. To measure the surface temperatures of the heated tube, the heated tube was instrumented with 5 E-type sheathed thermocouples outside the surface of the tube. Each thermocouple was brazed to the tube wall at an equal distance. To minimize end losses, the test tube was insulated with a mixture of polystyrene and Devcon five-minute epoxy. Calibrated Chromel-constantan thermocouples (E-type) were used to measure the surface of the sintered surface of the heated tube and pool temperatures. Power supplied to the test sections and the auxiliary heaters was controlled via separate powerstates. Test section power was measured by a wattmeter, providing heat flux accuracy of $\pm 80\ \mathrm{W/m^2}$ at $80000\ \mathrm{W/m^2}$. The data were logged by a recorder (180 mm hybrid recorder, AH 3000, chino). The wall thermocouple readings were averaged. In addition, experimental procedure is as follows. At the start of each run, the tube heater was used to vigorously boil the pool for one hour in order to degas the test liquid. The experiment system was started up with no power input. Then it was necessary to heat the pool at low power until it reached the saturation temperature. The power applied to the test section was then gradually increased by controlling the heater voltage. The voltage was increased in two to four volt increments until the initiation of boiling occurred, at which point ten volt increments were made up to the maximum attainable heat flux. The power was then decreased in the same fashion. At the current power setting, a comparison was made between the two average temperatures for each thermocouple after a heat flux decrement. The procedure was repeated until the temperature difference between the thermocouples was less than 0.1 K. The system was assumed to be at steady condition. It was permitted to stabilize for five minutes and the following data were recorded: heater voltage, vapor temperature, liquid pool temperature, and the temperatures on the sintered surface of the test heater.

Experimental parameters and plan

We used five experimental parameters of the sintered surface related to the characteristics of boiling heat transfer. The levels of each factor in the current study are shown in Table 1. The design parameters were named A, B, C, D, and E, for the sake of convenience. Five sintering factors were selected: sintering pressure (A, 1-2 atm), sintering time (B, 2-2.5 hr), sintering temperature (C, 850 to 900℃), heating rate (D, 5 to 20℃ /min), and particle size (E, 0.25 to 0.85 mm) under a nitrogen sintering atmosphere. Numerical analyses were performed to optimize the boiling heat transfer on the sintered surface of the copper tube using the $L_{15}(3^5)$ orthogonal array table for the design points, as shown in Table 2. In order to observe the effect of error sources on the boiling heat transfer, each experiment was repeated twice under

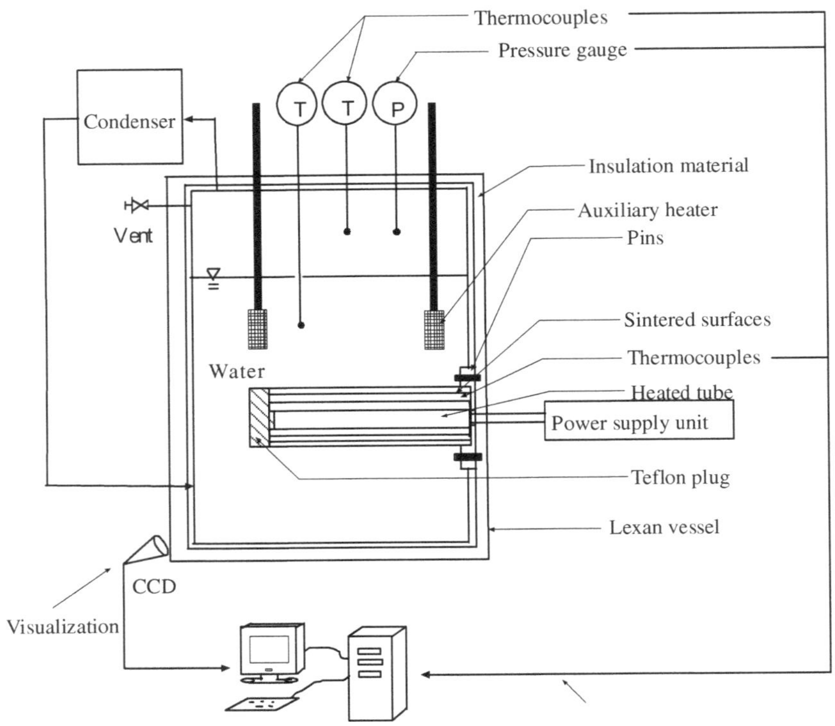

Figure 1. Schematic drawing of test apparatus.

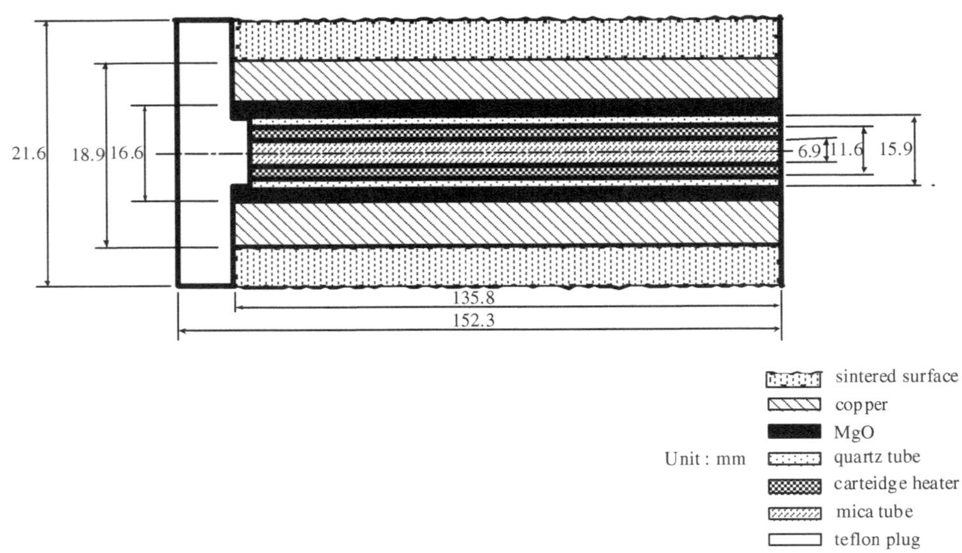

Figure 2. Schematic of test section.

the same conditions at different times. The SN (signal-to-noise) ratio can help designers to find out which levels of control factors are more efficient. The Taguchi method defines the SN ratio with the following equation:

$$SN = -10\log\left(\frac{1}{n}\sum_{i=1}^{n}\frac{1}{Y_i^2}\right) \qquad (1)$$

for the *n* observations, Y_i in each trial. In the Taguchi method, the experiment corresponding to optimum working conditions may not have been carried out during the whole period of experimentation. In such a case, the performance value corresponding to optimum working conditions can be predicted by utilizing the balanced characteristic of the orthogonal array (OA). For this aim, the additive model may be used (Phadke et al., 1983):

Table 1. Three-level cesign parameters in this study.

Symbol	Parameter	Unit	Level 1	Level 2	Level 3
A	sintering pressure	atm	1	1.5	2
B	sintering time	hr	2	2.3	2.5
C	sintering temperature	℃	850	880	900
D	heating rate	℃ /min	5	10	20
E	particle size	mm	0.85	0.35	0.25

Table 2. L_{15} (3^5) orthogonal array table for the Taguchi method.

Number of test	Design parameter					SN ratio
	A	B	C	D	E	
1	1	1	1	1	1	14.733
2	1	2	1	1	1	6.566
3	1	3	1	1	1	6.060
4	1	3	1	1	2	7.501
5	1	3	1	1	3	13.355
6	2	1	2	2	1	6.762
7	2	2	2	2	1	15.396
8	2	3	2	2	1	9.904
9	2	3	2	2	2	6.397
10	2	3	2	2	3	8.741
11	3	1	3	3	1	11.906
12	3	2	3	3	1	7.732
13	3	3	3	3	1	6.679
14	3	3	3	3	2	7.887
15	3	3	3	3	3	8.924

$$Y_i = \mu + x_i + e_i \qquad (2)$$

Because Equation 2 is a point estimation calculated by using experimental data to determine whether the results of the confirmation experiments are meaningful or not, the confidence interval can be obtained using the following formula:

$$Y_i \pm \sqrt{F_{\alpha;1,df_e} Mss_e n_e^{-1}}, \qquad (3)$$

The general formula for the effective number of replications is as follows:

$$n_e = \frac{N}{1+\sum(ED)}, \qquad (4)$$

If the experimental results are in percentages (%), before evaluating Equations 2 to 4, first, the omega transformation of the percentage values should be applied using the following equation and then the interested values are determined by reverse transformation using the same equation (Taguchi, 1987),

$$\Omega = -10 \log \left(\frac{1}{P} - 1\right) \qquad (5)$$

The interactive effects of the parameters were not taken into account in the theoretical analysis because some of confirmation experiments performed at the optimum conditions showed that they could be neglected. This is mainly due to the significant and main effects of these experimental conditions (the sintering pressure, sintering time, sintering temperature, heating rate, and particle size) on the boiling heat-transfer coefficient of the sintered surface.

Uncertainties and data reductions

The uncertainty in the experimental data was estimated using a propagation of error analysis. The uncertainty in the average wall superheat was estimated by taking into account the uncertainty in the average wall temperature.

The parameters required to make measurements in these pool boiling experiments include boiling wall temperature, the voltage drop *E* across the test heater and the current *I*. The power input to the heated tube can be obtained from the relation:

$$q = I \times E / A \qquad (6)$$

The front surface superheat ΔT can be obtained by a data acquisition system. Then, the boiling heat transfer coefficient can be estimated with the measured heat flux and averaged wall superheat as:

$$h = \frac{q}{\Delta T} \qquad (7)$$

Table 3. Ranges of experiments and measurement uncertainties.

Parameter	Range	Uncertainty
Heat flux supplied to test section, q (W/m^2)	41000	±0.5%
Wall superheat, (℃)	0.74-0.89	±7.5%
Heat transfer coefficient, h ($kW/m^2\ K$)	4.60-5.51	±10.6%
Sintering pressure (atm)	1-2	±5.6%
Sintering time (h)	2-2.5	±0.05%
Sintering temperature (℃)	850-900	±7.5%
Heating rate(℃ /min)	5-20	±8.9%
Particle size (mm)	0.25-0.85	±0.1%

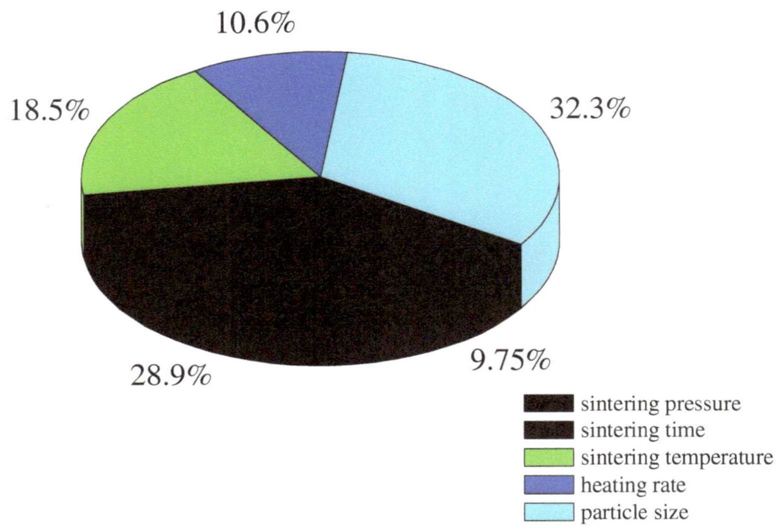

Figure 3. Contribution ratio on each factor.

RESULTS AND DISCUSSION

Repeated experiments showed that the data reproducibility was satisfactory. The ranges and measurement uncertainties for this study are listed in Table 3. Table 2 also shows the SN ratios calculated from the five tests. The effect of each sintering process parameter on the SN ratio at different levels can be separated out because the experimental design is orthogonal. Based on the results of SN ratio and ANOVA, the factorial effect and contribution ratio can be also calculated. The contribution ratio represents the effect of each factor on the boiling heat transfer coefficient. The percentage contribution is plotted in Figure 3. The effects of each factor on the boiling heat transfer coefficient are enumerated as 9.75 % for sintering pressure, 28.85% for sintering time, 18.47% for sintering temperature, 10.59% for heating rate, and 32.32% for particle size, as presented in Figure 3. Therefore, the particle size is seen to make the most major contribution to the overall performance.

Figure 4 shows the levels which provide optimal performance. The percentage contributions by each of the process parameters in the total sum of the squared deviations can be used to evaluate the importance of the process parameter change in relation to the quality characteristics. The numerical value of the maximum point of all levels on each factor marks the best value of that particular parameter, given in Table 4 for each parameter, and they resemble the optimum conditions in the range of the experimental conditions. Table 4 can be seen that an experiment corresponding to the optimum conditions is $A_3B_1C_3D_1E_2$. In addition, it should be noted that the heat transfer coefficient, 5.51 $kW/m^2\ K$, given in Table 4, is the result predicted by using Equation 2. The 98% significance level confidence interval of prediction is also given as 5.14-5.83 $kW/m^2\ K$ in Table 5, as predicted

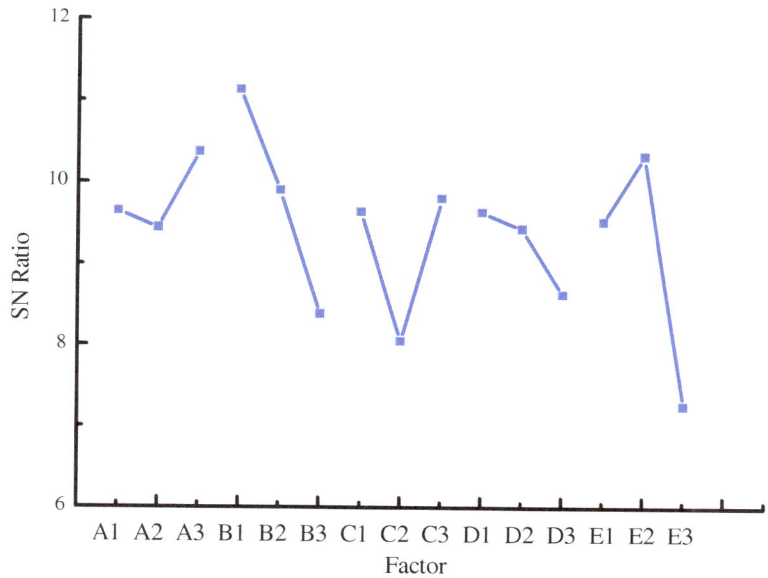

Figure 4. SN ratio on each factor.

Table 4. Optimum working conditions, and predicted and obtained values.

Parameter	Optimum conditions Optimum level	Optimum values
A	3	10.37
B	1	11.13
C	3	9.81
D	1	9.64
E	2	10.34

Predicted boiling heat transfer coefficient: 5.51 kW/m^2K; Predicted boiling heat transfer coefficient interval: 5.14-5.83 kW/m^2K; Obtained boiling heat transfer coefficient: 5.29 kW/m^2K.

by Equation 3. In order to test the predicted result, confirmation experiments were conducted twice at the optimum working conditions. Based on the fact that the heat transfer coefficients of 5.23 and 5.35 $kW/m^2\,K$, 5.29 $kW/m^2\,K$ are the averages obtained from the confirmation experiments at the same experimental conditions, as shown in Table 4, are well within the calculated confidence interval, it can be said that the experimental results are within ±4% in error. These results show that the interactive effects of the parameters are indeed negligible and also prove that the Taguchi method can be successfully applied to heat transfer experiments.

To expand the fundamental understanding of the pool boiling phenomena, flow visualization was made to observe the detailed fluid boiling characteristics of the present sintered surfaces. The flow visualization experiments were conducted using the strong-full light resource, a black background, a camcorder and a fast camera to record images of the pool boiling phenomena. The pictures are a small sample of that taken by the fast camera. The flow visualization showed how sintered surfaces affected the flow field and induced the boiling phenomena on the surface. Figure 5 shows typical photographs made by flow visualization of the present pool boiling on the sintered surfaces (Test 1, 3, 5, 7, 9, 11, 13, 15, and Test under the optimal condition) at a fixed heat flux of 41,000 W/m^2. The behavior of the unspecified other tests were similar. Starting from Figures 5 (b), (e), (g) and (h), one may find that bubbles rise from isolated nucleation sites. On the other hand, Figures 5 (a), (c), (d) and (f) indicate that because of the rough structure of the sintered surface, the active sites become very numerous due to plenty of additional tiny cavities in the sintered surfaces, especially for the test under the

Table 5. Optimum working conditions, and predicted and obtained values.

Parameters	Optimum condition	
	Optimum level	Optimum vales
A	3	10.37
B	1	11.13
C	3	9.81
D	1	9.64
E	2	10.34

Predicted boiling heat transfer coefficient: 5.51; Predicted boiling heat transfer coefficient interval: 5.14-5.83; Obtained boiling heat transfer coefficient: 5.29.

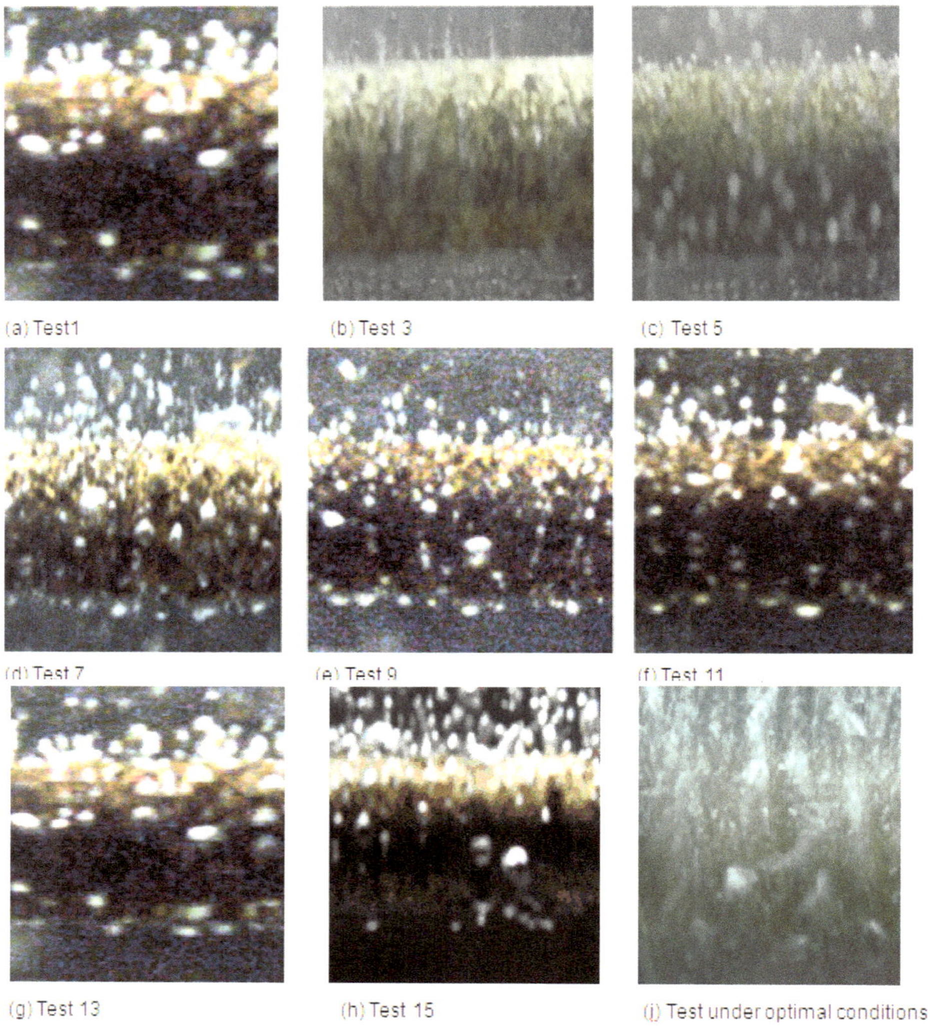

Figure 5. Photos of bubble formation in nucleate boiling under different test conditions, as shown in Table 1 and in the test under the optimal conditions.

optimal conditions (Figure 5 (i)). Looking at Figure 5 is apparent that the number of active nucleation sites generating bubbles is strongly influenced by the present surface condition. This indicates that the boiling phenomena of the sintered surface is of great importance and is responsible for improving the heat transfer performance. The results also contribute to our understanding of the best heat transfer effects for the sintered surface in the present study.

Conclusions

An optimized analysis for the experimental study of nucleate pool boiling heat transfer on sintered surfaces has been performed using the Taguchi approach. We proposed the optimum values of the design parameters for heat transfer from the sintered surface. In this study, the order of magnitude of the effects of each factor on the boiling heat transfer coefficient on the sintered surface is as follows: sintering pressure 9.75%, sintering time 28.85%, sintering temperature 18.47%, heating rate 10.59%, and particle size 32.32%. The results show that sintering time, sintering temperature, and particle size are the main factors that significantly influence the performance of the boiling heat-transfer surface discussed earlier. The contribution ratios of the sintering pressure and heating rate factors considered in this study also play important roles (about 10%) in the design process of a new boiling heat transfer surface. The optimum conditions of each factor are determined, and the reproducibility of these conditions has been verified by three analytical results. The flow visualization brought about a good understanding of the boiling phenomena for the present sintered surface. The findings of the present laboratory scale study will be useful for boiling heat transfer on sintered surfaces in heat–exchange processes used in a variety of industries.

ACKNOWLEDGEMENTS

The study was supported by the National Science Council of the Republic of China through grant No. NSC 98-2622-E-230-002-CC3. Special thanks to Mr. R.-F. Wu for his assistance in taking the experimental data.

Nomenclature

A	heat transfer surface area (m^2)
E	electrical voltage (V)
ED	effective number of degrees of freedom
e_i	random error in *i*th experiment
F_α	value of *F* table
h	heat transfer coefficient ($\mathrm{kW/m^2\ K}$)
I	electrical current (A)
MSS_e	mean square error
N	size of experiments
n_e	effective number of replications
P	percentage of product obtained experimentally
q	heat flux ($\mathrm{W/m^2}$)
SN	performance statistics
T	temperature (K)
ΔT	wall superheat, $\Delta T = T_w - T_{sat}$ (K)
Y_i	performance value of *i*th experiment
x_i	fixed effect of parameter level combination used in *i*th experiment
μ	overall mean of performance value
Ω	decibel value of percentage value subject to omega transformation (db)

Subscripts

df_e	degrees of freedom at level of α
sat	saturation
w	wall

REFERENCES

Bankoff SG (1958). Entrapment of gas in the spreading of a liquid over a rough surface. AIVhE J., 4: 24-26.

Bergles AE, Chyu MC (1982). Characteristics of nucleate pool boiling from porous metal coatings. J. Heat Transfer ASME Trans., 104: 279-285.

Bergles AE, Webb RL (1980). Bibliography on augmentation of convective heat and mass transfer. Previews Heat Mass Transfer, 6: 292-314.

Bilen K, Yapici S, Celik C (2001). A Taguchi approach for investigation of heat transfer from a surface equipped with rectangular blocks. Energy Conversion Manage., 42: 951-961.

Chang SH, Baek WP (2003). Understanding, predicting, and enhancing critical heat flux. The 10th Int. Topic Meeting on Nuclear Reactor Thermal Hydraulics (NURETH-10), Seoul, Korea, pp. 5-9.

Chyu MC, Bergles AE (1982). Characteristics of nucleate pool boiling from porous metallic coatings. Proc. 7th Int. Heat Transfer Conf., 6: 275.

Griffith P, Wallis JD (1960). The rate of surface conditions in nucleate boiling. Chem. Eng. Prog. Symp. Ser., 56: 49-63.

Li J, Wang S, Cai W, Zhang W (2010). Numerical study on air-side performance of an integrated fin and micro-channel heat exchanger. Appl. Thermal Eng., 30: 2738-2745.

Phadke MS, Kackar RN, Speeney DV, Grieco MJ (1983). Off-line quality control in integrated circuit fabrication using experimental design. Bell Syst. Technol. J., 62: 1273-1309.

Scurlock RG (1995). Enhanced boiling heat transfer surfaces. Cryogenics. 35: 233-237.

Taguchi G (1987). System of experimental design: engineering methods to optimize quality and minimize costs. Kraus International Publications, New York.

Taguchi G, Elsayed AE, Thomas CH (1989). Quality engineering in production system. McGraw-Hill, New York.

Tseng YS, Fu HH, Hung TC, Pei BS (2007). An optimal parametric design to improve chip cooling. Appl. Thermal Eng., 27: 1823-1831.

Webb RL (1981). The evolution of enhanced surface geometries for nucleate boiling. Heat Transfer Eng., 2: 46-69.

Webb RL (1994). Principles of enhanced heat transfer. John Wiley & Sons, Inc., New York, pp. 311-371.

Zeng M, Tang LH, Lin M, Wang QW (2010). Optimization of heat exchangers with vortex-generator fin by Taguchi method. Appl. Thermal Eng., 30: 1775-1783.

Does a physically reasonable solution of the Navier Stokes equations exist?

Asya S. Skal

P. O. Box 1836, Ariel 44837, Israel. E-mail: asyaskal@yahoo.com

According to the literature, the conservation of momentum equation needs to be coupled with the mass conservation equation. However, they cannot create a coupled system of equations of motion because they ignored third Newton law. The conservation of momentum equation is Newton's second law of motion, whereas conservation of mass belongs to kinematics that have no deal with forces at all. However, no one motion in nature can be described only by Newton's second law without Newton's third law (every action creates an equal and opposite reaction). Einstein (1905) referred to this as dynamic equilibrium. It is only half the task to construct the conservation of momentum equation of "action". The second equally important part is to find the equation of "reaction", which would satisfied flow problem. We will show that the system of the conservation of momentum and diffusion of momentum equations satisfies the dynamic equilibrium condition.

Key words: Static and dynamic equilibrium, momentum, Navier Stokes (NS) equations.

INTRODUCTION

The Navier Stokes (NS) equations were developed by Navier in 1831, and more rigorously by Stokes in 1845. According to the (literature) they can describe ocean currents, water flow in a pipe, flow around an airfoil (wing), motion of stars inside a galaxy, the analysis of the effects of pollution, and so on; we will show that they also deal with electrodynamics and plasma physics.

The NS equations are also of great interest in a purely mathematical sense; therefore the Clay Mathematics Institute calls them as one of the seven most important open problems in mathematics. They formulated the problem as follows (Constantin, 2001; Scheffer, 1976): The conservation of momentum coupled with supplemental equations (conservation of mass) and well formulated boundary conditions seem to model fluid motion accurately; even turbulent flows seem (on average) to agree with real world observations.

However, mathematicians cannot yet prove that in three dimensions, solutions always exist (existence), or that if they do exist they do not contain any infinities, singularities or discontinuities (smoothness). These are called the NS existence and smoothness problems. So the first of the entire question, which we present in title, needs to be answered: Does a physically reasonable solution of NS equations exist?

The main purpose of the present paper is to show that an answer for such formulated millennium problem does not exist. The NS equations consist of three foundational axioms of fluid dynamics, which are the conservation of mass, momentum and energy laws in physics. Such systems have five unknowns: scalar, pressure, temperature, vector, and velocity. However, it is convenient to view in fluid mechanics that conservation of momentum and mass equations can be uncoupled from the conservation of energy and solved for the four unknowns: a scalar pressure and a vector velocity without regards to the temperature variation. Such decoupling is possible if the viscosities are required to be constant.

The nonlinearity of NS equations are due to convective acceleration, which is an acceleration associated with the change in velocity over position. This was partly solved (equation with the one nonlinear term) in our previous papers (Skal, 1995, 2002).

However, there are two reasons why the momentum and mass conservation equations cannot be coupled. For incompressible fluid, the first reason is mathematical and the second more common reason is physical and valid for compressible and incompressible fluids.

THE MOMENTUM AND MASS CONSERVATION EQUATIONS CANNOT BE COUPLED TWICE FOR INCOMPRESSIBLE FLUID

For incompressible fluid the density for ρ = const. This reduces mass conservation equation for fluid with average velocity u:

$\nabla(\rho u) = -\partial\rho/\partial t$ (1)

To the equation of continuity:

$\nabla \cdot u = 0$ (2)

The conservation of momentum equation of motion for fluid is a relation equating the rate of change of momentum and the sum of all volume and surface forces acting on a fluid. During the derivation of momentum equation, we used the sum of all volume and surface forces acting on fluid. There are the external forces which act at distantly like gravity, the internal force which is due to fact of shearing stress called the viscous force. The latter take a form in common case as:

$$e_{ij} = \mu\left(\frac{\partial^2 u_{ij}}{\partial x_i \partial x_j} + \frac{\partial u_{ij}}{3\partial x_i}\right) \quad (3)$$

A simplification of the resulting flow equations is obtained when considering an incompressible flow of a Newtonian fluid. The assumption of incompressibility rules out diagonal term of stress tensor, which reduced to $\mu\ \nabla 2(u)$ by using mass-conservation equation. So here we have coupled the momentum and mass conservation equations for the first time and have obtained well known equation ρ ∂(u):

ρ ∂(u)/∂t + ∇ (ρ2u2)/(2ρ)+_ × (ρu)=v∇2(ρu)- ∇ [p + ρφ] (4a)

Ω = ∇ × u; ∇· Ω = 0 (4b)

where Ω [1/T] is the local vorticity, uρ is the momentum density, p [ML−1T−2] is pressure and φ[L2T−2] is the gravity potential, respectively.

Since this vector equation represent the system of three scalar equations and has four unknown variable (u, p), it is usually coupled a second time with the mass conservation equation, which is against the mathematical rules regardless of a statement that the conservation of mass is generally necessary.

THE PHYSICAL REASON WHY THE MOMENTUM AND MASS CONSERVATION EQUATIONS CANNOT BE COUPLED

The law of conservation of mass says that the mass of a closed system will remain constant, regardless of the processes acting inside the system. Conservation of linear momentum (also known as Newton's second law of motion) are physical laws that form the basis for classical mechanics. Whereas the conservation of momentums are dynamical equations, the conservation of mass is a kinematical equation. Kinematics deals with mere geometry of motion without reference to the applied forces, whereas dynamics deals with the applied forces that produce changes in the motion of the fluid. No one motion in nature can be described only by Newton's second law without Newton's third law (action had aroused reaction), that Einstein (Einstein, 1905), named dynamic equilibrium. How can conservation of momentum and mass equations satisfy a dynamic equilibrium condition if they belong to different branches of fluid mechanics? Therefore, one can say that the conservation of momentum equation alone (or even supplemented by the conservation of mass equation) is incomplete, because liquid would never flow in the pipe or river by forces presented in these equations.

So the millennium problem of a system of momentum and mass conservation equations has no mathematical solution because it has no physically reasonable solutions. The second purpose of this paper is to find the supplemented dynamical equation which satisfies Newton's third law and the dynamic equilibrium.

THE DIFFUSION OF MOMENTUM EQUATION SATISFIES NEWTON'S THIRD LAW

Momentum diffusion refers to the diffusion, or spread of momentum between particles (atoms or molecules) of matter. The rate of transport is governed by the viscosity of the fluid and the momentum gradient. Momentum transport in a fluid depends on the macroscopic momentum of molecules of the system.

If fluid is in motion, the molecules will posses a microscopic momentum in the moving x-direction of flow. If there is a variation in flow velocity, the transfer moving molecules possesses a greater momentum in the z-direction of flow and can transfer the excess momentum to their slower moving neighbors perpendicular to the flow in the y-direction and x-direction.

There are two friction mechanisms in fluid transport: the internal molecular shear friction and the drag friction. The shear stresses in fluid can be thought of as the transfer of momentum in the fluid.. The shear force is parallel to flow z-direction and belong to the equation of conservation of momentum (Newton's second law, "action"). However, this friction force inside a fluid in addition to all other driving forces would not make a fluid to flow and so is not responsible for the third Newton law.

Momentum diffusion can be seen as an important process responsible for Newton third law. In the case of the laminar flow of a liquid past a solid surface, momentum diffuses across the boundary layer which forms at the boundary where the solid meets the liquid.

The gradient in this case occurs between the liquid in contact with the surface, which does not move at all and has zero momentum (see no-slip boundary condition), and the liquid far away from the wall, which has momentum proportional to the speed at which it is flowing. This gradient of a velocity gives rise to the transfer of momentum toward the boundary, which causes drag friction force. For example, the liquid in the pipe is moving only because the liquid in a pipe exerts the force on the boundary of pipe due to that gradient. According to Newton's third law, the force exerts from liquid balanced by a friction force exerts by the boundary of a pipe.

One has two momentum diffusion mechanisms, one for shearing force inside the fluid and second for friction force on boundaries. The first force belongs to Newton's second law whereas the second force belongs to Newton's third.

In laminar flow, momentum is generated uniformly throughout the fluid and is transferred to the boundaries perpendicular to flow direction by molecular diffusion. A total force balance around the volume element states that the sum of drag forces must equal zero at steady state. This balance means that pressure drop is equal to the sum of the shearing force and the drag friction force on boundary.

Thus the system of equations of conservation of momentum and diffusion of momentum satisfied dynamical equilibrium. Exists or does not exist solution of such system can be proved by comparison of the experimental data with theory.

CONCLUSION

Just as in hydrostatics, we need to satisfy static equilibrium condition for forces, in the hydrodynamics, we need to construct our system so as to satisfy the dynamic equilibrium condition, which cannot be satisfied by coupling NS and mass conservation equations. Therefore the millennium problem formulated in such a way by Clay Mathematical Institute needs be reformulated.

ACKNOWLEDGEMENTS

The author would like to gratefully acknowledge the useful comments and conversation with Dr. Igor Grebnev, and also very grateful for help and assistance of Dr. Lechem Israel, Dr. Amber Avigail, and Mr. Marc DVer.

REFERENCES

Constantin P (2001). Some open problems and research directions in the mathematical study of fluid dynamics, in Mathematics Unlimited2001 and Beyond, Springer Verlag, Berlin, pp. 353-360.

Einstein A (1905). 'Investigations on the theory of the Brownian movement', Annalen der Physik, 17: 891-921.

Scheffer V (1976). Turbulence and Hausdorff dimension, in Turbulence and the NavierStokes Equations, Lecture Notes in Math. 565, Springer Verlag, Berlin, 941: 12.

Skal AS (2002). 'Equivalence between fundamental equations of elasticity and conductivity in a magnetic field". Math. Phys. Eng. Sci., 458: 2099-2117.

Low temperature synthesis of La_2O_3 and CrO_2 by Sol – Gel process

A. Bahari[1]*, A. Anasari[2] and Z. Rahmani[3]

[1]Department of Physics, University of Mazandaran, Babolsar, Iran.
[2]Department of Physics, Islamic Azad University, Sari Branch, Sari, Iran.
[3] Department of Physic, Islamic Azad University, Centeral Tehran Branch, Iran.

Some issues such as tunneling, leakage current and light atom penetration through the film are threatening the ultra thin SiO_2 be as a good dielectric for future industrial and electronic devices and in ceramic technologies. A series of experiments to synthesis La_2O_3 as well as CrO_2 has been done at low temperature with using sol-gel method. The nano structural properties of La_2O_3/CrO_2 equipped with 5 to 40 nm size are investigated. The obtained results show the potential of La_2O_3/CrO_2 for not only for lowering the interface – state density, but also as a good dielectric for the future of nano electronic. These structures have been studied by using XRD (x ray diffraction) technique and X- powder method.

Key words: Nano structures, La_2O_3, Gate dielectric, x-powder method and sol-gel method.

INTRODUCTION

The gate SiO_2 dielectric thickness in current complementary metal oxide semiconductor (CMOS) transistors is less than 1 nm. It is mostly interface and so thin film cannot obviously prevent leakage, tunneling currents and boron diffusion (Morgen et al., 2006; Bahari et al., 2005; Morgen et al., 2005; Bahari et al., 2006, 2006b; Morgen et al., 2007; Bahari et al., 2008). The higher K- gate dielectric materials can thus be introduced as the alternative gate dielectrics for future of CMIS (I: insulator) devices. La_2O_3 has largest band gap of the rare earth oxides at 4.3 eV, while also having the lowest lattice energy, with very high dielectric constant, $\varepsilon = 27$. It is widely used in industry as well as in the research laboratory. Moreover, the La_2O_3/CrO_2 gate stacks with high electrical permittivity show an amorphous structure of low temperature which can be used instead of ultrathin silicon dioxide gate dielectric in current CMOS devices (Duan et al., 2008; Alexazder et al., 2006; Dercz et al., 2006) due to its excellent chemical properties (Dercz et al., 2006; Zhu, 1998; Kurzweg et al., 1998; Betz et al., 1999) and independent La_2O_3 network within the CrO_2 network. However, La_2O_3 has p-type semi-conducting properties because its resistivity decreases with an increase in temperature, average room temperature resistivity is 10 kΩ·cm (Kale et al., 2005).

In this work, we have synthesized La_2O_3/CrO_2 which is suitable for CMIS. Different crystalline forms of lanthanum oxide have been prepared. At low temperatures, La_2O_3 has an A-M_2O_3 hexagonal crystal structure (Kale et al., 2005). Its crystalline phase can be used for the potential applications of the ceramic – mesoporous structures, as an ingredient for the manufacture of piezoelectric and thermoelectric materials, automobile exhaust-gas converters contain La_2O_3 (Cao et al., 2005). La_2O_3 is also used in X-ray imaging intensifying screens, phophors as well as dielectric and conductive ceramics. We have demonstrated to synthesis of hexagonal crystal phase (h-La_2O_3) at room temperature by using Sol-gel method. Some researchers (Wells, 1984; Wyckoff, 1963; Vinogradova et al., 2004; Manoilova, 2004) believe that the La^{3+} metal atoms are surrounded by a 7 coordinate group of O^{2-}atoms, while the oxygen ions are in an octahedral shape around the metal atom, we see there is one oxygen ion above one of the octahedral faces. In contrast to low temperature case, at high temperatures the La_2O_3 converts to a C-M_2O_3 cubic crystal structure. The La^{3+} ion is surrounded by a 6 coordinate group of O^{2-} ions. However, La_2O_3 in ferroelectric materials content has still needed to be studied due to its structural collapse during formation of the mesoporous phase and its phase transformation from hexagonal crystal phase

*Corresponding author. E-mail: a.bahari@umz.ac.ir.

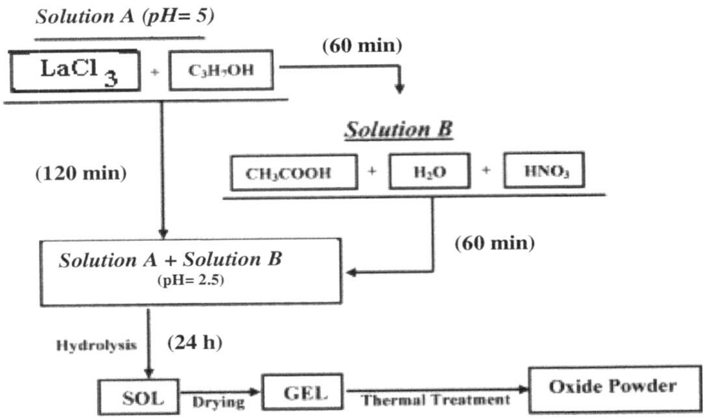

Figure 1. Preparation steps of La_2O_3 powders. Similar procedure is for CrO_2.

to cubic crystal phase (Pajak et al., 2003; Duan et al.,2008; Alexazder et al., 2006; Dercz et al., 2006).

For this reason, we synthesize CrO_2 nanoparticles in parallel to La_2O_3 nanoparticles in this structure and the obtained results indicate the potential applications of the present combination.

Experimental procedure and details

Lanthanum dioxide was prepared by the hydrolysis and condensation of $LaCl_3$ used as precursor under acidic conditions. In this process, a 0.1 M solution of $LaCl_3$ is sprayed onto a preheated substrate. We got two ways to control hydrolysis rate which can be viewed as occurring in hydrolysis followed by dehydration:

$2\ LaCl_3 + 3\ H_2O \rightarrow La(OH)_3 + 3\ HCl$
$2\ La(OH)_3 + heat \rightarrow La_2O_3 + 3\ H_2O$

On the other hand, hexagonal La_2O_3 involves precipitation of nominal $La(OH)_3$ from aqueous solution using a combination of 2.5% NH_3 and the surfactant sodium dodecyl sulfate followed by heating and stirring for 24 h at 80°C (Vinogradova et al., 2004):

$2\ LaCl_3 + 3\ H_2O + 3\ NH_3 \rightarrow La(OH)_3 + 3\ NH_4Cl$
$LaCl_3 \cdot 3H_2O \rightarrow La_2O_3$

Other routes include:

$2\ La_2S_3 + 3\ CO_2 \rightarrow 2\ La_2O_3 + 3\ CS_2$
$2\ La_2(SO_4)_3 + heat \rightarrow 2\ La_2O_3 + 6\ SO_3$

Thus the kinetic is conditioned and it is becoming possible. The water amount necessary to the hydrolysis can be chemically controlled with the hydrolysis in order to decrease its reactivity. The precursor can then be hydrolyzed with a small amount of water produced *in situ* by an esterification reaction resulting of the two points: Depending on the pH of the compound, different crystal structures can be obtained. La_2O_3 is hygroscopic; under atmosphere, lanthanum oxide absorbs moisture over time and converts to lanthanum hydroxide (Manoilova, 2004). It is worth noting that there are some reactions as well:

$$C_2H_5OH + CH_3COOH \Leftrightarrow H_2O + CH_2COOC_2H_5$$
$$C_2H_7OH + CH_3COOH \Leftrightarrow H_2O + CH_3COOC_2H_7$$

We have chosen the Sol–gel process to obtain a gel and privileging at maximum of connection formation La -O- La. The overall hydrolysis and condensation reactions are illustrated below:

Hydrolysis

$$\mathrm{La(OR)_3OR} + H_2O \rightarrow \mathrm{La(OR)_3OH + ROH}$$

Condensation

$$\mathrm{La(OR)_3OH + La(OR)_3OR \rightarrow (OR)_3La\text{-}O\text{-}La(OR)_3 + ROH}$$

Or

$$\mathrm{La(OR)_3OH + La(OR)_3OH \rightarrow (OR)_3La\text{-}O\text{-}La(OR)_3} + \mathrm{H_2}O$$

Where $OR = \mathrm{OC}_3H_7$. The chemically controlled condensation of lanthanum alkoxide leads to stable colloidal solutions of monodispersed lanthania nanoparticles.

The La_2O_3 sol was prepared from a lanthanum alkoxide precursor (Betz et al., 1999; Kale et al., 2005; Cao et al., 2005; Wells, 1984; Wyckoff, 1963) lanthanum propoxide (Aldrich), acetic acid (sigma-aldrich) was used as a chelating agent to lanthanum propoxide and isopropanol (sigma-aldrich) as a solvent. To reduce the viscosity of the alkoxide and reactivity with moisture, the alkoxide solution was prepared by stirring lanthanum propoxide and isopropanol at room temperature in the molar ratio 1 La:15 isopropanol. Catalyst solution was with the molar composition of 1.0 H_2O:0.6 HNO_3:7.5 isopropanol was prepared using distilled water, nitric acid (sigma- A ldrich) and isopropanol. Acetic acid was added drop – wise to the stirred alkoxide solution until the molar ratio of acetic acid to La became 2.0, and the solution was kept stirring for 2 h to let complete the reaction between alkoxide and acetic acid. The catalyst solution was mixed with the aforementioned solution to produce a lanthania sol until the molar ratio of La:CH_3COOH:HNO_3:H_2O:C_3H_7OH became 1.0;2.0:1.2:2.0:30 and was then kept stirring for 2 h. The scheme of preparation is illustrated in Figure 1. Similarly, we can synthesize the nano scale CrO_2 and add it into La_2O_3 (Figures 2 and 3).

The properties of Lanthania forms of higher symmetry are very often preferable to cubic one. CrO_2 is usually used as the stabilization component of higher symmetry lanthania (Betz et al., 1999; Cao et al., 2005; Wells, 1984; Wyckoff, 1963; Vinogradova et al., 2004; Manoilova, 2004). Figure 2 shows the XRD patterns of La_2O_3 without CrO_2. As can be seen in Figure 2 (c) peaks at 13°, 24°, 29 to 31° and 44° attributed to the (111), (200), (220) and (311) diffraction planes of the hexagonal lanthania and some cubic lanthania peak at 50 to 80°C. Figure 3 displays the XRD pattern La_2O_3 with CrO_2 (the presence of CrO_2 (50 wt%) in La_2O_3). The new peaks in comparison to Figure 2, attributed to CrO_2 phases. Phase beside of cubic, hexagonal of La_2O_3 and cubic forms of CrO_2 phase are stabled. The size of nanoparticles is in the range from 20 to 60 nm as found with using x- powder method (Dercz et al., 2006; Zhu, 1998; Kurzweg et al., 1998; Betz et al., 1999; Kale et al., 2005; Cao et al., 2005) and shown in Figures 4 and 5. The amorphous structure is obtained at 300°C which can be used for gate dielectric of nano electronic devices. This amorphous structure has become to nano crystalline structure with heating the sample (Figure 2b, T = 500°C).

Conclusion

Lanthanum oxide (La_2O_3), also known as "lanthana" is usually supplied as an odourless white powder. It is

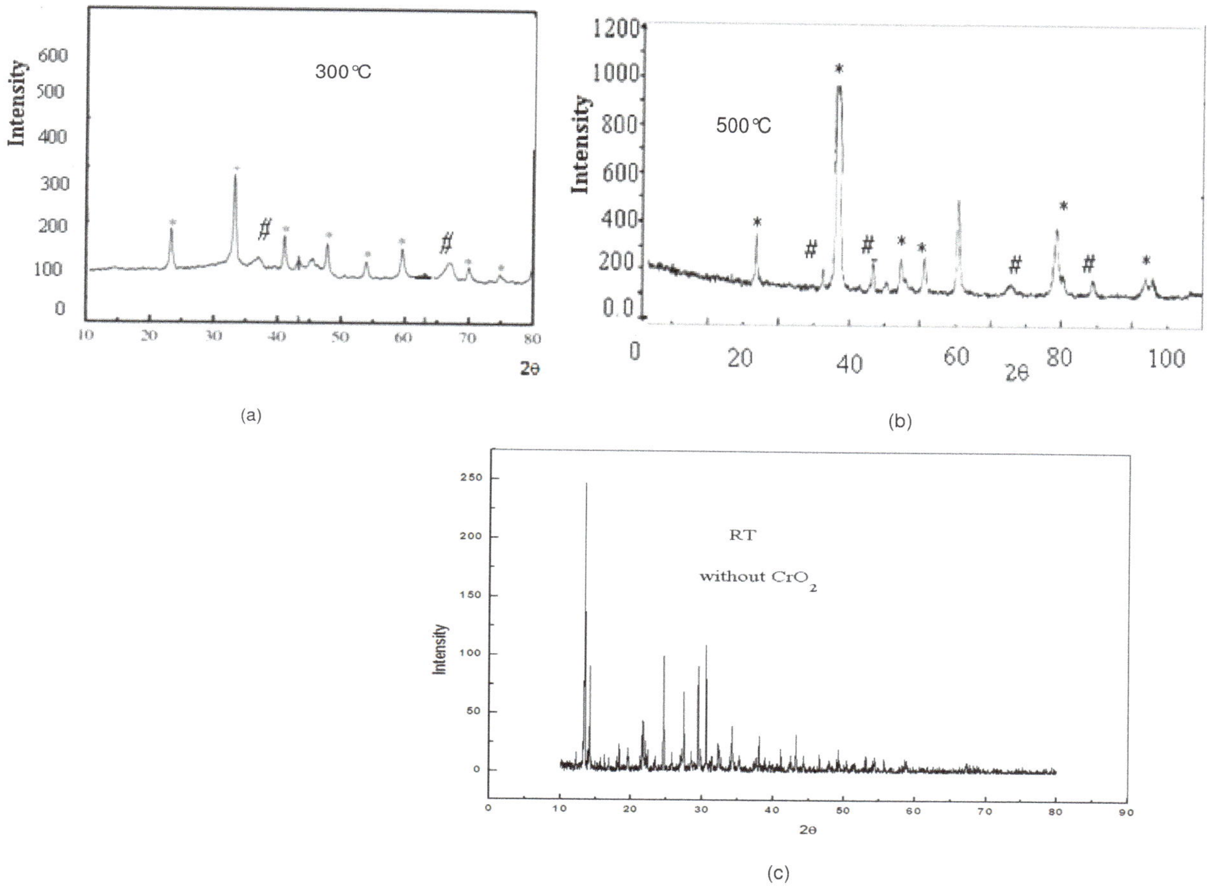

Figure 2. (a) The XRD pattern of nanoparticles at 300 °C without CrO_2. * Indicates the hexagonal crystal phase and # shows cubic crystal phase. (b) The XRD pattern of nanoparticles at 500 °C without CrO_2 and (c) The XRD pattern of nanoparticles at RT without CrO_2.

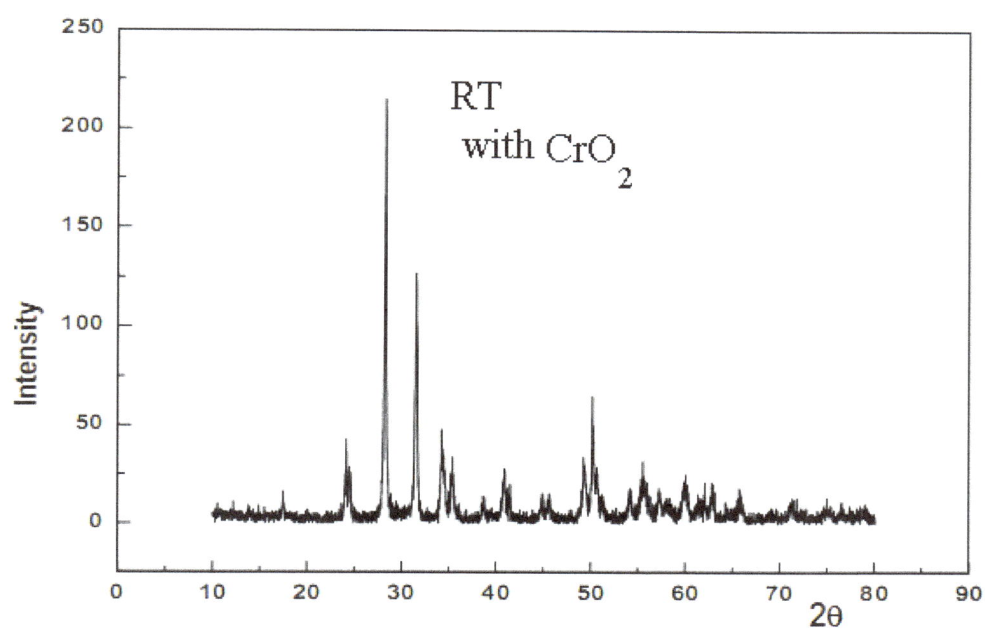

Figure 3. The XRD pattern of nanoparticles at RT with CrO_2 (50% wt La_2O_3 and 50% wt CrO_2).

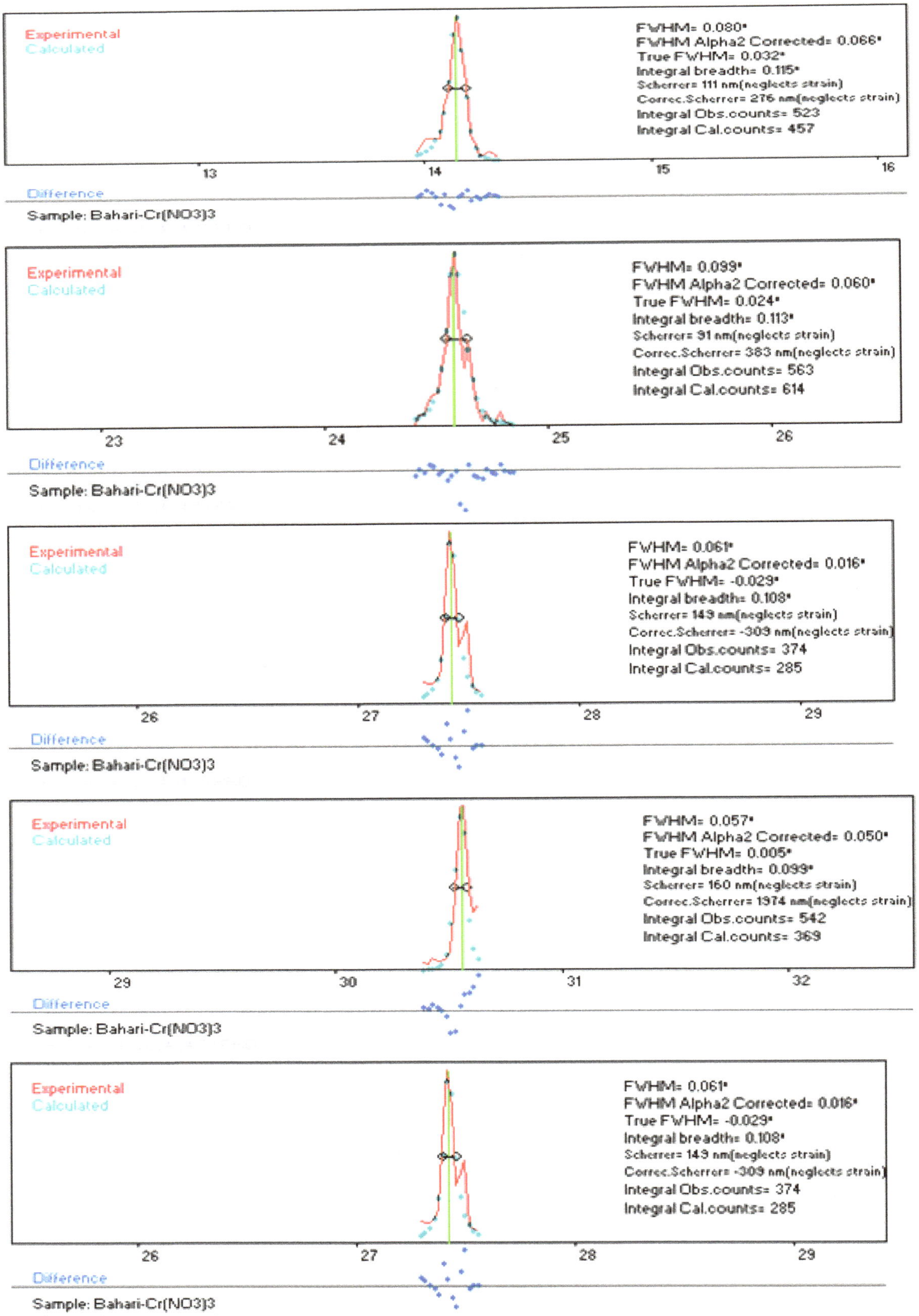

Figure 4. The size of nanoparticles correspond with Figure 2; and are determined using x- powder method.

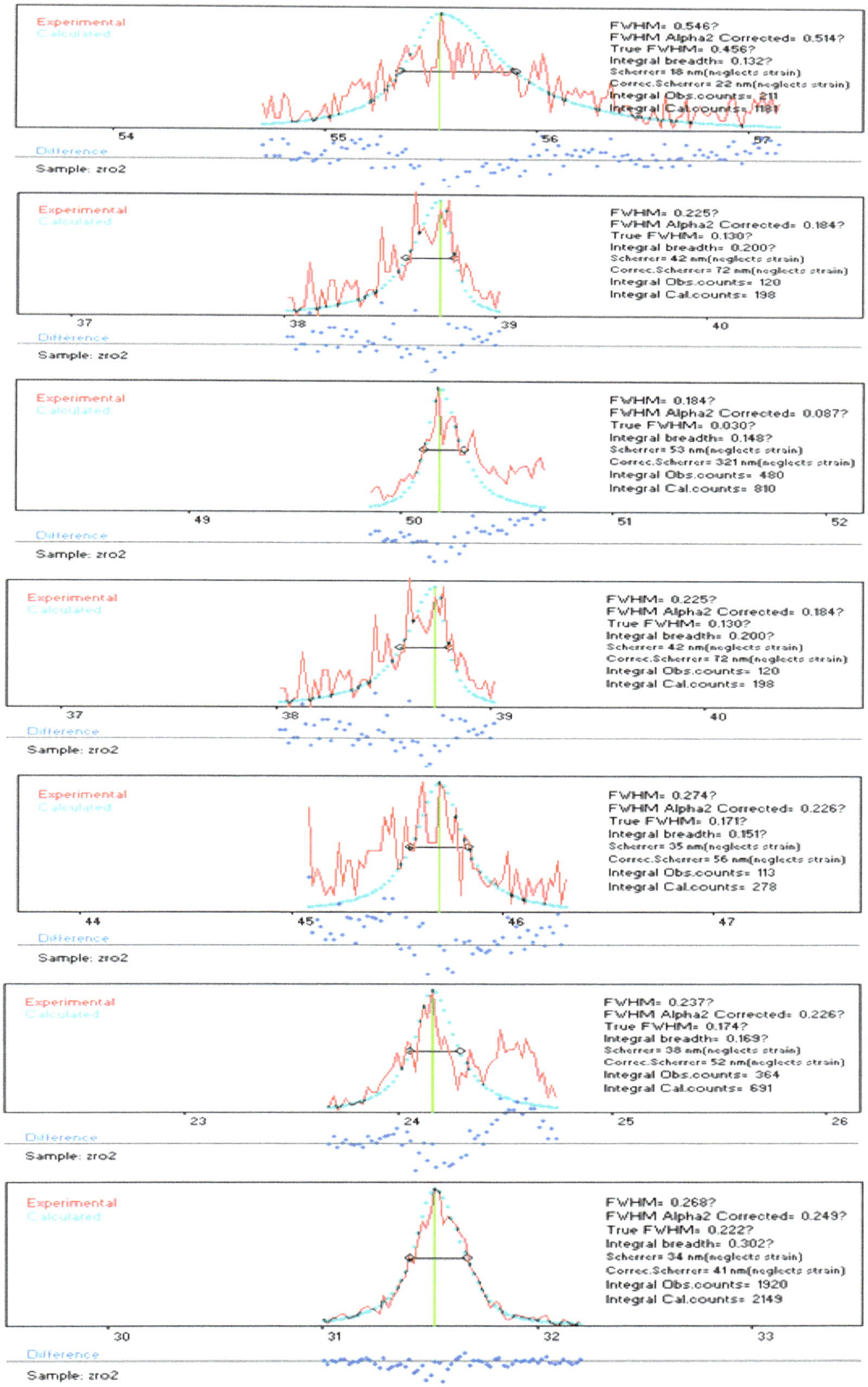

Figure 5. The size of nanoparticles correspond with Figure 3; and are determined using x- powder method.

slightly soluble in water and soluble in acids. La_2O_3 has been used to make optical glasses, to which this oxide confers increased density, refractive index and hardness. La_2O_3 in CrO_2 content have been synthesized at different temperatures. The fraction of the cubic La_2O_3 decreases at low temperature, whilst it increases at 500℃. It means that we could get the progressive phase transformation from metastable hexagonal La_2O_3 to stable cubic La_2O_3 during heating. In the XRD patterns, one can see the presence of cubic, hexagonal of La_2O_3 and cubic CrO_2 phases. In this spectrum, the intensity cubic "lanthania" is so small component which indicates this phase is interoperated into "lanthania" phases. The peak broadening at 500℃ (respect to room temperature) indicated the bigger size of the crystallites in the growth direction.

The amorphous structure of sample (at 300℃) can be used as a good gate dielectric of nano electronic devices.

REFERENCES

Alexazder VI, Lyseuuo SV, Baranova SV (2006). thermally stable materials based on mesostructured sulfated Lanthania. Mesopor. Mater., 91: 254-2600.

Alexazder VI, Lyseuuo SV, Baranova SV (2006). Thermally stable materials based on mesostructured sulfated Lanthania, Microporous Mesoporous Mater., 91: 254-2600.

Bahari A, Morgen P, Li ZS (2008). Ultra thin silicon nitride films on Si(100) studied with core level photoemission. Surf. Sci., 602: 2315-2324.

Bahari A, Robenhagen U, Morgen P (2005). Growth of Ultra thin Silicon Nitride on Si(111) at low temperatures. Phys. Rev. B., 72: 205323-205329.

Morgen P, Bahari A, Rao MG, Li ZS (2005). Roads to Ultra thin Silicon Oxide. J. Vacuum Technol. A., 23: 201-207.

Bahari A, Morgen P, Li ZS, Pederson K (2006). Growth of a Stacked Silicon Nitride /Silicon Oxide Dielectric on Si (100). J. Vacuum Sci. Technol., 24: 2119-2123.

Bahari A, Morgen P, Li ZS (2006). Valence Bend Studies of the Formation Ultra thin Pure Silicon Nitride Films on Si(100). Surf. Sci., 600: 2966-2971.

Betz U, Sturm A, Loeffler JF, Wagner W, Wiedenmann A, Hahn H (1999). Low-temperature isothermal sintering and microstructural characterization of nanocrystalline Lanthania ceramics using small angle neutron scattering. NanoStruct. Mater., 12: 689-692.

Cao J, Liu H, Zheng M, Chang X, Ma X, Zhang A, Xu Q (2005). Controllable syntheses of hexagonal and lamellar mesostructured lanthanum oxide. Mater. Lett., 59: 408-416.

Duan G, Zhang C, Li A, Yang X, Lu L, Wang X (2008). Preparation and characterization of mesoporous Lanthania made by using a poly (methyl methacrylate) tamplate. Nanoscale Res. Lett., 3: 118-122.

Dercz G, Prusik K, Pajak L (2006). Structure investigations of commercial Lanthania ceramic powder, J. Achiev. Mater. Manuf. Eng., 18: 259-262.

Duan G, Zhang C, Li A, Yang X, Lu L, Wang X (2008). Preparation and characterization of mesoporous Lanthania made by using a poly (methyl methacrylate) tamplate. Nanoscale Res. Lett., 3: 118-122.

Kale S, Jadhav K, Patil P, Gujar T, Lokhande C (2005). Characterizations of spray-deposited Lanthanum oxide (La_2O_3) thin films. Mater. Lett., 59: 3007-3024.

Kurzweg W, Heimann RB, Troczynski T, Waymann ML (1998). Development of plasma sprayed bioceramic coatings with bond coats based on Titania and Lanthania. Bomaterials, 19: 1507-1511.

Manoilova OV (2004). Surface Acidity and Basicity of La_2O_3, LaOCl, and $LaCl_3$ Characterized by IR Spectroscopy, TPD, and DFT Calculations. J. Phys. Chem. B., 108: 15770-15776.

Morgen P, Bahari AP (2006). Functional properties of Nano structured Material. Springer, pp. 229-257.

Morgen P, Bahari A, Pedersen K, Li Z (2007). Plasma assisted growth of ultra thin nitrides on Si surfaces under ultra high vacuum conditions. J. Phys., 86: 12019-12038.

Pajak L, Formanek B, Dercz G (2003). Dispersion analysis of NiAl-TiC-Al_2O_3 composite powder. Proceedings of 12th Scientific International Conference. Achiev. Mech. Mater. Eng. AMME'2003, Gliwice-Zakopane, pp. 723-726.

Vinogradova N, Dmitruk L, Petrova O (2004). Glass Transition and Crystallization of Glasses Based on Rare-Earth Borates. Glass Phys. Chem., 30: 1-15.

Wells AF (1984). Structural Inorganic Chemistry. Oxford: Clarendon Press, p. 546.

Wyckoff RWG (1963). Crystal Structures: Inorganic Compounds RXn, RnMX2, RnMX3. New York: Interscience Publishers.

Zhu WZ (1998). Effect of Cubic Phase on the Kinetics of the Isothermal Hexagonal to Cubic Transformation in La_2O_3 (3 mol%Y_2O_3) Ceramics. Ceram. Int., 24: 35-43.

Two-dimensional particle modeling of submicrometer ZnO MESFET based on an ensemble Monte Carlo calculation including five-valley band structure model

H. Arabshahi[1,2]

[1]Department of Physics, Ferdowsi University of Mashahd, Mashhad, Iran
[2]Department of Physics, Payame Nour University of Fariman, Fariman, Iran. E-mail: arabshahi@um.ac.ir.

A Monte Carlo method has been developed for the study of electron transport properties in ZnO MESFET for high field, using a five-valley conduction band model. The effects of upper valleys on the characteristics of ZnO MESFETs have been investigated. The following scattering mechanisms, that is impurity, polar optical phonon, acoustic phonon, alloy and piezoelectric are included in the calculation. Ionized impurity scattering has been treated beyond the born approximation using the phase-shift analysis. The simulation results show that on the drain side of the gate region, hot electrons attained enough energy to be scattered into the upper satellite conduction valleys. Approximately 17% of the electrons occupy the higher valleys (mainly *U* and *M* valley). The simulated device geometries and doping are matched to the nominal parameters described for the experimental structures as closely as possible, and the predicted drain current and other electrical characteristics for the simulated device including upper valleys show much closer agreement with the available experimental data.

Key words: Particle modeling, submicrometer, ionized impurity, polar optical phonon.

INTRODUCTION

In recent years, devices made from ZnO have attracted considerable interest because of its potential for short-wavelength photonic devices and high-power, high-frequency electronic devices (Allen et al., 2010). Up to now mostly metal-insulator-semiconductor structures (MISFETs) are used for those purposes due to the simple realization of high-*k* insulators by *rf*-sputtering techniques (Grundmann et al., 2010). However, the key advantage of metal-semiconductor FETs (MESFETs) is the higher mobility of carriers in the channel, as compared to MISFETs. Since the carriers in the channel of a MISFET are accelerated toward the semiconductor/insulator interface by the gate electric field where they suffer interface scattering, their mobility is less than half of the mobility of bulk material (Nemec et al., 2010). As the depletion region in MESFETs keeps the carriers off the gate interface, their mobility is close to that of bulk material. The higher mobility leads to a higher current, transconductance and transit frequency of the device (Arabshahi et al., 2007). In this article, we present a two dimensional ensemble Monte Carlo simulation of electron transport in ZnO MESFET (Cameron et al., 1983). Our current approach employs a valley conduction band model to investigate steady-state electron transport in this device (Arabshahi, 2007). However, the momentum space treatment is three dimensional, and the scattering events consider all three dimensions. Specifically, our model includes the five lowest valleys of the conduction band with nonparabolicity (Cameron et al., 1982). The device geometries and transport model are described in "model, device and simulations", after which the simulation results are thus provided. Subsequently, details of the device fabrication and simulation model which is used in the characteristic of the device are presented, after which details of the effects of upper valley in ZnO MESFET and the results obtained are interpreted.

MODEL, DEVICE AND SIMULATIONS

Our self-consistent MonteCarlo simulation was performed using an analytical band structure model consisting of five non-parabolic ellipsoidal valleys. The pseudo potential band structure shows the conduction band minimum to be located at the Γ point (Γ_1), and lowest energy conduction band satellite valleys to occur at the *U* point (located about two thirds of the way between the *L* and *M* symmetry points). Higher conduction band valleys are located at the Γ point (Γ_3), at the M point, and at the *K* point (Figure 1). In our Monte Carlo simulation, the two different Γ valleys, the six equivalent *U* valleys, the three equivalent *M* valleys and the two equivalent *K* valleys are represented by ellipsoidal, non-parabolic dispersion relationships of the following form (Jacoboni et al., 1989).

$$E(k)[1+\alpha_i E(k)] = \frac{\hbar^2}{2}\left[\frac{k_x^2+k_y^2}{m_\perp^*}+\frac{k_z^2}{m_\parallel^*}\right] \quad (1)$$

where $m_\perp^*$ and $m_\parallel^*$ are the transverse and longitudinal effective masses at the band edge and α_i is the non-parabolicity coefficient of the *i*-th valley.

Material parameters such as nonparabolicities, valley separations, and transverse and longitudinal effective masses are obtained either from experimental results or from pseudopotential calculations (Table 1). A constant time step discretization scheme is chosen in the Monte Carlo procedure, which allows us to track the time evolution of the electronic transport as well as to determine the position of all sample electrons at given instances of time. Consequently, the electron concentration as a function of space and time can be determined and Poisson's equation can be solved to obtain the self-consistent electric field. The modeled structure is divided into three regions, as indicated. Electron particles are initially distributed keeping all regions charge neutral. The location of the source and the drain implants and the top and back buffer layer are marked (Arabshahi et al., 2007).

In order to minimize the statistical fluctuations, always associated with the stochastic Monte Carlo method, we choose 20000 electron particles for the simulation and a time step of 10 fs for the readjustment of the electric field. A charge assignment scheme is also used to suppress the fluctuations. Each particle represents a cloud of electrons. The charge of the particle is assigned to its nearest four background mesh points proportionally to the position of the particle in the cell. To make the charge assignment more efficient, field cell size used for the central region is 30 nm^2 (horizontal × vertical), but that in the high doped source and drain implants is finer (10 nm^2).

Electrons in the bulk are scattered by ionized impurities and by bulk acoustic and non-polar optical phonon modes. Intervalley scattering by the absorption and emission of long wavelength acoustic and optic phonons have also been considered in the model. As described in detail, model devices are built up as a series of joined rectangular regions, with the electric field cell sizes matched along the join between each region. Each region can consist of multiple layers of different alloy composition and doping/compensation density. The ZnO MESFET can be described simply by three regions (Figure 1), representing source and drain doping implants and a central region containing the supply layers.

Figure 1 shows a schematic of the modeled ZnO MESFET. The overall device length is 3.3 µm in the *x*-direction and the device has a 0.3 µm gate length and 0.5 µm source and drain length. The source and drain have Ohmic contacts and gate is in Schottky contact in 1 eV to represent the contact potential at Au/Pt. The source and drain regions are doped to 5×10^{23} m^{-3} electron concentration and the top and down buffer layers are doped to 2×10^{23} and 1×10^{22} m^{-3} electron concentration, respectively. The geometry of the device to be simulated is specified at the beginning of the simulation by defining it in an *x*-*y* plane as a set of joined rectangular regions, each with uniform doping and other material parameters, and a set of contact regions. All physical quantities are assumed constant in the *z* direction (device width).

RESULTS

Figure 2 compares the instantaneous distribution of electrons throughout the device in the steady-state for different gate and drain biases at room temperature. When a positive potential V_{ds} is applied to the drain, electrons flow from source to drain, giving a current I_{ds} from source to drain. The depletion region of the Schottky barrier restricts the current path to the lower part of the channel and the buffer layer. At zero drain bias (Figure 2a), the depletion layer beneath the gate has a symmetric shape. At zero gate and drain bias, the depletion region corresponds to that associated with the built-in potential of the Schottky barrier, but is larger for the case of gate bias V_{gs}=-1 V shown here. As V_{ds} is increased from zero, the depletion layer becomes asymmetric in shape since the potential difference between the gate and the channel is greater at the drain end than the source end of the gate. The channel is more constricted at the drain end of the gate and the field along the channel is also higher in that region.

As the drain-source voltage is increased further, the field at the drain end of the gate approaches the value at which the electron velocity along the channel saturates. Beyond this value, which corresponds to the threshold voltage in the drain current-voltage characteristics, further increase in drain-source voltage does not substantially increase the drain-source current. The length of the region of the channel over which the electrons are in velocity saturation increases and there is some carrier accumulation within the channel. The depletion layer edge at the drain end of the gate also moves closer to the drain as the drain-source voltage increases. Note also that the transition between the gate depletion region and the charge neutral bulk is far from sharp on the drain side. At a sufficiently negative gate potential the depletion layer punches through to the high-resistively buffer layer, and the source and drain electrodes are connected only by leakage paths within the buffer layer and substrate (Figure 2b).

An important factor in the high frequency performance of the MESFET is the time taken for electrons to traverse the region of the device beneath the gate, and the performance is enhanced by a high average electron velocity through that region. The spatial distribution of hot electrons throughout the device for each valley at V_{gs}=-1 V and V_{ds}=30 V for room temperature operation is shown in Figure 3. Electrons are seen to exist in the upper valleys only to the right of the high field region, which exists on the drain side of the gate, because it is only there that the electrons have attained enough energy to be scattered into the satellite conduction valleys. Also note there is an injection of electrons from the channel into the buffer layer; a process which is eventually

Table 1. Important parameters used in our simulations for ZnO material (Foutz et al., 1997).

Parameter	ZnO
Density (kgm^{-3}) (Allen et al., 2010)	3200
Sound velocity (ms^{-1}) (Arabshahi et al., 2007)	1373
Static relative permitivity ε_0	9.7
Optical relative permitivity ε_∞	6.5
Deformation potential (eV)	15
Polar optical phonon energy (meV)	120
Energy gap (eV) (Nemec et al., 2010)	3.43
Intervalley deformation potential (10^7 eVm^{-1})	0.356
Nonparabolocity (eV^{-1}), (Arabshahi et al., 2007)	3.39 0.323

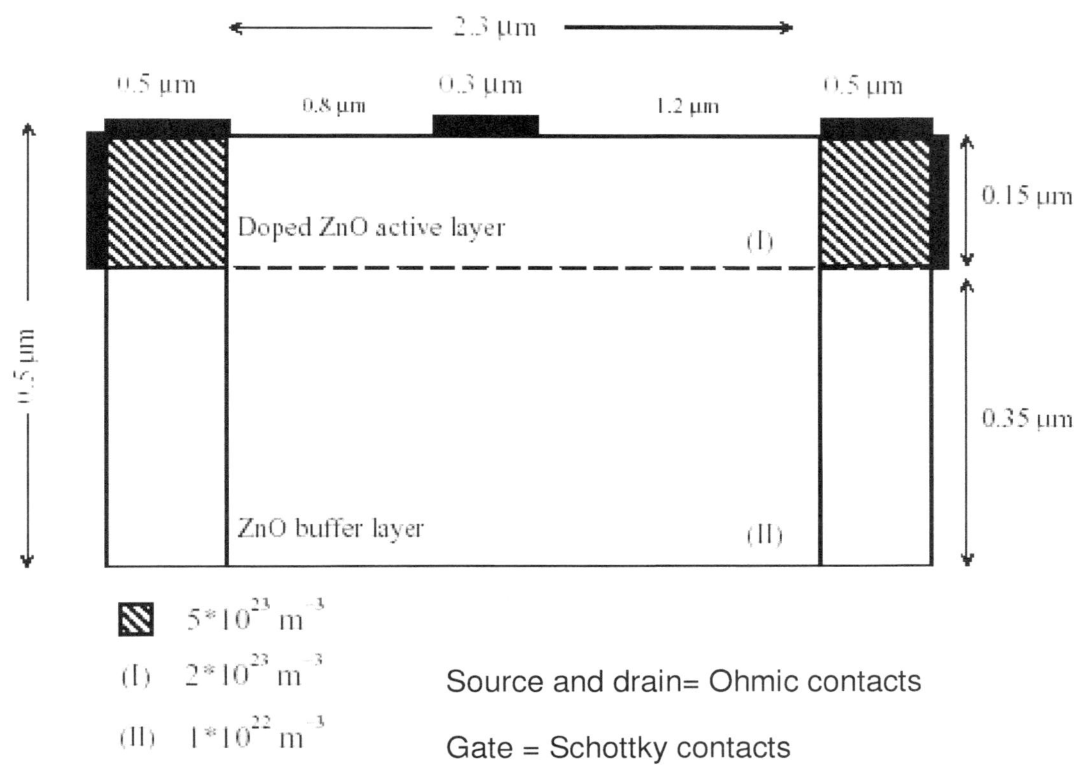

Figure 1. The two-dimensional model of ZnO MESFET.

opposed by the electric field created by the resulting negative space charge in the buffer layer. Figure 3 also shows that the distribution of electrons occupying the upper valleys extends a significant way towards the drain region where the electric field is much lower. This is a result of the finite time that it takes for phonon scattering to return the electrons to the Γ-valley.

Figure 4 shows the valley electron occupancies throughout the device. It can be seen that significant electron transfer to the upper valleys only begins to occur under the gate. Approximately 17% of the electrons occupy the higher valleys (mainly *U* and *M* valleys) in the vicinity of the gate which is similar to the valley occupancy ratio in the active layer of the n^+-n-n^+ diode. Figure 5 shows various microscopic properties of the device when the source-drain bias is 30 V and the gate voltage is -1 V; specifically the longitudinal electric field, the Γ-valley band profile, the electron kinetic energy, the

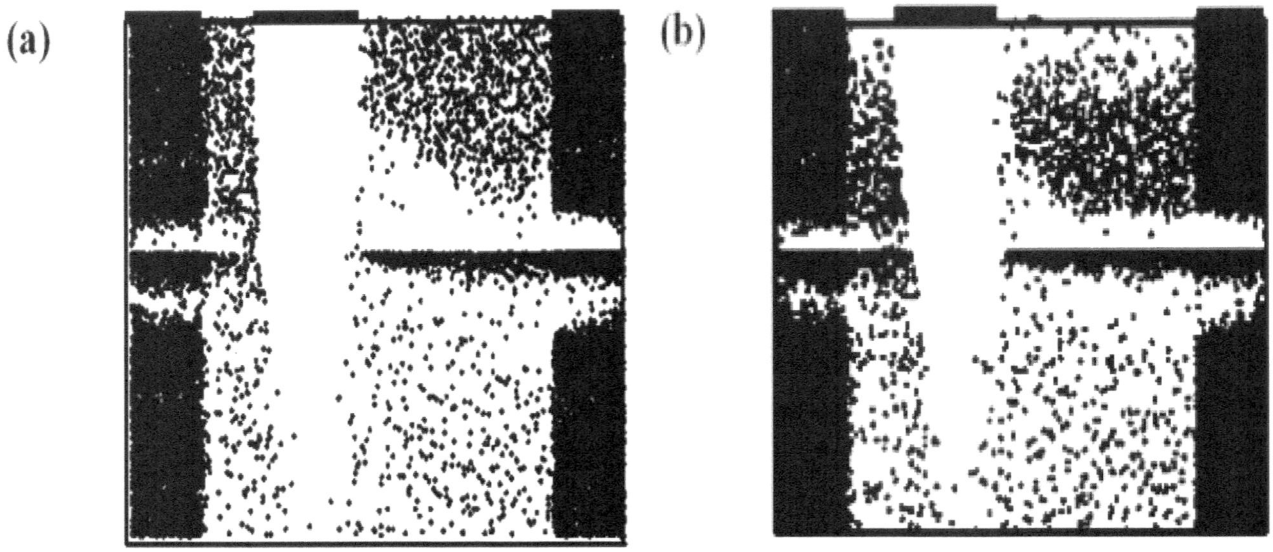

Figure 2. Depletion layer profiles for different drain and gate biases for ZnO MESFET at room temperature.

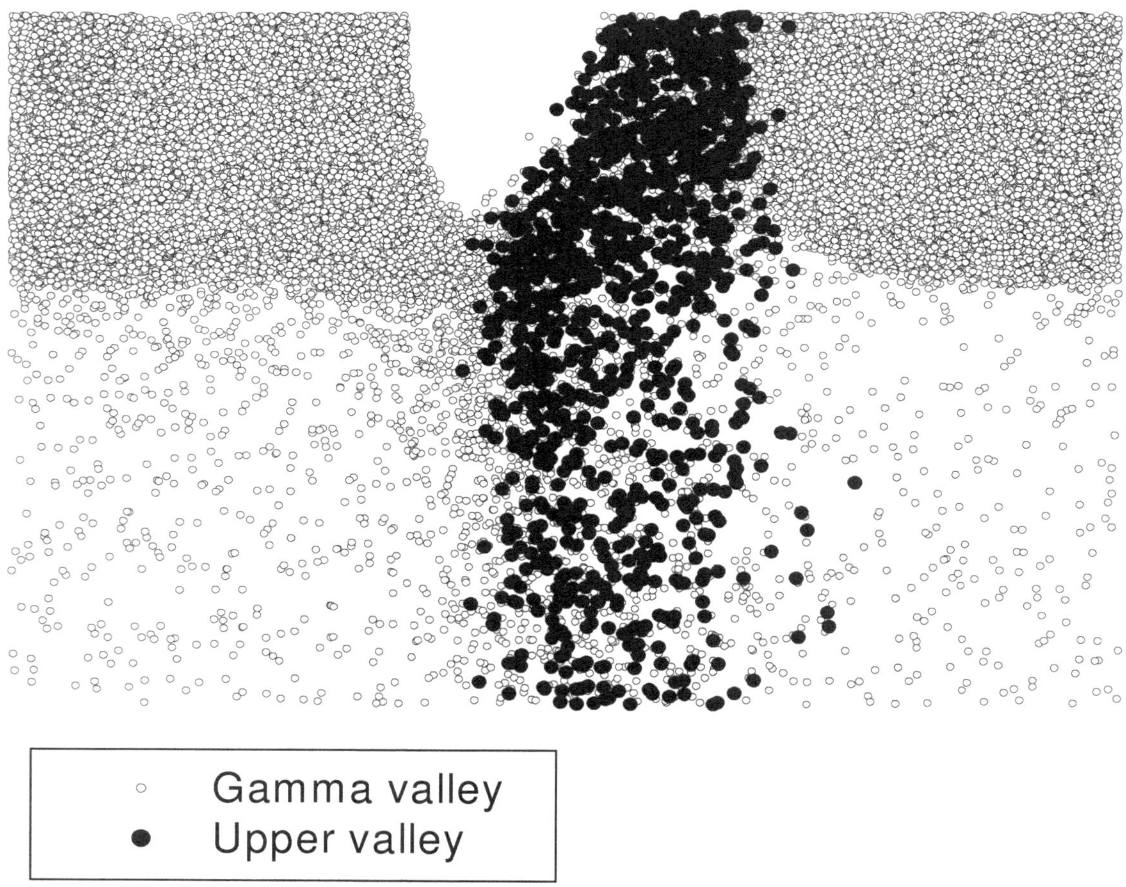

Figure 3. The distribution of hot electrons at room temperature for V_{gs}=-1 V_{ds}=30 V in central Γ valley and upper valleys.

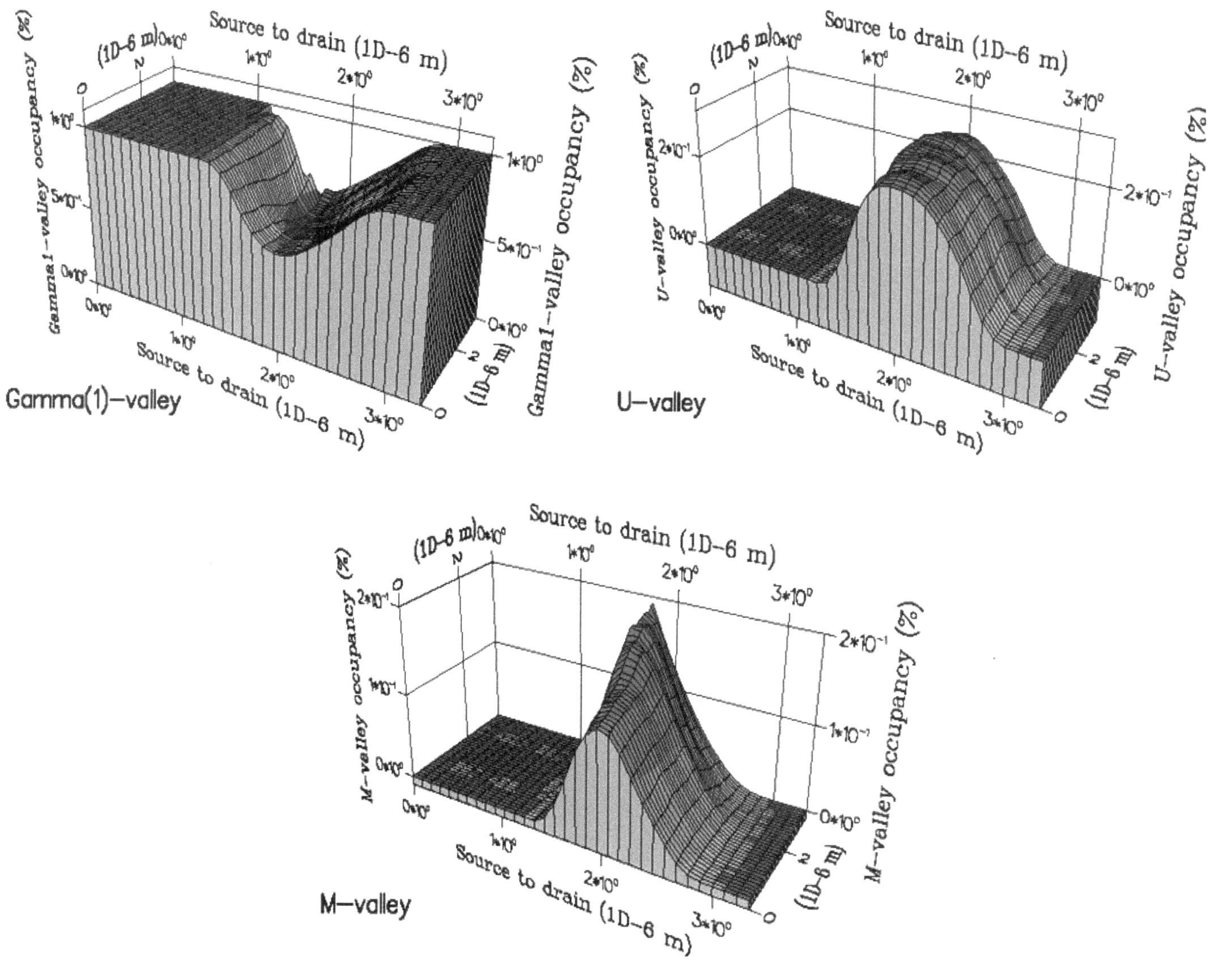

Figure 4. The valley occupation ratios for the central Γ_1-valley and upper valleys when the source-drain bias is 30 V and the gate voltage is -1 V at T=300 K.

average drift velocity and the total electron density as a function of distance from the source. The longitudinal electric field plotted in Figure 5a shows the high electric field in the region under the gate, which has been referred to earlier. Related to this is the Γ-valley band profile throughout the device in Figure 5b. Note almost all the drain-source potential is dropped within the gate-drain region, leaving a flat potential profile near the source and drain. As electrons move towards the drain, they lose potential energy and gain sufficient kinetic energy to transfer to the upper conduction valleys where their drift velocity is reduced.

The variations of average electron kinetic energy and average drift velocity throughout the simulated device are shown in Figures 5c and d, respectively. The average electron velocity reaches about 2×10^5 ms^{-1} and then declines towards the drain. The steep decrease in the average kinetic energy on the drain side of the gate is due to the transfer of electrons to the upper valleys. The electron density through the device is shown in Figure 5e. The gate depletion region is clearly seen where the electron density is several orders of magnitude lower than it is near the source and drain.

The drain current is obtained by counting the net charge flow through the drain contact. Figures 6a and b show the calculated drain current versus drain-source voltage at different gate biases for temperatures of 300 and 420 K. The simulated characteristics at 300 K show good saturation behavior with a knee voltage around 15 to 20 V and a saturation drain current of about 1200 mA/mm for V_{gs}=0 V. The high drain current density is encouraging for the use of ZnO for high-power applications.

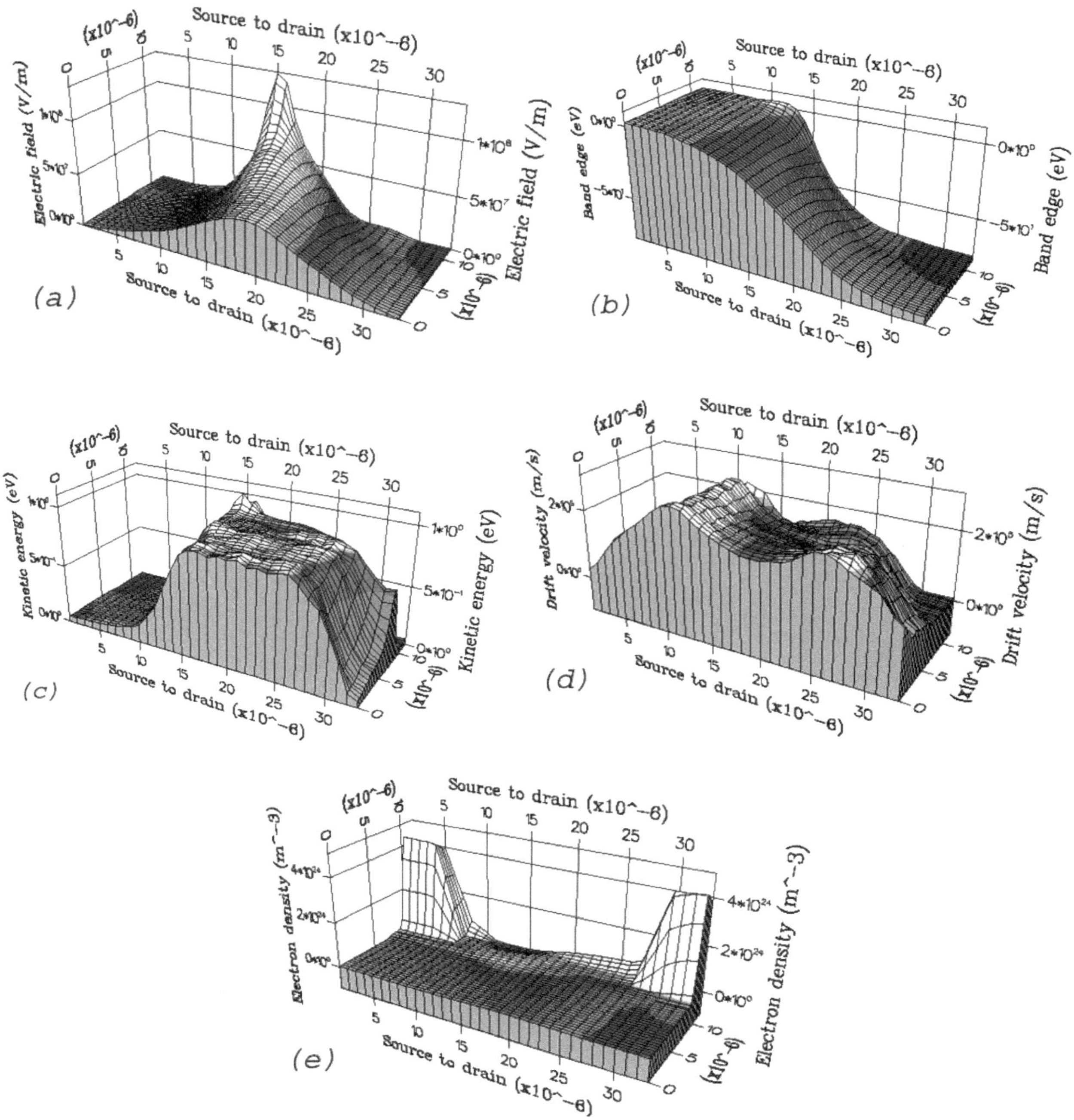

Figure 5. Three dimensional distribution of electron data recorded through the simulation of the GaN MESFET when the source-drain bias is 30 V and the gate voltage is -1 V at room temperature. This figure shows: (a) Longitudinal electric field, (b) Γ-valley conduction band profile, (c) Average kinetic energy, (d) Drift velocity and (e) Electron density. At V_{ds}= 80 V essentially the whole drain current flows entirely through the buffer.

From Figure 6a it is clear that the device is not completely pinched-off even at large negative gate bias (V_{gs}=-6 V) which is due to strong electron injection into the buffer layer at high electric fields. An increasing fraction of the drain current flows through the buffer as the drain voltage increases.

To obtain some idea of the effect of high temperature on ZnO MESFETs, simulations were carried out at T= 420 K, keeping the other device parameters unchanged. The *I-V* curves obtained are shown in Figure 6b.

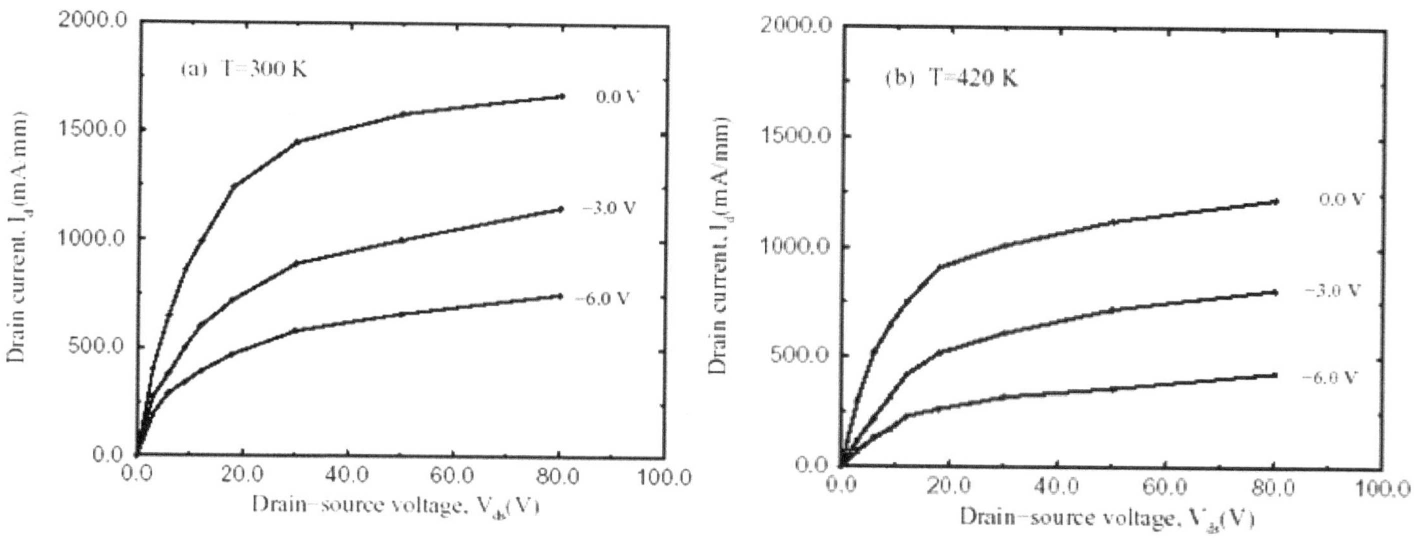

Figure 6. (a) The simulated drain current versus drain voltage at T= 300 K, (b) The simulated drain current versus drain voltage at T= 420 K.

Comparing the *I-V* curves at T= 300 and 420 K, it can be seen that the drain current is somewhat lower at the higher temperature, due to increased phonon scattering, but the effect is not major.

Figure 6 shows the general shape of the simulated *I-V* characteristics is similar to the experimental measurements but the predicted drain currents are higher. This discrepancy could be explained by electron trapping effects which are thought to play an important role in the electrical properties of ZnO devices due to the current limitations of the material technology. On applying drain bias, electrons passing from source to drain will have a certain probability of being captured. Emission of trapped electrons will also occur with a certain probability and in the steady-state there will be a density of trapped electrons determined by the balance between capture and emission. Hence, for any given electron there will be additional processes of electron capture and re-emission during electron transport and the effective transit time will be increased. In addition the trapped electrons cause space charge which acts to reduce the drain current. The transconductance of the MESFET is given by (Moglestue, 1993)

$$g_m = \frac{\Delta I_d}{\Delta V_g}\Big|_{V_d} \qquad (2)$$

and is calculated from Figure 6a to be about 150 mS/mm at 18 V drain bias and -1 V gate voltage for room temperature. When the drain bias is increased to 30 V at the same gate voltage , the transconductance increases approximately to 220 mS/mm. In comparison, good GaAs MESFETs without a gate recess have a transconductance around 110 mS/mm. The higher value of transconductance in simulated ZnO MESFET is related to a higher drain current.

Conclusions

In summary, Monte Carlo simulations of electron transport in ZnO MESFET for temperatures greater than 300 K, using an approximate band structure comprised of five analytic conduction band valleys yield higher valleys are important, because more than 17% of the electrons occupy the higher valleys (mainly *U* and *M* valley) which will produce a serious reduction in the drain current and consequently the output power of ZnO MESFET. Our calculations indicate that the high temperature output drain current is limited by strong polar optical phonon scattering.

REFERENCES

Allen MW, Swartz CH, Myers TH, Veal TD, Durbin SM (2010). Bulk transport measurements in ZnO: The effect of surface electron layers. Phys. Rev., 81: 075211.

Arabshahi H, Benam MR, Salahi B (2007). A Shock-Capturing Upwind Disceritization Method For Characterization of SiC MESFETs. Mod. Phys. Lett. B., 21: 1715.

Arabshahi H (2007). Monte Carlo Modeling of Hot Electron Transport in Bulk AlAs, AlGaAs and GaAs at Room Temperature. Mod. Phys. Lett. B., 21: 199.

Cameron DC, Irving LD, Whitehouse CR (1983). Comparison of High and Low Field Electron Transport in AlGaN, AlN and GaN. Thin Solid

Films, pp. 103-161.
Cameron DC, Irving LD, Whitehouse CR (1982). Comparison of Low Field Electron Transport in ZnO and GaN Structures for High-Power and high Temperature Device Modelling, Electron. Lett., pp. 18-534.
Foutz BE, Eastman LF, Bhapkar UV, Shur M (1997). Discretization Method of Hydrodynamic for simulation of ZnO MESFETs. Appl. Phys. Lett., 70: 2849.
Grundmann M, Frenzel H, Lajn A, Lorenz M, Schein F, Wenckstern H (2010). Transparent semiconducting oxides: Materials and devices. Physica Status Solidi (a), 207: 1269.
Jacoboni C, Lugli P (1989). The Monte Carlo Method for semiconductor and Device Simulation. Springer-Verlag.
Moglestue C (1993). Monte Carlo Simulation of Semiconductor Devices. Chapman and Hall.
Nemec H, Rochford, J, Taratula O, Sundstrom (2010). Influence of the Electron-Cation Interaction on Electron Mobility in Dye-Sensitized ZnO and TiO_2 Nanocrystals: A Study Using Ultrafast Terahertz Spectroscopy. Phys. Rev. Lett., 104: 197401.

Efficient enhancement of polymer light emitting diode by inserting hole blocking layer

M. Y. Lim[1], W. M. M. Yunus[1]*, Z. A. Talib[1] and A. Kassim[2]

[1]Department of Physics, Faculty of Science, Universiti Putra Malaysia, 43400 UPM Serdang, Selangor, Malaysia.
[2]Department of Chemistry, Faculty of Science, University Putra Malaysia, 43400 UPM Serdang, Selangor, Malaysia.

The triplet layer for the PLED was fabricated with ITO / PVK (77 nm) / CdS (115 nm) / Alq_3 (84 nm) / Al (300 nm). The PLED devices were inserted in the CdS layer between the 77 nm PVK and 84 nm Alq_3 absorbing the ambient light to reduce the reflection of the device. The maximum wavelength and luminescence intensity increase with the thickness of CdS. This indicated that the maximum wavelength shifted to a higher wavelength when the thickness of CdS increases. With the insertion of 115 nm of CdS layer between the PVK and Alq_3, a lower turn on voltage can be obtained at 5.4 eV.

Key words: Cadmium sulphide, photoluminescence, energy band gap, electrical conductivity, optical reflectance.

INTRODUCTION

In 1990, a Cambridge group of friends announced a conduction of polymer-based LED (PLED) involving aromatic, π conjugated species. The field of PLEDs has grown rapidly and has now entered industrialization. This new field has recorded enormous progress made in the improvements of color gamut, luminance efficiency and device reliability. The organic systems and the polymeric materials in particular, have the distinctive advantages that are easy to process.

In 1990, Burroughes et al. (1999) reported a very-high-efficiency green-light-emitting polymer LEDs (PLEDs) with conjugated polymer of poly (p-phenylene vinylene) (PPV) shown in Figure 2. The device used PPV as a precursor coated on the ITO by ion beam sputtering. Aluminum (Al) worked as a cathode by thermal evaporation. Since then, the conjugated polymer has been given considerable attention and intensive studies have been done on the PLEDs. Braun and Heeger from the University of California at Santa Barbara have an interest in the electroluminescence from conjugated polymer materials.

Further study in PLEDs has focused more on phosphorescent emissive materials, like Iridium complexes such as Tris [2-(p-tolyl) pyridine] and Iridium (III) [Ir $(mppy)_3$]. The electrophosphorescence converts the electrical energy in an OLED into light in a highly efficient manner (Kang et al., 2007; Tsutsui et al., 1999). The internal quantum efficiency of the devices has been improved to approximately 25% by using these phosphorescent materials. This is due to the decay radioactivity of both the singlet and triplet excitons compared to a standard PLED, where only the singlet states will contribute to emission of light.

Recently, a number of researches have been carried out by using organic materials as hole blocking layers, such as: N, N'-dicarbazolyl-4,4'-biphenyl (Peng et al., 2006), bathcuproine (Ikai et al., 2001; Adachi et al., 2001; Kwong et al., 2001; Kolosov et al., 2002), 4-biphenylphenolate aluminium (III) bis (2-methyl-8-quinolinato-4-phenylphenolate (Kim et al., 1999; Jae et al., 2007), 1,8-napthalimides (Vadim et al., 2003) and polyquinolines (Chandra et al., 1980). A good example is seen in devices reported by Jae et al. (2007), where

*Corresponding author. E-mail: mylim82@yahoo.com

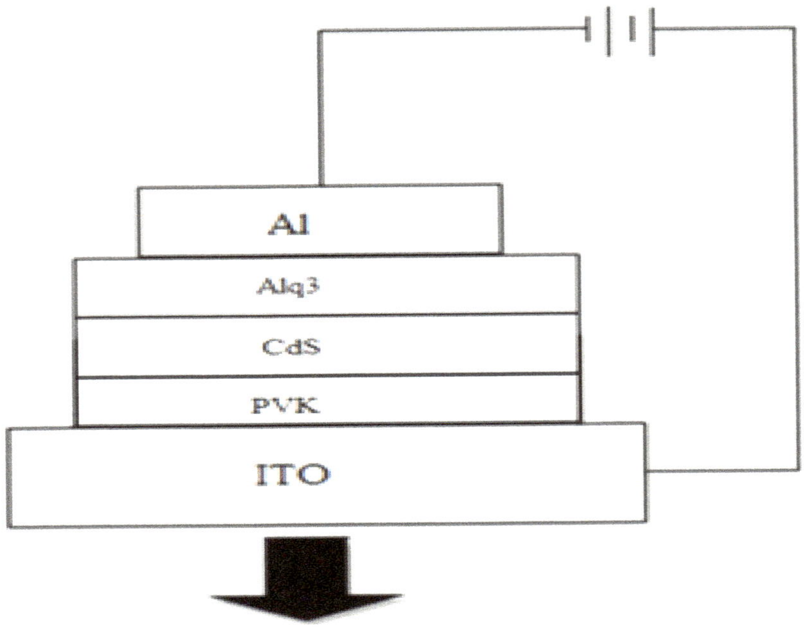

Figure 1. The structure of OLED device.

silane and triazine moieties, 2,4-diphenyl-6-(49-triphenylsilanylbiphenyl- 4-yl)-1,3,5-triazine (DTBT) were used as hole blocking materials,and N,N9-dicarbazolyl-4, 49-biphenyl (CBP) doped with *fac*-tris(2- phenylpyridine) iridium $[Ir(ppy)_3]$ showed a maximum external quantum efficiency of 17.5% with a maximum power efficiency of 47.8 lm W^{-1}. The insertion of the HBL will solve the problem of excitons leakage, in which efficiency blocks the diffusion of holes from the luminescent layer into the ETL layer (Alex et al., 2004).

Cadmium sulfide is in the hexagonal structure and it has a direct band gap semiconductor with a band gap of 2.42 eV. It has useful properties for optoelectronics, being used in both photosensitive and photovoltaic devices. Thin films of CdS are required in components, such as a photoresistor and solar cells (Eih-Zhe et al., 2003), nonlinear integrated optical devices and light emitting diodes (LEDs) or laser heterostructures for emission in the visible spectral range (Johnson et al., 1995; Peng et al., 2006). It has been reported that some research used CdS nano particles as a active layer for light emission (Ikai et al., 2001). Chemically synthesized cadmium sulfide (CdS) nano particles are deposited on Si substrate and they exhibit electroluminescence, corresponding to different process treatment. Thus, optimum thickness of the CdS is very important to improve the performance of the OLEDs. Therefore, in this paper, the effect of CdS thickness is reported on the electrical conductivity, turn on voltage and photoluminescence of the CdS as a hole blocking layer in OLED device.

EXPERIMENTALS

In this experiment, Alq_3 was used as the ETL and PVK as the HTL, while Indium Tin Oxide (ITO) was used as the anode aluminium (Al) and cathode, respectively. Devices with a structure of ITO / PVK (77 nm) / CdS (T nm) / Alq_3 (84 nm) / Al (300 nm) were fabricated, where T was chosen as 75, 90, 105, 115 and 134 nm. The PVK and Alq_3 compounds were purchased from Sigma-Aldrich. The PVK was deposited on a subtract by using deep coating method to transfer it onto indium tin oxide (ITO) glass substrates with a sheet resistance of about 15 Ω, while the Alq_3 and CdS were deposited by thermal evaporation at room temperature in a high vacuum chamber (Figure 1). The ITO glass was immersed in ultrasonic bath with acetone for 10 min. Then, the ITO glass was rinsed in deionized water for 10 min and then, it was blown dry with nitrogen gas. This procedure was used to remove organic contamination and particles from the ITO surface.

Current-voltage measurement was carried out using current-voltage source (Keithley, 2400). The thickness of the layers was measured by disk profile machine (Tencor P-12). The photoluminescence (PL) of the films was measured using Spectrofluorometer (Oceon Optics 2000 FLG). Spectrophotometer and Micro Autolab Type III Potentiostat were used for measuring the energy band gap and cyclic voltametry study.

RESULTS AND DISCUSSION

Figure 3 showed that the different thicknesses of CdS were inserted between the PVK and Alq_3 to enhance the

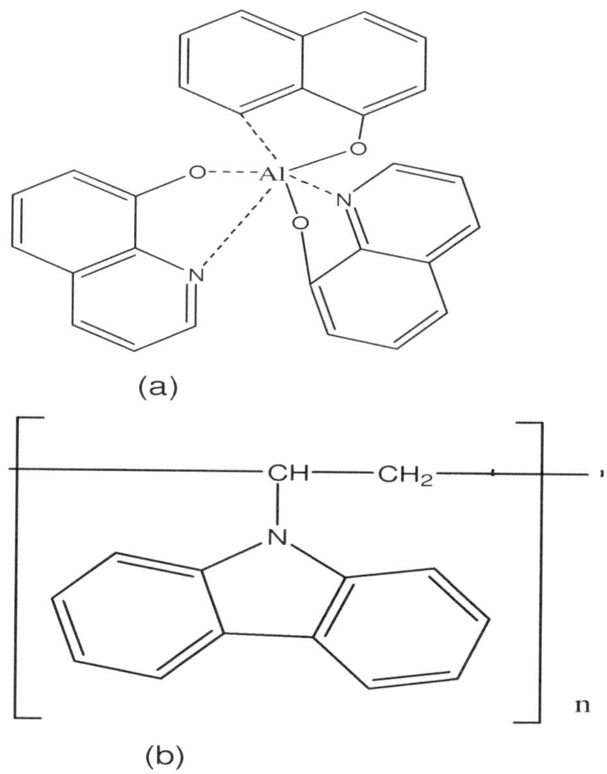

Figure 2. Molecular structure of (a) Alq_3 and (b) PVK.

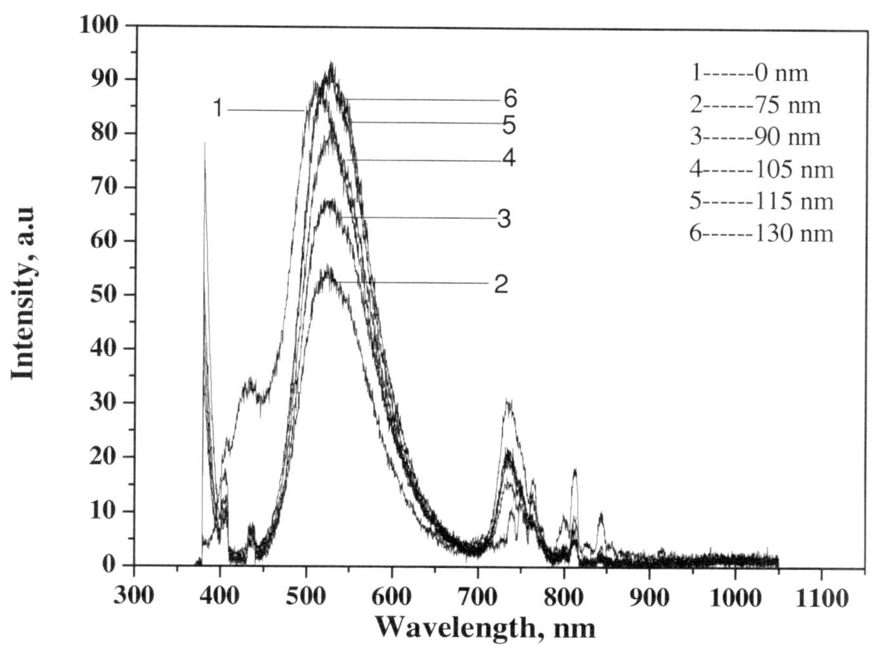

Figure 3. Photoluminescence spectrum for different thickness of CdS with 77 nm PVK and 84 nm Alq_3.

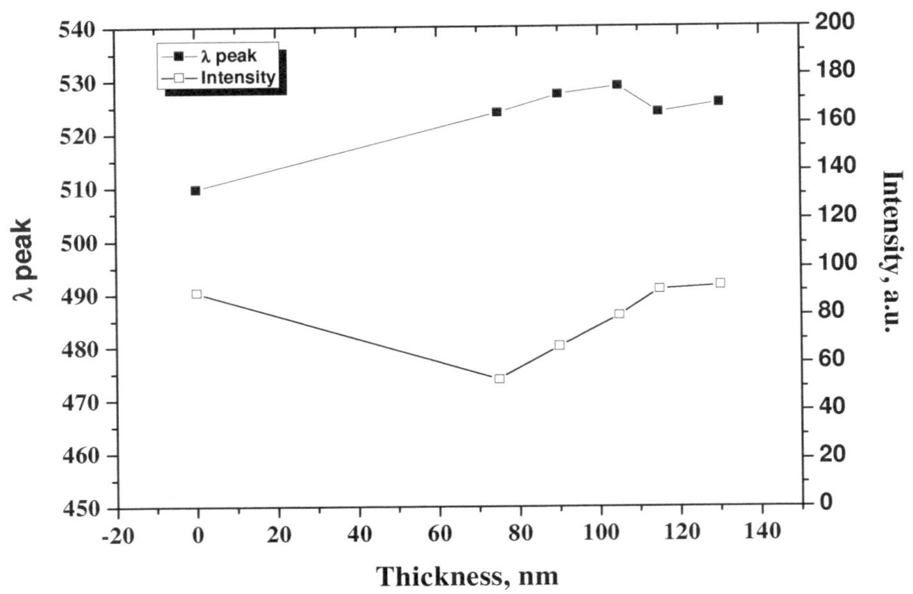

Figure 4. λ peak versus intensity for different thickness of CdS with 77 nm PVK and 84 nm Alq_3.

luminescence intensity of the PLED devices. Triplet layers of PLED devices were prepared with ITO / PVK (77 nm) / CdS (T nm) / Alq_3 (84 nm) / Al (300 nm), where T was chosen as 75, 90, 105, 115 and 130 nm. The maximum wavelength for the PLED device, without the CdS layer, was around 530 nm. However, the maximum wavelength shifted to a higher wavelength (550 nm), when the CdS was inserted between the PVK and Alq_3. This may be due to the fact that the PVK can be used as a light emitting layer in the PLED device, while the 550 nm is attributed to the excimer emission of PVK. The relationship between the λ peak and the luminescence intensity for the triplet layers PLED is shown in Figure 4. The maximum wavelength and luminescence intensity increase with the thickness of CdS. This indicated that the maximum wavelength shifted to a higher wavelength, when the thickness of CdS increased. When the film was excited by 345 nm light, the maximum emission was observed at 550 nm, confirming an efficient energy transfer from the PVK excimer to CdS. As a result, Forster-type energy transfer is expected to occur at the interface of PVK and CdS. In addition, the hole transport barrier between the host PVK and the CdS blocks holes, directly quenches at the cathode, hence increasing the probabilities of charge trapping and recombination on the emitting layer, which leads to high efficiency.

This is very important to minimize the reflectance of the PLED, but also to simultaneously maximize its emission through the optimization of its transmittance. The PLED devices insert the CdS layer between the 77 nm PVK and the 84 nm Alq_3, which absorb the ambient light to reduce the reflection of device. From Figures 5 and 6, the CdS with 115 nm has the lowest intensity reflectance light which is 6 a.u. The reduction of the reflectance of the device is done by introducing an insertion of the layer between the HTL and ETL. This is an approach that can also enhance the extraction of light from the device since CdS is a good photoconductivity which absorbs light.

Figure 7 shows the I-V characterization for the different thickness of CdS insertion between the PVK and Alq_3. The thicknesses of the CdS are from the range of 75 to 130 nm. It can be found that the current initially increase with the increase of CdS thickness until 105 nm. This implies that once the CdS molecules are introduced into the devices, the CdS molecules will trap the hole and reduced the mobility of the hole. Further increase in the thickness of the CdS, until 115 nm, shows a decrease in the trend of the current. There are some pre-trapped electrons in the emitting layer, when the voltage applied to the electron helps the injection of the hole into the emitting layer. Thus, with higher CdS thickness, the number of trapped electrons increased.

To determine the turn-on voltage of each multi layer PLED devices, the corresponding Fowler-Nordheim plot log (I/V^2) versus 1/V, in Figure 8, yields a straight line which implies that the measured anode current follows the model of Fowler-Nordheim model and can be defined as a field emission current. Generally, field emission is characterized using the theory of Fowler and Nordheim. Figure 9 showed that the turn-on voltage without the

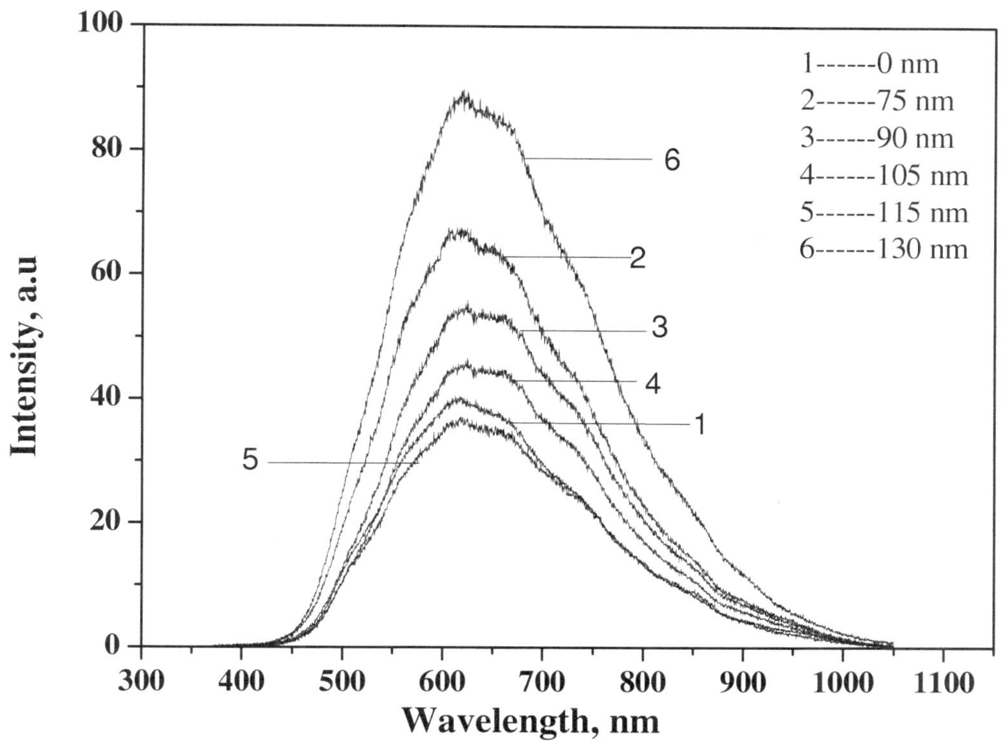

Figure 5. The reflectance light for different thickness of CdS with 77 nm PVK and 84 nm Alq_3.

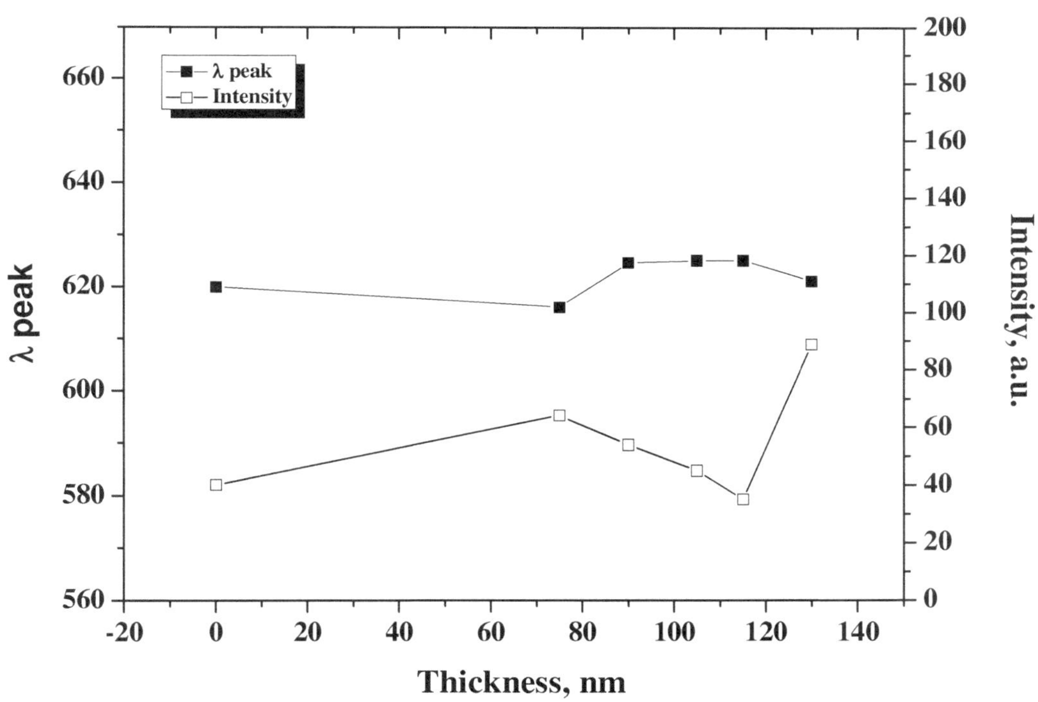

Figure 6. The reflectance light for different thickness of CdS with 77 nm PVK and 84 nm Alq_3.

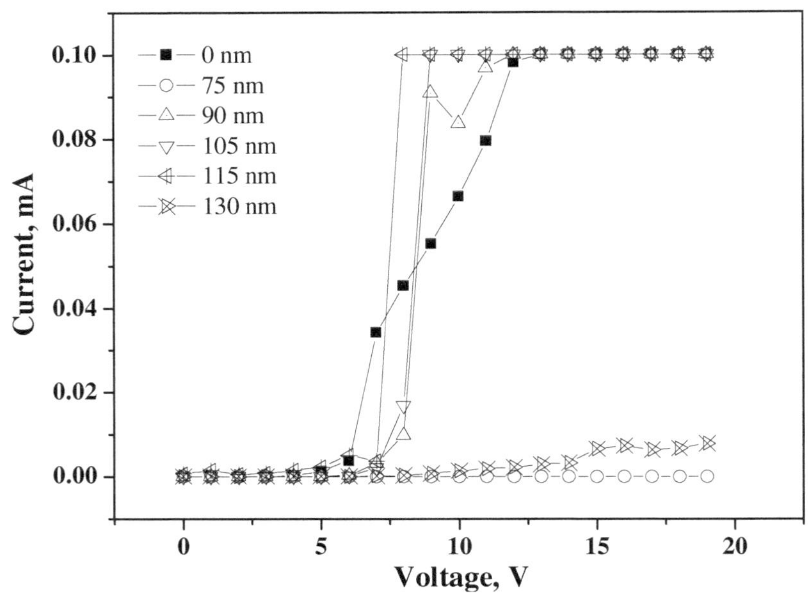

Figure 7. I-V characteristic for different thickness of CdS with 77 nm PVK and 84 nm Alq_3.

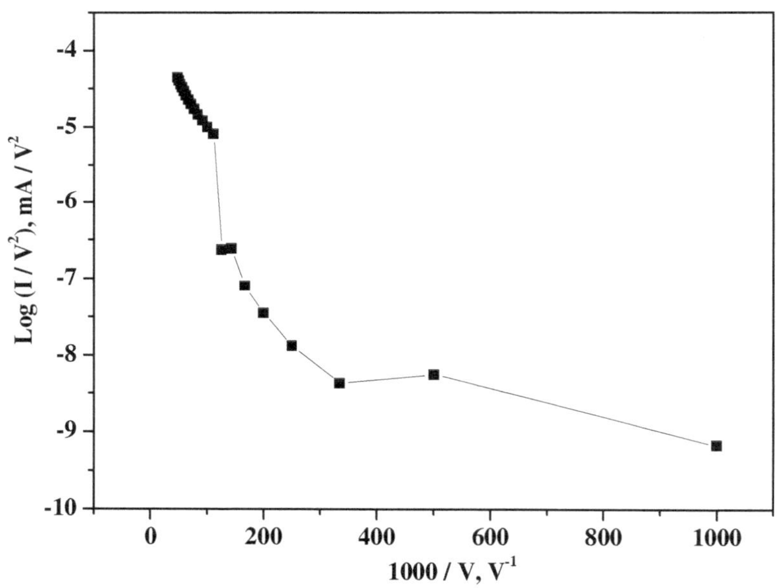

Figure 8. The corresponding Fowler-Nordheim plot for 84 nm Alq_3, 115 nm CdS and 77 nm PVK PLED device.

CdS layer is 5.9 V; however, by inserting 115 nm of CdS layer between PVK and Alq_3, a lower turn-on voltage can be obtained at 5.4 V. The decrease in the turn-on voltage is due to the decrease in the trap density which means lower voltage is required to send the electron to interface PVK: Alq_3. As a conclusion, the optimum thickness of CdS will enhance the injection of the hole.

Figure 10 depicts the log-log scale I-V characteristics of the multi layers PLED devices with the 84 nm Alq_3, 115 nm CdS and 77 nm PVK. The I-V curve shows three

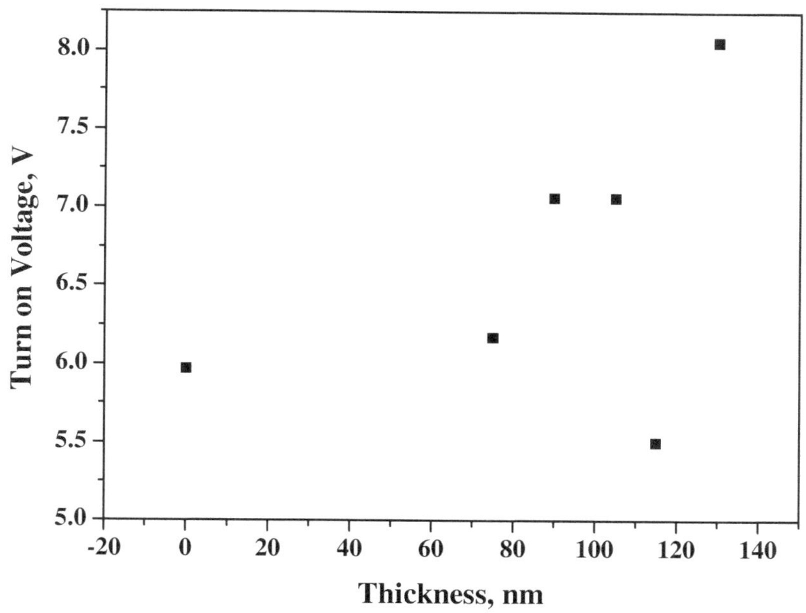

Figure 9. Turn-on voltage for OLED devices with different thickness of CdS with 77 nm PVK and 84 nm Alq_3.

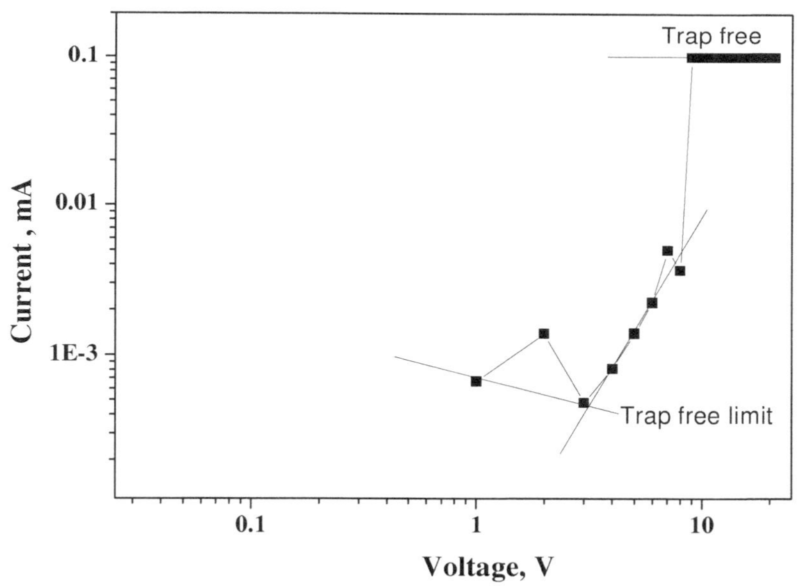

Figure 10. I-V characteristic on a log-log scale for the devices with 84 nm Alq_3, 115 nm CdS and 77 nm PVK PLED.

distinct regions, characteristic of a one-carrier single set of traps. These regions include:

(1) The Ohmic region, where the current density is proportional to the voltage;
(2) SCLC trap limited region;
(3) Trap-filled limit, where the onset trap-filled voltage (VTFL) results in the density of traps; and

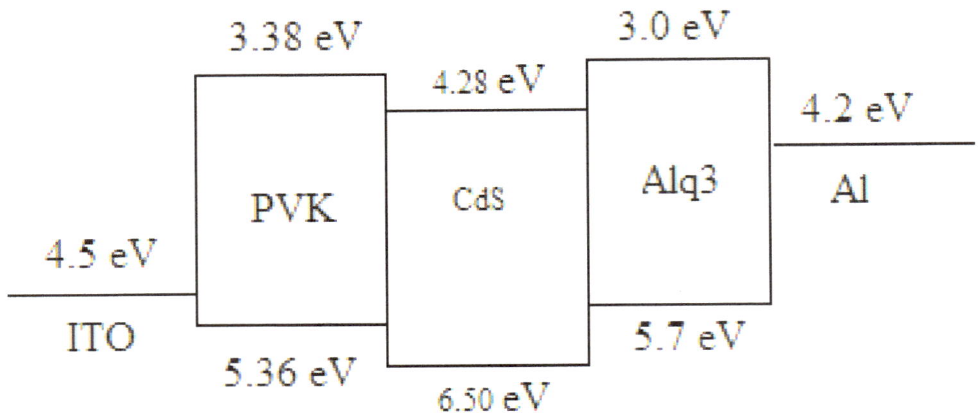

Figure 11. Energy band diagram of ITO / NPB (55 nm) / CdS (115 nm) / Alq_3 (84 nm) / Al (300 nm) device.

(4) The final region, in which the current is only limited by the space charge and is free from the influence of traps.

With the CdS layer found in the electric field region, where only a part of the deep traps become filled, the increase of current with increasing electric field is fast, resulting in the power law dependence with a power larger than 2. This is the so-called trap-filled limited region, where ionic defects can create the relevant deep-trap energy levels in the band gap.

To improve the efficiency of the PLED, inserting a layer between the HTL and ETL is needed in order for the luminescence values to result at a low current density. Figure 11 shows a proposed energy level diagram of the PLED device.

The lower HOMO level of CdS (6.50 eV), than those of Alq_3 and PVK, leads to better confinement of holes within the emitting layer. This is due to the efficient electron injection from the ETL into the emitting layer. The mobility electron in the material is higher and it increases the quantum efficiency of the device. This is in good agreement with the findings of the literature, which stipulates that the quantum efficiency improves if a material with high triplet energy and low HOMO level is used as the HBL.

Conclusions

The OLEDs with the ITO/ PVK (77 nm) / CdS (130 nm) / Alq_3 (84 nm) /Al have been successfully fabricated using thermal evaporation and dip coating method. The device with 115 nm of CdS produces the lowest current turn-on voltage which is 5.4 V and gives the highest intensity of the light emission.

ACKNOWLEDGEMENTS

The authors gratefully acknowledge the financial support received from Malaysia Government through Fundamental Research Fund (FRGS 01-11-08-664FR/5523664). More so, Mei Yee is grateful for the financial support of NSF scholarship (2009 to 2010).

REFERENCES

Adachi C, Kwong R, Djurovichi P, Adamovichi V, Baldo M, Thompson ME, Forrest SR (2001). Endothermic energy transfer: A mechanism for generating very efficient high-energy phosphorescent emission in organic materials. Appl. Phy. Lett., 79: 2082-2085.

Alex CJ, Luo Z, Seo J, Cheung N, Lee LP, Sands TD (2004). Heterogeneous integration of CdS filters with GaN LEDs for fluorescence detection microsystems. Sensors Actuators, 111: 1-7.

Baldo MA, Lamansky S, Burroes PE, Thompson ME, Forrest SR (1999). Very high-efficiency green organic light-emitting devices based on electrophosphorescence. Appl. Phys. Lett., 75(1): 4-6.

Baldo MA, O'Brien DF, You Y, Shoustikov A, Sibley S, Thompson ME, Forrest SR (1998). Highly eficient phosphorescent emission from organic electroluminescent devices, 395: 151- 154.

Braun D, Heeger AJ (1991). Visible light emission from semiconducting polymer diodes. Applied Physics Letters 58:1982-1984.

Burroughes JH, Bradley DDC, Brown AR, Marks RN, Mackay K, Friend RH, Burns PL and Holmes AB (1990). Light-emitting diodes based on conjugated polymers, 347: 539-541.

Chandra S, Pandley RK, Agrawal RC (1980). Solar Energy Conversion by Photoelectrochemical Cells using Chemical-bath-deposited CdS Films. J. Phy. D: Appl. Phys., 13: 1757-1760.

Dai CM, Horng L, Hsieh WF, Shih YT (1992). High orientation CdS thin films grown by pulsed laser and thermal evaporation. J. Vac. Sci. Technol. A, 10: 484-488.

Eih-Zhe L, Ching-Fuh L, Ting-Wien S, Wu-Ping H, Hsing-Hung H (2003). Light emitting diodes on Si. Proceedings of SPIE, 4996: 46-56.

Ikai M, Tokito S, Sakamoto Y, Suzuki T, Taga Y (2001). Highly efficient phosphorescence from organic light-emitting devices with an exciton-block laye Appl. Phys. Lett., 79: 156- 158.

Jae WK, Deug SL, Hyung DP, Young SP, Ji WK, Won IKJ, Kyung MY, Kyoungmoon G, Se HK, Jang JK (2007). Silane and triazine containing hone and exciton blocking material for high efficiency phosphorescent organic light emitting diodes. J. Mat. Chem., 17: 3714-3719.

Johnson GE, McGrane KM, Stolka M (1995). Electroluminescence from single layer molecularly doped polymer films. Pure Appl. Chem., 67: 175-182

Kang JW, Lee SH, Park HD, Jeong WI, Yoo KM, Park YS, Ikai M, Tokito S, Sakamoto Y, Suzuki T, Taga Y (2001). Highly efficient phosphorescence from organic light-emitting devices with an exciton-block layer Appl. Phys. Lett., 79: 156- 158.

Kim JJ (2007). Low roll-off of efficiency at high current density in phosphorescent organic light emitting diodes. Appl. Phys. Lett., 90: 223508-223511.

Kim JS, Friend RH, Cacialli F (1999). Improved operational stability of polyfluorene based organic light-emitting diodes with plasma-treated indium-tin oxide anodes. Appl. Phys. Lett., 74: 3084-3086.

Kolosov D, Adamovichi V, Djurovichi P, Thompson ME, Adachi C (2002). 1,8-Naphthalimides in Phosphorescent Organic LEDs: The Interplay between Dopant, Exciplex, and Host Emission. J. Am. Chem. Soc., 124: 9944-9945.

Kwong RC, Nugent MR, Michalski L, Ngo T, Rajan K, Tung YJ, Weaver MS, Zhou TX, Hack M, Thompson ME, Forrest SR, Brown JJ (2001). High operational stability of electrophosphorescent devices. Appl. Phys. Lett., 81: 162-164.

Peng HJ, Zhu XL, Sun JX, Yu XM, Wong M, Kwok HS (2006). Efficiency improvement of phosphorescent organic light-emitting diodes using semitransparent Ag as anode. Appl. Phys. Lett., 88: 033509-033511.

Tsutsui T, Yang MJ, Yahiro M, Nakamra K, Watanabe T, Tsuji T, Fukuda Y, Wakimoto T, Miyaguchi S (1999). High Quantum Efficiency in Organic Light-Emitting Devices with Iridium-Complex as a Triplet Emissive Center. Japanese J. Appl. Phys., 38: L1502-L1504.

Vadim I, Adamovich, Steven R, Cordero A, Peter I, Djurovich, Arnold T, Mark E, Thompson, Brian W, Andrade D, Stephen R (2003). Forrest.New charge-carrier blocking materials for high efficiency OLEDs. Organic Electron., 4: 77-87.

.

An investigation of the effect of hall current and rotational parameter on dissipative fluid flow past a vertical semi-infinite plate

J. G. Abuga*, M. Kinyanjui and J. K. Sigey

Department of Mathematics and Statistics, Jkuat, P.O.BOX 62000, Nairobi,Kenya.

Magneto hydrodynamics (MHD) stokes problem for a vertical semi-infinite plate in a dissipative rotating fluid with hall current has been considered. The partial differential equations governing the problem are framed and then solved using numerical methods of implicit finite difference approximations. An analysis of effects of parameter on the velocity (both primary and secondary) profiles and temperature distribution profiles are shown graphically and the results are discussed.

Key words: Hall current, dissipative fluid, rotational parameter.

INTRODUCTION

Considerable progress has been made recently in the general theory of rotating fluids due to its various applications especially in the geological sciences. The application of rotating fluid are in oceanic, atmospheric and in cosmic fluid dynamics and solar physics (Kinyanjui and Uppal, 1998). Takhar and Soundalgekar (1976) did as study on the viscous dissipation effects on heat transfer in boundary layer flow past a semi-infinite horizontal flat plate and in 1977 they studied a forced and free convective flow past a semi- infinite vertical plate and also the MHD and heat transfer over a semi-infinite plate under a transverse magnetic field. Some of their other works include dissipation effects of MHD flow past a semi-infinite vertical plate, MHD free convection flow past a vertical semi-infinite plate with a uniform free-stream in 1985 and MHD free convection flow past a semi-infinite plate with uniform heat flux also in 1985.

Soundalgekar et al. (1979 a) studied free convection effects on MHD stokes problem for a vertical plate and they discovered that skin friction increased owing to a greater heating of the plate. Soundalgekar et al. (1979b) studied the finite difference analysis of free convection effect on stokes problem for a vertical plate in dissipative fluid. Soundalgekar et al. (1981) studied the MHD stokes problem for a vertical infinite plate with variable temperature. Chartuverdi (1996) studied the finite difference study of MHD stokes problem for a vertical infinite plate in a dissipative heat generating fluid with hall and Ion-slip current. Takhar and Soundalgekar, (1997) studied the forced and free convective flow past a semi-infinite vertical plate and MHD and heat transfer over a semi-infinite plate under a transverse magnetic field.

Kinyanjui and Uppal (1998) studied the MHD stokes problem for a vertical infinite plate in a dissipative rotating fluid with hall current and they later investigated the effect of both hall and Ion-slip currents on the flow of heat generating rotating fluid system. They observed that for an Eckert value of 0.02, there was a decrease in the primary velocity profile with an increase in rotational parameter but in the case of secondary velocity profiles, there was initially a decrease and as the distance from the plate increased, the secondary velocity profile increased. They also observed that an increase in hall parameter has no effect on the temperature profile but an increase in time causes an increase in the temperature profiles. Kinyanjui et al. (1999) studied the finite difference analysis of free convection effects on MHD problem for a vertical plate in a dissipative rotating fluid system with constant heat flux and hall current and also did a finite difference analysis of MHD stokes problem for a vertical infinite plate in a dissipative fluid with constant heat and hall current.

Kinyanjui et al. (2001) studied Magnetohydrodynamic free convection heat and mass transfer of a heat

*Corresponding author. E-mail: jadeabuga@yahoo.com.

generating fluid past an impulsively started infinite vertical porous plate with hall current and radiation absorption MHD stokes free convection flow past an infinite vertical porous plate subjected to constant heat flux with ion-slip current and radiation absorption was studied (Kwanza and Uppal, 2003).

Chamkha (2004) studied unsteady MHD convective heat and mass transfer past a semi-infinite vertical permeable moving plate with heat absorption. The presence of heat absorption (thermal sink) effects had the tendency to reduce the fluid temperature. This caused the thermal buoyancy effects to decrease resulting in a net reduction in the fluid velocity.

Hazem (2008) studied the effect of hall current on transient hydromagnetic Couette–Poiseuille flow of a viscoelastic fluid with heat transfer. Palani et al. (2009) studied MHD flow past a semi-infinite vertical plate with mass transfer. It observed that the magnetic parameter M has a retarding effect on velocity (Palani et al., 2009). Sweet et al. (2011) studied the analytical solution for the unsteady MHD flow of a viscous fluid between moving parallel plates. They discovered that an increase in the fluid density corresponded to a decrease in the velocity in the y-direction and that the initial velocity in the x-direction decreased with an increase in the fluid density.

In spite of all these studies, effect of hall current on a dissipative rotating fluid past a vertical semi-infinite plate has received little attention. Hence, the main objective of the present study is to investigate the MHD stokes problem for a vertical semi-infinite plate in a dissipative rotating fluid with hall current.

MATHEMATICAL ANALYSIS

We consider the flow of a viscous incompressible magneto hydrodynamics (MHD) free-convection heat generating fluid past an impulsively started semi-infinite vertical plate. The plate is jerked into motion in its own plane with constant velocity.

Since the motion of the boundary is in the x-direction then it may be assumed that the motion of the fluid will also be in that direction. Assume that a strong magnetic field of uniform strength H_0 is applied perpendicular to the direction of the flow as shown in Figure 1.

It is assumed that the induced magnetic field is negligible such that $H = (0,0,H_o)$. The assumption is justified because the magnetic Reynold's number of a partially ionized fluid is very small. The system is considered to be rotating with a uniform angular velocity Ω about the z-axis is taken normal to the plate. Since the plate is semi which -infinite in length, the variables are functions of z^+ and t^+ only. At time $t^+ > 0$, the plate start moving impulsively in its own plane with constant velocity v_0 and its temperature is instantaneously increased or decreased to $T^+{}_w$ which is constantly maintained later. The temperature of the fluid and the plate are assumed to be the same initially.

The equation of conservation of electric charge $\nabla . J = 0$ gives j_{z^+} = constant where $J = (j_{x^+}, j_{y^+}, j_{z^+})$, this constant equal to zero, $j_{z^+} = 0$ at the plate which is non-conducting thus $j_{z^+} = 0$ everywhere in the flow. The generalized Ohm's law must be modified to include Hall current as follows:

$$\vec{J} + \frac{\varpi_e \tau_e (\vec{J} \times \vec{H})}{H_0} = \sigma\left(\vec{E} + \mu_e \vec{q} \times \vec{H} + \frac{1}{e\eta_e} \nabla p \right) \quad (1)$$

Where $\sigma, \mu_e, \tau_e, e, \eta_e, P_e$ are electrical conductivity, magnetic permeability, cyclotron frequency, collision time, electric charge, the number density of electrons and the electron pressure respectively.

It is assumed that $\varpi_e \tau_e \le 1$. The applied magnetic field is assumed to be zero and the pressure gradient may be neglected. The effect of Ionslip and thermoelectric are neglected. Thus equation 1 becomes:

$$(j_{x^*}, j_{y^*}) + \frac{\varpi_e \tau_e (j_y . H_0, -j_x . H_0)}{H_0} = \sigma\mu_e (v^* H_0, -u^* H_0) \quad (2)$$

Equating x^+ and y^+ component in the above equation yields:

$$j_x + \varpi_e \tau_e j_y = \sigma\mu_e v^* H_0 \quad (3)$$

$$j_y + \varpi_e \tau_e j_x = \sigma\mu_e u^* H_0 \quad (4)$$

Calculating j_{x^+} and j_{y^+} we have:

$$j_{x^+} = \frac{\sigma\mu_0 H_o [v^+ + mu^+]}{1+m^2} \quad (5)$$

$$j_{y^+} = \frac{\sigma\mu_0 H_o [mv^+ + u^+]}{1+m^2} \quad (6)$$

When the effect of rotation is considered, the coriolis force has to be included in the momentum equation. Considering a rotating frame of reference with a uniform angular velocity Ω, the equations of motion become:

$$\frac{\partial u^+}{\partial t} - 2\Omega v^+ = \frac{\upsilon \partial^2 u^+}{\partial z^2} + g\beta^* (T - T^+{}_\infty) + \frac{\mu_0 H_0}{\rho} j_{y^+} \quad (7)$$

$$\frac{\partial v^+}{\partial t} - 2\Omega v^+ = \frac{\upsilon \partial^2 u^+}{\partial z^2} - \frac{\mu_0 H_0}{\rho} j_{x^+} \quad (8)$$

$$\frac{\partial T^+}{\partial t} = \frac{K}{\rho C_p} \frac{\partial^2 T^+}{\partial z^{+2}} + \frac{v}{C_p}\left[\left(\frac{\partial V^+}{\partial z^+} \right)^2 + \left(\frac{\partial u^+}{\partial z^+} \right)^2 \right] \quad (9)$$

where g is the acceleration due to gravity, β^* is the volumetric

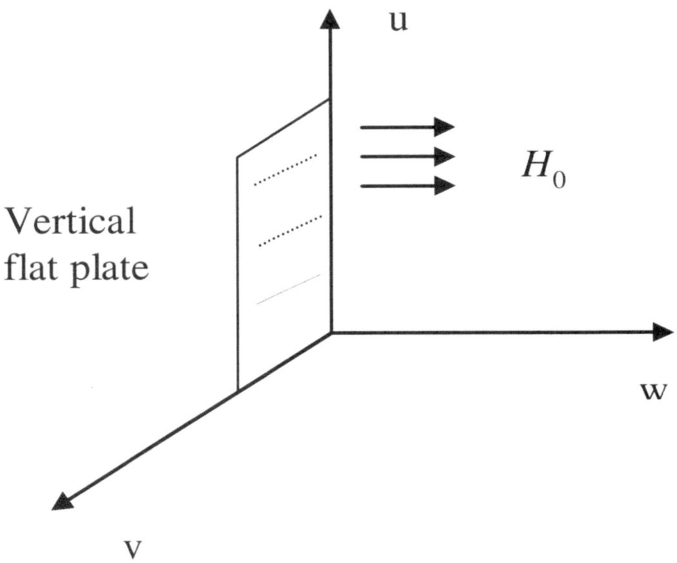

Figure 1. The flow configuration with co-ordinate system.

coefficient of thermal expansion, T^+, T^+_∞ are the temperature in the boundary layer and free- stream respectively, the fluid density , k is the thermal conductivity, C_p is the specific heat at a constant pressure, v is the kinematic viscosity, J_{x+}, J_{Y+} are the current density components and u^+, v^+ are the components in the X^+ and Y^+ directions.

The non-dimensional quantities are:

$$t=\frac{t^+u_o^{\ 2}}{\nu},\quad z=\frac{z^+u_0}{\nu},\quad u=\frac{u^+}{u_o},\quad v=\frac{v^+}{u_o},$$

$$x=\frac{x^+u_0^{\ 2}}{\nu},\quad y=\frac{y^+u_0^{\ 2}}{\nu},\quad Ec=\frac{U_o^{\ 2}}{Cp\left(T^*_s-T^*_\infty\right)},$$

$$\theta=\frac{\left(T^+-T^+_\infty\right)}{\left(T_s^{\ +}-T_\infty^{\ +}\right)},\ \Pr=\frac{\upsilon\rho C_p}{K}=\frac{\mu C_p}{K},\ Gr=\frac{\upsilon g\beta\left[T^*_s-T^*_\infty\right]}{U_o^{\ 3}}$$

$$Er=\frac{\Omega\upsilon}{U^2_{\ 0}},\ M_l=\sqrt{\frac{\sigma\mu^2{}_eH^2{}_o\upsilon}{\mu U^2{}_o/\upsilon}}=\sqrt{\frac{\text{Magnetic force}}{\text{inertia force}}}$$

Where

Gr- Grashof number
Pr- Prandtl number
Er- Rotational parameter
M^2- Magnetic parameter
Ec- Eckert number

Using the non-dimensional parameters, the governing equations of the fluid flow can be expressed in non-dimensional form as:

$$\frac{\partial\vec{q}}{\partial t}=\frac{\partial^2\vec{q}}{\partial z^2}+Gr\theta-M^2\vec{q} \tag{10}$$

$$\Pr\frac{\partial\theta}{\partial t}=\frac{\partial^2\theta}{\partial^2 t}+\Pr Ec\left[\left(\frac{\partial\vec{q}}{\partial z}\right)\left(\frac{\partial q}{\partial z}\right)\right] \tag{11}$$

The initial conditions at t< 0 take the form

$$q(z,t)=0\qquad \theta(z,t)=0$$

The initial conditions at $t\geq 0$ take the form

$$q(0,t)=1\qquad \theta(0,t)=1$$

The boundary conditions as $z\longrightarrow\infty$ take the form

$$q(\infty,t)=0\qquad \theta(\infty,t)=0$$

SOLUTION OF THE PROBLEM OF FINITE-DIFFERENCE METHOD

We solve our system of equations using the finite-difference method. In the finite-difference method, our Equations 10 and 11 take the form:

$$\frac{q(i,j+1)-q(i,j)}{\Delta t}=\frac{q(i+1,j)-2q(i,j)+q(i-1,j)}{\Delta z^2}-m^2q+Gr\theta(i,j) \tag{12}$$

$$\Pr\frac{\theta(i,j+1)-\theta(i,j)}{\Delta t}=\frac{\theta(i+1,j)-2\theta(i,j)+\theta(i-1,j)}{(\Delta z)^2}+\Pr Ec\left(\frac{q(i+1,j)-q(i,j)}{\Delta z}\right)\left(\frac{\vec{q}(i+1,j)-\vec{q}(i,j)}{\Delta z}\right) \tag{13}$$

The index i refers to z and j to time t and $\Delta z=0.1$

DISCUSSION

The results obtained are represented in form of graphs. The Prandtl number Pr is taken to be 0.71 (corresponding to air) and the magnetic parameter M^2 is taken to be 5.0 signifying a strong magnetic field. Gr > 0 (= 5) corresponds to cooling of the plate by free convection currents since the plate is at higher temperature than the surrounding and Gr < 0 (= -5) to heating of the plate by free convection currents since the plate is at a lower temperature that the surrounding.

In the presence of cooling of the plate by free convectional currents that is, when the Grashoff's number is greater than zero (equal to five) from Figures 2 and 3 we note that for Ec=0.01, an increase in the rotational parameter leads to a decrease in both the primary and secondary velocity profiles and for Ec=0.02, in Hall parameter for Ec = 0.01 and Ec = 0.02 leads to an increase both the primary velocity profiles and the secondary velocity profiles. We also observe that an increase in time for both Ec = 0.01 and Ec = 0.02 causes an increase in both the primary and secondary velocity. From Figure 4, the temperature velocity profiles for Gr >0, we observe that an increase in the rotational parameter

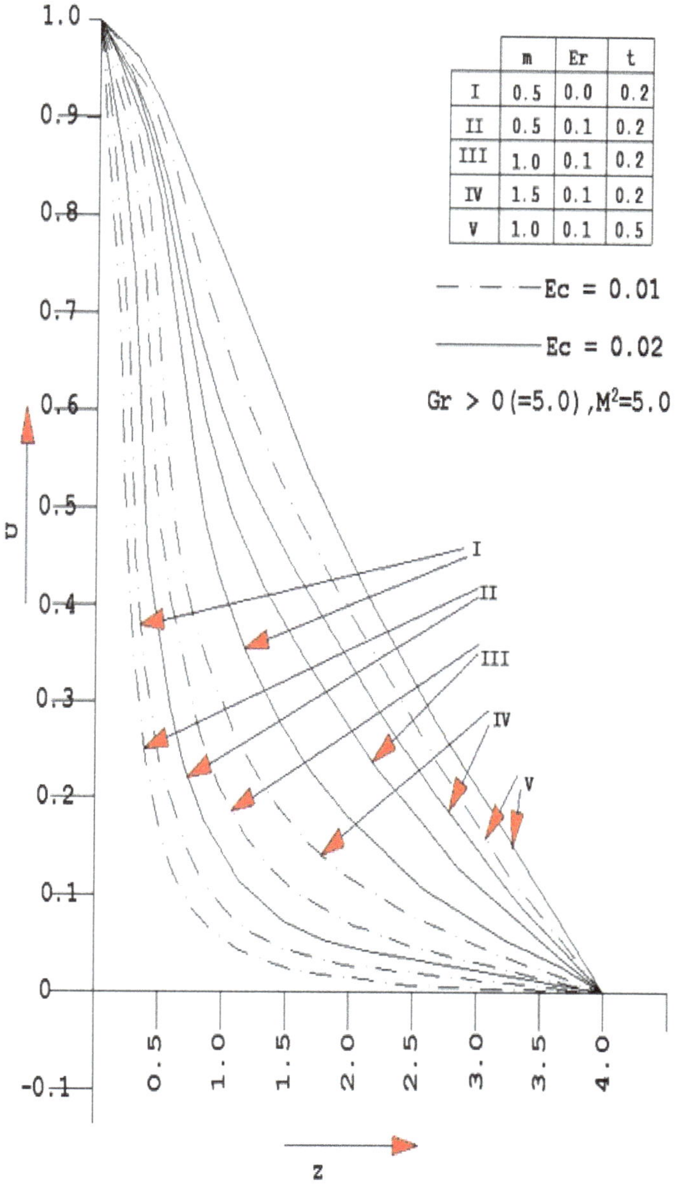

Figure 2. Primary velocity profiles for case of cooling plate by free convectional currents strong magnetic field with Eckert numbers of 0.01 and 0.02.

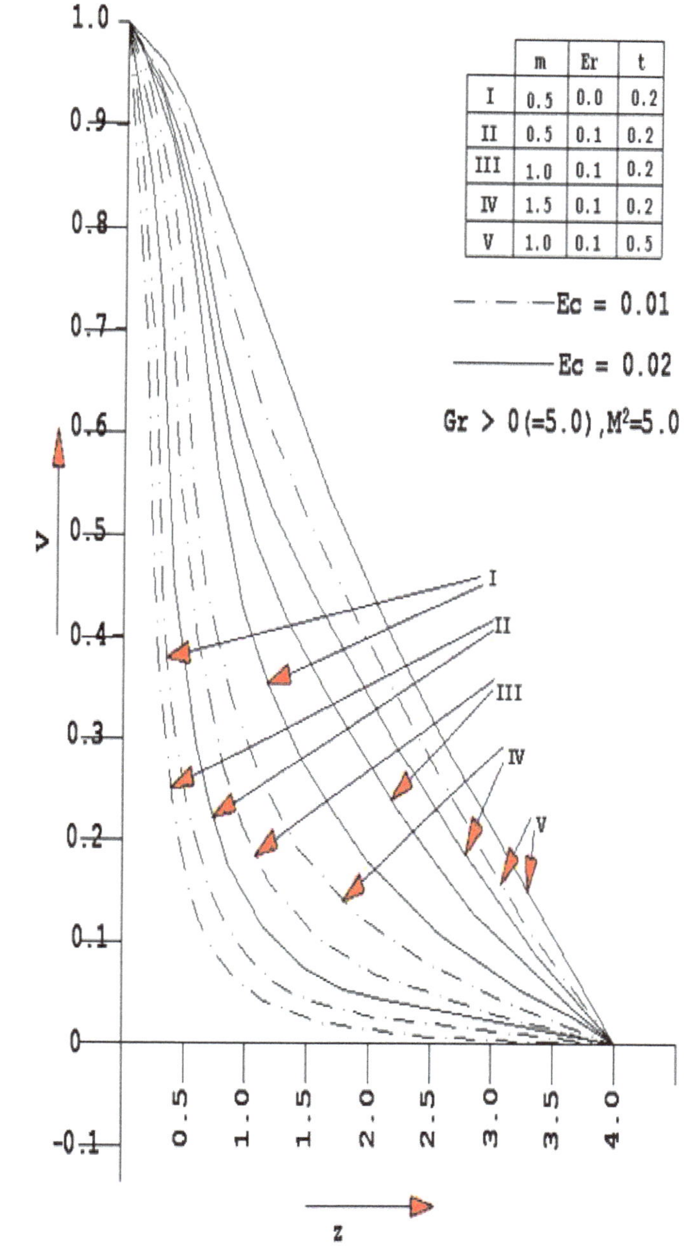

Figure 3. Secondary velocity profiles for case of cooling plate by free convectional currents strong magnetic field with Eckert numbers of 0.01 and 0.02.

Er, leads to an decrease in the temperature profiles for Ec = 0.01 but there is no change for Ec = 0.02 in the temperature profiles and an increase in Hall parameter m for both Ec = 0.01 and Ec = 0.02 causes no change in the temperature profiles. An increase in time causes an increase in the temperature profiles for both Ec = 0.01 and Ec = 0.02.

In case 2, in the presence of heating of plate by free-convection current for the case of Gr < 0 (= -5) from Figures 5 and 6 we observe that for Ec=0.01, an increase in the rotational parameter leads to a decrease in both the primary and secondary velocity profiles. For Ec=0.02, an increase in the rotational parameter leads to an increase in both the primary and secondary velocity profiles and an increase in hall parameter for Ec = 0.01 and Ec = 0.02 leads to an increase both the primary velocity profiles and the secondary velocity profiles. An increase in time for both Ec = 0.01 and Ec = 0.02 causes an increase in both the primary and secondary velocity.

From Figure 7, we observe that change in the rotational

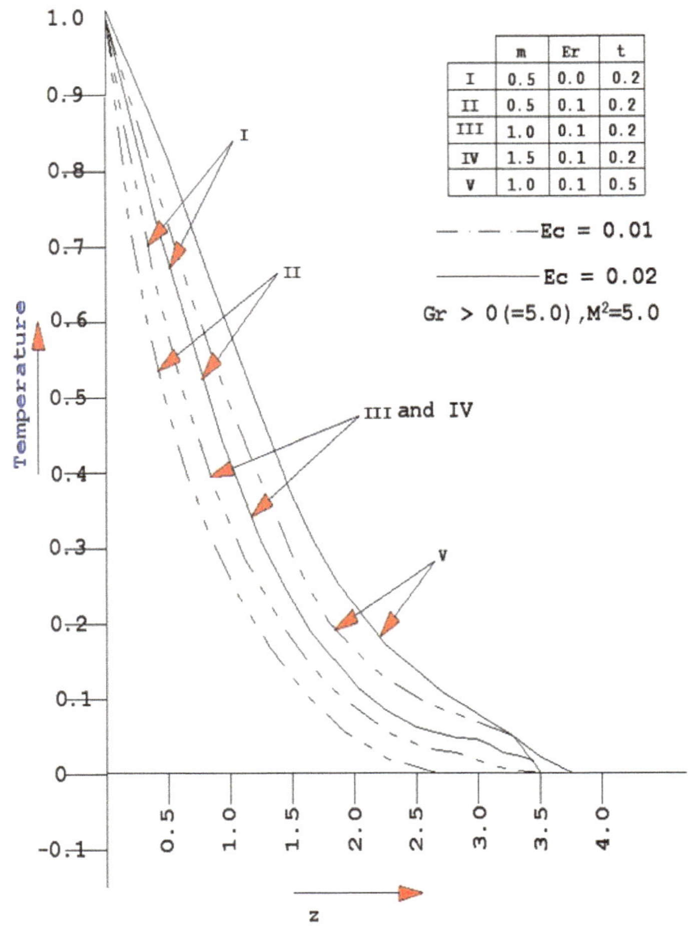

Figure 4. Temperature distribution profiles for case of cooling plate by free convectional currents strong magnetic field with Eckert numbers of 0.01 and 0.02.

parameter causes no change in the temperature profiles. An increase in the Hall parameter causes no change in the temperature profile and also an increase in time causes a decrease in the temperature profiles.

Conclusion

This paper considered the flow of a viscous incompressible magneto hydrodynamics (MHD) free-convection heat generating fluid past an impulsively started semi-infinite vertical plate. The governing equations were non-dimensionalized and then transformed using implicit-finite difference method. The following points were concluded:

1) An increase in hall parameter for both cooling of the plate by free convection currents and heating of the plate by free convection currents led to an increase in the velocity profiles.

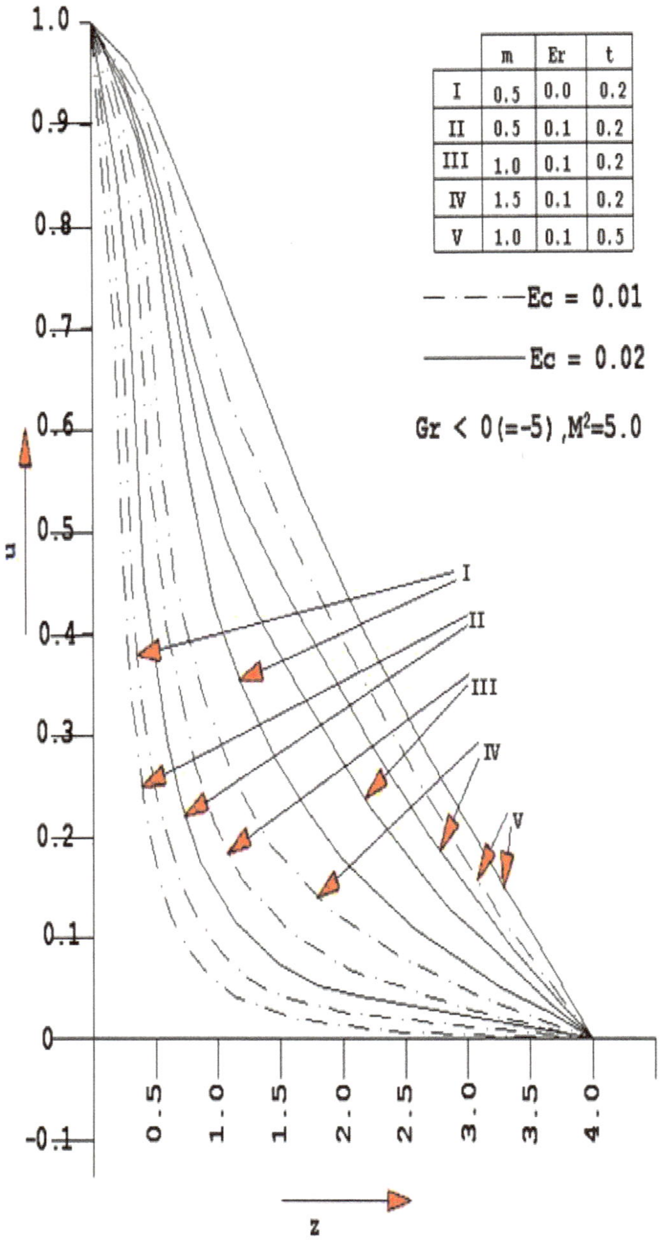

Figure 5. Primary velocity profiles for the case of heating plate by free convectional currents strong magnetic field with Eckert numbers of 0.01 and 0.02.

2) An increase in hall parameter for both cooling of the plate by free convection currents and heating of the plate by free convection currents has no effect on the temperature profiles.

3) An increase in rotational parameter led to a decrease in the velocity profiles when the Eckert number was 0.01 and an increase in velocity profiles when the Eckert number was 0.02.

4) An increase in rotational parameter had no effect on temperature profiles except in the case of cooling the

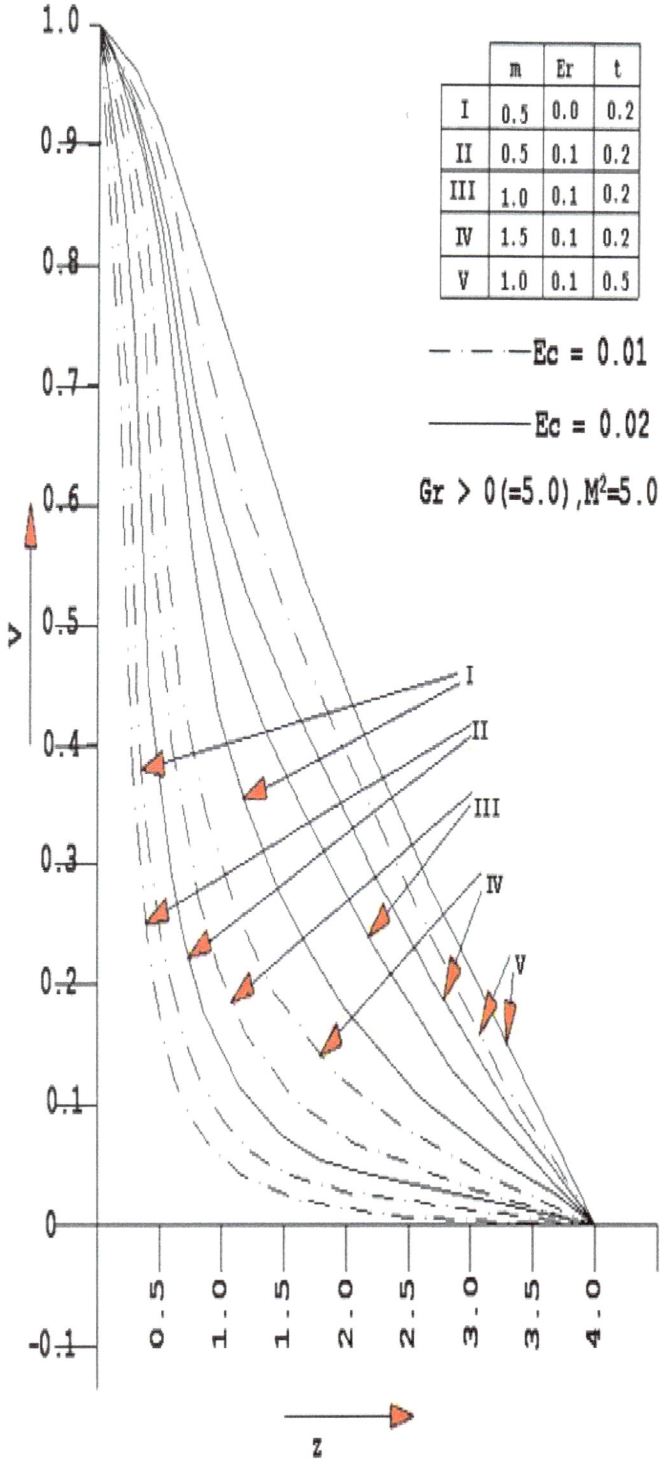

Figure 6. Secondary velocity profile for the case of heating plate by free convectional currents strong magnetic field with Eckert numbers of 0.01 and 0.02.

plate when the Eckert number was 0.01. In this case, it caused a decrease in the temperature profiles.

5) An increase in the time led to an increase velocity

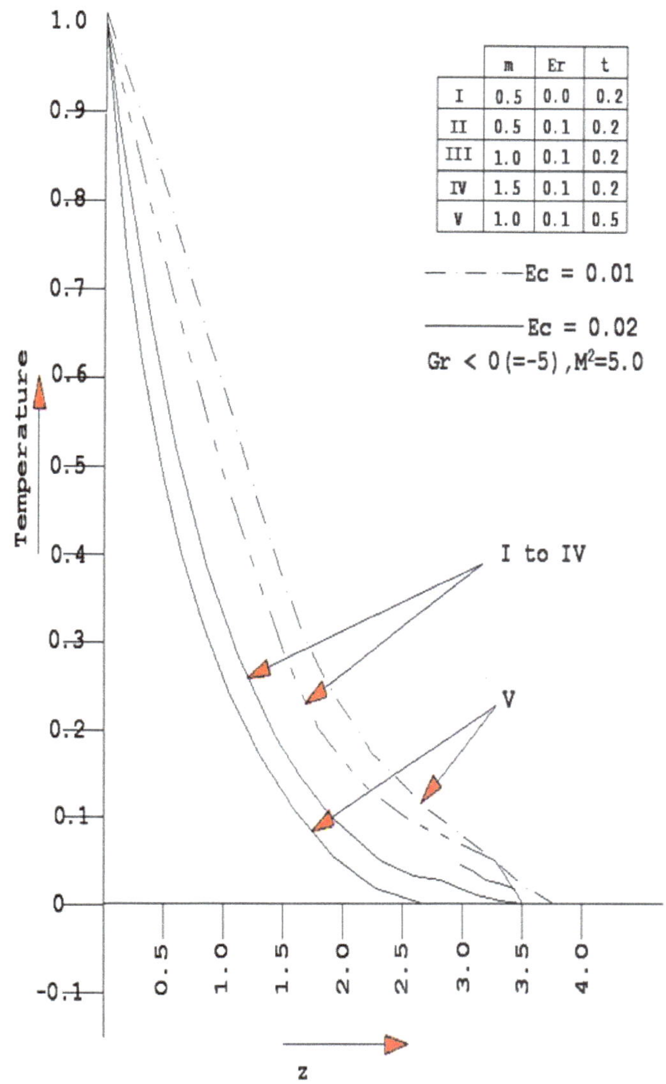

Figure 7. Temperature distribution profiles for the case of heating plate by free convectional currents strong magnetic field with Eckert numbers of 0.01 and 0.02.

(both primary and secondary) profiles in the case cooling the plate by free convection currents and led to a decrease in velocity profiles in the case of heating the plate by free convection currents.

It is hoped that the results will be useful for applications and that they can also be used for comparison with other problems dealing with Hall current and rotational parameter which might be more complicated .It is also hoped that the results can serve as a complement to other studies.

REFERENCES

Chamkha JA (2004). Unsteady MHD convective heat and mass transfer

past a semi-infinite vertical permeable moving plate with heat absorption. Int. J. Eng. Sci., 42: 217–230.
Chartuverdi N (1996). Finite difference study of MHD stokes for a vertical infinite plate in a dissipative heat generating fluid with hall and Ion-slip current. Botswana J. Technol., 12: 21-31.
Hazem AA (2008). Effect of Hall current on transient hydromagnetic Couette–Poiseuille flow of a viscoelastic fluid with heat transfer. Appl. Math. Model., 32(4): 375-388.
Kwanza KM, Uppal SM (2003). MHD stokes free convection flow past an infinite vertical porous plate subjected to constant heat flux with ion-slip current and radiation absorption. Far East J. Appl. Math., 12: 105-131.
Kinyanjui CN, Uppal SM (1998). MHD stokes problem for a vertical infinite plate in a dissipative rotating fluid with Hall current. J. Magneto hydrodyn. Plasma Res., 8(1): 15-30.
Kinyanjui M, Chartuverdi N, Uppal SM (1999). Finite difference analysis of free convection effects on MHD problem for a vertical plate in a dissipative rotating fluid system with constant heat flux and hall current. J. Magneto hydrodyn. Plasma Res., 8(2/3): 191-224.
Kinyanjui M, Kwanza JK, Uppal SM (1999). Finite difference analysis on the MHD stokes for a vertical plate in a dissipative rotating fluid system with constant heat flux and Hall current. J. Magneto hydrodyn. Plasma Res., 8(4): 301-319.
Kinyanjui M, Chartuverdi N, Uppal SM (2001). Magnetohydrodynamic free convection heat and mass transfer of a heat generating fluid past an impulsively started infinite vertical porous plate with Hall current and radiation absorption. Energy Convers. Manage., 42(8): 917-931.
Palani G, Srikanth U (2009). MHD flow past a semi-Infinite vertical plate with mass transfer Nonlinear Analysis: Model. Control, 14(3): 345–356.
Sweet E, Vajravelu K, Robert A, Van Gorder PI (2011) Analytical solution for the unsteady MHD flow of a viscous fluid between moving parallel plates. Commun. Nonlin. Sci. Numer. Simul., 16: 266–273.
Soundalgekar VM, Bhat JP, Mohiuddin M (1979a). Finite difference analysis of free convection effects on stokes problem for a vertical plate in a dissipative fluid with constant heat flux. Int. J. Eng. Sci., 17(12): 1283-1288.
Soundalgekar VM, Gupta SK, Aranake RN (1979b). Free convection effects on MHD Stokes problem for a vertical plate. Nuclear Eng. Des., 51(3): 403-407.
Soundalgekar VM, Takhar HS (1977). Forced and free convective flow past a semi-infinite vertical plate, Israel J. Technol., 15: 368-372.
Soundalgekar VM, Patil MR, Jahagirdar MD (1981). MHD stokes problem for a vertical infinite plate with variable temperature. Nuclear Eng. Des., 64(1): 39-42.
Takhar HS, Soundalgekar VM (1976). Viscous dissipation effects on heat transfer in boundary layer flow past a semi-infinite horizontal flat plate. J. Flügwissenschaften, 24: 75-77.
Takhar HS, Soundalgekar VM (1997). Effects of viscous dissipation on heat transfer in an oscillating flow past a flat plate. Appl. Sci. Res., 33: 101-111.
Takhar HS, Soundalgekar VM (1985). MHD free convection flow past a vertical semi-infinite plate with a uniform heat flux. Mech. Res. Commun., 12: 155-162.

Atomic force microscopy (AFM) and X-ray diffraction (XRD) studies of $BiFeO_3$ nanoparticles with sol–gel method

A. Bahari[1]* and N. Noori[2]

[1]Department of Physics, University of Mazandaran, Babolsar, Iran.
[2]Department of Physics, Islamic Azad University, Sari Branch, Sari, Iran.

Recently, there has been considerable interest in studying nanostructural properties of Bismuth (Bi) as a good electrode of electro-analysis, shape–memory alloys and photo-typical multiferric devices. We have thus demonstrated a series of experiments to synthesis Bi and $BiFeO_3$ (BFO) nanoparticles using sol–gel method at ambient pressure and study their nanostructural properties using tapping mode of atomic force microscopy (AFM) and X-ray diffraction (XRD) techniques. These techniques lend themselves very well to investigate the structure of nanoparticles and to map out properties and structure on surfaces of Bi and multiferroic BFO. The present work shows that, all nanoparticles are crystallized by annealing at 500°C. The narrower and sharper peaks in 10% Bi respect to 15% Bi in BFO, indicate the compositional changes of BFO. High purity BFO-phase powders with particle size of about 70 nm have been successfully synthesized by the sol–gel process, followed by leaching in diluted nitric acid.

Key words: Bismuth, nanoparticles, X-ray diffraction (XRD), atomic force microscopy (AFM) techniques.

INTRODUCTION

Bismuth, a semi metallic element, has rhombohedral structure (its atoms are covalently bonded,), whilst BFO is an insulator which has different electrical polarization. The domain walls themselves can conduct electricity at room temperature. This domain wall is virtually a two-dimensional sheet through the material, because they are so small and can be moved. Due to the sensitivity of Bi to its environment and less carriers on the surface, it has great promise for future electronics, shape–memory alloys and phototypical multiferric which simultaneously exhibit two or more ferric properties (anti-ferromagnetism and ferro-electricity) (Ikemoto et al., 2007; Kim et al., 2010; Kharissova et al., 2009; Kim et al., 2009; Jun et al., 2010). Moreover, the stress in BFO can cause a spontaneous change in orientation or crystal structure and flip back again, and produce regions separated by domain walls. The Bi nanoparticles in BFO, can also control its properties by reducing the stress and changing orientations (Jun et al., 2005; Lu et al., 2006; Saito et al., 2010). The main reasons are due to weak covalent bonds of Bi atoms and formation of layer structure in rhombohedral Bi. In fact, the interlayer interactions cause semi-metallic behavior and weaken the covalent bonds of Bi.

Atomic force microscopy (AFM) and X-ray diffraction (XRD) techniques are rightly considered routine tools for studying the nanoparticles properties of Bi, as well as in the development of advanced materials, devices and processes analysis. The obtained results indicate that, Bi nanoparticles present in BFO exhibit a transformation from nanocrystalline rhombohedral structures to smaller amorphous particles at a thickness of about 10 nm with decreasing size. As Ikemoto et al. (2007) found that, the amorphous nanoparticles are semiconducting, and

*Corresponding author. E-mail: a.bahari@umz.ac.ir.

covalent interactions between Bi atoms can be increased with reducing the nanoparticles size. Many researchers (Kim et al., 2010; Kharissova et al., 2009; Kim et al., 2009; Jun et al., 2010) have synthesized nanoparticles of BFO with the thermal decomposition of the inorganic complex Bismuth hexacyanoferrate (III) tetrahydrate, self-catalyzed fast reaction process, self-catalyzed fast reaction process and rapid liquid-phase sintering (Navarro et al., 2010), we have demonstrated a series of experiments to synthesize nanoparticles of Bi and amorphous BFO which expected to be two key elements for the future of nano-electronic devices.

EXPERIMENTAL PROCEDURES AND DETAILS

The sol-gel process is commonly applied to synthesis such BFO materials, owing to its several advantages such as low temperature processing and the ability to prepare materials in various shapes, compared with the conventional preparation procedures of glass and ceramics (Qi et al., 2005; Chen et al., 2005). In this work, we prepare BFO by using hydrolysis procedure of $Bi(NO_3)_3$ and $Fe(NO_3)_3$ - which is transformed to rhombohedral phase by heating it at 500℃. It obviously depends on the preparation procedures and BFO content in the combination. By adding more Bi to BFO, the obtained powder trend to crystalline structure. For the preparation of BFO- Bi gel (Figure 1) follows, such that 0.02 mol $Bi(NO_3)_3$. $5H_2O$ and 0.02 mol $Fe(NO_3)_3$. $9H_2O$ are made as the prepared solution and transferred into the 30% dilute nitric acid. This transparent solution is then evaporated in a bath at 500℃ and the BFO with Bi nanoparticles are finally formed.

Furthermore, Tetraethyl-Orthosilicate (TEOS) (Merk, > 99) was hydrolyzed in that ethanol. Ethanol acts as a mutual solvent. TEOS, in ethanol was hydrolyzed with water containing acetic acid at room temperature (RT) for 30 min. The solution was then mixed with bismuth nitrate ($Bi(NO_3)_3$.) and iron nitrate ($Fe(NO_3)_3$) at 0℃ in specific molar ratio to obtain various content of BFO. After 30 min of stirring at room temperature, the sol was vibrated for 20 min in ultrasonic bath to deconglomerate particles and then relaxed at room temperature for 30 min. The sol was stirred at 60℃, until it became a gel and removed ethanol (about 24 h). After gelatin samples were dried at 60℃ to remove water and acetic acid and leave a transparent lump depending on BFO content. After that, the lump samples were milled with mortar and calcinated at 500℃. The thermal gradient during experiments procedure was 5°/min and the samples were put in oven during 2 h at calcinations temperature stated earlier. The procedure is shown in Figure 1. BFO is synthesized at the calcinations temperature of 500℃. The composition, structure and surface morphology of the BFO-Bi powder were investigated by XRD and AFM techniques.

EXPERIMENTAL RESULTS

In this work, AFM is used to map different topographic features as shown in Figures 2 and 3. This allows insight in the nanoscale structure and the corresponding properties changes which relates to crystallographic and compositional changes.

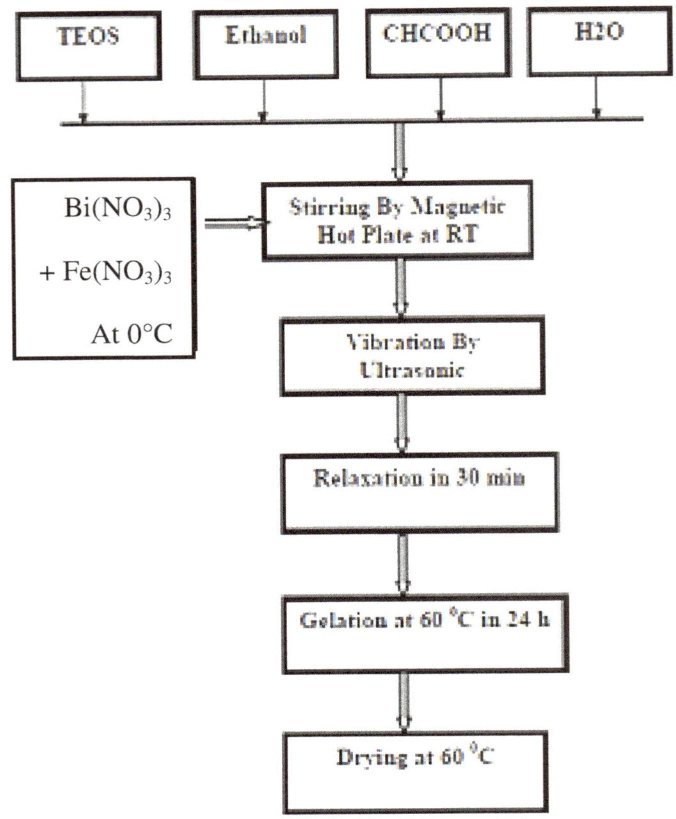

Figure 1. Flowchart of sol–gel process.

The AFM images obviously show much more detailed grain and nanostructure analysis than XRD allows. In the topographic image in Figure 3 bright colors are high structures while dark colors are low structures, due to different orientations of Bi grains. These different colors indicate different magnetic polarization as found in Ikemoto et al. (2007) and Kim et al. (2010), where weak contrasts at the background with many large pores mean a weak magnetic field. By measuring the nanoparticles size, different grain sizes areas are found in the outer layer (the range of nanoparticle's size is in 100 to 150 nm). It means different magnetic response and different amounts of pores. This leads to the assumption that, the pores can change the nanostructural properties of Bi as a good.

In parallel to AFM technique, patterns in the XRD spectra Figures 4 and 5, show the BFO powders with 0.005 M nitric acid and 10 or 15% Bi which annealed at 500℃. XRD patterns are measured on a (GBC-MMA 007 (2000)) X-ray diffractometer and recorded with (K_α (Cu), 1.54 Å), 0.020 step size, where the speed was (10°/min) radiation over a 2θ range of 10 to 60°. The range of Bi – nanoparticle's size is between 6 and 75 nm as

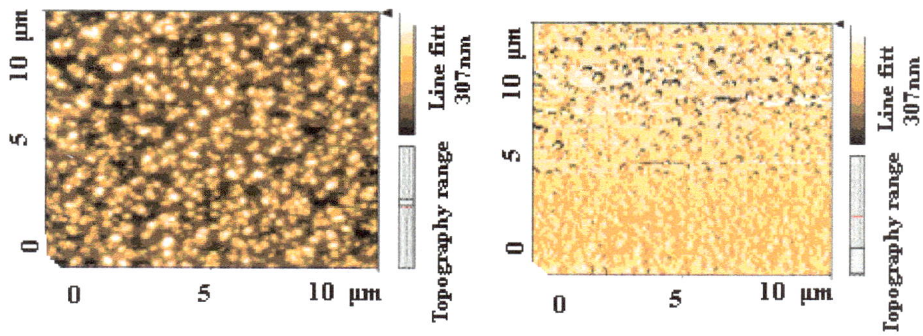

Figure 2. 2D (D: dimension) AFM microstructure of the BFO phase powder ceramics with 10% (left) and 15% (right) Bi at 10 µm x 10 µm size .

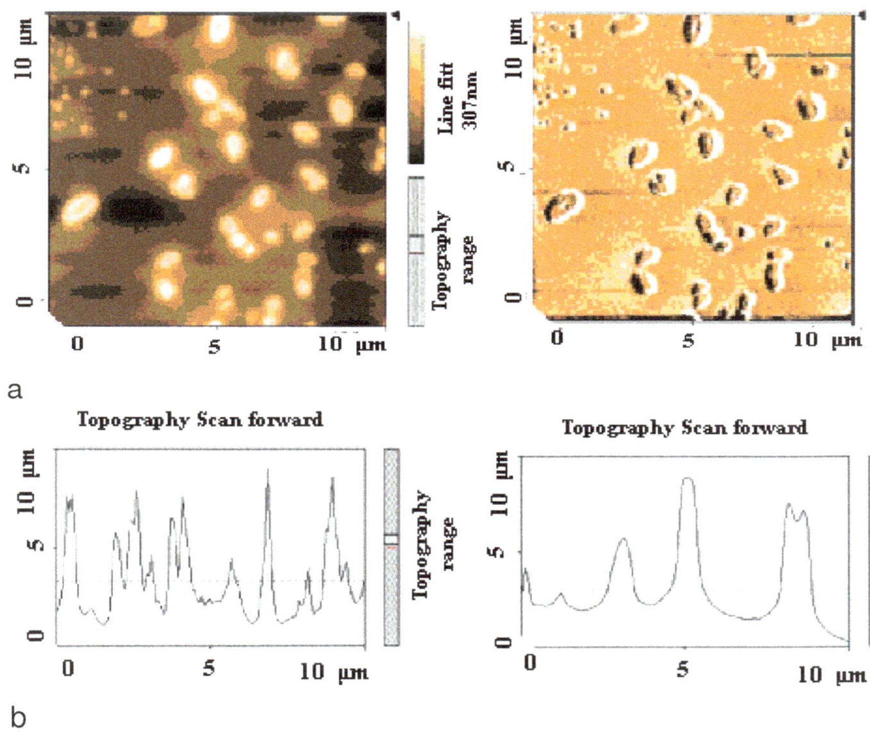

Figure 3. (a) 2D- AFM microstructure of the BFO phase powder ceramics with 10% (left) and 15% (right) Bi. Below the images are the profile (height) of the Bi and SFO respectively. The Bi nanoparticles look like dimer atoms in Si(100) 2 x 2 (Bahari et al., 2006a; Morgen et al., 2007; Bahari et al., 2008b; Bahari et al., 2008c; Bahari et al., 2009c; Bahari et al., 2009d). (b) AFM microstructure topography of the BFO phase powder ceramics with 10% (left) and 15% (right) Bi at 10 x 10 µm size.

determined by using X-powder method (Figure 6) agrees with the small value of average grain size which has been deduced from Scherer's equation (Mott et al., 1979):

$$D = \frac{K\lambda}{\beta c \cos\theta}$$

Where D is the crystallite size of nanoparticle, k_{α} is a constant (0.94), λ is the wavelength of X-ray ($Cu_{k\alpha}$ = 1.5406 Å), β is the true half- peak width, and θ is the half diffraction angle of the center of the peak in degree.

Strongest peaks at 22 and 32° indicate $BiFeO_3$:B(102) and B(110). $Bi_{36}Fe_{24}O_{57}$ phases are broad, which is due

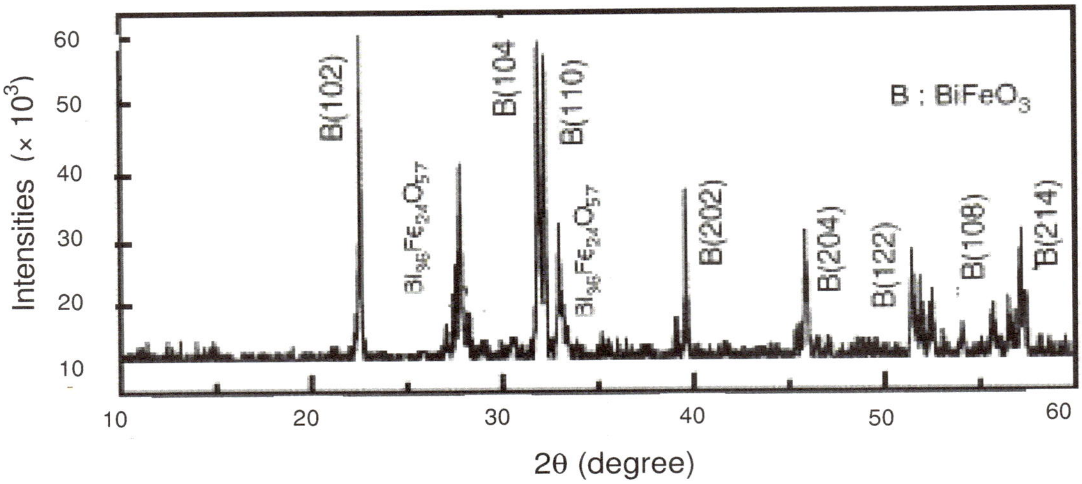

Figure 4. XRD patterns of the BFO powders with 0.005 M nitric acid and 10% Bi annealed at 500℃. The other small peaks are due to some Bi nanoparticles in the content which made non linear background.

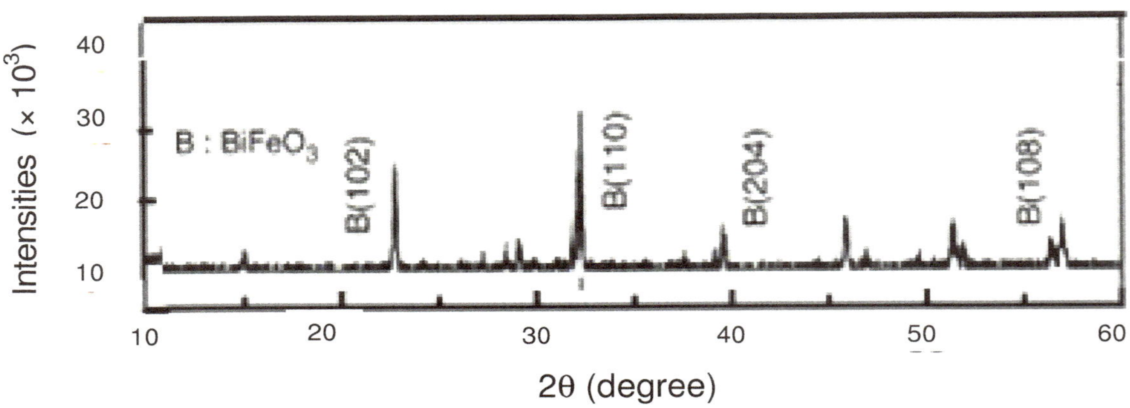

Figure 5. XRD patterns of the BFO powders with 0.005 M nitric acid and 15% Bi annealed.

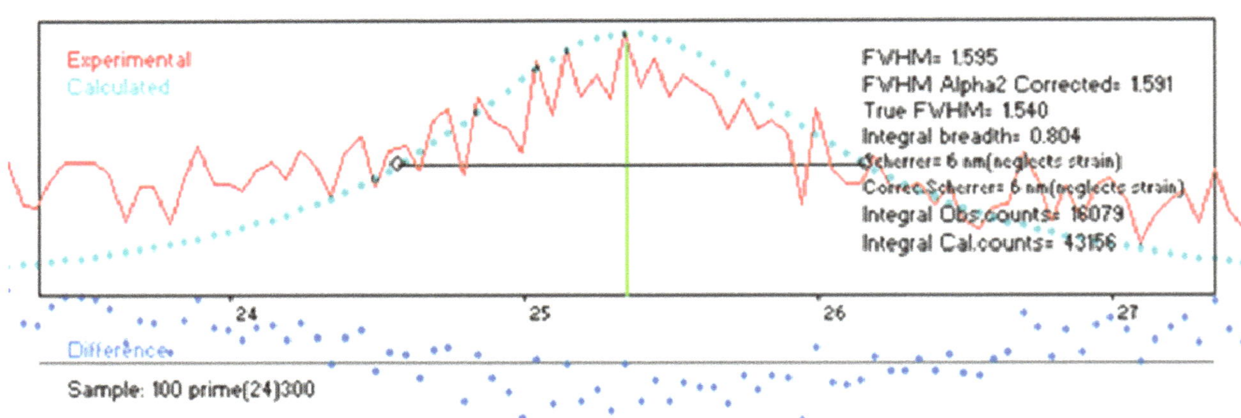

Figure 6. X–powder s patterns of the BFO powders with 0.005 M nitric acid and 15% Bi annealed at 500℃. The nanoparticle size is about 6 nm, as found by using X-powder method.

to small size of the Bi nanoparticles. Few broad peaks are obtained around 30, 35, and 48° in the background of XRD spectra, which are due to small size of Bi nanoparticles. B (104) peak is completely removed after adding 5% more of Bi nanoparticles in BFO. It could be attributed to weak bonds in this orientation and changeable crystalline state.

DISCUSSION

It is well-known that sol-gel method is a simple way to synthesize nanoparticle material. As stated earlier, Bi is a semi metallic element and its metallic behavior is so important in many industrial applications in catalysis, as a component in alloys with low melting point and polymetallic nanocomposites, for example Sn-Bi, $Bi_{1-x}CO_x$, Mn_5Bi_{95} and radiation sensors, among many other uses (Kim et al., 2010). In its nanostructural forms, as found in the present work, Bi has been rightly considered in a series of recent monographs, reviews, patents and a lot of experimental articles (Kim et al., 2010; Kharissova et al., 2009; Kim et al., 2009; Jun et al., 2010). By looking at XRD patterns and AFM images, we think the weakness of covalent bonds within the layer; reduce the inter-layer correlation in small Bi nanoparticles as found in Ikemoto et al. (2007).

Crystal structures and morphologies of BFO powders at every fabrication step were investigated by XRD diffractometer and AFM techniques. In order to eliminate the unwanted impurity phases, both BFO powders have been leached with diluted nitric acids (0.05 mol) for 30 min, which is known to be an effective method to eliminate the impurity phases from BFO powders prepared by a solid state reaction method (Bahari et al., 2006a) and also for getting better AFM images, as shown in Figures 2 through 3. Figures 4 and 5 show the XRD patterns of BFO powders annealed in air, followed by leaching in nitric acids with different Bi concentrations. Reflection peaks related with $Bi_{36}Fe_{24}O_{57}$ phase, were nearly disappeared after leaching in nitric acid solution. Based on these results, the peaks in the present work become a little broader and the intensities decrease with more amount of Bi nanoparticles and annealing at 500°C. It indicates that the crystallization phase is changed to an amorphous one.

Figures 4 and 5 show XRD patterns of the sol–gel synthesized BFO powders crystallized at 500°C for 30 min in air atmospheres with 10 and 15% of Bi, respectively. It was found that the majority of both powders were a rhombohedrally distorted perovskite BFO phase, as shown in Figures 4 and 5. Besides the formation of BFO phase, the formation of the impurity phases, such as Bi36Fe24O57, are confirmed from the XRD analysis. In the case of crystallization in air, the BFO powder contained a small amount of non-perovskite phases, such as $Bi_{36}Fe_{24}O_{57}$.

The aforestated figures show Bi – predominant phase in the BFO content is in z-direction. These nano-wires are not homogonously distributed on the BFO surface due to their interlayer interactions. From Mott (1979), the Bi atoms are weakly bonded in the background direction with the three neighbors of the next layer. The interlayer interactions cause semi metallic behavior of Bi and weaken the covalent bonds.

Conclusion

On the basis of the earlier stated results, it could be proposed that the presence of different sizes of Bi nanoparticles on top of the sample, causes the formation of two different magnetic properties of nanoparticles due to their different nanoparticles size. Nanoparticles of Bi described in this work, offer a new medium for synthesizing BFO nanostructures, for studying their unique electronic properties including finite-size effects, and for realizing large positive magneto resistance.

REFERENCES

Bahari A, Morgen P, Li Z (2006). Valence Bend Studies of the Formation Ultra thin Pure Silicon Nitride Films on Si(100). Surf. Sci., 600: 2966-2971.

Bahari A, Morgen P, Li Z (2008). Ultra thin silicon nitride films on Si(100) studied with core level photoemission. Surf. Sci., 602: 2315-2324.

Bahari A, Dlshadmanesh M (2008). Photoemission measurements of Ultrathin SiO_2 film at low take-off angles. IJNN, 4: 31-38.

Bahari A, Zokai M (2009). Nanostructural properties of Bismuth. WASJ, 7: 1-5.

Bahari A (2009). An alternative method for determining carbon nano tube band energy, Acta Phys. Polonica A., 115: 622-625.

Chen H, Li Z, Wu Z, Zhang Z (2005). A novel route to prepare and characterize Sn–Bi nanoparticles. J. Alloys Comp., 17: 282-285.

Ikemoto H, Yoshida S, Goyou A (2007). X–ray diffraction study of Bi nano particles. e–J. Surf. Sci. Nano Technol., 5: 110–120.

Jo Y, Jang K, Park J, Kim H, Hur N, Park S, Cheong S (2007). Measurements of the pressure and magnetic field dependence of multiferroic $DyMn_2O_5$. Phys. Rev. B., 76: 012406-012410.

Kharissova O, Osorio M, Kharisov B, Yacamán M Méndez U (2010). A comparison of bismuth nanoforms obtained in vacuum and air by microwave heating of bismuth powder., Mater. Chem. Phys., 121: 489–496.

Kim J, Kim Y, Connor P, Irvine J, Bae J, Zhou W (2009). Structural, thermal and electrochemical properties of the layered perovskite $SmBaCo_2O_5$d, a potential cathode material for intermediate-temperature solid oxide fuel cells. J. Power Sour., 194: 704-711.

Kim J, Kim S, Kim W (2009). Sol–gel synthesis and properties of multiferroic BiFeO3 powders. J. Eur. Ceram. Soc., 29: 2903-3104.

Kumar M, Palkar V, Srinivas K, Suryanarayana S (2000). Appl. Phys. Lett., 76: 2764-2773.

Lu J, Günther A, Schrettle F, Mayr F, Krohns S, Lunkenheimer P, Pimenov A, Travkin VD, Mukhin AA, Loidl A (2010). On the room temperature multiferroic $BiFeO_3$: Magnetic, dielectric and thermal properties. Eur. Phys. J., 75: 451–460.

Morgen P, Bahari A, Pedersen K, Li Z (2007). Plasma assisted growth of ultrathin nitrides on Si surfaces under ultrahigh vacuum conditions. J. Phys., 86: 12019-12038.

Mott NF, Davis EA (1997). Electronic progress in non–crystalline materials. Clarendon press, Oxford, p. 413.

Navarro M, Lagarrigue L, De Paoli M, Carbonio E, Go´mez I (2010). A new method of synthesis of $BiFeO_3$ prepared by thermal decomposition of $Bi[Fe(CN)_6].4H_2O$. J. Therm. Anal. Calorim., 102: 655–660.

Qi X, Wei M, Lin Y, Jia Q, Zhi D, Dho J, Blamire M, MacManus-Driscoll J (2005). High-resolution x-ray diffraction and transmission electron microscopy of multiferroic $BiFeO_3$ films. Appl. Phys. Lett., 86: 071913-071916.

Saito K, Ulyanenkov A, Grossmann V, Ress H, Bruegemann L, Ohta H, Kurosawa T, Ueki S, Funakubo X (2006). Structural Characterization of $BiFeO_3$ Thin Films by Reciprocal Space Mapping. J. Appl. Phys., 45: 7311-7314.

Effect of geopathic stress zone on human body voltage and skin resistance

N. P. Dharmadhikari[1]*, D. C. Meshram[2], S. D. Kulkarni[3], A. G. Kharat[4] and S. S. Pimplikar[5]

[1]Department of Applied Physics, Sinhgad Academy of Engineering, Kondhwa (Bk.) Pune-48, (MS) India.
[2]Department of Geology, University of Pune, Pune-7, (MS), India.
[3]Department of Environmental Science, New Arts Commerce and Sciences College, Ahmednagar-.414001, (MS), India.
[4]Department of Civil Engineering, Sinhgad Academy of Engineering, Kondhwa (Bk.) Pune-48, (MS), India.
[5]Department of Civil Engineering, Maharashtra Institute of Technology, Pune-38, (MS), India.

Geopathic stress zones are places on the surface of the Earth known to cause health problem. Empirical knowledge of the existence of geopathic stress is probably as old as mankind. The aim of this study is to evaluate the effect of geopathic stress on human body. It further tries to scientifically evaluate dowsing phenomenon by measuring body voltage, skin resistance, using V-20 bio voltmeter, GSR-2 biofeedback system(R) respectively in and around the pre-detected geopathic stress zones on the Mumbai-Pune express way and some residential areas of Pune city (India). Healthy male volunteers were subjected to the stress in prospective study. The observations indicate a change in body voltage, skin resistance in geopathic stress zones as compared to non-stress zone.

Key words: Geopathic stress, skin resistance, body voltage.

INTRODUCTION

The occurrence of stress as a factor in ill health is now widely accepted and its effects on the body are relatively very well-documented. Geopathic stress is always neglected and considered as a causal syndrome to ill health and has been always ignored because most of the people are unaware of its occurrence in the selected location (Freshwater, 1997). In ancient times, people were aware of geopathic stress associated with specific areas and were very particular in selecting location for civilizations and sites for construction of any structure and dwelling. Those people used to carryout elaborate tests to consider suitability and sustainability of site for structures required for dwelling. The archaic method for selecting a site for dwelling considered the presence of bushes, color of land, presence of water bodies and the development of trees. Housing animals on the proposed site and observing their behavior for certain period was also one of the tests (Bradna, 2002).

In contrast, the modern world with growing population, construction companies hardly give any choice for selection of a site for their clients, hence the opportunity to select site for construction of any structure and dwelling by an architect is also minimum (Saunders, 2003). In majority of the cases the role of architect ceases, the moment a structure is completed. This results in the occupant occupying the built structure ,to slowly get a feel of the building and he start comparing it with the previous site in term of health, compatibility with relatives and others, and progress in careers, business and other social activities. There are numerous disease of which aetiology is not known, prominent among these are the disease related to human body because of impact of Geopathic Stress (Derek, 1994).Energies from earth at specific location that have ability to change the normal body function may be termed as geopathic stress (Kharat, 2000) or in other words geopathic stress is the effect of detrimental earth radiation and electromagnetic radiation on the health of human body (Gerhard, 2005). Geopathic stress are as a result of certain natural or spontaneous potentials generated in the subsurface due to movement of groundwater and its mechanical friction as it flows through fissures, faults, joints and lineaments etc (Geldart, 1976; Morrison et al., 2004).

These potentials are associated with electric currents in

*Corresponding author. E-mail: sejalrohit2007@gmail.com

the ground and are observed to rise vertically from the groundwater to the surface of the earth and above it, to a distance of 220 km (Rolf, 2005). The subterranean groundwater flows at various depths and is the most important source that generates the geopathic stress zone (Creightmore, 1997; Thurnell-Read, 2006). In the ancient time, location of subterranean groundwater was done with the help of dowsing rod which is an old unscientific and well-known practice. Many of the spas were discovered by dowsing method. Professor Benedict explains dowsing as a phenomenon of combination of Bio-physics. According to him the bipolar (positive and negative) sides of the body are closed in an 'emanation stream' through the dowsing rod. The characteristic 'turning of the rod' occurs at the moment the dowser walks over the geopathic stress zone (Bachler, 1989). Though this technique has been very successful based on its statistical success ratios, and is also simple, fast and inexpensive; there is a continued skepticism as regards its acceptance by the scientific community (Betz, 1995). Also dowsing is a very crude technique and does not give any numerical value or a reading; hence there is need to study scientific way in which effect of geopathic stress on human body can be quantified.

Extensive work has been published by various researchers to understand the effect of geopathic stress on the built environment (Bachler, 1989; Thomas, 1993; Gregory, 2002). Possible influences attributed to geopathic stress phenomena have been widely reported by the mass media, albeit without scientific proof. Apparently, geopathic stress does not only influence humans but all kind of animals, plants, fungi and bacteria (Gerhard, 2008; Dubrov, 2008; Von Pohl, 1993; Gak, 2008). The effects of geopathic stress on human body have not yet been proven by scientifically accepted techniques except a few (Dharmadhikari, 2010). The existence of the phenomenon is archaic. At present, electronic devices such as bio-voltmeter and GSR-2 biofeedback system can easily measure the changes happening in human body. This paper attempts to correlate the changes in human body voltage and skin resistance on location of GS and non-GS zones. It further tries to scientifically evaluate dowsing phenomenon.

Human body voltage and skin resistance

The human system contains tissues, the electrical properties of which depend on the ions dissolved in water molecules that have electric dipoles. The cells in the tissues maintain a high concentration of potassium (K^-) and sodium (Na^+) ions, exhibiting an active transport mechanism, termed as the sodium-potassium pump. The nerve impulse is a wave of electro-chemical activity. When large number of tissues contract, a large electrical signal is produced through various motor points on the body surface. The human body reacts to the nervous tension and stress through skin, this is termed as galvanic skin resistance (GSR). The GSR is a reflection of variation in sweat gland activity and pore size, which are controlled by the sympathetic nervous system. GSR and body voltage levels get changed when human body becomes excited, frightened or disturbed.

The system activates chemical and physical changes all through the human body. "Reversal theory" has shown in biofeedback research, that meditation and relaxation procedures cause a rise in skin resistance. It has therefore been assumed that high and low skin resistance correlate directly with relaxation and stress respectively, and that a high resistance indicates a pleasant relaxed state of mind, whereas low resistance indicates tension. The aim of this study is to detect the effect of geopathic stress (GS) on human body by measuring body voltage, skin resistance using V-20bio-voltmeter(R), GSR2-biofeedback system(R) in and around the predetected geopathic stress location on the Mumbai-Pune expressway and some residential areas of Pune city (India).

EXPERIMENTAL DETAILS

Geopathic stress zones were identified with the help of dowsing. Five locations were selected for this study. The presence of groundwater vein was confirmed by resistivity technique. Over each location, body voltage (Figure 1) and tone intensity of GSR-2 biofeedback system was measured (Figure 4). On each location, healthy volunteers of age group 19 to 36 years were tested. The readings were observed and suitably plotted (Figures 2 and 3). Each subject was tested for more than 20 min on each location. For skin resistance, measurement was studied by recording tone intensity of GSR-2 biofeedback system. The GSR-2 biofeedback system was turned on and the fingers were placed on the sensing plates. The dial was then turned very slowly until a tone was heard, that is output of GSR-2 biofeedback system was recorded using sound level meter (SL-4010) in dB. With respect to time spent by each volunteer on geopathic stress zone, the tone intensity of GSR unit was recorded in calm and relax position (Figures 5 and 6). After 20 min, the stress zone GSR-2 biofeedback system was shut off automatically as soon as the fingers were removed from the sensing plates. For body voltage measurement, one end of the instrument was grounded in the earth and the other end, which has the golden plate electrode, was kept in the hand of the person. The body voltage was measured.

Observations

Readings of the all volunteers show that body voltage increased on geopathic stress zone (Figure 1). It was further observed that with the increase in age body voltage on geopathic stress zone was much higher (Figures 2 and 3).

Observation of skin resistance measured by GSR-2 biofeedback system, shows that the tone intensity changes on geopathic stress zone as compared to normal zone(Figure 4). It was further observed that tone intensity increases with age of the volunteers (Figure 5 and 6).

Average body voltage and tone intensity measurement for 14 different age group candidates were tested by paired t-test and ANOVA f-test statistically inside and outside geopathic stress zone. The result shows highly significant change in body voltage (Tables 1 and 2), and significant change seen for tone intensity of GSR-2

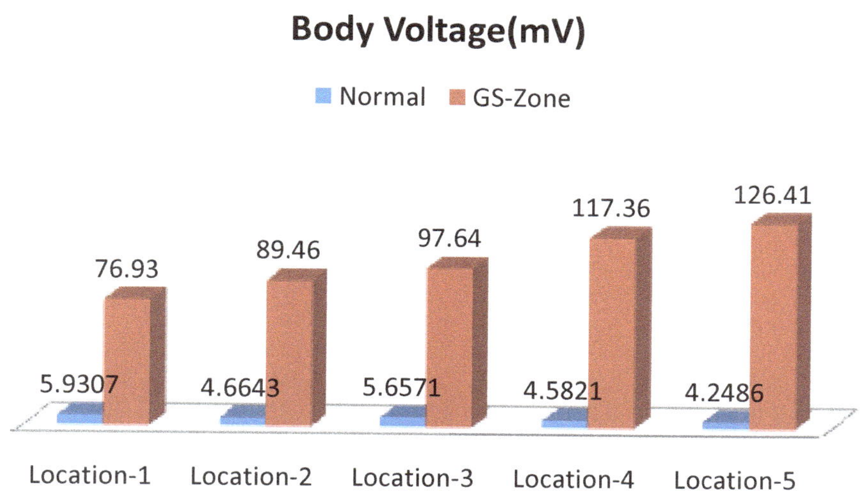

Figure 1. Location wise variation in human body voltage (mV).

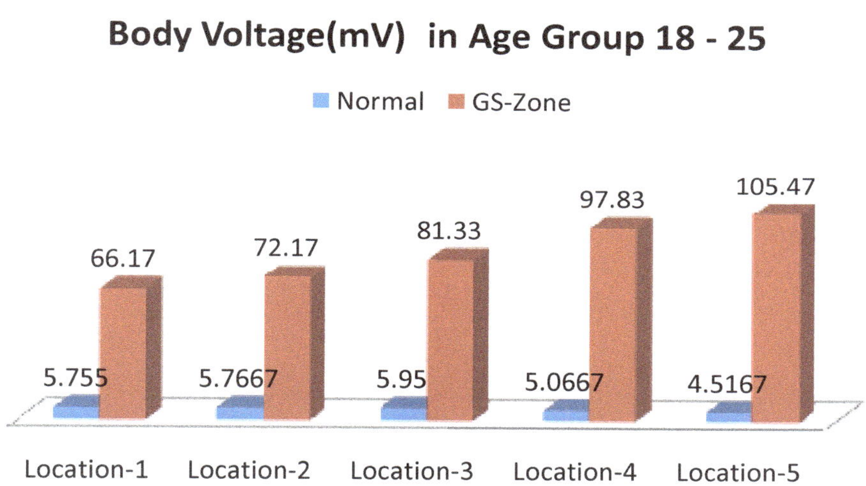

Figure 2. Variation in body voltage (mV) for different location in age 18-25 year group.

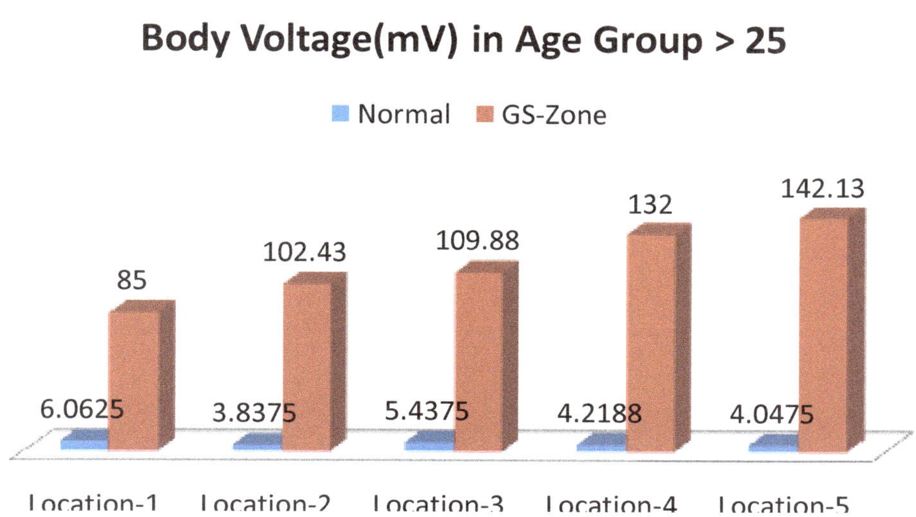

Figure 3. Variation in body voltage for different location in age above 25 year group.

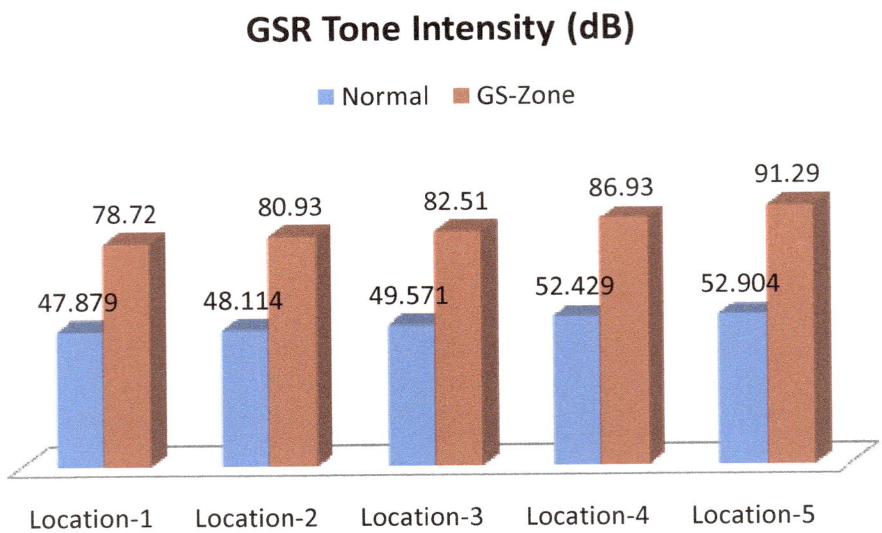

Figure 4. Location wise variation in GSR tone intensity (dB).

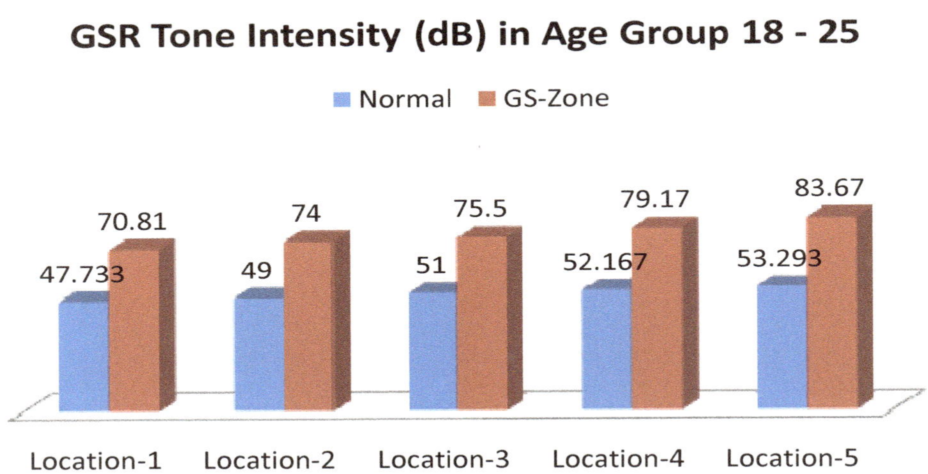

Figure 5. Variation in GSR tone intensity for different location in age 18-25 year group.

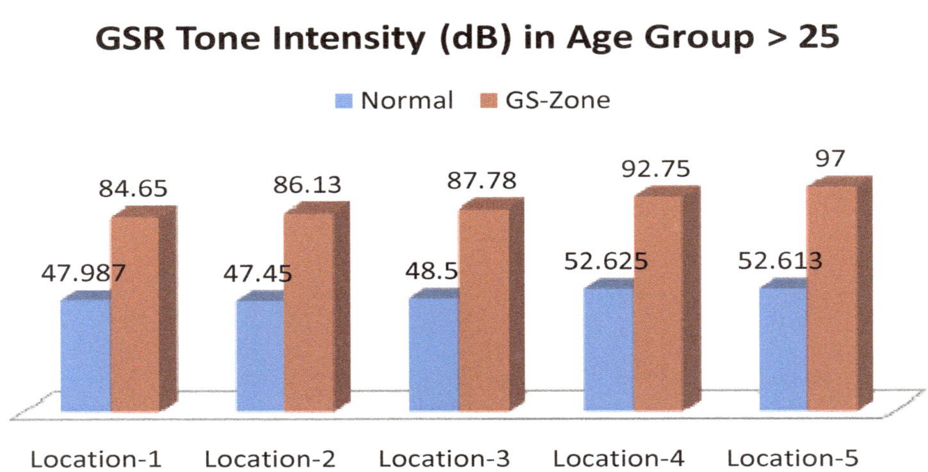

Figure 6. Variation in GSR tone intensity for different location in age above 25 year group.

Table 1. Location wise mean, SD body voltage, t-test and P value.

Location number	Location	No. of candidates tested	Mean voltage	SD	Paired t-test	P-value
1	Normal Zone GS Zone	14 14	5.9307 76.93	1.39686 12.964	20.924	P<0.001 [HS]
2	Normal Zone GS Zone	14 14	4.6643 89.46	1.84500 18.943	16.342	P<0.001 [HS]
3	Normal Zone GS Zone	14 14	5.6571 97.64	1.67961 18.838	17.926	P<0.001 [HS]
4	Normal Zone GS Zone	14 14	4.5821 117.36	2.02424 21.539	18.967	P<0.001 [HS]
5	Normal Zone GS Zone	14 14	4.2486 126.41	1.65309 23.094	19.793	P<0.001 [HS]

HS = Highly significant.

Table 2. Normal zones/GS Zone ANOVA F-test.

Location no.	Location	No. of candidates	Mean voltage (mV)	SD	ANOVA F-test	P-value
1	Normal	14	5.9307	1.39686		
2	Normal	14	4.6643	1.84500		
3	Normal	14	5.6571	1.67961	2.506	0.051 [NS]
4	Normal	14	4.5821	2.02424		
5	Normal	14	4.2486	1.65309		
1	GS Zone	14	76.93	12.964		
2	GS Zone	14	89.46	18.943		
3	GS Zone	14	97.64	18.838	15.235	P<0.001 [HS]
4	GS Zone	14	117.36	21.539		
5	GS Zone	14	126.41	23.094		

NS = Not significant, HS = Highly significant.

biofeedback system (Table 4 and 5). Paired and unpaired t-test for different age groups (18 to 25 and 26 to 37 years) shows that agewise, body voltage (Table 3) and tone intensity of GSR-2 biofeedback system (Table 6) change significantly. Out of five locations, location number two, four and five had the body voltages which were highly significant as compared to normal zone.

RESULTS AND DISCUSSION

It is seen that the energy emitted by the earth at specific surface locations, affects the normal human body voltage and skin resistance. It is observed that on geopathic stress zone, the body voltage increases and skin resistance decreases as compared to the non- geopathic stress zone. As per Dr. Apter, if tone of GSR unit gets increased then skin resistance gets decreased, this means that the person becomes upset or agitated. It can be said that the human body is affected adversely by the geopathic stress zone. One of the arguments may be that as age increases, the effect of the geopathic stress zone also causes changes in the human electrical system. This may be one of the causes for induction of degenerative diseases like heart attack, cancer, etc as the age increases.

Conclusion

The work tries to bring out that the geopathic stress zone has some sort of subtle energy which affects the human system. This energy can be easily detected by dowsing. The subjectivity in dowsing can be removed by measuring changes in body voltage and electrical skin resistance. Thus, it can be concluded that the archaic method of using the human system to detect

Table 3. Age and location wise paired t-test and p-value.

Age group (in years)	Location	No. of candidate	Mean voltage (mV)	SD	Paired t-test	P-value
			Location - 1			
18 - 25	Normal Zone	6 6	5.7550 66.17	1.24922 10.068	16.262	P<0.001 HS
26 - 37	Normal Zone	8 8	6.0625 85.00	1.56929 8.124	26.296	P<0.001 HS
Diff. Body Voltage (mV) vol	Unpaired t = 3.921 , P=0.002 Sig					
			Location - 2			
18 - 25	Normal Zone	6 6	5.7667 72.17	1.87581 9.579	16.695	P<0.001 HS
26 - 37	Normal Zone	8 8	3.8375 102.43	1.40909 12.346	24.971	P<0.001 HS
Diff. Body Voltage (mV) vol	Unpaired t = 5.642 , P<0.001 HS					
			Location - 3			
18 - 25	Normal Zone	6 6	5.9500 81.33	1.55145 6.919	27.192	P<0.001 HS
26 - 37	Normal Zone	8 8	5.4375 109.88	1.84153 15.028	19.138	P<0.001 HS
Diff. Body Voltage(mV) vol	Unpaired t = 4.277 , P=0.001 Sig					
			Location - 4			
18 - 25	Normal Zone	6 6	5.0667 97.83	.83106 12.608	17.546	P<0.001 HS
26 - 37	Normal Zone	8 8	4.2188 132.00	2.60082 13.277	25.629	P<0.001 HS
Diff. Body Voltage (mV) vol	Unpaired t = 4.755 , P<0.001 HS					
			Location - 5			
18 - 25	Normal Zone	6 6	4.5167 105.47	1.12324 14.839	16.459	P<0.001 HS
26 - 37	Normal Zone	8 8	4.0475 142.13	2.01644 13.228	31.457	P<0.001 HS
Diff. Body Voltage (mV) vol	Unpaired t = 5.069 , P0.001 HS					

HS = Highly significant.

subterranean features such as groundwater disturbs the body voltage and skin resistance. People occupying such location for a long time are subject to the subtle changes of body voltage and skin resistance. It may be further concluded that ill health can be induced on geopathic stress zones.

Table 4. Location wise mean tone intensity of GSR (db), t-test and P value.

Location no.	Location	No. of candidates	Mean tone intensity of GSR (dB)	SD	Paired t-test	P-value
1	Normal Zone	14	47.879	5.3760	9.948	P<0.001 HS
	GS Zone	14	78.72	9.435		
2	Normal Zone	14	48.114	5.3716	11.923	P<0.001 HS
	GS Zone	14	80.93	7.691		
3	Normal Zone	14	49.571	5.7206	11.486	P<0.001 HS
	GS Zone	14	82.51	7.160		
4	Normal Zone	14	52.429	5.6223	11.954	P<0.001 HS
	GS Zone	14	86.93	8.380		
5	Normal Zone	14	52.904	7.3207	13.008	P<0.001 HS
	GS Zone	14	91.29	7.800		

HS = Highly significant.

Table 5. Normal zones/GS Zone ANOVA F-test and p-value.

Location no.	Location	No. of candidates	Mean tone intensity of GSR (dB)	SD	ANOVA F-test	P-value
1	Normal Zone	14	47.879	5.3760		
2	Normal Zone	14	48.114	5.3716		
3	Normal Zone	14	49.571	5.7206	2.232	0.075 NS
4	Normal Zone	14	52.429	5.6223		
5	Normal Zone	14	52.904	7.3207		
1	GS Zone	14	78.72	9.435		
2	GS Zone	14	80.93	7.691		
3	GS Zone	14	82.51	7.160	5.357	
4	GS Zone	14	86.93	8.380		0.001 Sig
5	GS Zone	14	91.29	7.800		

NS=Not significant, Sig.= Significant.

Table 6. Age and location wise mean tone intensity of GSR and its ,SD value, paired t-test, p-value.

Age group (in years)	Location	No. of candidates	Mean tone intensity of GSR (dB)	SD	Paired t-test	P-value
			Location – 1			
18 - 25	Normal Zone	6	47.733	3.4419	12.020	P<0.001 HS
	GS Zone	6	70.81	6.580		
26 - 37	Normal Zone	8	47.987	6.7217	8.652	P<0.001 HS
	GS Zone	8	84.65	6.370		
Diff. Tone Intensity of GSR (dB)	Unpaired t = 2.608 , P=0.023 Sig					

Table 6. Contd.

			Location – 2			
18 - 25	Normal Zone	6	49.000	3.8471	15.309	P<0.001 HS
	GS Zone	6	74.00	5.329		
26 - 37	Normal Zone	8	47.450	6.4682	11.288	P<0.001 HS
	GS Zone	8	86.13	4.190		
Diff. Tone Intensity of GSR (dB)	Unpaired t = 3.230 , P=0.007 Sig					
			Location – 3			
18 - 25	Normal Zone	6	51.000	4.3818	11.202	P<0.001 HS
	GS Zone	6	75.50	4.037		
26 - 37	Normal Zone	8	48.500	6.6332	11.949	P<0.001 HS
	GS Zone	8	87.78	3.127		
Diff. Tone Intensity of GSR (dB)	Unpaired t = 3.464 , P=0.005 Sig					
			Location – 4			
18 - 25	Normal Zone	6	52.167	4.9160	11.211	P<0.001 HS
	GS Zone	6	79.17	5.193		
26 - 37	Normal	8	52.625	6.4296	10.954	P<0.001 HS
	GS Zone	8	92.75	4.559		
Diff. Tone Intensity of GSR (dB)	Unpaired t = 2.768 , P=0.017 Sig					
			Location – 5			
18 - 25	Normal Zone	6	53.293	6.7850	24.717	P<0.001 HS
	GS Zone	6	83.67	4.131		
26 - 37	Normal Zone	8	52.613	8.1497	11.289	P<0.001 HS
	GS Zone	8	97.00	3.703		
Diff. tone intensity of GSR (dB)	Unpaired t = 2.978 , P=0.012 Sig					

HS = Highly significant.

REFERENCES

Bachler K (1989). Earth Radiation. Word Asters Ltd, ISBN 095141510 7: 34-35.

Betz HD (1995). Unconventional Water Detection: Field Test of the Dowsing Technique in Dry Zones, Part I & II. J. Sci. Explor., 9: 1- 2.

Bradna L (2002). The Influence of Hydro pathogenic zones on Drivers. Narendra Prakashan, pp. 38–43.

Derek JC (1994). The Effect of Geopathic Stress on Building Occupants. Renew. Energy, pp. 993-996.

Dharmadhikari NP (2010). Effect of Geopathic Stress on Human Heart Rate and Blood Pressure. Indian J. Sci. Technol., 3(1): 54-57.

Dubrov AP (2008). Geopathic Zones and Oncological Diseases. Druskininkai, pp. 42-44.

Freshwater D (1997). Geopathic Stress. Complement. Ther. Nurs. Midwifery, 3: 160-162.

Gak EZ (2008). About Nature Influence Geophysics Earth's Fields Anomalies on the living systems. Druskininkai, pp. 88-91.

Gerhard WH (2005). Biomedical Evidence of Influence of Geopathic Zones on the Human Body: Scientifically Traceable Effects and Ways of Harmonization. Forsch Komplementärmed Klass Naturheilkd Karger, pp. 315-327.

Gerhard WH (2008). Geopathic Stress Zone and Their Influences on Human Organism. Druskininkai, pp. 8-27.

Gregory AS (2002). Geopathic Zones and The Iron Stake Method-No-2. A Dowser's Series, pp. 12-20.

Thurnell-Read J (2006). Geopathic Stress and Subtle Energy. Life-Work Potential, (UK), pp. 11-24.

Kharat AG (2000). Empirical and Theoretical Investigation on Built Environment, Ph.D.Thesis, University of Pune, p. 107.

Morrison F, Gasperikova E, Washbourne J (2004). The Berkeley Course in Applied Geophysics. http://appliedgeophysics.berkely.edu:7057.
Rolf G (2005). Are you sleeping in a safe place? 130 Gipsy Hill London SE19, 7th edition, ISBN 0951 4017 0X. Dulwich Health, pp. 9-10.
Saunders T (2003). Health hazards and electromagnetic fields. Complement. Ther. Nurs. Midwifery, 9: 191–197.
Telford WM, Geldart LP (1976). Applied Geophysics. Cambridge University Press, NY, p. 859.
Thomas JM (1993). Noxious (Geopathic) Fields Are Damaging To Your Health Erie Pennsylvania U.S.A. 16511, 10. Effects of Harmful Radiations and Noxious Rays., A series of Research paper by Noted Scientists and Authors, American Society of Dowsers, INC.Danville Vt. 05828, pp. 4-10.
Von PGF (1993). Earth currents: Causative factors of cancer and other diseases. Frech- Verlag, Stuttgart, ISBN 13:978-3772494024, pp. 7-17.

Electrical characterization of in-house fabricated polysilicon micro-gap for yeast concentration measurement

Th. S. Dhahi[1]*, U. Hashim[1], N. M. Ahmed[2], Md. Eaqub Ali[2] and T. Nazwa[2]

[1]Institute of Nano Electronic Engineering, University Malaysia Perlis (UniMAP), Malaysia.
[2]School of Physics, University Sains Malaysia (USM), Malaysia.

Fabrication and electrical characterization of micro-gaps and their properties are discussed with their applications in electrochemical sensors and bimolecular detection. To understand the relationship between the biosensor and micro-structure we have carried out the fabrication and characterization of micro-gap structures for yeast concentration measurement. In this paper two photo mask designs are used. The first mask is for defining the lateral micro-gap and the second one is for the aluminum electrode pattern. Lateral micro-gaps are formed using polysilicon and Al as contact pads. Conventional photolithography techniques are used to fabricate the micro-gap. The electrical measurements were carried out using a Semiconductor Parameter Analyzer to measure the change in capacitance across the micro-gap with the yeast solution drop put across the gap. The measured value of capacitance for each yeast concentration is presented as a function of the gap size. The results indicate that these micro-gaps could possibly be used for yeast sensor applications.

Key words: Micro-gap, conductivity, yeast detection, capacity, electrochemical sensor.

INTRODUCTION

The objective of the present work is to explore the possibility of using polysilicon micro-gaps for yeast concentration measurement. Micro- and nano-fabrication techniques are progressively being applied to create ultra-miniature sensors for characterizing and quantifying bio-molecules. The reduction in sensor size can result in lower materials cost, reduced weight, and lower power consumption, which are the key factors driving newer opportunities for sensors in the marketplace. Micro- and nano-sensors of highly reduced power consumption are very suitable for integration into wireless communication devices to enable widespread distributed monitoring and control. Very low-power micro- and nano-sensors would also be beneficial for use as battery-operated handheld or wearable sensors. Logical and promising sensing application areas for micro- and nano-sensors include medical (e.g., blood analysis, patient monitoring and diagnostic testing), bio warfare detection, genetic analysis, drug discovery, food inspection/testing, environmental monitoring, and industrial chemical process monitoring. Micro- and nano-gap structures fabricated using semiconductors and materials have been used quite often for such sensors. In this work polysilicon (Poly-Si) is used for micro-gap fabrication.

Poly-Si is known to be compatible with high temperature processing and interfaces very well with normal semiconductor processing steps. As a gate electrode, it has also been proven to be more reliable than aluminum (O'Mara et al., 1990). It can also be deposited conformably over steep topography. Doped polysilicon thin films are used in emitter structures in bipolar circuits and in resistors. Polysilicon is a key component for integrated circuits. At the component level, polysilicon has long been used as the conducting gate material in metal–oxide–semiconductor field-effect transistor (MOSFET) and complementary metal oxide semiconductor (CMOS) technologies. For such applications it is deposited using low-pressure chemical-vapor

*Corresponding author. E-mail: sthikra@yahoo.com

deposition (LPCVD) at high temperatures. In terms of chemical resistance, polysilicon is very much like single crystal silicon. The oxidation rate of undoped polysilicon is typically between that of (100)- and (111)-oriented single crystal silicon. For temperatures below 1,000℃, heavily phosphorus-doped polysilicon oxidizes at a rate significantly higher than undoped polysilicon (Bhushan and Bharat, 2004).

The resistivity of polysilicon can be modified using conventional dopants for silicon processing. Phosphorous, the most commonly used dopant in polysilicon, diffuses significantly faster in polysilicon than in single crystal silicon, due primarily to enhanced diffusion rates along grain boundaries. The diffusivity in polysilicon thin films (that is, small equiaxed grains) is about $1x10^{-12}$ cm^2/s (Irudayaraj and Reh, 2008).

The electrical characteristics of a poly-Si thin film depend on its doping. Poly-Si is more resistive than single-crystal silicon for any given level of doping primarily because the grain boundaries in poly-Si offer additional scattering sites to reduce carrier mobility. Common dopants for polysilicon include arsenic, phosphorus, and boron. Polysilicon is usually deposited undoped and the dopants are introduced later on (James, 2005).

Yeast extract is the common name for various forms of processed yeast products made by extracting the cell contents (removing the cell walls); they are used as food additives or flavourings, or as nutrients for bacterial culture media. They are often used to create savory flavors and umami taste sensations. Monosodium glutamate (MSG) is used for umami, but has no flavor. Yeast extract, like MSG, often contains free glutamic acid. Yeast extracts in liquid form can be dried to a light paste or a dry powder. Glutamic acid in yeast extracts are produced from an acid-base fermentation cycle, only found in some yeasts, typically yeasts used in the baking of breads.

Yeast extract is the water-soluble portion of autolyzed yeast. The autolysis is carefully controlled to preserve naturally-occurring B-complex vitamins. Yeast extract is prepared and standardized for bacteriological use and cell cultures, and is an excellent stimulator of bacterial growth. Yeast extract is generally employed in the concentration of 0.3 - 0.5%. Yeast extract is typically prepared by growing baker's yeast, *Sacharomyces* spp., in a carbohydrate-rich plant medium. The yeast is harvested, washed, and resuspended in water, where it undergoes autolysis, that is, self-digestion using the yeast's enzymes. Yeast extract is the total soluble portion of this autolytic action. The autolytic activity is stopped by a heating step. The resulting yeast extract is filtered clear and dried into a powder by spray drying. Yeast extract has been successful in culture media for bacterial studies in milk and other dairy products. Several media containing yeast extract have been recommended for cell culture applications (Chan et al., 1998). The capacitance offered by micro- and nano-gap structures depends on the electrode geometry. It increases with the electrode area, increases with the electrode spacing, and also depends on the nature of the material present between the two electrodes.

In this work, the electrical measurements were carried out using a Semiconductor Parameter Analyzer to check the capacitance across the micro-gap, which changes with the yeast solution drop put across the gap. The measured values of capacitance for each yeast concentration used are presented as a function of the gap size in the micro-device structure. The fabricated micro-gaps were used for yeast sensors.

RESEARCH METHODOLOGY

Mask design

In this work, Si wafer is used as the substrate to fabricate a micro-gap biosensor. The starting material used here is a p-type, 100 mm diameter. Two photomask processes are used to fabricate the micro-gap using conventional photolithography and polysilicon dry-etching techniques. Commerical chrome mask is used for better pattern quality. The photomasks are designed using AutoCAD and then printed onto a chrome glass surface.

Figure 1 is the first mask for microgap electrode formation with a length and width of 5000 and 2500 µm, respectively.

Six sets of angular length S_d (600 to 1100 µm, with a difference of 100 µm) has been chosen in the present study to check the best angle for micro-gap formation after the etching process. S_d refers to the dimension for the side angle of the design for micro-gap formation as shown in Figure 1(a) and (b), which shows the actual arrangement of the device design on the chrome mask. It consists of 160 dies with 6 different designs.

Figure 2 is a schematic device design of mask 2 with 1500 µm length and 1500 µm width. The distance between two squares is 3500 µm.

Micro-structure fabrication

The proposed process steps for the Al electrode with polysilicon micro-gap fabrication are starting by cleaning the Si wafer before depositing a 150 nm SiO_2 substrate on the Si wafer by using plasma-enhanced chemical vapor deposition (PECVD) equipment. Then, a low-pressure chemical-vapor deposition (LPCVD) machine is used to deposit the 600 nm polysilicon layer to perform a layer of 135 nm Al substrate as a hard mask to avoid damage to the polysilicon layer during the etching/ RIE process (Thikra et al., 2011). Next, in the photolithography process, a layer of positive photo-resist 1200 nm is first applied to the Al surface, and then exposed to ultraviolet light through mask 1 as shown in Figure 3(d). After developing, only the unexposed resist will remain, and then the wet-etching process of Al layer is performed before removing the resist. After that, the dry-etching process for the polysilicon layer is applied to fabricate the micro-gap for the micro-structure, then a layer of 135nm Al substrate is deposited before the resist coating process. After exposing mask 2, the layer of the resist is developed as in Figure 3(g), and then the wet-etching process of the Al substrate is performed before removing the resist. Finally a structure of the Al electrode with a polysilicon micro-gap is obtained as shown in Figure 3(h).

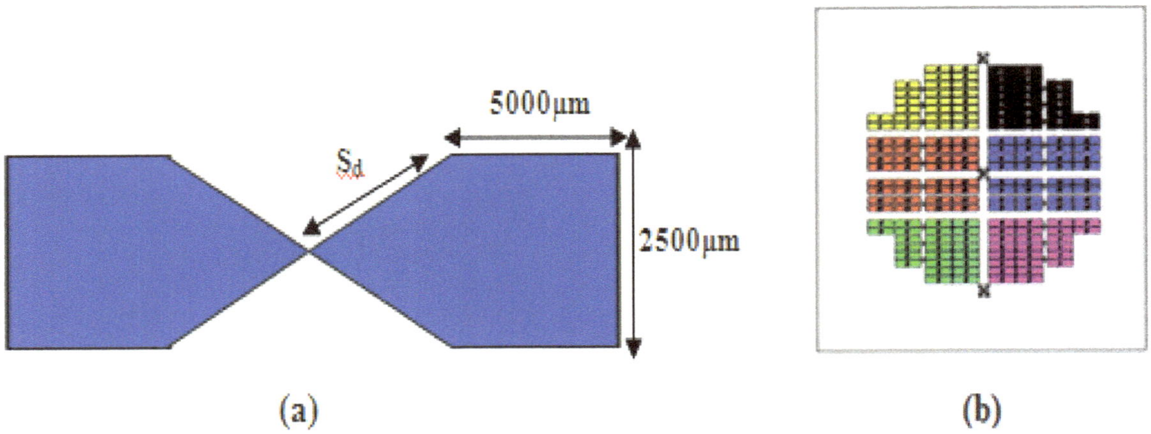

Figure 1. (a) Design specification of the first mask, and (b) Schematic design of the actual mask on chrome glass.

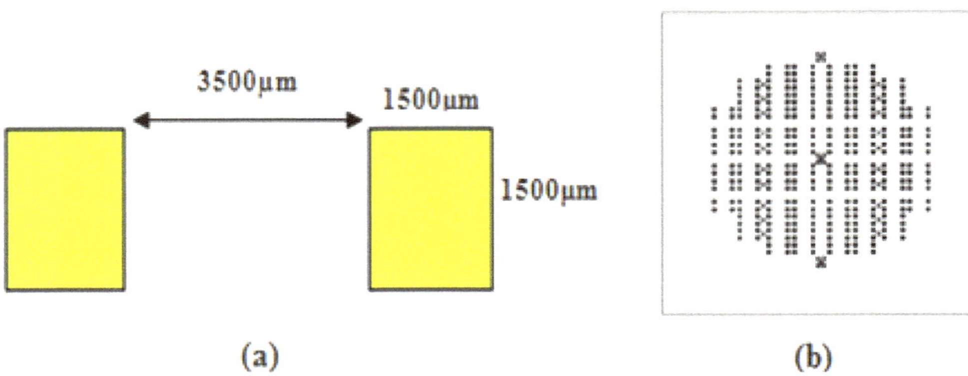

Figure 2. (a) Design specification for mask2, and (b) Schematic for mask2 on chrome glass.

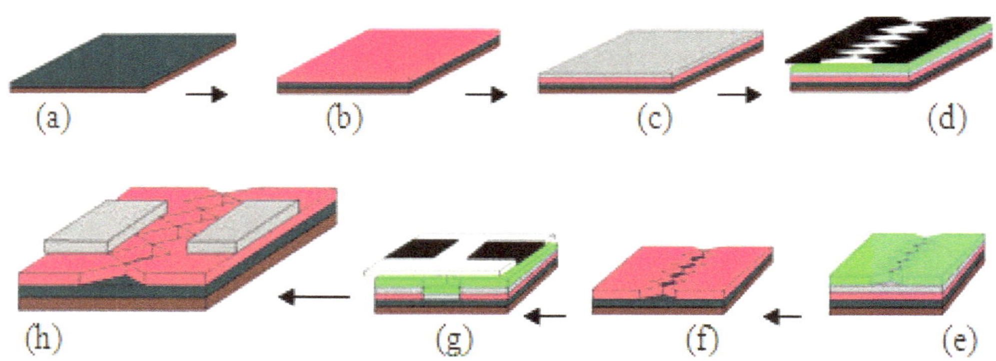

Figure 3. Polysilicon micro-gap structure fabrication process flow.

Micro-gap measurement and characterization

The fabricated micro-gap structures were characterized by the 3D Nano Profiler and a scanning electron microscope (SEM). Figure 4 shows the final device structures used for the electrical characterization using the Semiconductor Parameter Analyzer. Probing is done using large size Al pad electrodes on the device. The micro-gap biosensor will thus act as a capacitor between two polysilicon structures connected to the Al pads.

Yeast solution preparation and detection

In this work, the yeast-DI water solutions were tested using

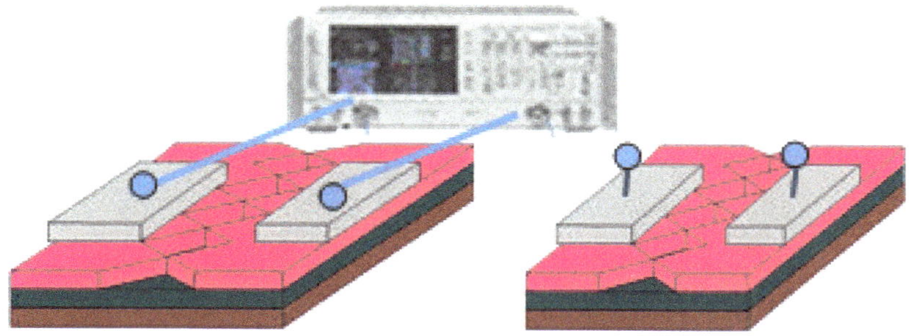

Figure 4. Capacitance measurements taken using a dielectric analyzer.

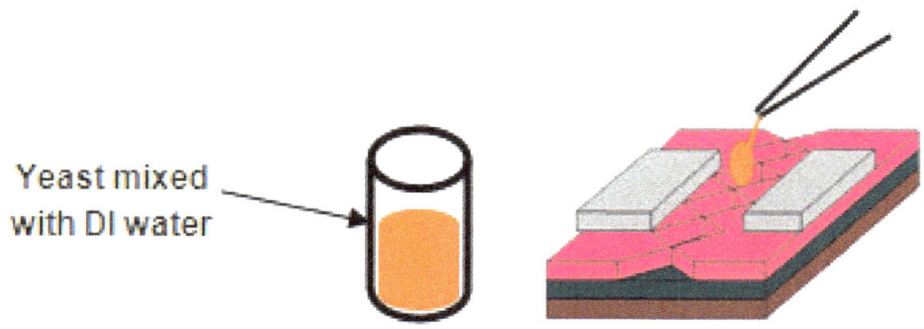

Figure 5. The probe process on the micro-gap structure.

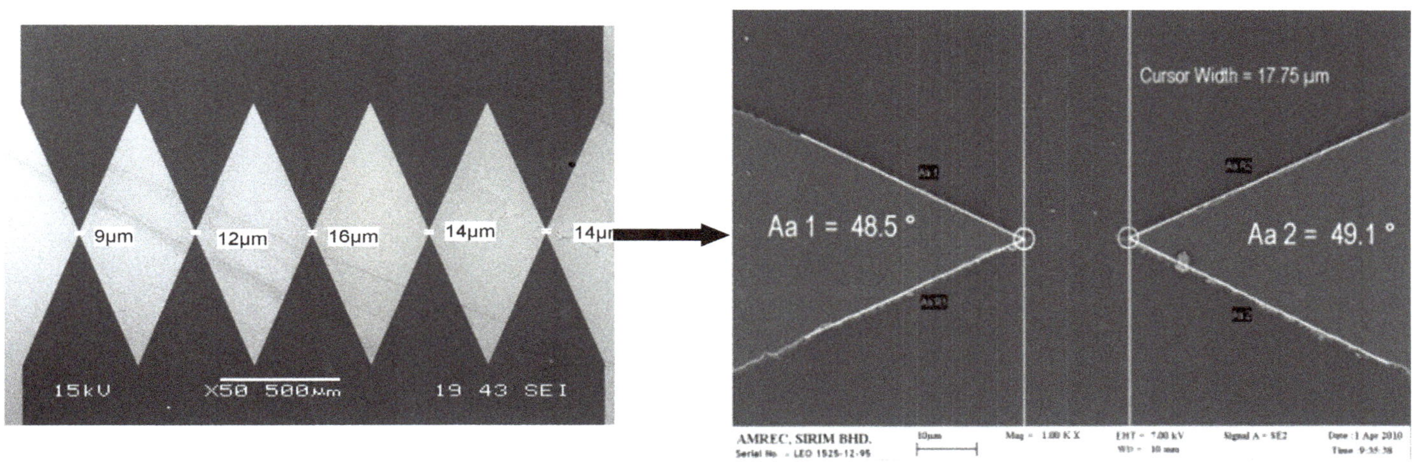

Figure 6. SEM image of polysilicon micro-gap measurement.

micro-gap structures for the capacitance measurements. The micro-gap is cleaned using DI water followed by dry spinning, then baking at less than 100ºC for a few minutes. After that, the yeast solution is dropped across the micro-gap as shown in Figure 5, to measure the change in the capacity as will be shown in the results and discussion.

RESULTS AND DISCUSSION

Figures 6 and 7 show the SEM and 3D nano profiler images for the polysilicon micro-gap after fabrication processes, where it can be seen quite clearly that the size of the gap is approximately between 9-18 µm.

Capacitance-voltage (C-V) characterizations of micro-gap electrodes were performed at room temperature by a two-point probe using a Keithley 4200 semiconductor characterization system.

Figure 8 shows that for the applied voltage from -3 to +3 V, capacitance is almost constant at 127E-12 F. That happens because the charges between the two plates do not have enough energy to create the depletion region, and it would be increasing when probed by any contact

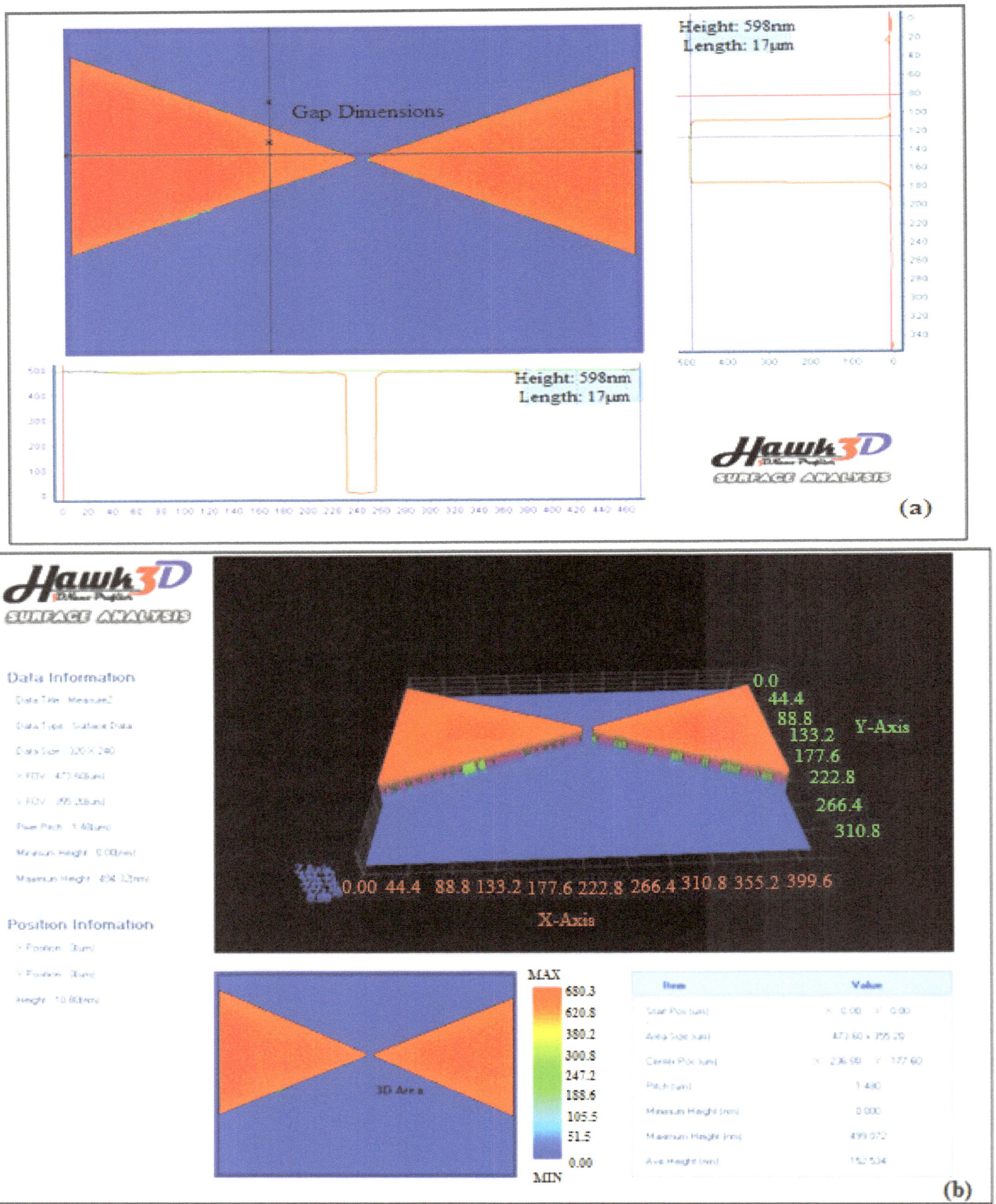

Figure 7. 3D Nano profile measurements for: (a) The gap dimensions, (b) The microstructure surface section.

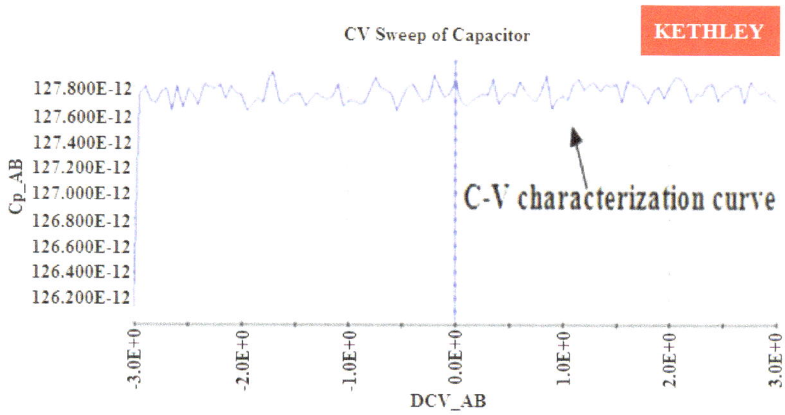

Figure 8. Capacitance –Voltage curve for polysilicon micro-gap structure.

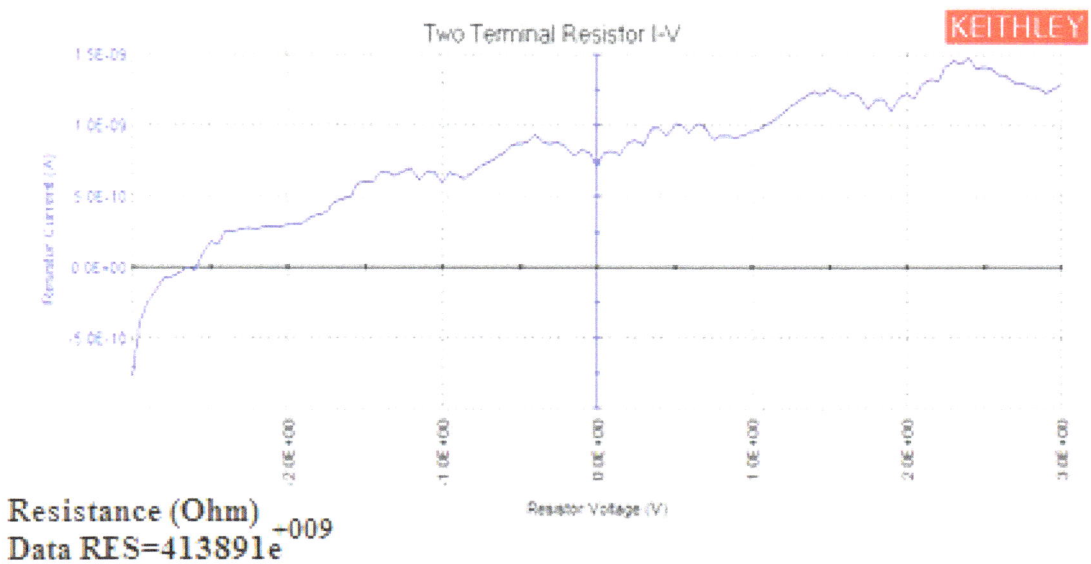

Figure 9. I-V curve for the polysilicon micro-gap structure.

polymer. The Keithley 4200-Simiconductor Parameter Analyzer has built-in support for measuring current-voltage (I-V) characterization as well. The I-V curve in Figure 9 shows that for the voltage applied from -5 to +5 V, the resistance is very high at 4.13891^{+9}Ohms, because the capacitance will decrease in this range of voltage and allow passing the current with a small value of -5E-10 to1E-9 Amperes. C-V and I-V measurements confirm the capacitance nature of the micro-gap fabricated in this work (Ikonomou and Schneider, 2001).

Measurement and analysis of dielectric properties (capacity value) for the yeast solution is carried out by using WINDETA software. Different types of parameters are set at a value list for frequency and AC volt, start condition, list order of independent variables and end condition. These parameters will generate the result graph that was used in this research for analysis. From the frequency - capacitance curve results, we can determine the value of capacitance that would be used to analyze what yeast solutions give the best performances among the others. The experimental method of this research is based on a capacitance measurement technique using DI water and a yeast solution. For this measurement, different yeast solution concentrations, which are 0.1, 0.3, 0.5, 0.7 and 0.9 mg, are used with various gap sizes using a constant voltage at 0.1V. The range of frequency that was used is around 10^0-10^2 Hz; the results display almost exactly the same behavior but different capacitance values.

Figure 10 presents the electric capacity versus frequency results for various gaps using DI water. The results show that the capacitance increases with the size

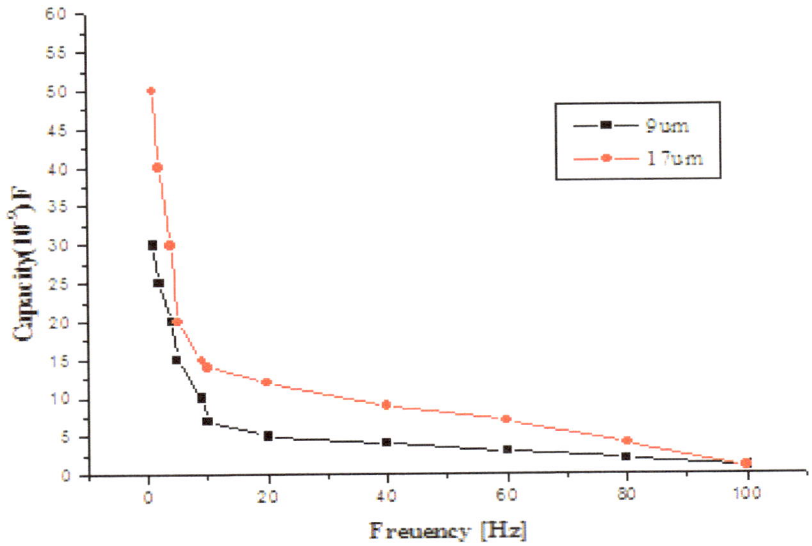

Figure 10. The capacitance measurement for DI water at gap size 9 and 17 μm.

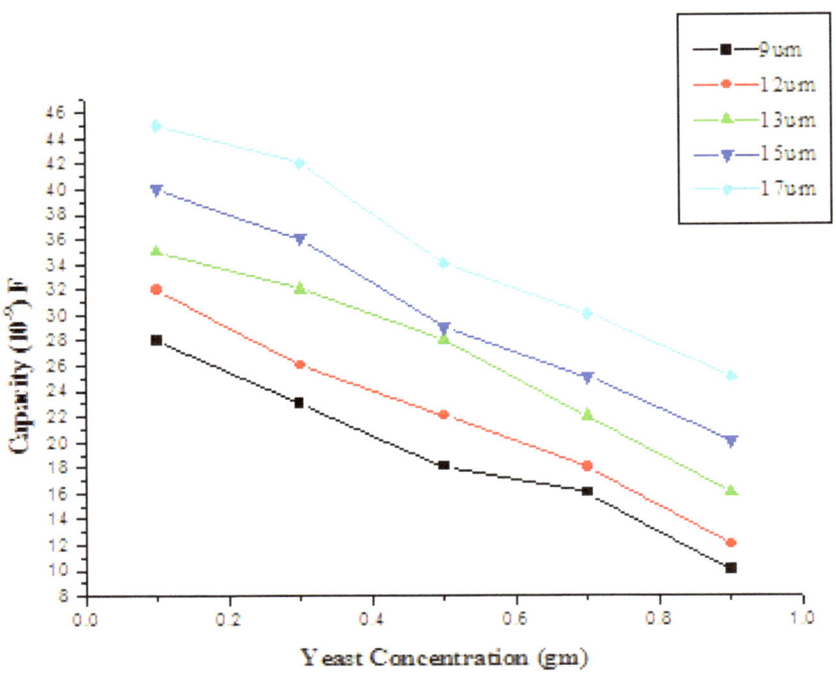

Figure 11. The capacity measurement with the yeast concentration for different gap sizes.

The most characteristic feature of the obtained results is a positive charge increment, which shows the capacity of polysilicon, measured at positive increases when a strong electric field is applied. The curve behavior is almost the same just the value of the capacitance has changed. This is because the bigger gap size has many charges inside the insulator compared to the smaller gap.

According to the theory, the permittivity of a substance is a characteristic that describes how it affects any electric field set up in it. A high permittivity tends to reduce any electric field present.

Figure 11 shows that the capacity value decreases with the increase in yeast concentration. We can see that the gap size can affect the capacity value, in that where the

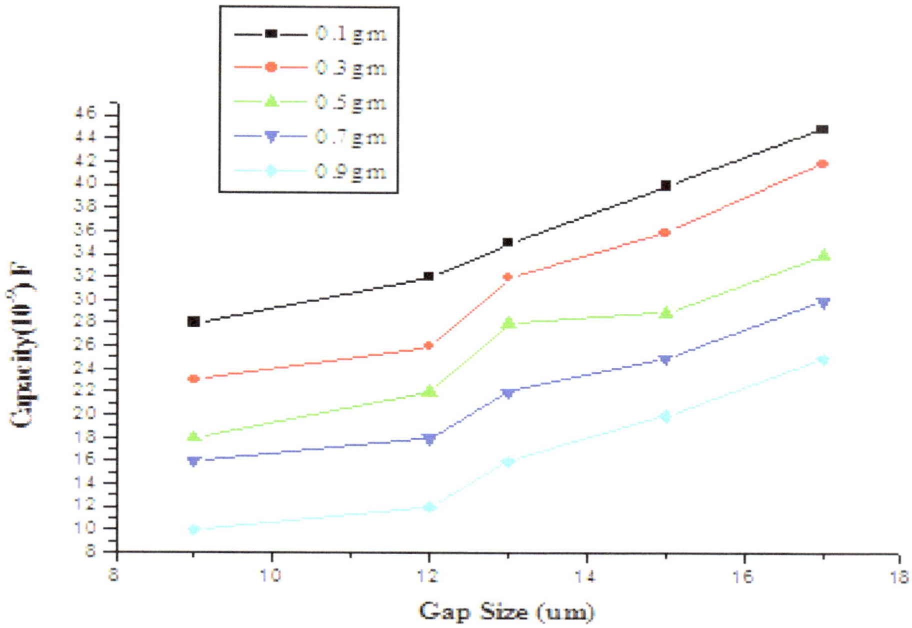

Figure 12. The capacity measurement with the gap size for various yeast concentrations.

smaller gap size decreases the capacity value, the conductivity becomes high. As a result, the resistivity decreases. Figure 12 shows the reaction of the capacity measurement as a function of gap size (µm) at different concentrations of yeast, using 0.1 V as the applied voltage, whereas the capacity value recorded decreased with the increase in the yeast concentration. Gap sizes are proportional to the value of capacity. When the gap size is bigger, the value of the capacity is higher. The larger capacity value means that a better contact is produced between the conductor and insulator, which means the yeast concentration that we used, has a good effect on the insulator. If the air gap between the electrodes is replaced by an insulating material (dielectric), the current that flows between the plates will increase to a new value, where the increase in the current is caused by a property of the insulating material known as permittivity, and the introduction of the insulating material therefore increases the capacitance between the two electrodes (Suchodolskis et al., 2011), and decreases the conductivity. The lower capacity value means that less contact is produced between the conductor and insulator, which mean the yeast con-centration that we used is still not good enough, and the dielectric fields that are produced between the insulator and conductor are not strong (Thikra et al., 2010).

The most characteristic feature of the obtained results is a positive charge increment, which shows the capacity of polysilicon, measured at positive increases when a strong electric field is applied. The graphs are almost all the same for all the gap sizes, the only difference is that the value of the capacitance has changed. This is because the bigger gap sizes have many charges inside the insulator compared to the smaller gap. By referring to theory, the permittivity of a substance is a characteristic that describes how it affects any electric field set up in it. A high permittivity tends to reduce any electric field present, then decreases the conductivity value and increases the resistivity. We can increase the capacitance of a capacitor by increasing the permittivity of the dielectric material (Liu and Zhao, 2007).

Conclusion

The capacitance as well as the conductivity was measured for the polysilicon micro-gap structure, using a Semiconductor Parameter Analyzer. Two chrome masks are used to fabricate the micro-gap structure. I-V and C-V curves are carried out using the Dielectric Analyzer to check the resistivity of the micro-gap device. The electrochemical potential of different yeast concentration values were made by measuring associated capacity, where the capacitance values decrease with the yeast concentration level and increase with the size of the gap.

ACKNOWLEDGMENTS

We are grateful for fruitful discussions with our collaborators at the Institute of Nano Electronic Engineering (INEE) at University Malaysia Perlis (UniMAP). This work was supported by INEE at UniMAP, through the Nano Technology project. The views expressed in this publication are those of the authors and do not necessarily reflect the official view of the funding agencies on the subject.

REFERENCES

Bhushan B, Bharat (2004). Springer Handbook of Nanotechnology. 978-3-540-01218-4.

Chan L, Greenfield PF, Reid S (1998). Optimizing fed-batch production of recombinant proteins using the baculovirus expression vector system. Biotechnol. Bio. Eng., 59: 178-188.

Ikonomou B, Schneider A (2001). Design of efficient medium for insect cell growth and recombinant protein production. *In vitro* Cell Dev. Biol. Anim., 37: 549-559.

Irudayaraj J, Reh C (2008). Nondestructive Testing of Food Quality, 4, 978-0-8138-2885-5.

James JA (2005). Micro Electro Mechanical System Design. Dekker Mechanical Engineering, 0824758242/0-8247-5824-2.

Liu Y, Zhao Z (2007). Design and Fabrication a 3-D Nano-gap Electrode Grid Array Biosensor, pp. 753-758.

O'Mara WC, Herring RB, Hunt LP (1990). Handbook of Semiconductor Silicon Technology, 978-0-8155-1237-0.

Suchodolskis A, Stirke A, Timonina A, Ramanaviciene A, Ramanavicius A (2011). Baker's yeast transformation studies by Atomic Force Microscopy, Adv. Sci. Lett., 4: 171–173,

Thikra SD, Hashim U, Ahmed NM, Mat Tai (2010). A review on the Electrochemical Sensors and Biosensors Composed of Nanogaps as Sensing Material. J. Optoelectronics Adv. Mater. (JOAM), 12(29): 1857-1862.

Thikra SD, Hashim U, Ahmed NM (2011). Fabrication and Characterization of 50 nm Silicon Nanogap Structure. J. Sci. Adv. Mater. (SAM), 3(2): 1–6.

The effect of mesophilic and thermophilic temperature on bonny light crude oil degradation in Niger Delta Area of Nigeria using *Pseudomonas* sp.

Ukpaka C. P.

Department of Chemical/Petrochemical Engineering, Rivers State University of Science and Technology, Nkpolu, P. M. B. 5080, Port Harcourt, Nigeria. E-mail: chukwuemeka24@yahoo.com.

In this study the effect of mesophilic and thermophilic temperature on Bonny light crude oil degradation was tested on the freshwater and salt water media. To this end the pH, optical density, refractor index, specific gravity, density, microbial population and viscosity were determined using experiment analysis. Experiments were performed using environmentally relevant concentrations of the test substances in the batch reactor. The mesophilic temperature is within the range of 18-35°C and thermophilic is within the range of 35-75°C; the research was conducted within the temperature range of 27-80°C, it is seen that a temperature range of 27-35°C (mesophilic) there was increase in microbial population, decrease in Bonny light crude oil concentration, decrease in pH, increase in optical density for both fresh and salt water. Under the thermophilic temperature (35-75°C) the characteristics of the process is the same as that of mesophilic temperature. This paper can be found useful in monitoring and predicting the degradation of Bonny light crude oil upon the influence of mesophilic and thermophilic temperature using *Pseudomonas* sp. in Niger Delta Area of Nigeria.

Key words: Mesophilic, thermophilic, temperature, crude oil, degradation, *Pseudomonas* sp.

INTRODUCTION

Crude oil contamination has caused critical environmental and health defects and increasing attention has been made for developing and implementing innovative technology for clearing up the contaminant as well as considering some possible factors that influence the bioremediation process such as temperature, pH and other physicochemical parameters. The rates of uptake and mineralization of organic compounds by microbes depends on the concentration of the compound as well as the operating condition of the system (Ukpaka, 2007). Since the fate of hydrocarbon degradation is largely determined by the local environmental conditions which influence the microbial growth and enzymatic (Mckenna and Kallio, 2004; Millkin et al., 2004) activities. This research was conducted to explore the effect of temperature on biodegradation of crude oil (petroleum hydrocarbon) using Bonny light.

The significance of this study lies on the need to known how temperature affects *Pseudomonas* sp. activity during the degradation of Bonny light crude oil in Niger Delta Area of Nigeria. The relative resistance of heavy oil to degradation at low temperature will be presented in the next paper as well as having a proper understanding on how to manipulate temperature in order to reduce inhibitory factors on crude oil degradation using *Pseudomonas* sp. where studied by various research groups (David, 2002; Atlas, 2004, 2005).

Crude oil is a complex mixture also known as petroleum; it is a dark viscous fluid which contains a complex mixture of paraffin, alicyclic and aromatic hydrocarbons. These hydrocarbons are compounds of carbon and hydrocarbon only (Atlas, 2000; Anderson, 2008; Brennan and Withgott, 2005; DPR, 2002). Crude oil was obtained in porous sedimentary or any other permeable reservoir rock as liquid at atmospheric pressure; the crude oil is made of hydrocarbon, non hydrocarbon, metallic substance and other impurities

Abbreviations: THC, Total hydrocarbon content; **IPS,** Institute of Pollution Studies; BOD, biochemical oxygen demand.

Table 1. Analysis results on the crude oil, salt and fresh water.

Sample	Surface tension (N/m)	Viscosity (cp)	Density (g/cm^3)
Crude oil	0.036	34.3	0.925
Salt water	0.0727	4.10	1000
Fresh water	0.0716	6.04	938

(Coulson and Richardson, 1994; Davis and dewiest, 2006; Foster, 2002; Gibson and Subramanian, 2005; Leahy and Colwell 2006; Haramaya et al., 1999; and Gibson, 2004).

Investigation conducted revealed that most chemical reactions, a lot of biochemical reactions are greatly influenced by the temperature which may enhance the contact between the enzyme and substance resulting in enzyme-substrate complex, this is possible to certain temperature range (FMEM, 2006; Kallio, 2002; Kiel, 2008; Kobayshi and Rittman, 2000). Depending on the type of microbes, the temperature can be considered in two phases such as; mesophilic types of microbes that survive in a temperature range of 18 -35°C and thermophilic within the temperature range of 35-75°C (Ogoni et al., 2001; Okoh, 2003; Ogoni, 2003; Odu et al., 2004; Okpokwasili, 2006; Spiri and Stighani, 2006; Westlake et al., 2003; Van de-Linden and Thijsse, 2005; and Zobell, 2002).

The main aim of this study is to direct the effect of temperature on *Pseudomonas* sp. during bio-degradation of Bonny light crude oil. The study was conducted in salt and fresh water media and the physicochemical parameters of the salt, and fresh water, crude oil sample and microbial species were measured and recorded.

MATERIALS AND METHODS

Collection of samples/methodology/procedure

The various samples used for the investigation were collected in Port Harcourt City in Niger Delta Area of Nigeria. The physical properties of the samples were determined as shown in Table 1 to confirm their effect on messophilic and thermophilic temperature on Bonny light crude oil degradation in Niger Delta Area of Nigeria using *Pseudomonas* sp. The surface tension, viscosities and densities were measured with the acid of stlimaizu ST-1 wilhelong-type surface tension meter, cannon-penske capacillary type viscometer and hygrometers respectively.

The salt water sample was collected from the Eagle Island River of Mgbousimini Community and the fresh water sample was collected from the Omuihuechi stream located in Omuihuechi village in Aluu, Ikwere, Rivers State of Nigeria.

The water samples (fresh and salt water) were weighed to 3 kg and transferred into 2 different plastic buckets, then labeled fresh and salt water respectively. Crude oil concentration of 270 ml by volume was used to contaminate the water samples. The physicochemical parameters of the crud oil were determined at different temperatures, and also the microbiological parameters and the extent of biodegradation of the crude with time.

Microbial culture of 100 ml concentration by volume was introduced into the 2 contaminated samples. Physicochemical parameters, total hydrocarbon content (THC), PH, density, retractive index, specific gravity and viscosity were analysed. Microbiological parameters including, the THC analyses was carried out at the Institute of Pollution Studies (IPS) laboratory, Rivers State University of Science and Technology.

Apparatus and reagents

The following apparatus, reagents and media was used for the investigation such as, apparatus: cylindrical flask, burette, pipette, thermometer, conical flask, measuring cylinder, spectrophotometer, APLAB turbidity meter, crucibles, biochemical oxygen demand (BOD) bottles 300 ml, BOD incubator, colony counter and petri dishes.

Reagents and media

Crude oil, cultural medium, salt and fresh water, dilute hydrochloric acid, nutrient agar, concentration H_2SO_4, 0.025 N sodium thiosulphate solution, $MnSO_4$ solution, starch indicator, ferric chloride, calcium chloride solution, $AgNO_3$ solution (N/35.45) and barium chloride solution.

The petroleum hydrocarbon (crude oil) was treated using ex-situ treatment in which the water samples (fresh and salt water) were transferred into plastic buckets with the introduction of the microbial culture. Petroleum hydrocarbon (crude oil) was collected from Reservoir fluids laboratory Port Harcourt and stored in a clean sterile container. Microbial culture was obtained form the micro biological sciences laboratory in the Rivers State University of Science and Technology, Port Harcourt.

Analysis of microbial parameters

Preparation of bacteria culture for biodegradation

Bacteria culture was prepared inoculating culture of *Pseudomonas* sp.; isolated into a 500 ml of sterile nutrient broth medium. The inoculated medium was incubated at 37°C for 24 h, and served as culture of biodegradation of crude oil.

Use of bacteria population for monitoring biodegradation

Populations of heterotrophic bacteria and hydrocarbon utilizing bacteria here enumerated in the water samples were used as indices for monitoring the biodegradation. The ten-fold serial dilution method was applied for enumeration of bacterial (Harigan and McCane, 1990) In this method, 1 g of crude oil was added to 10 g of standard fresh water and CaCl as (diluent) to give 10^{-1} dilution. Further serial dilutions were made by transferring 1 ml of the 10^{-1} dilutions to 9 ml diluent up to 10^{-3} dilution from the 10^{-3} dilution, 0.1ml aliquots of samples was introduced on to the surface of sterile dishes. The innocula spread plated using a sterile beat glass rod and the inoculated plates were incubated at 37°C for 24 h. Colonies that developed were counted and recorded, and taken

Table 2. Analysis results on the pH; viscosity density refractor index and specific gravity of Bonny light crude oil at different operating temperatures.

Temperature (°C)	pH	Viscosity	Density (g/ml)	Refractor index	Specific gravity
27	6.08	2.36	0.899	1.388	0.771
35	6.14	2.14	0.895	1.368	0.813
45	6.36	2.02	0.782	1.313	0.763
60	6.38	1.72	0.741	1.107	0.685
80	6.38	1.69	0.710	1.101	0.671
Control 37	6.40	2.70	0.92	1.55	0.823

as the population of bacteria in colony forming unit per gram (cfu/ml) water. This procedure was repeated for different temperature range; 27, 35, 45, 60 and 80°C.

Evaluation of the pH

In this study, the different pH levels of 4 to 10 were examined. To maintain the pH; citrate-phosphate buffer (pH 4 and 5), phosphate buffer (pH 7 and 8), and carbonate-bicarbonate buffer (pH 9 and 10) were used to measure the pH of the crude. Electrometric method was used. An equipment with an electrode was inserted into the crude and the pH level was determined from the experimental set-up as well as a buffer solution was used to increase and maintain the sensitivity of the probe.

Temperature

The experiment was conducted under various operating temperatures, in other to determine the effect of the different temperature ranges on biodegradation of petroleum hydrocarbon using *Pseudomonas* sp. The following operational temperatures were used such as 27, 35, 45, 60 and 80°C. The mesophilic temperature is < 35°C whereas the themophilic is < 80°C > 35°C. The *Pseudomonas* sp. were isolated and identified as well cultured to with stand the operating temperature.

Total hydrocarbon content (THC)

The (extraction/spectrophotometer) method was used. A known volume of the sample was well agitated and poured into a separation funnel. A known quantity of sodium chloride ($NaCl_2$) was add to the prevent emulsification. 50 ml of xylene was added to different samples and then shaken properly to rinse the container before transferring into the separation funnel. The funnel was corked and shaken vigorously for about 1 min. The mixture was allowed to stand for separation. The sample portion was run-off by opening the tap and then the extract transferred into a 100 ml centrifuge tube by passing it through a filter paper containing 1 g of sodium sulphate. The extraction process was repeated with another 50 ml of xylene. The xylem layer was then collected into same centrifuge tube containing the first extract.

The separating funnel was rinsed with 10 ml xylene before it was collected into the centrifuge tube. The extract was centrifuge for 15bmin at 1500 rpm and transferred to a standard curve with a light path of 10 mm. The spectrophotometer was standardized and sample readings taken. THC concentration was calculated with reference to the standard curve and multiplication by the appropriate dilution factor. Detection limit is 0.01 mg/L.

RESULTS AND DISCUSSION

The results obtained from the investigation are presented in tables and figures. The viscosity surface tension and density of the crude oil, fresh and salt water wee analyzed in the laboratory and the results obtained recorded.

Table 1 illustrates the analysis result on the crude oil, salt and fresh water characteristics in terms of surface tension, viscosity and density. The following results were obtained, for surface tension crude oil its value is given as 0.036N/m, salt water value is 0.0727N/m and fresh water value is 0.0716N/m; for viscosity we have 34.3cp for crude oil sample, 4.10cp for salt water sample and 6.04cp for fresh water sample, whereas the following parameters were obtained for density crude oil is 0.925g/cm^3, salt water is 1000g/cm3 and fresh water is 938g/cm3 as presented in Table 1.

Results presented in Table 2 illustrate the characteristics of pH, viscosity, density, refractor index and specific gravity at various operating temperature. Increase in pH values were obtained with increase in temperature until a constant pH was obtained within the temperature range of 60-90°C, as well at control temperature of 37°C its pH value is given as 6.40. Decrease in viscosity was observed with increase in temperature. The variation on the pH and viscosity can be attributed to variation in temperature with control value of 2.70cp at a control temperature of 37°C as shown in Table 2. Similarly, the density decreases with increase in temperature as well its control value is given as 0.92g/ml at a control temperature of 37°C. Decrease in the refractor index was observed with increase in temperature as well a control value of 1.55 was obtained at control temperature of 37°C. An increase in specific gravity was obtained with the temperature range of 27 to 35°C and a sudden decrease in the specific gravity was observed above 35°C with specific gravity control value of 0.823 at a constant control temperature of 37°C as presented in Table 2.

Table 3 illustrates the analysis results on the petroleum hydrocarbon degradation at various operating temperature for both salt and fresh water as well as the characteristics of pH, optical density and microbial

Table 3. Analysis results on the petroleum hydrocarbon degradation at various temperatures for both salt and fresh water (pH, optical density and microbial population).

S/No	Temperature (°C)	Time (days)	Salt water			Fresh water		
			pH	Optical density	Microbial population (cfu/ml) × 10^6	pH	Optical density	Microbial opulation (cfu/ml) × 10^6
1.	27	2	8	0.5	1.5	8.00	1.50	1.50
2.		4	8	1.20	2.0	7.50	1.50	1.50
3.		6	8	1.50	10.0	7.75	2.75	3.50
4.		8	8	11.10	14.5	7.63	2.50	6.25
5.		10	8	7.00	12.5	7.80	1.75	5.75
Control		-	8	0.75	2.75	8.00	0.50	1.00
1.	35	2	7.00	1.25	2.00	7.6	1.25	1.50
2.		4	7.50	1.68	2.75	7. 75	1.50	2.00
3.		6	7.75	11.80	6.25	7.90	3.50	11.00
4.		8	6.50	12.85	8.50	7.80	3.75	12.00
5.		10	6.38	6.75	8.25	7.0	3.00	10.00
Control		-	8.00	2.50	2.75	8.0	1.09	0.50
1.	45	2	7.75	1.25	1.50	7.6	4.50	7.00
2.		4	7.50	2.50	3.5	7. 0	8.50	10.00
3.		6	7.00	4.00	6.00	7.0	13.50	13.50
4.		8	6.50	9.0	6.25	7.0	13.50	15.00
5.		10	6.25	9.0	5.75	7.0	12.50	14.00
Control		-	8.00	0.00	1.50	8.0	3.00	3.00
1.	60	2	7.75	1.25	1.50	8.0	1.25	1.50
2.		4	7.75	1.38	1.93	8.0	1.40	1.68
3.		6	7.75	1.34	1.72	7.0	1.35	1.72
4.		8	7.75	2.20	1.83	7.0	2.25	1.84
5.		10	7.75	1.77	1.78	7.0	1.98	1.76
Control		-	8.00	1.00	1.50	8.0	1.05	1.50
1.	80	2	8.0	1.67	2.0	8.0	1.67	2.00
2.		4	8.0	1.92	1.83	8.0	1.92	1.76
3.		6	8.0	2.15	1.72	8.0	2.15	1.64
4.		8	8.0	2.40	1.50	8.0	2.40	1.45
5.		10	80	2.48	0.50	8.0	2.48	0.78
Control		-	8.0	1.60	2.0	8.0	1.60	2.00

population in the different bioreactor. A constant pH value was obtained at temperature of 27°C with increase in optical density and microbial population for salt water environment. The fresh water characteristics of pH values decreases at 4days, increase at 6 days decrease at 8 days and increases at 10 days, whereas the optical density remain constant within 2-4 days, increases at 6 days and decreases from 8 to 10 days with the control value of pH = 8.00 and optical density = 0.50. Increase in microbial population was observed with increase in time. The variation in the microbial population can be attributed to the variation in time for fresh water environment.

Considering the characteristics of the system at temperature of 35°C we have the pH values within the range of 6.38 to 7.75, optical density values within the range of 1.25 to 12.85 and the microbial population within the range of 2 ×106 cf/ml to 8.50 × 106 cfu/ml for salt water environment. The fresh water environment characteristics is given as pH values within the range of 7.0 to 7.90, optical density 1.25 to 3.75 and the microbial population is within the range of 1.50 × 106 cfu/ml to 12.00 × 106 cfu/ml. At temperature of 45°C for 10 days investigation yielded the following results pH value is within the range of 6.25 to 7.75, optical density value within the range of 1.25 to 9.0 and microbial population value within 1.50 × 106cfu/ml to 6.25 × 106 cfu/ml. Fresh water value for pH is within the range of 7.0 to 7.6, optical density value within the range of 4.50 to 13.50 and microbial population value is within the range of 7.0 × 106cfu/ml to 15.00 × 106cfu/ml as presented in Table. 3.

At temperature of 60°C for 10 days intensive studied on petroleum hydrocarbon degradation, the following characteristics value were obtained for the pH, optical density and microbial population. For pH values are

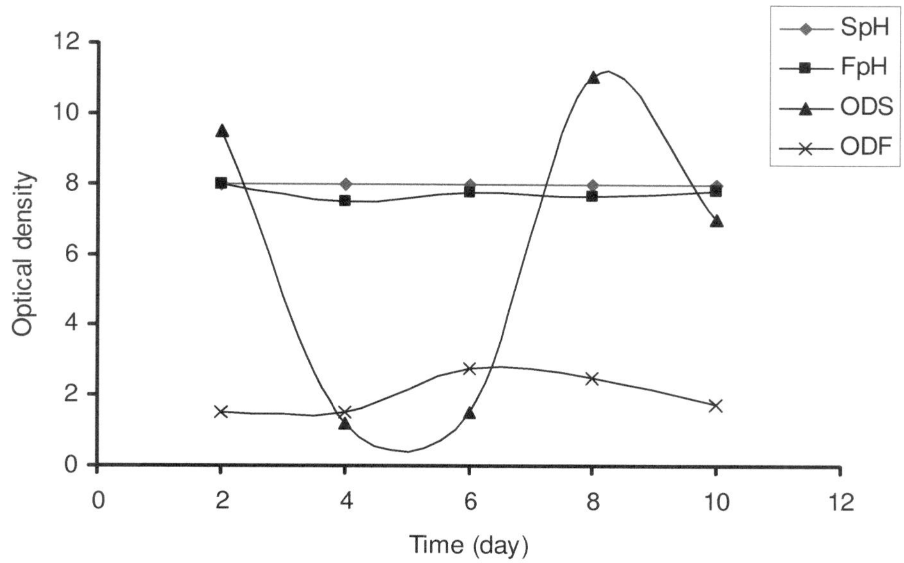

Figure 1. Graph of optical density and pH versus time for temperature at 27°C. Where SpH and FpH (pH of salt and fresh water), ODS and ODF (dissolved oxygen of salt and fresh water).

constant from 2–10 days, whereas the optical density values are within the range of 1.25 to 2.20 and the microbial population is within the range of 1.50 × 106 cfu/ml to 1.93 × 106 cfu/ml for salt water environment. In fresh water environmental the following results were obtained, for pH value within the range of 7.0 to 8.0, optical density value within the range of 1.25 to 2.25 and microbial population within the range of 1.50 × 106 cfu/ml to 1.84 × 106 cfu/ml as presented in Table 3.

Finally, at temperature of 80oC for 10days investigation revealed pH value was constant at 8.0, optical density value within the range of 1.67 to 2.48 and microbial population within the range of 0.5 × 106 cfu/ml for salt water environment as presented in Table 3. In the fresh water environment the pH value was constant at 8.00 whereas the optical density value is within the range of 1.67 to 2.48 and the microbial population is within the range of 0.78 × 10+cfu/ml to 2.00 × 106 cfu/ml as presented in Table 3.

The optical density and pH values on petroleum hydrocarbon degradation upon the influence of temperature were investigated. Results obtained revealed increase and decrease in optical density of petroleum hydrocarbon degradation in salt water medium with increase and decrease on petroleum hydrocarbon degradation in fresh water medium. The variation in the optical density concentration can be attributed to the composition of crude oil, salt and fresh water as well as the variation in time as, shown in Figure 1. Similarly, the pH value of salt water medium remain constant for the period of days under investigation whereas the pH value of fresh water medium decreases and increase with increase in time. These variations can be attributed to the characteristics of crude oil composition, salt and fresh water composition and other environmental factors as shown in Figure 1.

In Figure 2, it is seen that optical density increases with increase in time; on till the maximum optical density was achieved, then a decrease in optical density was observed with increase in time for both salt and fresh water media. The variation in optical density values can be attributed to the variation in time as well as the chemical composition of the samples under investigation. Similarly, the pH values increased and later decrease with increase in time. The variation can be attributed to the characteristics of crude oil composition, salt and fresh water composition, temperature and other environmental factors as shown in Figure 2.

The pH value on the biodegradation of petroleum hydrocarbon on mesophilic and thermophilic temperature can influence the optical density. From Figure 3, it is seen that the pH value of salt water medium decreases with increase in time, but for fresh water medium, a decrease in 1day investigation was observed and after that the pH value remain constant. These characteristics can be attributed to the composition of samples used for the investigation. The variation in the pH values can be attributed as well to the variation in time. Similarly, the optical density for both salt and fresh water media increase with increase in time on till the optical density was achieved, which yielded decrease in optical density values as shown in Figure 3.

Results presented in Figure 4 shows increase in optical density for both salt and fresh water on till the maximum optical density was achieved with increase in time. A decrease in optical density was observed after the

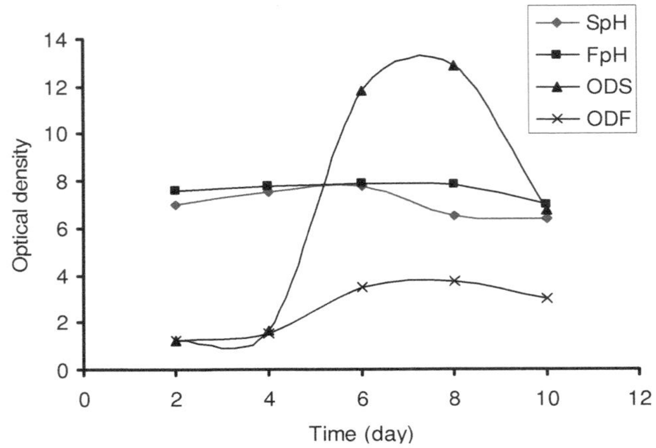

Figure 2. Graph of optical density and pH versus time for temperature at 35°

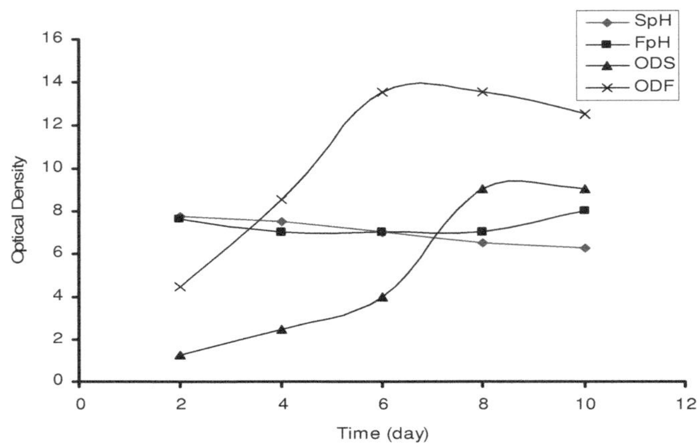

Figure 3. Graph of optical density and pH versus time for temperature at 45℃.

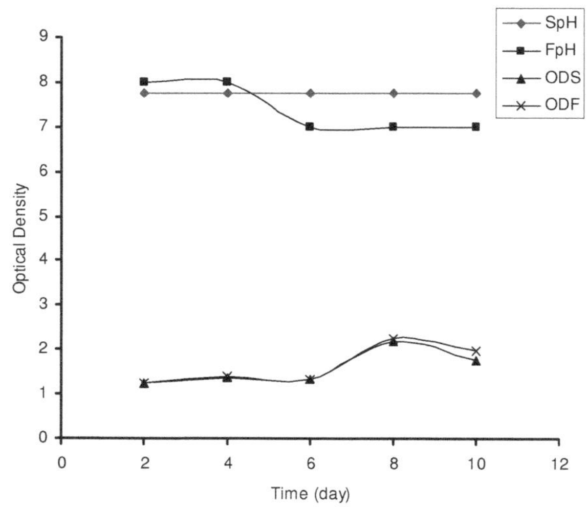

Figure 4. Graph of optical density and pH versus time for temperature at 60℃.

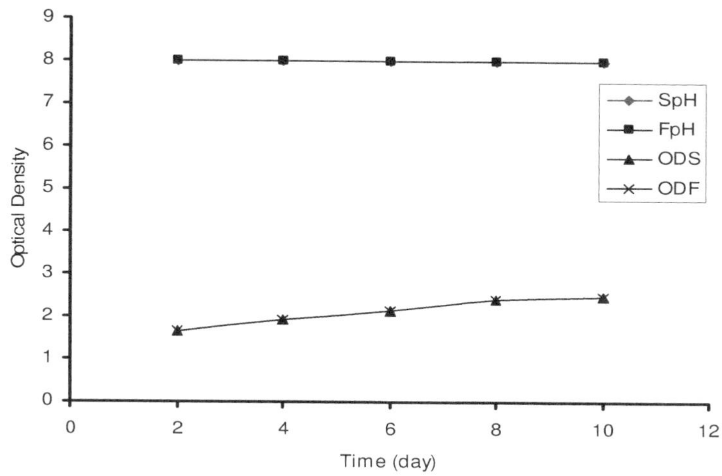

Figure 5. Graph of optical density and pH versus time for temperature at 80°C.

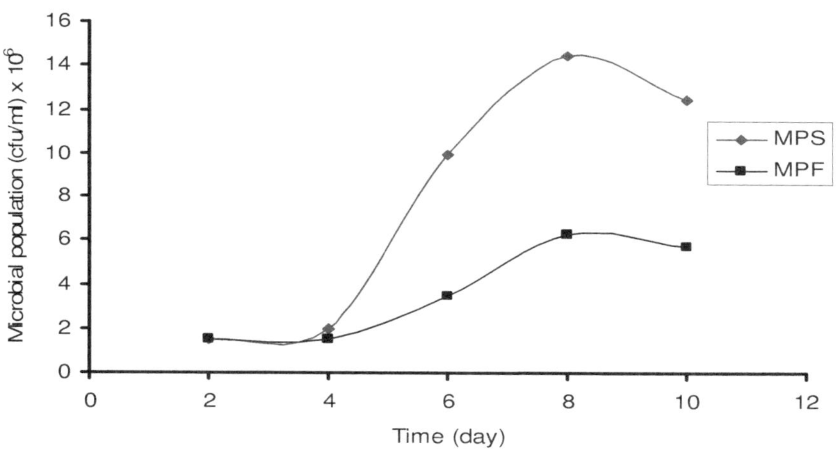

Figure 6. Graph of microbial population versus time of temperature at 27°C. Where, MPS and MPF (microbial population of salt and fresh water).

maximum optical density with increase in time. The variation in the optical density can be attributed to the variation in time. Similarly, the pH value was examined and results obtained illustrate constant pH value for salt water medium and slight variation in pH value of fresh water as shown in Figure 4.

The optical density and the pH values on petroleum hydrocarbon degradation upon the influence of temperature was investigation. From Figure 5, it is seen that the optical density increases with increase in time as well with constant optical density parameters for both fresh and salt water. The observation was unique and different from other operating temperature. The temperature inhibits the microbial growth of the *Pseudomonas* sp. which causes less impact on degradation process or slows the rate of degradation of petroleum hydrocarbon. In case there is slight microbial activity in both salt and fresh water system. Similarly, from Figure 5, it is seen that the pH increases with increase in time as well with constant pH values for both fresh and salt water environment.

The microbial population of *Pseudomonas* sp. identified, and isolated from the experimental set-up was evaluated per day for a constant temperature. It is seen that increase in microbial population from 9 to 10days as shown in Figure 6, for both salt and fresh water system. The variation in microbial population of *Pseudomonas* sp. can be attributed to variation in time as well as the characteristics of the crude oil, salt water and fresh water. An intercept on microbial population was observed within 2 to 3 days as shown on Figure 6.

In Figure 7, it is seen that a lag, progressive, stationary and decline phase was observed at different time interval. Increase in microbial population was observed with

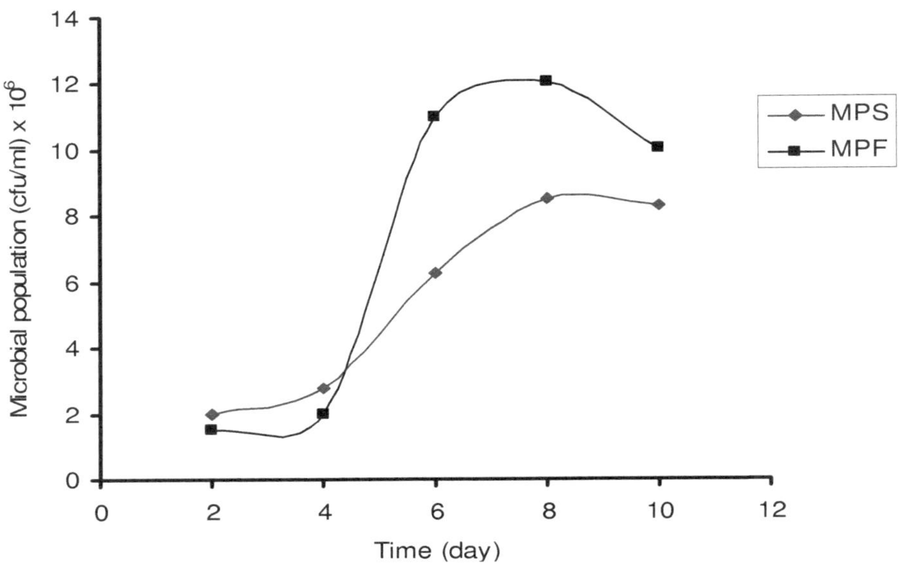

Figure 7. Graph of microbial population versus time for temperature at 35°C.

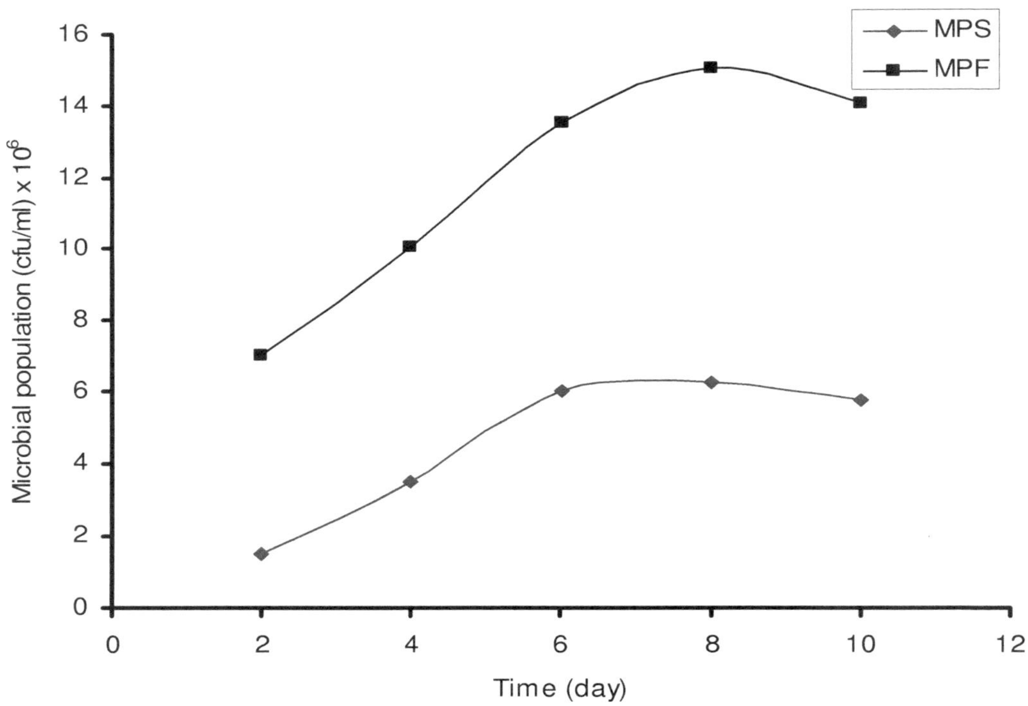

Figure 8. Graph of microbial population versus time for temperature at 45°C.

increase in time for both salt and fresh water media. The microbial population in salt water medium was higher than microbial population in fresh water medium. This implies that the characteristics of the salt water medium favours the *Pseudomonas* sp. used for the investigation. The variation in the microbial population can be attributed to variation in time and crude oil concentration and other functional parameters.

The graph of microbial population against time for operating temperature of 45°C is shown in Figure 8. Results obtained revealed increase in microbial population for both salt and fresh water environment with

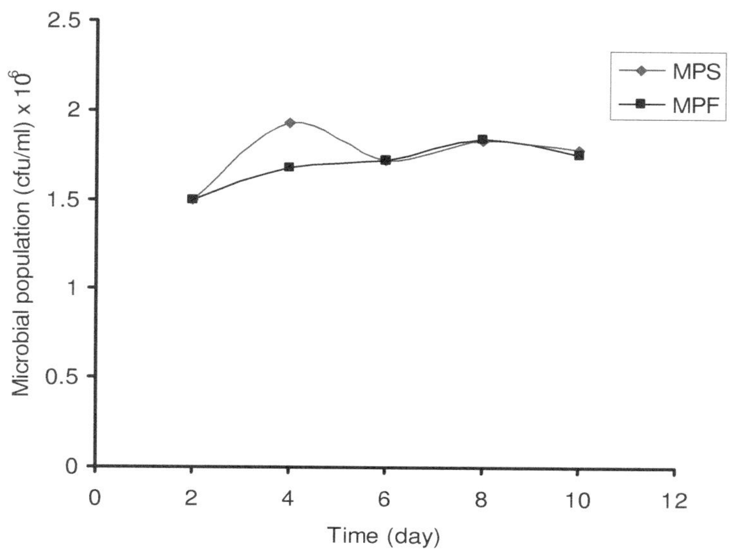

Figure 9. Graph of microbial population versus time for temperature at 60°C.

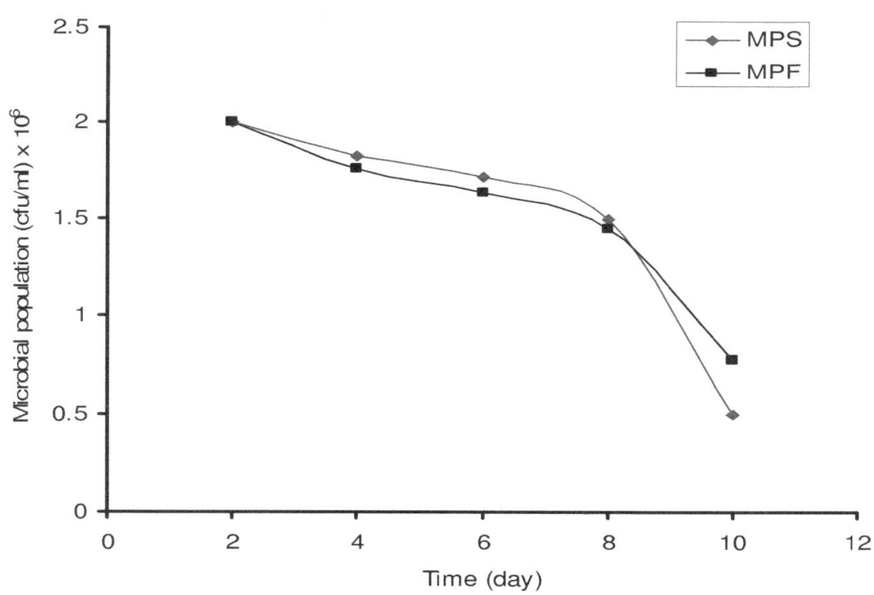

Figure 10. Graph of microbial population versus time for temperature at 80°C.

increase in time. In this case there is no lag phase, but progressive, stationary and decline phase was observed. The variation in microbial population of *Pseudomonas* sp. can be attributed to variation in time as well as the characteristic and chemical composition of the crude oil, salt water and fresh water media. Optimum microbial population for *Pseudomonas* sp. was experienced on 6 and 8 days for fresh and salt water environment.

In Figure 9, it is seen that the microbial population of *Pseudomonas* sp. in fresh water medium increases with increase in time. Progressive, stationary and decline phase observed with no lag phase. Similarly, for salt water environment, the microbial population increased and later decrease with time. It is seen that on the salt water environment various lag phases was observed. This can be attributed to increase in temperature, chemical composition of the crude oil, fresh water and salt water environment as well as the type of microorganism used for the investigation.

Results presented in Figure 10, illustrates decrease in microbial population with increase in time. Only decline phase was experienced, this sudden decline by the

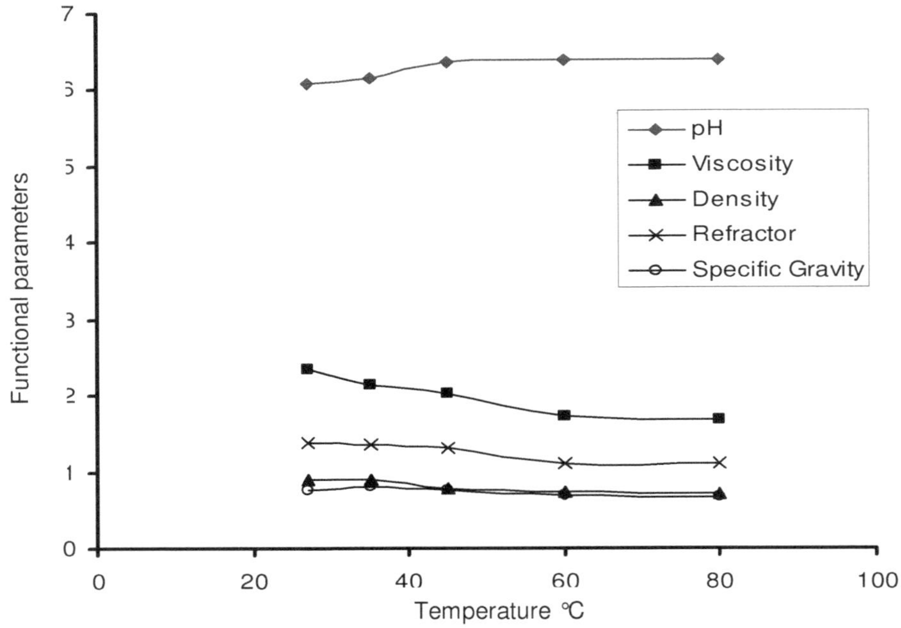

Figure 11. Graph of functional parameters versus temperature.

Pseudomonas sp. can be attributed to the operating temperature, crude oil composition, salt and fresh characteristics and other environmental factors. In this case no lag, progressive and stationary phase was observed, indicating the effect of temperature on *Pseudomonas* sp. as well as non-effectiveness of biodegradation of petroleum hydrocarbon in salt and fresh water environment.

Figure 11 illustrates the relationship among functional parameters with temperature. From Figure 11, it is seen that the viscosity, density, refractor index and specific gravity parameters decreases with increase in temperature as well as increase in pH. The variation in the functional parameters can be attributed to variation in temperature.

Conclusion

The following conclusions were drawn from the investigation such as:

(i)The effect of temperature on microbial activity was examined at different operating temperatures and the degradation process was faster at mesophilic stage than the mesophilic stage. This can be attributed to the effect on the physicochemical properties of crude oil and the aqueous media.

(ii) The effect of temperature on pH of salt and fresh water environment was examined at different operating temperatures. The characteristics of pH influence the rapid growth of the microorganism in the batch bio-reactor, the process influence the degradation of crude oil at both mesophilic and themophilic conditions.

(iii) The effect of temperature on optical density was examined at different operating temperature for both salt and fresh water environment. The variation on the optical density can be attributed to the variation in temperature as well influence the degradation of crude oil in both salt and fresh water.

(iv) The best operating temperature to enhance rapid remediation of polluted environment with crude oil and other contaminants using *Pseudomonas* sp. for mesophilic it is within the range of 20-35°C and thermophilic is within the range of 35-60°C.

(v) *Pseudomonas* sp. is capable of biodegradating petroleum hydrocarbon in both salt and fresh water environment provided the operating conditions are favourable.

REFERENCES

Anderson B (2008). Report on the soils of the Niger Delta Special Area Niger Delta Development Board, pp. 17-50.

Atlas RM, Bertha R (2000). Degradation and Mineralization of Petroleum by two bacterial Isolated from crude oil, Microbial Rev., 15: 173-196.

Atlas RM (2005). Effects of Temperature and crude oil composition on Petroleum Biodegradation. Available:http://www.pubmedcentral.nih.gov/picrender.fegi?artd = 187194&blobname=applmicrooo(2008June21st).

Atlas RM (2004). Microbial Degradation of Petroleum Hydrocarbons, an Environmental Perspective, Microbial. Rev., 10: 180-209.

Brennan S, Withgott J (2005). Environment The Science behind the stores. Pearson Education. Inc. USA, pp. 306-472.

Coulson JM, Richardson JF (1994). Chemical and Biochemical

Reactors and Process control vol. 3, 3rd Edition, Butternorth Heinmann, pp. 342-353.
David JB (2002). Petroleum Microbiology. *Elselier*, New York, pp.107-209.
Davis SN, Deweist RJM (2006). Hydrogeology John Wiley and Sons, Inc. New York, pp.1-500.
DPR (Department of Petroleum Resources) (2002) Engineering Guidelines and Standards for the Petroleum Industry in Nigeria, pp. 10-36.
Foster JW (2002). Hydrocarbons as Substrates for Microorganism. Antonie coastal Waters. biotechnology Bioeng. 14:297-308.
Van Leeuwenhoek J. Microbiol. Serol., 28: 241-274.
FMEM (Federal Ministry of Environment) (2006). National Guidelines and Standards for Industrial Effluents, Gaseous Emissions and Hazardous Wastes Management in Nigeria, pp. 59-66.
Gibson DT, Subramanian V. (2005). Microbial Degradation of Aromatic Hydrocarbons, in Microbial Degradation of Organic Compounds (Ed. D.T. Gibson) Marcel Dekker, New York, pp. 181-252.
Gibson DT (2004). The Microbial Oxidation of Aromatic Hydrocarbons. CRC Crit. Rev. Microbial. pp.1: 199-223.
Haramaya S, Suguira K, Ishihara M, Shimauchi T (1999). Physiochemical Properties and biodegradability of Crude Oil. Environ. Sci. Technol., 31: 45-51.
Leahy JG, Colwell RR (2006). Microbial Degradation of Hydrocarbons in the Environment, Appl. Environ. Microbial., 54: 305-315.
Kallio RE (2002). Microbial Transformation of Alkanes. In D. Perlman (Ed.), Fermentation advances. Academic Press Inc. New York, pp. 635-648
Kiel G (2008). Environmental Engineering McGraw-Hill Publishing Company, England, pp. 72-213.
Kobayshi H, Rittman BE (2000). Microbial Removal of Hazardous Compounds. Environ. Sci. Technol., 19: 470-481.
Millkin BA, Philips GT, Steart JE (2004). Effect of Environment Parameter on Bacterial Degradation of Bulk Oil, Crude Oil and Hydrocarbon. Appl. Microbial., 2: 315-322.
Mckenna EJ, Kallio RE (2004). Biology of Hydrocarbons Annu. Rev. Microbiol., 19: 183-206.
Odu CT, Nwoboshi LC, Esuruoso OF (2004). Environmental Study of the Nigerian Agil Oil Company Operation Areas in Proceedings of the International Seminar on the Petroleum Industry and the Nigeria Environment, Kaduna, Nigeria, pp. 274-283.
Ogoni HA (2003). Topics in Biochemical Engineering vol.1, 1st Edition, Divine Printers Port Harcourt. pp. 1-58.
Ogoni HA, Oboho EO, Amadi SA (2001).Determination of the Optimum Temperature and PH for the Biodegradation of Petroleum Hydrocarbon Mixture, African J. Environ. Stud., 1(1-2): 40-58.
Okoh AI (2003). Biodegradation of Bonny Light Crude Oil in Soil Microcosm by some Bacterial Strains Isolated from Crude Oil flow Stations Saver Pits in Nigeria. African J. Biotech., 12(5): 104-108.
Okpokwasili GC (2006). Biodegradation of Petroleum Hydrocarbons by Bacterial, Abstract of 14th, Annual Conference and 3rd African Regional Network Microbiology. (1996) Unical, Nigeria, pp. 57-69.
Spiri TG, Stighani WM (2006). Chemistry of the Environment. 2nd edition Prentice Hall Inc. New Jersey, USA , pp. 371-810.
Westlake DWS, Jobson A, Phillippe R, Cook FD (2003). Biodegradability and Crude Oil composition Can. J. Microbio., 20: 915-928.
Ukpaka CP (2007). Modelling the microbial thermal kinetics system is Biodegradation of crude oil. J. Modelling, Simulation and Control (AMSE), 67(2): 40-59.
Van der LAC, Thijsse GJE (2005). The Mechanism Adv. Ezzymol. 27: 469-548.
ZoBell CE (2002). Action of Microorganisms on Hydrocarbons. Bacterial Rev., 10: 1-49.

Influence of finishing temperature on the mechanical properties of conventional hot rolled steel bar

S. A. Balogun, G. I. Lawal*, O. I. Sekunowo and S. O. Adeosun

Department of Metallurgical and Materials Engineering, University of Lagos, Nigeria.

The conventional mill operators are faced with challenges of producing reinforcing steel bars of adequate strength characteristics. This stems from the inability to effectively control and monitor process variables such as temperature, strain-rate and cooling. These parameters collectively determine the microstructure and invariably the mechanical properties of the rolled bars. However, the finishing temperature appears to be more important than the temperature at the roughing stand. Often, the type and extent of in-process cooling method and duration of each rolling cycle largely determine the finishing temperature. Thus, installation of temperature monitoring devices along the rolling line will furnish prompt thermal information necessary for both process conditions and product properties optimization respectively. Detailed temperature tracking of a conventional rolling operation was carried out with a view to establish the range of finishing temperature that give rise to improved strength characteristics of steel bars produced, taking cognizance of the rolling stock chemistry. Improvement in the bars mechanical properties was achieved within 840 to 860°C finishing temperature and rolling stock chemistry in the range of 0.21 to 0.23% C, 0.20 to 0.25% Si and 0.50 to 0.60% Mn.

Key words: Rolling, temperature, microstructure, steel, strength.

INTRODUCTION

In hot rolling of steel, temperature is the dominant parameter controlling the kinetics of metallurgical phenomena such as flow stress, strain-rate and recrystallization (both static and dynamic). The mechanical properties of the final product are determined by a complex sequence of microstructural changes conferred by thermal variations (Pereloma, 2001; Richard, 2001). Temperature also aids the softening mechanism by which rolling stocks (billet) are prevented from brittle fracture due to work hardening effect of the rolling forces.

There exist in hot rolling, a correlation between strain-rate and the flow stress. The working temperature however, influences both phenomena such that increase in temperature usually results in increased strain rate, which eventually affects the austenite grain size (Choi, 2002). Unless by means of innovative cooling, the austenite grain size at the last pass solely determines the product grain size and mechanical properties.

Most mills in developing nations of the world still operate on the basis of conventional rolling which is devoid of modern facilities offered by controlled rolling (Obikwelu, 1987). Usually, conventional mill operations are not executed along with the necessary temperature monitoring with a view to controlling the evolved microstructure. Hence, steel bars produced through conventional rolling often exhibit abysmally low mechanical properties (Lamberterie, 2006; Saroj, 2000). Control of inter-stand temperature such that the desired initial austenite grain size is achieved at the last stand is imperative. This will ensure that appropriate phase transformation of the right grain size, morphology and texture is obtained during cooling of the bars.

The above notwithstanding, control of temperature during rolling is more important at the finishing than at the roughing stage (Laasraoui and Jonas, 2007). Choi (2002) also established that property sensitive parameters of hot rolled steel bar depend largely on the finishing temperatures.

The idea was employed (Barrett and Wilshire, 2002) in the early 1980s in the production of ferritic hot rolled interstitial free steel to eliminate temperature control problems. This was accomplished by reducing the

* Corresponding author. E-mail: gilawal@yahoo.com.

Table 1. Chemical composition analysis of rolling stocks (billets).

Sample ID	Elements (%)												
	C	Si	S	P	Mn	Ni	Cr	Sn	Mo	V	Cu	Fe	Ceq
A	0.194	0.167	0.039	0.025	0.856	0.146	0.178	0.038	0.029	0.006	0.344	97.978	0.42
B	0.220	0.199	0.046	0.032	0.501	0.101	0.104	0.036	0.015	0.003	0.216	98.454	0.35
C	0.164	0.123	0.046	0.027	0.768	0.137	0.149	0.037	0.017	0.002	0.318	98.336	0.36
D	0.308	0.258	0.050	0.028	0.684	0.117	0.147	0.037	0.015	0.002	0.342	98.012	0.49
E	0.211	0.246	0.039	0.028	0.506	0.112	0.148	0.035	0.013	0.002	0.306	98.354	0.36
F	0.172	0.113	0.046	0.019	0.697	0.105	0.136	0.035	0.019	0.001	0.249	98.686	0.34
G	0.231	0.250	0.055	0.034	0.602	0.102	0.120	0.034	0.020	0.002	0.274	98.276	0.38

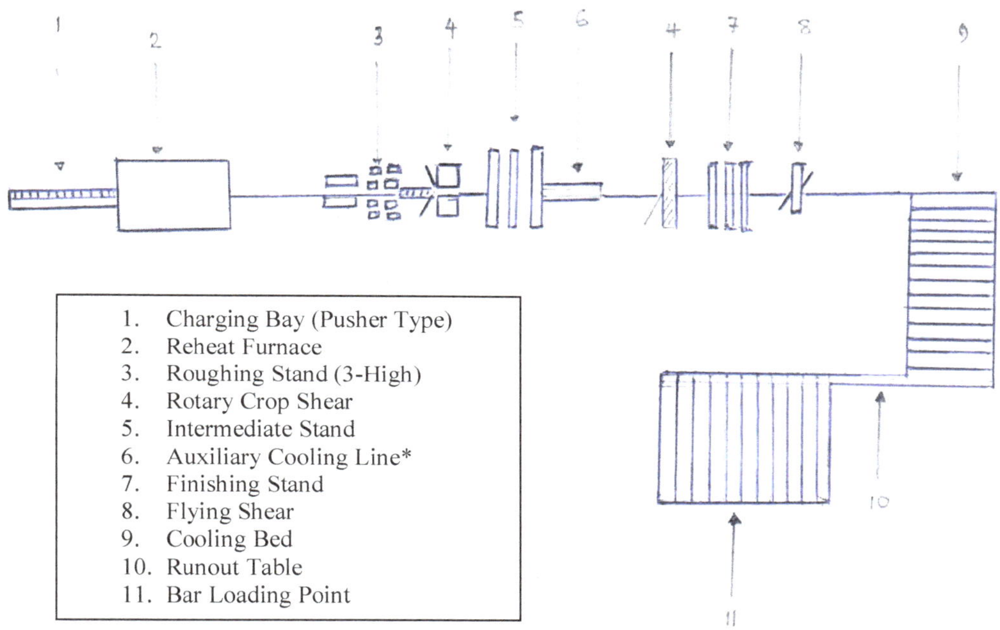

Figure 1. A conventional bar mill configuration.

Figure 1: A Conventional Bar Mill Configuration

finishing temperature from the conventional 1030 to 810°C. Given that fine-grained product is usually desired, the best practice is to ensure a much lower working temperature at the last pass. This will drastically reduce grain growth during cooling. In this study attempt is made to obtain thermal variations data through monitoring of a conventional mill operation and evaluate its effect on the strength characteristics of the bars taking cognizance of the rolling stock chemistry.

EXPERIMENTAL PROCEDURE

Materials

Billets (100 x 100 x 1600 mm) of chemical composition as shown in Table 1 were charged into the re-heat furnace and heated to the rolling temperatures in the range 1200 – 1220°C. They were then rolled into 12 mm diameter high-yield bars. One hundred and twenty billets were rolled in each of the seven rolling cycles monitored.

Temperature tracking (TT)

Using a Jenway digital pyrometer model 220k, monitoring of the process temperature was carried out at each of the critical points where high temperature deformation occured namely roughing, intermediate and finishing stands respectively. As illustrated in Figure 1, the points at which temperature readings were obtained include the reheat furnace, roughing, intermediate and finishing stands respectively and at the cooling bed. Bar samples of 12 mm were obtained at the end of each rolling cycle for mechanical and microstructural analyses.

Mechanical property tests

In carrying out both hardness and tensile evaluation properties on the bars, the entire test specimens were prepared according to the

Table 2. Temperature tracking data (TTD).

Rolling cycle	Reheat oven to (°C)	Temperature at the stands (°C)			Cooling bed to (°C)
		Roughing, T_r	Intermediate, T_i	Finishing, T_f	
A	1217	1085	1013	872	792
B	1215	1075	998	848	762
C	1218	1094	1026	893	817
D	1209	1074	1005	864	786
E	1215	1078	1003	858	774
F	1216	1087	1016	879	800
G	1214	1076	1000	853	768

Table 3. Hardness of bars produced by conventional rolling.

Sample ID	Finishing temperature (°C)	Hardness (HRB)	Carbon (%)
A	872	72.7	0.194
B	848	84.3	0.220
C	893	62.1	0.164
D	864	93.2	0.308
E	858	87.1	0.211
F	879	67.8	0.172
G	853	86.4	0.231

British standard (BS 18). Relevant clauses of the Nigerian Industrial Standards (NIS 117-42/50HD 2004) were also complied with. The test specimens hardness values were evaluated using the 'B' scale Rockwell hardness machine model United TB-II. An Instron electro-mechanical testing system model 3369 was used to obtain the test specimens yield and tensile strength characteristics.

Microstructural analysis

Test specimens were ground on a water-lubricated grinding machine using silicon carbide abrasive papers grades 240, 320, 400 and 600 grits. Final polishing of the specimens was effected with 0.5 µm chromic oxide powder. The surfaces so obtained were etched in 2% Nital solution and left for 30 s then, rinsed with water. The microstructural features of the specimens were examined under a metallurgical inverter microscope model FEROX PL at ×100 magnification.

RESULTS AND DISCUSSION

Results of the temperature tracking experiment are shown in Table 2 while listed in Table 3 are values of micro-hardness induced in the steel bars after air-cooling. Using the data obtained during tensile test on specimens, relevant tensile data are computed and presented in Table 4. The microstructural features of test specimens are shown on Figure 2 (a-g) Variations in the functional mechanical characteristics in terms of ultimate tensile strength, yield strength and hardness exhibited by test specimens in relation to the finishing temperatures are illustrated in Figures 3 and 4. The results of yield property behaviour with variation in carbon content are shown in Figures 5 and 6.

Figure 3 shows the variation of reheat temperatures of billets used during the rolling cycles. The values were almost the same, 1216± 2°C. The finishing temperatures, T_f however, vary widely and in the range 848.2°C to 893.4°C.This gives a variation of 45.2°C, which is high enough to induce microstructural transformations during the air-cooling of the bars.

Wide variations in finishing temperatures can be attributed to the combination of two factors namely, in-process cooling and speed of rolling. The combination of direct and indirect cooling of rolling stock along with other heat sensitive devices of the mill facility impact on the finishing temperature. Other than technology, internal state of the rolling stock (British Standards Specification: (BS 4449 (1988); Curtin and Dewald, 2005) in term of cleanness affects the extent of strain hardening suffered during rolling hence, the speed of rolling. Similarly, large amount of strain hardening occasioned by inclusions usually give rise to delay in material flow (Curtin and Dewald, 2005).

The duration of rolling cycle in this study was in the range of 90-105 seconds culminating in different finishing temperatures. Generally, property sensitive parameters of hot rolled steel bars depend largely on the finishing temperatures (Choi, 2002). Similarly, flow stress; strain, strain rate and recrystallisation (static and dynamic) are influenced by temperature during hot rolling. However, temperature prediction is more important during finishing

Table 4. Calculated true stress and true strain.

A: Lo=23.14, A=19.46 mm^2		B: Lo=23.53, A=20.11 mm^2		C: Lo=23.36, A=19.32 mm^2		D: Lo=26.21, A=20.67 mm^2		E: Lo=23.37, A=20.19 mm^2		F: Lo=24.92, A=21.73 mm^2		G: Lo=22.38, A=19.71 mm^2	
True stress (MPa)	True strain (ε)	True stress (MPa)	True strain (ε)	True stress (MPa)	True strain (ε)	True stress (MPa)	True strain (ε)	True stress (MPa)	True strain, (ε)	True stress (MPa)	True strain (ε)	True stress (MPa)	True strain (ε)
210.7	0.02	256.6	0.03	56.0	0.01	179.8	0.02	101.0	0.01	99.0	0.01	157.4	0.01
266.8	0.03	336.1	0.04	107.2	0.01	225.8	0.03	153.1	0.02	189.2	0.02	260.8	0.02
321.8	0.04	366.6	0.05	158.9	0.02	296.4	0.04	256.1	0.03	285.3	0.03	315.1	0.03
379.4	0.05	444.9	0.06	215.3	0.02	354.1	0.05	339.5	0.04	336.0	0.04	367.4	0.03
448.1	0.09	488.3	0.09	267.1	0.03	435.9	0.06	417.6	0.05	414.7	0.08	442.2	0.04
518.0	0.11	585.0	0.12	323.5	0.04	500.5	0.07	482.1	0.09	498.4	0.11	499.2	0.05
535.4	0.12	634.5	0.15	396.1	0.06	559.8	0.10	586.0	0.12	553.6	0.15	553.6	0.09
586.8	0.16	652.2	0.16	507.1	0.12	672.9	0.12	651.3	0.16	591.5	0.18	648.4	0.12
614.3	0.19	692.6	0.19	604.6	0.22	748.8	0.16	696.1	0.20	623.0	0.22	749.6	0.20
618.8	0.20	712.1	0.22	616.7	0.25	801.5	0.20	711.4	0.23	635.5	0.25	773.2	0.24
596.4	0.23	728.2	0.25	588.3	0.28	806.9	0.21	680.6	0.26	645.7	0.28	745.8	0.27
531.5	0.25	607.5	0.30	480.1	0.30	676.2	0.27	563.3	0.29	484.9	0.35	624.6	0.30
298.0	0.25	361.8	0.30	268.9	0.31	381.9	0.28	328.1	0.29	290.8	0.35	344.5	0.30

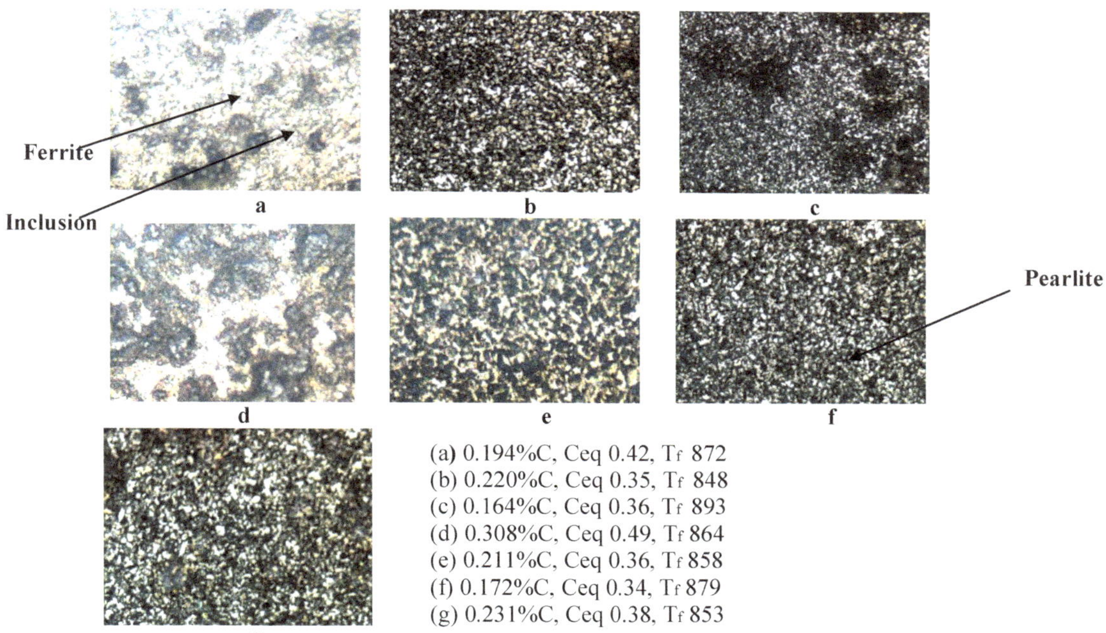

Figure 2. Micrographs of air-cooled rolled bar samples.

than roughing (Laasraoui and Jonas, 2007).

Microstructural observation

The micrographs on plates A-G showed two major phases, ferrite and pearlite including large pod-like non-metallic inclusions. The phases are distributed approximately in the ratio 0.3, 0.6 and 0.1 volume fractions respectively. Accurate prediction of microstructural evolution (Singh, 2007) in various carbon steels during hot rolling is one of the challenges steel millers are grappling with. This is because the finishing temperature is closely related to visco-plastic flow of the

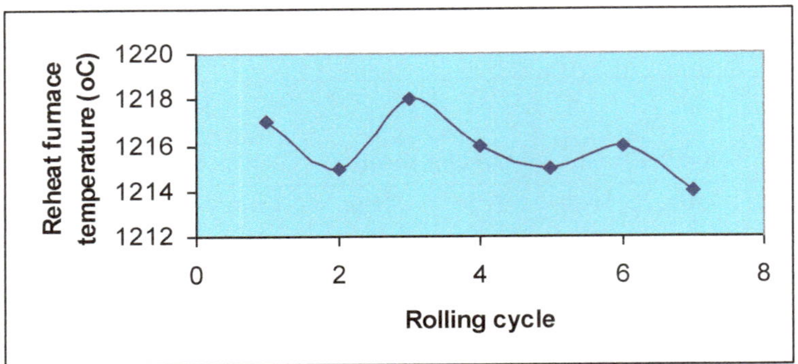

Figure 3. Variation of reheating temperature of rolling stock.

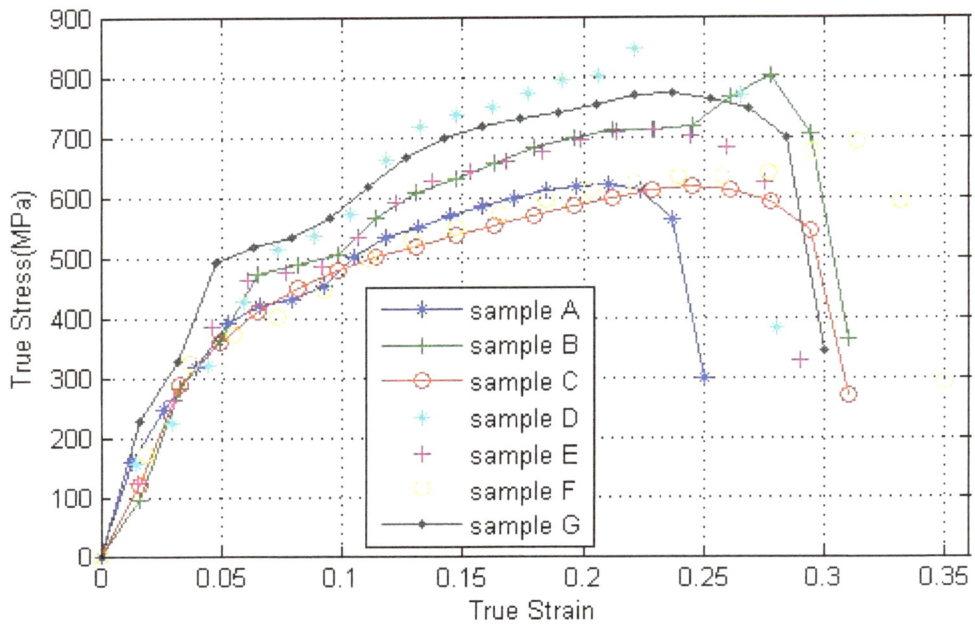

Figure 4. True stress against true strain.

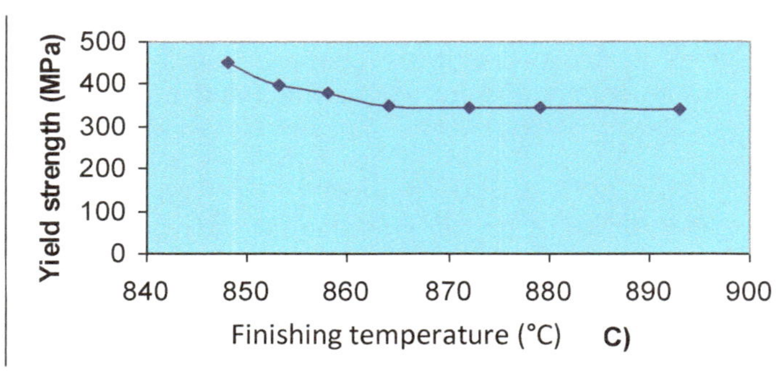

Figure 5. Yield strength against finishing temperature.

rolling stock, which affects the austenite grain size. Prior austenite grain size determines to a large extent the size of grain in the final product and hence its mechanical properties.

The rolling cycles monitored in this study had their finishing temperatures between 864 and 893.4°C, which is about 150°C above the lower critical point, 721°C. Hence, about 14 s elapsed before the start of

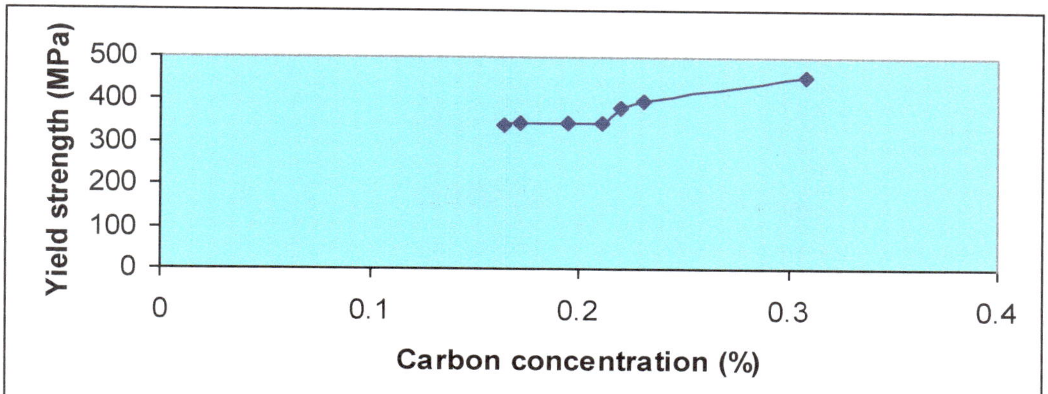

Figure 6. Variations of yield property with carbon concentration.

transformation. The delay resulted in the formation of coarse pearlite in test specimens at finishing temperatures of 872, 893 and 879°C respectively (Figure 2a, c, f).

However, the degree of coarseness of the pearlite reduces with decreasing finishing temperatures (Figure 2(e) T_f 858°C, (g) T_f 853°C and (b) T_f 848°C). Coarse pearlite formed at the nose of TTT curve just below A_1 line exhibits high strength but poor ductility (Singh, 2007). This accounts for the low yield stress exhibited by all test specimens except specimen G (452.8MPa) as shown in Figures 5 and 6.

Ultimate tensile strength

The effects of microstructures in concert with other relevant parameters such as temperature, composition and cooling regime manifested in the flow curves of Figure 4. Specimen D exhibited the highest UTS, 806.9 MPa (Figure 4) mainly due to its relatively high carbon concentration, 0.30% coupled with high concentrations of inclusions (Ceq 0.49), which coalesced along with pearlite grains. This type of structure impedes dislocation mobility thereby requiring higher stress to cause plastic deformation. However, the bar is not recommended for application as reinforcement due to its abysmally low modulus of elasticity, 13493.6 MPa.

The non-metallic inclusions observed are a combination of tramp elements and slag that could not be removed at the refining stage of the steel from which the rolling stock was produced. Melting of most inclusions is not feasible during the reheating of rolling stocks in the furnace, which operates around 1200°C. This is because basic slags that are mainly compounds of silica and magnesia are highly refractory (Obikwelu, 1987). Such inclusions merely deform along the direction of rolling thus conferring directional properties on the rolled bars. Deformability of inclusions during hot working of steel influences the final properties of the product (Lai, 2007). Deformed inclusions also distort normal grain boundary arrangements that are potential barriers to dislocations motion.

Specimen G exhibited a good combination of UTS, 773.2 MPa and elastic modulus, 19157.9 MPa being the highest amongst all specimens tested. This may be attributed to two factors namely fairly low finishing temperature 855°C, which gave rise to fine grained pearlite under air-cooling and low concentrations of inclusions, Ceq 0.38. The combination of these factors favours modest strain hardening both during elastic and plastic deformations.

Yield strength

Figure 5 shows the variations of test specimens yield strength at different finishing temperatures. Specimens A-F exhibited low yield strengths in the range 380.8 to 396 MPa. The yield strength of sample G, 452.8 MPa is comparable to local and international specifications, which are 420 MPa (NIS), 460 MPa (BS) and 500 MPa (ASTM).

The yield point phenomenon common in steel, aluminium and copper, is associated with small amounts of interstitial or substitutional impurities (Curtin and Dewald, 2005). This partly accounts for the observed substantial ductility in low carbon steels having interstitial carbon concentrations between 0.1 and 0.25%maximum. It has been shown (Lai, 2007; Maunder and Charles 2006; 1968) that almost complete removal of carbon and nitrogen from low carbon steel by wet-hydrogen treatment will remove the yield point phenomenon. However, only about 0.001% of either of these elements is required for a reappearance of the yield point.

Gradual increase in yield strength of test specimens occurred from 0.15 – 0.19% carbon (Figure 6). However, sharp increase in yield strength was observed between 0.19 and 0.23% carbon with corresponding finishing temperature in the range 848 to 858°C. Above this temperature range (Figure 5) and irrespective of the carbon composition, the yield strength dropped drastically. From this observation, it can be inferred that the type of microstructure developed at finishing

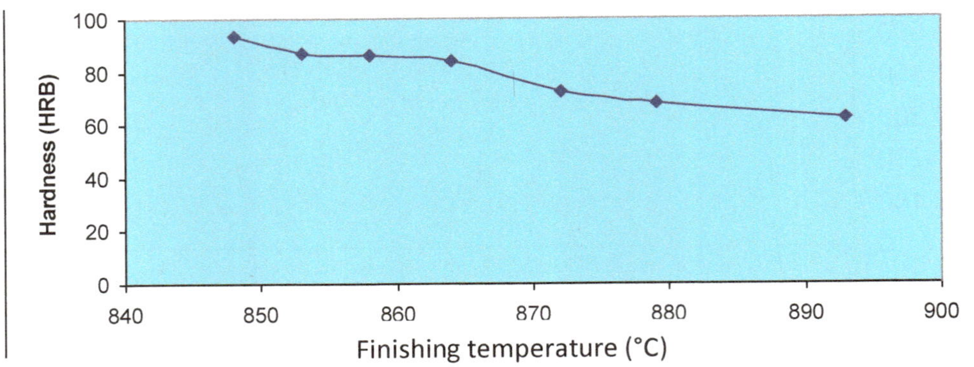

Figure 7. Variations of micro-hardness with finishing temperature.

temperatures greatly influenced the yield values obtained. Though specimen D has 0.30% carbon yet, it exhibited merely yield strength of 344.8 MPa. Yielding phenomenon in low carbon steel peaked between 0.20 and 0.30% of carbon. Beyond this range, the ductility of carbon steel is impaired. Apart from weldability criterion, this may be another basis for the BS 4449 specification of 0.25% carbon maximum in billets/ingots employed in hot rolling of construction steel bars.

Hardness of conventional rolled bars

Hardness exhibited by the specimens varies with the rolling stock carbon concentrations (Figure 7). Under natural air-cooling as the case in this study and at such finishing temperature range 848 to 893°C, the hardness measured must have been induced entirely by cementite rather than martensite. This is because martensite could not have formed given the prevailing processing conditions. Relevance of adequate surface hardness required in reinforcing bars concerns the ribs, which are meant to offer resistance to slip of the bar member within the structure. Free slip of bars should not be greater than 0.2 mm in a pull-out test (British Standards Specification: (BS 4449 (1988). The ribs must therefore exhibit sufficient bond strength in order to function effectively.

Conclusion

The influence of process factor such as finishing temperature, on the mechanical properties of hot rolled steel bar in a conventional mill has been investigated. From the analyses of results obtained, it can be concluded that in hot rolling the type and extent of in-process cooling method and duration of each rolling cycle largely determine the finishing temperature. Given that under conventional rolling whereby the bars are air-cooled, it is imperative that finishing temperature be just above A1 point in order to avoid the formation of coarse pearlite.

Finishing temperatures in the range 840 to 860°C gave rise to fine grain microstructures that resulted in improved strength. Close monitoring of inter stand temperature will furnish prompt information regarding the extent of cooling required. This will ensure that rolled products get to cooling bed with temperature profile amenable to desirable microstructural evolution under air-cooling. Similarly, the rolling stock chemistry and soundness in term of level of cleanness exhibits a complementary influence on the strength characteristics of the steel bar.

REFERENCES

Barrett CJ, Wilshire B (2002). "The Production of Ferritically Hot Rolled Interstial Free Steel on Modem Hot Strip Mill." J. Mater. Process. Technol., 122(1): 56–62.

British Standards Specification (BS 4449) (1988). "Carbon steel bars for the reinforcement of concrete", pp. 6-9.

Choi Y (2002). "Integrated Model for Thermo-Mechanical Controlled Process in Bar Rolling." Mater. Process. Technol., 125–126: 678–688. Sept.

Curtin WA, Dewald M (2005). "Multiscale Modeling of Dislocation / Grain Boundary Interactions." Focus Session: Multiscale Phenomena in Solid Interactions, March.

Laasraoui A, Jonas JJ (2007)."Prediction of Temperature Distribution Flow Stress and Microstructure during Multipass Hot Rolling of Steel Plate and Strips." ISIJ Int., 31: 95–105.

Lai GY (2007). "The Effect of Austenitising Temperature on the Microstructure and Mechanical Properties of as-quenched 4340 Steel." Metallurg. Mater. Trans. B, June, 5(7): 1663–1670.

Lamberterie B (2006). "Recent Evolutions and Trends in the Steel Rolling Industry." Inaugural Steel Rolling Conference, Paris, Aug, pp. 311–318.

Obikwelu DCN (1987). "Metallurgical Consideration in the Optimization of Mechanical Properties of Rolled Products." Seminar Paper Presented at the Metallurgical and Research Department, Delta Steel Company, Aladja, Warri, Nigeria.

Perelom EV (2001). "Strain-Induced Precipitation Behaviour in Hot Rolled Strip Steel." Mater. Sci. Eng., A299: 27–37.

Richard S (2001). "Method for Controlling the Temperature of Steel on a Continuous Hot Rolling Mode." Metallurg. Trans. B, 22: 121–128.

Saroj K (2000)."Optimal Temperature Tracking for Accelerated Cooling Processes in Hot Rolling of Steel." J. Dyn. Control, 7(4): 327–340. Nov.

Modeling the biosorption of crude oil in water using plantain pseudo stem as the adsorbent

Okoromah Candy Ozioma and Ukpaka C. Peter

Department of Chemical/Petrochemical Engineering, Rivers State University of Science and Technology, Nkpolu, P.M.B. 5080, Port Harcourt, Nigeria.

Biosorption is a technique that can be used for the removal of pollutants from waters. A variety of biomaterials are known to bind these pollutants, including bacteria, fungi, algae, and industrial and agricultural wastes. In this work, the biosoption abilities of plantain pseudo stem towards crude oil are emphasized. Experimental and theoretical approach was used to monitor and predict the rate of biosorption using plantain pseudo stem obtained from Erema town in Ogba/Egbema/Nndoni Local Government Area of Nigeria. The rate of adsorption of Bonny light crude oil by plantain pseudo stem as adsorbent of different thicknesses were used for the process for which it was observed that the adsorbent of 1 cm thickness agreed very closely with theoretical data. Thus, this article revealed that plantain stem is useful in enhancing effective remediation of oil polluted water environment.

Key words: Modeling, biosorption, crude oil, water, plantain pseudo stem, adsorbent.

INTRODUCTION

The major problem associated with oil exploitation is the spillage of petroleum hydrocarbon during transportation, offshore drilling and even during routine maintenance. Crude oil is a complex mixture of hydrocarbons and other organic compounds, including some organometallic constituents' (Butler and Mason, 1997; Mandelbaum et al., 1997; Reddy and Chinthamreddy, 1999; Page and Page, 2002). Investigation conducted by various research groups revealed that the Petroleum hydrocarbon contains hundreds and thousands of aliphatic compounds as well as branches based on the reaction in which aromatic hydrocarbons formed are toxic to living organisms (Prince, 1993; Wang et al., 1998). Crude oil spillage can thus be defined as the accidental discharge of petroleum hydrocarbons into the environment and this spillage can damage ecosystems, including plants and animals, and contaminate water for drinking and other purposes (Virkutyte et al., 2002; Toyoda and Inagaki, 2003; Carberry and Wik, 2001; Pamukeu et al., 2004; Reed et al., 1995). When the contamination is deposited in water environment, fishes can suffocate by the thick sludge of oil formed on the water surface, and bottom-dwelling fish can develop liver disease, as well as reproductive and growth problems (Safe drinking water formulation, 2000). Also, plants that grow on water and near the water can be harmed by oil pollution, which when spited on plant leaves can block the sunlight needed for photosynthesis, thus impeding growth as well leading to the death of the plant. Investigation conducted revealed that oil spillage results in closed beaches and harbors' high level of contaminants (Desehamps et al., 2003; Jacman et al., 2001; FRTS, 2000; Shapiro and Probstein, 1993; USEPA, 1999). It in like manner, affects fishing and hunting, which is especially detrimental for people who rely on hunting and fishing, such as many rural communities. Water sources that are intended to provide drinking water can become contaminated (Ukpaka, 2005, 2006, 2007, 2008, 2010; Wei et al., 2003; Teas et al.,

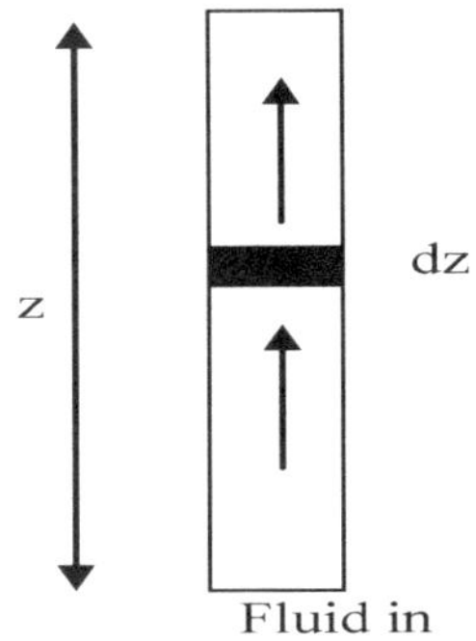

Figure 1. Pore size in the adsorbent.

2001; USDOD, 1994; USEPA, 2000).

There are some areas that are more sensitive to oil pollution than others, for example, coral reefs, mangroves and marshes are more sensitive than sandy beaches and sea-grass beds. Currents and wind can sometimes carry the oil far from the spill location, causing damage to the vegetation, birds and marine animals. Crude oil constitutes a potential threat to humans especially, when fat soluble components may accumulate in the organs of animals and may be enriched in the food chain such that prolonged exposure and high oil concentration may cause the development of liver and kidney disease, possible damage to the bone marrow and an increased risk of cancer. One of the most notable oil spills in the US and even in the world was in March, 1989, when a large tanker, Exxon Valdez spilled thousands of tons of crude oil (Paine et al., 1996; Albaiges et al., 2006; Butter and Mason, 1997; Gottz et al., 2001; Acar and Alshawabkeh, 1993; Cerniglia, 1997; Morgan and Watkinsons, 1992; Pamucks, 1997; Prince, 1993).

Engineering methods has been put in place using different approach in solving the pollution problem due to this almost inevitable occurrence, difference means of removing, reducing and remediating these problems caused by petroleum hydrocarbons spills were developed. These methods can be grouped into physiochemical remediation and bioremediation (Jain et al., 2011; Reynold et al., 2001; O'Connor et al., 2003; Ribeiro et al., 2005; Van Cauwenberghe, 1997; Roszak and Colwell, 1987; Colwell et al., 1978).

This work engulfs effective, efficient and economic preliminary means of remediating water bodies by biosorption of petroleum hydrocarbons using pseudo-stem of plantain as a biosorbent or adsorbent. It also gives a head-start for further research on ways to improve the process. This project will demonstrate: how adsorbent enhances bioremediation, the effect of size of adsorbent on bioremediation, development of a model that describes the rate of adsorption of crude oil in water using plantain pseudo stem as adsorbent, the effect of moisture on biosorption, determination of the physical properties of adsorbents relevant to this work, determination of the physical properties of crude oil relevant to this work, development of a model of rate of adsorption of crude oil, determine the constants from experiment and test the model for its deviation from experimental data. This project will be carried out within the above stated framework due to funds insufficiency, limited time specified for the project to be carried out and equipment available.

MATERIALS AND METHODS

Developing the model

The mathematical model for this research work was developed by considering an adsorbent of area (A) and thickness (Z) as well as assuming the pore arrangement within the adsorbent to be like cylindrical pipes of constant volume arranged parallel to each other and adsorption occurring due to bulk flow and diffusion, taking into consideration only axial gradient in volume of the substance (in this case oil) to be adsorbed, the rate of adsorption is given in the Figure 1.

Applying the principal of conservation of mass which states that:

$$\left[\frac{\begin{array}{l}\textit{Accumulation of}\\ \textit{a substabce 'A' within}\\ \textit{'A' a system}\end{array}}{\textit{Time}}\right] = \left[\frac{\begin{array}{c}\textit{Inflow of substance}\\ \textit{'A'}\end{array}}{\textit{Time}}\right] - \left[\frac{\begin{array}{c}\textit{Outflow of substance}\\ \textit{'A'}\end{array}}{\textit{Time}}\right] + \left[\frac{\begin{array}{l}\textit{Amount of substance}\\ \textit{generated}\end{array}}{\textit{consumed}}\right] \quad (1)$$

where 'A' is total mass/moles or component mass/moles.

Analyzing each term in the Equation (1) for the material balance of oil and representing oil as component a, thus:

$$\frac{\begin{array}{l}\textit{Accumulation of}\\ \textit{component 'a' within the system}\end{array}}{\textit{Time}} = \frac{\partial m_a}{\partial t} \quad (2)$$

$$\frac{\textit{Inflow of a quantity 'a'}}{\textit{Time}} = M_a + (-D\nu A)\frac{\partial \frac{m_a}{mrv}}{\partial z} \quad (3)$$

$$\frac{\textit{Outflow of a quantity S}}{\textit{Time}} = m_a + \frac{\partial m_a}{\partial z} + (-D\nu)A\frac{\partial \frac{m_a}{mrv}}{\partial z} + \frac{\partial}{\partial z}\left((-Dv)A\frac{\partial \frac{m_a}{mrv}}{\partial z}\right) \quad (4)$$

$$\frac{\textit{Amount of a quantity S generated}}{\textit{consumed within the system}} = 0 \quad (5)$$

The last term above is zero since no reaction occurred during the time of observation of the process. Substituting the Equations, (2), (3), (4), and (5) into Equation (1) gives:

$$\frac{\partial m_a}{\partial z} = m_a + (-Dv)\ \frac{\partial m_a}{\partial z}\ - \left(m_a + \frac{\partial m_a}{\partial z} dz + (-DvA)\, A \frac{\partial \frac{m_a}{mrv}}{\partial z} + \frac{\partial}{\partial z}\left((-Dv)A\ \frac{\partial \frac{m_a}{mrv}}{\partial z} \right) \partial z \right) \quad (6)$$

Equation (6) can be written as;

$$\frac{\partial m_a}{\partial z} = - \left(\frac{\partial m_a}{\partial z} + \frac{\partial}{\partial z}\left((-Dv)A\ \frac{\partial \frac{m_a}{mrv}}{\partial z} \right) \right) \quad (7)$$

From Equation (7) considering Mr and V as constant as well as insignificance to the process, therefore Equation (7) can be written as;

$$\frac{\partial m_a}{\partial z} = - \left(\frac{\partial m_a}{\partial z} + \frac{(-Dv)A}{mrv} \frac{\partial^2 m_a}{\partial z^2} \right) \quad (8)$$

The above expression, Equation (8) is the model for the rate of adsorption crude oil in water using a adsorbent (plantain pseudo stem).

where, Ma = mass of crude oil (a) uptake (g), -Dv = the diffusivity of adsorbent (plantain stem) (m^2s^{-1}), Mr = the molecular weight of crude oil (a) (g/gmol), A = the surface area of adsorbent (m^2), *V* = the volume the adsorbent (m^3).

The developed expression is based on oil adsorbed by the adsorbent that is mass balance for water was not taken into consideration because it is assumed that only oil is adsorbed as the adsorbent, plantain stem has a high affinity for just oil, that is, the only fluid flowing into the adsorbent, which is consider as the control volume, is oil.

Solving the model

The model was further resolved by considering the following operations:

Let $$\frac{(-Dv)A}{mrv} = q \quad (9)$$

Therefore substituting Equation (9) into Equation (7) we have:

$$\frac{\partial m_a}{\partial z} = - \left(\frac{\partial m_a}{\partial z} dz + q \frac{\partial^2 m_a}{\partial z^2} \partial z \right) \quad (10)$$

Putting $M_a = f(z)g(t)$ as the solution of the above model thus

$$\frac{\partial m_a}{\partial z} = f\ (z) \frac{\partial g}{\partial t} \quad (11)$$

$$\frac{\partial m_a}{\partial z} = g\ (t) \frac{\partial f}{\partial z} \quad (12)$$

$$\frac{\partial^2 m_a}{\partial z^2} = g\ (t) \frac{\partial^2 f}{\partial z^2} \quad (13)$$

Substituting Equations (11), (12) and (13) into Equation (10) becomes

$$f\ (z)\ \frac{\partial g}{\partial t} = -\ g\ (t)\ \frac{\partial f}{\partial z}\ -\ g\ (t)\ q\ \frac{\partial^2 f}{\partial z^2} \quad (14)$$

Dividing through Equation (14) by $f(z)g\ (t)$ we have

$$\frac{1}{g}\ \frac{\partial g}{\partial t} = -\ \frac{1}{f}\ \frac{\partial f}{\partial z}\ - q\ \frac{1}{f}\frac{\partial^2 f}{\partial z^2}\ = k \quad (15)$$

Equation (15) can be expressed as

$$\frac{1}{g}\ \frac{\partial g}{\partial t} = k\ and\ -\ \frac{1}{f}\ \frac{\partial f}{\partial z}\ - q\ \frac{1}{f}\frac{\partial^2 f}{\partial z^2}\ = k \quad (16)$$

Therefore Equation (15) can be written as

$$\frac{\partial g}{\partial t} = k \partial t \quad (17)$$

Integrating both sides of the Equation (17) we have

$$\int_0^g \frac{\partial g}{\partial t} = \int_0^t k\, \partial t \quad (18)$$

$$\text{Ing} = kt + d \quad (19)$$

$$g = e^{kt+d} \quad (20)$$

$$g = C^{ekt} \quad (21)$$

In the same vein, Equation (16) can be expressed as

$$-\ \frac{1}{f}\ \frac{\partial g}{\partial t}\ - a\ \frac{1}{f}\ \frac{\partial^2 f}{\partial z^2}\ = k \quad (22)$$

$$-\ \frac{1}{f}\left(\frac{\partial f}{\partial z} + q\ \frac{\partial^2 f}{\partial z^2} \right) = k \quad (23)$$

Putting $f\ = e^{hz}$ as the solution of the Equation (23), then

$$\frac{\partial f}{\partial z}\ = he^{hz} \quad (24)$$

$$\frac{\partial^2 f}{\partial z} = h^2 e^{hz} \quad (25)$$

Substituting Equations (24) and (25) into Equation (23) and resolving gives

$$(he^{hz} + qh^2 e^{hz}) = -ke^{hz} \quad (26)$$

Further rearrangement of Equation (26) yields

$$e^{hz}(h + qh^2) = -ke^{hz} \quad (27)$$

Canceling the exponential terms on both sides gives the quadratic equation below

$$(h + qh^2) = -k \quad (28)$$

$$qh^2 + h + k = 0 \quad (29)$$

Using the formula for quadratic equations to solve Equation (29), gives

$$h = \frac{-1 \pm \sqrt{1 - 4qk}}{2q} \quad (30)$$

Substitute the value of h into f to obtain the solution to the second order differential equation thus,

$$f = P_e\left(\frac{-1 \pm \sqrt{1-4qk}}{2q}\right)z + N_e\left(\frac{-1 - \sqrt{1-4qk}}{2q}\right)z \quad (31)$$

Substitute Equations (31) and (23) into the expression for *Ma* gives

$$Ma = Ce^{kt}\left(P_e\left(\frac{-1 \pm \sqrt{1-4qk}}{2q}\right)z + N_e\left(\frac{-1 - \sqrt{1-4qk}}{2q}\right)z\right) \quad (32)$$

where CP=A and CP=B, thus Equation (32) becomes

$$Ma = e^{kt}\left(A_e\left(\frac{-1 + \sqrt{1-4qk}}{2q}\right)z + B_e\left(\frac{-1 - \sqrt{1-4qk}}{2q}\right)z\right) \quad (33)$$

Equation (33) is the solution to the model for the rate of adsorption of crude oil using plantain pseudo stem as adsorbent

Equipment/materials used

The materials used for this experiment includes: Weighing balance, Plantain pseudo stem, Plastic cans, measuring cup, tooth pick, stainless steel bowl, Bonny light crude oil, Machete, cutter, detergent, wipes, desiccators, oven, and water.

Sample collection and preparation

The plantain pseudo stem was collected from a plantain plantation in Erema town in Ogba/Egbema/Ndoni Local Government Area (ONELGA) in Rivers State. The stem was first cut from the living tree laterally with the aid of a machete in the form of a log. The log-like stem's outer back (impermeable part) was peeled, further cut through its middle, longitudinally, and then an area of 9 cm^3 (3 × 3 cm) and thickness of 1 cm^3 was measured out with a ruler and cut. Four other samples of equal volume (3 × 3 × 1 cm) were cut using the same procedure as stated above. Each sample was labeled 1 to 5.

The above steps were carried out for adsorbents of thickness of 2 and 3 cm, given a total number of 15 samples of the sorbent realized and used for the experiment, whereby adsorbents of thickness of 1, 2, and 3 cm had 5 samples each. Due to the nature of the stem, the samples were held in place with tooth picks. 15 empty plastic cans were weighed using an electronic weigh balance and each labeled accordingly based on the samples obtained stating from sample 1 for 1cm thickness, sample 2 for 1 cm thickness, down to sample 5 for 3 cm thickness. The adsorbent were placed in their corresponding cans and the gross weights were obtained. The weight of the empty measuring cup and crude oil (bonny light) of 25ml was also obtained

Adsorption process

Five bowls were filled with water and polluted with 25 ml of crude oil each. The five samples of 1 cm thickness were placed all at the same time in the individual bowls. Using a stop watch to monitor the process, sample 1 was removed and placed in its plastic can after 30 s, sample 2 after 1 min, down to sample 5 which was removed after 2 min 30 s. The new weights of the samples were obtained. The five bowls were washed with detergent, filled with water, polluted with 25 ml crude and the process repeated for samples of thickness 2 and 3 cm.

Moisture content

Seven (7) samples of thickness 2 cm and area 9 cm^3 were obtained as explained under the subheading of sample collection and preparation. An oven was preheated to 70°C and all seven samples were placed on a metal tray, and then put into the oven to dry. Sample 1 was removed after 20 min, and placed in a desiccator to cool and dry before weighing again. This step is repeated for samples 2 and 3 after which the remaining four samples (4 and 5) were removed after every 6 min. After cooling and weighing the samples, the 7 samples were placed all at the same time in 7 bowls of water polluted with 25 ml of crude oil for 2 min. All samples were removed at the same time and weighed again.

RESULTS AND DISCUSSION

The results obtained from the research work are presented in Figures and Tables. Tables 1, 2 and 3 contain few physical properties relevant to this study. These properties include the mass of all 15 samples and the mass of oil adsorbed by each with respect the time of contact with the oil. In Table 1 the 5th sample, the sample with longest contact time, adsorbed oil about its own weight with the difference between its weight and the oil adsorbed being (5.96 - 5.12 = 0.84 g). This is pretty impressive and promising as it can be said that the adsorbent has the ability to utilize its entire volume for the sole purpose and aim of the work, which is adsorbing crude oil in water.

Table 1. The rate of adsorption of crude oil using 1 cm thickness adsorbent.

Weight of empty can (g)	Weight of can + adsorbent (g)	Weight of adsorbent (g)	Adsorption time (s)	Weight of can + Adsorbent + oil (g)	Weight of oil (g)
36.04	43.04	7.00	30	45.13	2.09
39.54	46.12	6.58	60	48.74	2.62
40.60	47.70	7.10	90	50.98	3.28
13.57	19.29	5.72	120	23.39	4.10
20.60	26.56	5.96	150	31.68	5.12

Table 2. The rate of adsorption of crude oil using 2 cm thickness adsorbent.

Weight of empty can (g)	Weight of can + adsorbent (g)	Weight of adsorbent (g)	Adsorption time (s)	Weight of can + adsorbent+ oil (g)	Weight of oil (g)
36.04	49.69	13.65	30	56.80	7.11
39.54	54.68	15.14	60	59.50	4.82
40.60	53.46	12.86	90	58.77	5.31
13.57	28.11	14.54	120	35.12	7.01
20.60	28.11	15.08	150	33.80	5.69

Table 3. The rate of adsorption of crude oil using 3 cm thickness adsorbent.

Weight of empty can (g)	Weight of can + adsorbent (g)	Weight of adsorbent (g)	Adsorption time (s)	Weight of can + adsorbent+ oil (g)	Weight of oil (g)
36.04	52.36	16.32	30	66.39	4.72
39.54	54.38	14.84	60	66.39	12.01
40.60	58.71	18.11	90	21.02	2.91
13.57	29.30	15.73	120	37.39	8.09
20.60	38.64	18.04	150	49.75	11.11

On the other hand, Table 4 shows the effect of moisture content on the rate of adsorption of crude oil using 2 cm thickness adsorbent at various drying times. The samples were exposed to drying within a time limit of 30 to 150 s

The result presented in Figure 2 illustrates the relationship between the mass of oil adsorbed and the time taken for adsorption process using adsorbent of 1 cm thickness. The variation in the mass adsorbed can be attributed to the variation in time as well as the adsorbent thickness as presented in Figure 2. Increase in the rate of oil adsorbed was observed with increase in time, that is the rate of adsorption is directly proportional to time.

From Figure 3, it is seen that the mass of oil adsorbed decreases with increase in time for a period of 30 to 60 s and later increase with increase in time. The variation in the mass of oil adsorbed can be attributed to the variation in time as well as adsorbent thickness.

Figure 4 illustrates the relationship between the mass of oil adsorbed by the adsorbent per unit time. Increase in the rate of adsorption of oil with the time period of >30 s <60 s was observed and later decreased with the time range of >60 s <90 s and finally increased with increase in time. The variation in mass of oil adsorbed can be attributed to the variation in the time as well as the adsorbent thickness.

A linear curve was observed within the time range of 20 to 40 min and later a decrease in mass of oil adsorbed was experienced with increase in time. The mass of oil adsorbed can be attributed to the variation in time as well as adsorbent thickness as presented in Figure 5.

Determining the constants of the model

Recalling Equation (33) which states that;

$$Ma = e^{kt}\left(A_e\left(\frac{-1+\sqrt{1-4qk}}{2q}\right)z + B_e\left(\frac{-1-\sqrt{1-4qk}}{2q}\right)z\right)$$

Table 4. Illustrates the effect of moisture content on the rate of adsorption upon the influence of drying.

Weight of empty can (g)	Weight of can+ adsorbent before drying (g)	Weight of adsorbent (g)	Drying time (min)	Weight of can + adsorbent after drying (g)	Moisture content (g)	Weight of can + adsorbent+ oil (g)	Weight of oil (g)
36.04	56.93	20.89	20	45.02	11.91	49.02	4.00
39.54	55.42	15.88	40	42.55	12.87	46.53	3.98
40.60	55.14	14.54	60	40.91	14.23	43.89	2.98
13.57	28.72	14.97	66	14.52	14.20	17.50	3.30
20.60	40.26	19.66	72	23.25	17.01	26.35	3.10
12.45	28.86	16.41	74	13.67	15.19	16.76	3.09
13.65	28.78	15.13	80	13.66	15.12	16.64	2.98

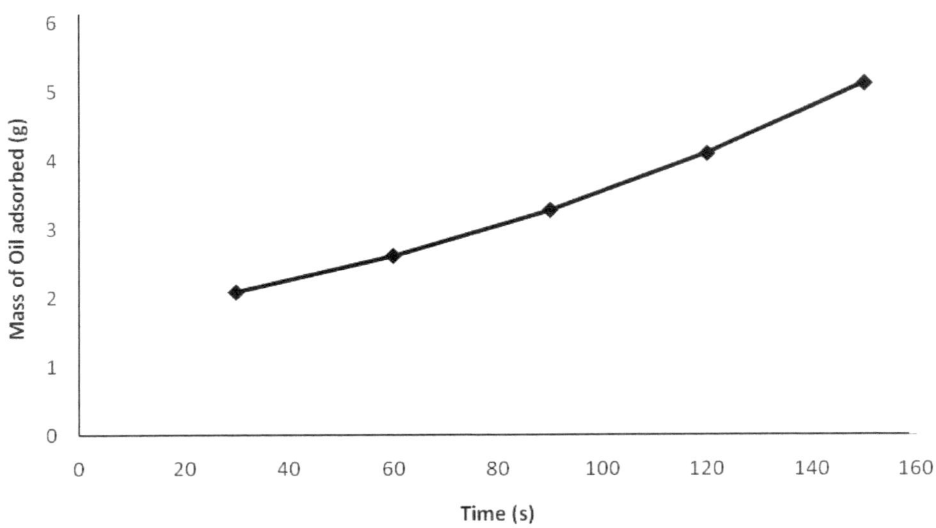

Figure 2. Graph of the relationship between mass adsorbed and the time taking for the adsorption process using adsorbent of 1 cm thickness.

Using the result data in Table 1 to determine the constants in the developed model shown in Equation (33) gives values of some functional parameters as shown in Table 5.

Substituting the result presented in Table 5 into the Equation (33) and resolving mathematically gives,

$$2.09 = e^{30k}\left(A_e^{\;0.01}\left(\frac{-1+\sqrt{1-4qk}}{2q}\right)+B_e^{\;0.01}\left(\frac{-1-\sqrt{1-4qk}}{2q}\right)\right) \quad (34)$$

$$2.62 = e^{60k}\left(A_e^{\;0.01}\left(\frac{-1+\sqrt{1-4qk}}{2q}\right)+B_e^{\;0.01}\left(\frac{-1-\sqrt{1-4qk}}{2q}\right)\right) \quad (35)$$

Dividing Equation (35) by (34) gives,

$$\frac{2.62}{2.09} = \frac{e^{60k}}{e^{30k}} \quad (36)$$

Resolving Equation (36) mathematically we have

$$1.2536 = e^{30k} \quad (37)$$

Taking the natural log of both sides of the Equation (37) gives,

$$0.2260 = 30k \quad (38)$$

Therefore, Equation (37) becomes

$$k = \frac{0.2260}{30}$$

$k = 0.0075$

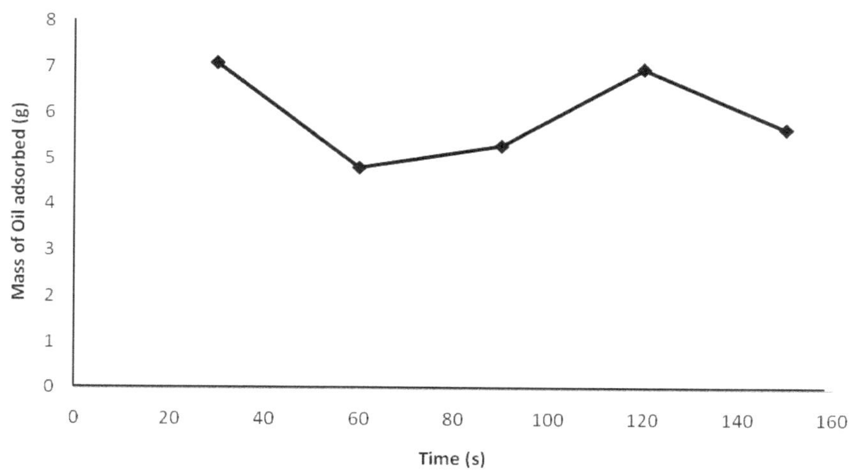

Figure 3. Graph of the relationship between mass adsorbed and the time taking for the adsorption process using adsorbent of 2 cm thickness.

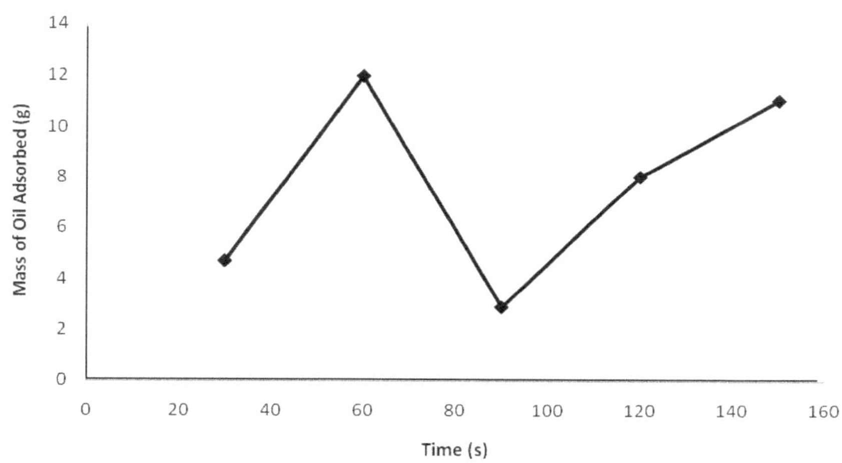

Figure 4. Graph of the relationship between mass adsorbed and the time taking for the adsorption process using adsorbent of 3 cm thickness

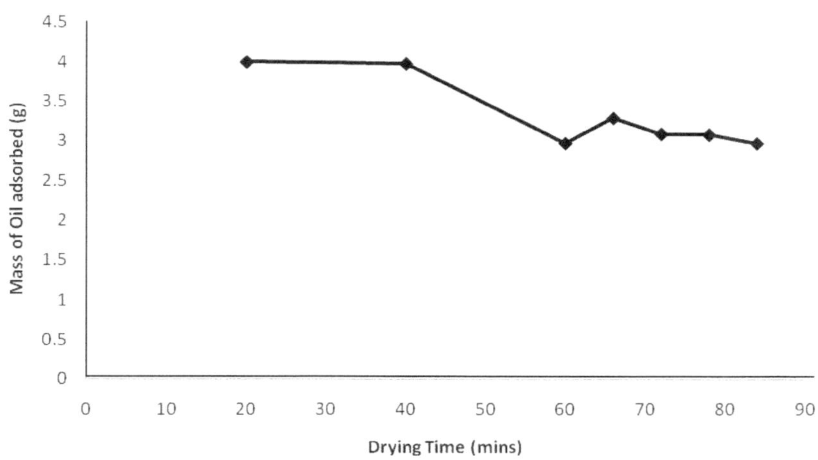

Figure 5. Graph of the relationship between mass adsorbed versus the drying time for a adsorbent of 2 cm thickness.

Table 5. Values of some functional parameters.

Component	Ma (g)	t (s)	z (m)
Sample 1	2.09	30	0.01
Sample 2	2.62	60	0.01

Table 6. Values for functional parameters for various thicknesses of adsorbent.

Functional parameter	Values of functional	parameters for different thicknesses	
Z (m)	0.01	0.02	0.03
M (g)	3.442	5.988	7.768
T (s)	90	90	90

To check if the value of 'k' is correct then

$$\frac{2.09}{e^{30k}} = \frac{2.62}{e^{60k}} = \frac{3.28}{e^{90k}} = \frac{4.10}{e^{120k}} = \frac{5.12}{e^{150k}}$$

$$\frac{2.09}{e^{30k}} = 1.6689$$

$$\frac{2.62}{e^{60k}} = 1.6706$$

$$\frac{3.28}{e^{120k}} = 1.6700$$

$$\frac{4.10}{e^{120k}} = 1.6669$$

$$\frac{5.12}{e^{150k}} = 1.6622$$

These values are acceptable as errors are expected due to the assumption made during the development of the model. To determine other functional parameters that govern the rate of adsorption of adsorbent in oil, Tables 2 and 3 are required. Due to the irregularities in both data the average will be used instead. Note that this will bring about errors as approximated values (the averages) are being used.

Substituting the values in the Table 6 into Equation (33) gives,

$$3.442 = e^{0.0075(90)}\left(A_e^{\ 0.01\left(\frac{-1+\sqrt{1-4qk}}{2q}\right)} + B_e^{\ 0.01\left(\frac{-1-\sqrt{1-4qk}}{2q}\right)}\right) \quad (39)$$

$$1.7525 = \left(A_e^{\ 0.01\left(\frac{-1+\sqrt{1-4qk}}{2q}\right)} + B_e^{\ 0.01\left(\frac{-1-\sqrt{1-4qk}}{2q}\right)}\right) \quad (40)$$

$$1.7525 = (Ae^{0.01} + Be^{0.01b}) \quad (41)$$

$$5.988 = e^{0.0075(90)}\left(A_e^{\ 0.02\left(\frac{-1+\sqrt{1-4qk}}{2q}\right)} + B_e^{\ 0.02\left(\frac{-1-\sqrt{1-4qk}}{2q}\right)}\right) \quad (42)$$

$$3.049 = \left(A_e^{\ 0.02\left(\frac{-1+\sqrt{1-4qk}}{2q}\right)} + B_e^{\ 0.02\left(\frac{-1-\sqrt{1-4qk}}{2q}\right)}\right) \quad (43)$$

$$3049 = (Ae^{0.02a} + Be^{0.02b}) \quad (44)$$

$$7.768\ e^{0.0075(90)} = \left(A_e^{\ 0.03\left(\frac{-1+\sqrt{1-4qk}}{2q}\right)} + B_e^{\ 0.03\left(\frac{-1-\sqrt{1-4qk}}{2q}\right)}\right) \quad (45)$$

$$3.9551 = \left(A_e^{\ 0.03\left(\frac{-1+\sqrt{1-4qk}}{2q}\right)} + B_e^{\ 0.03\left(\frac{-1-\sqrt{1-4qk}}{2q}\right)}\right) \quad (46)$$

$$3.9551 = (Ae^{0.03a} + Be^{0.03b}) \quad (47)$$

where

$$a = \left(\frac{-1+\sqrt{1-4qk}}{2q}\right) \quad (48)$$

$$b = \left(\frac{-1+\sqrt{1-4qk}}{2q}\right) \quad (49)$$

Multiplying Equation (41) by $e^{0.02a}$, gives

$$1.7525e^{0.02a} = (Ae^{0.03a} + Be^{(0.02a\ +\ 0.01b)}) \quad (50)$$

Multiplying Equation (44) by $e^{0.01a}$, gives

$$3.049e^{0.01b} = (Ae^{0.03a} + Be^{(0.01a\ +\ 0.02b)}) \quad (51)$$

Resolving Equations (41) and (44) simultaneously gives,

$$3.049e^{0.01a} - 1.7525e^{0.02a} = B\,(e^{0.01a + 0.02b} - e^{(0.02a + 0.01b)}) \quad (52)$$

Making B subject of the formula from Equation (52) gives,

$$B = \frac{3.049\,e^{0.01a} - 1.7525e^{0.02a}}{(e^{0.01a + 0.02b)} - e^{(0.02a + 0.01b)})} \quad (53)$$

Substituting Equation (53) into Equation (47) gives,

$$3.9551 = \left(A\,e^{0.03a} + \frac{3.049\,e^{0.01a} - 1.7525\,e^{0.02a}}{\left(e^{(0.01a + 0.02b) - e(0.02a + 0.01b)}\right)}\,e^{0.03b} \right) \quad (54)$$

Making A subject of formula in Equation (54) gives,

$$3.9551 = \left(e^{(0.01a + 0.02b)} - e^{(0.02a + 0.01b)} = \left(Ae^{0.03a}\left(e^{(0.01a + 0.02b)}\right)^{-} e^{(0.02a + 0.01b)} + (3.049\,e^{0.01a} - 1.7525e^{0.02a})\,e^{0.03b}\right) \quad (55)$$

$$A = \frac{3.9551\left(e^{(0.01a + 0.02b)} - e^{(0.02a + 0.01b)} - (3.049e^{0.01a + 0.03b}) - 1.7525\,e^{0.02a + 0.30b}\right)}{\left(^{(0.04a + 0.02b)}\;e - {}^{(0.05a + 0.01b)}\right)} \quad (56)$$

Substituting the values of A and B obtained above into Equation (41)

$$1.7525 = \frac{3.9551\left(e^{(0.01a + 0.02b)} - e^{(0.02a + 0.01b)} - (3.049e^{0.01a + 0.03b}) - 1.7525\,e^{0.02a + 0.30b}\right)}{\left(^{(0.04a + 0.02b)}\;e - {}^{(0.05a + 0.01b)}\right)}$$

$$e^{0.01a} + \frac{3.049\,e^{0.01a} - 1.7525e^{0.02a}}{\left(e^{(0.01a + 0.02b)} - e^{(0.02a + 0.01b)}\right)}\,e^{0.01b} \quad (57)$$

$$1.7525\left(e^{(0.01a + 0.02b)} - e^{(0.02a + 0.01b)}\right) = 3.9551\left(e^{(0.01a + 0.02b)} - e^{(0.02a + 0.01b)}\right) \quad (58)$$

$$- \left(3.049\,e^{(0.01a + 0.03b)} - 1.7525\,e^{(0.02a + 0.03b)}\right)e^{0.01a} + \left(3.049\;e^{(0.01a + 0.02b)} - 1.7525\,e^{(0.02a + 0.01b)}\right)$$

$$1.7525\left(e^{(0.04a + 0.02b)} - e^{(0.05a + 0.01b)}\right) = 3.9551\left(e^{(0.02a + 0.02b)} - e^{(0.03a + 0.01b)}\right)$$

$$- \left(3.049\,e^{(0.02a + 0.03b)} - 1.7525\,e^{(0.03a + 0.03b)}\right) + (3.049\,e^{0.04a + 0.01b} - 1.7525e^{0.05a + 0.01b}) \quad (59)$$

Rearranging Equation (59) by collecting like coefficient terms gives

$$1.7525\left(e^{(0.04a + 0.02b)} - e^{(0.05a + 0.01b)}\right) = 3.9551\left(e^{(0.05a + 0.01b)} - e^{(0.03a + 0.03b)}\right)$$
$$= 3.9551\left(e^{(0.02a + 0.02b)} - e^{(0.03a + 0.01b)}\right) + 3.049\left(e^{0.04a + 0.01b} - e^{0.02a + 0.03b}\right) \quad (60)$$

Further simplifying of Equation (60) yields

$$1.7525\left(e^{(0.04a + 0.02b)} - e^{(0.03a + 0.03b)}\right) = 3.9551\left(e^{(0.02a + 0.02b)} - e^{(0.03a + 0.01b)}\right)$$
$$+ 3.049\left(e^{(0.04a + 0.01b)} - e^{(0.02a + 0.03b)}\right) \quad (61)$$

$$(0.04a + 0.02b) = \frac{-0.04 + 0.04\sqrt{1 - 4q\,(0.075)}}{2q} + \frac{-0.02 - 0.02\sqrt{1 - 4q\,(0.075)}}{2q} \quad (62)$$

$$(0.04a + 0.02b) = \frac{-0.03 + 0.01\sqrt{1 - 0.03q}}{2q} \quad (63)$$

$$(0.03a + 0.03b) = \frac{-0.03}{q} \quad (64)$$

$$(0.02a + 0.02b) = \frac{-0.02}{q} \quad (65)$$

$$(0.03a + 0.01b) = \frac{-0.02 + 0.01\sqrt{1 - 0.03q}}{q} \quad (66)$$

$$(0.04a + 0.01b) = \frac{-0.05 + 0.03\sqrt{1 - 0.03q}}{2q} \quad (67)$$

$$(0.02a + 0.03b) = \frac{-0.05 + 0.01\sqrt{1 - 0.03q}}{2q} \quad (68)$$

Substituting Equations (68) into Equation (60) gives

$$1.7525\left(e^{\left(\frac{-0.03 + 0.01\sqrt{1 - 0.03q}}{q}\right)} - e^{\frac{-0.03}{q}}\right) = 3.9551\left(e^{\left(\frac{-0.02}{q}\right)} - e^{\left(\frac{-0.02 + 0.01\sqrt{1 - 0.03q}}{q}\right)}\right)$$

$$3.049\left(e^{\left(\frac{-0.05 + 0.03\sqrt{1 - 0.03q}}{2q}\right)} - e^{\left(\frac{-0.05 - 0.01\sqrt{1 - 0.03q}}{2q}\right)}\right) \quad (69)$$

$$1.7525\, e^{\frac{-0.03}{q}}\left(e^{\left(\frac{0.01\sqrt{1 - 0.03q}}{q}\right)} - 1\right) = 3.9551 e^{\left(\frac{-0.02}{q}\right)}\left(1 - e^{\left(\frac{0.01\sqrt{1 - 0.03q}}{q}\right)}\right)$$

$$+ 3.049 e^{\left(\frac{-0.05}{2q}\right)}\left(e^{\left(\frac{0.03\sqrt{1 - 0.03q}}{2q}\right)} - e^{\frac{-0.01\sqrt{1 - 0.03q}}{2q}}\right) \quad (70)$$

$$f(q) = 1.7525 e^{\frac{-0.03}{q}}\left(1 - e^{\left(\frac{0.01\sqrt{1 - 0.03q}}{q}\right)}\right) + 3.9551\, e^{\left(\frac{-0.02}{q}\right)}\left(1 - e^{\left(\frac{-0.02 + 0.01\sqrt{1 - 0.03q}}{q}\right)}\right)$$

$$f(q) = 1.7525 e^{\frac{-0.03}{q}}\left(1 - e^{\left(\frac{0.01\sqrt{1 - 0.03q}}{q}\right)}\right) + 3.9551\, e^{\left(\frac{-0.02}{q}\right)}\left(1 - e^{\left(\frac{-0.02 + 0.01\sqrt{1 - 0.03q}}{q}\right)}\right)$$

$$+ 3.049 e^{\left(\frac{-0.05}{2q}\right)}\left(e^{\frac{0.03\sqrt{1 - 0.03q}}{2q}} - e^{\frac{-0.01\sqrt{1 - 0.03q}}{2q}}\right) \quad (71)$$

Collecting like terms from Equation (71) and simplifying gives

$$f(q) = \left(1.7525 e^{\frac{-0.03}{q}} + 3.9551\, e^{\left(\frac{-0.02}{q}\right)}\right)\left(1 - e^{\left(\frac{0.01\sqrt{1 - 0.03q}}{2q}\right)}\right) + 3.049\, e^{\left(\frac{-0.025}{q}\right)}$$

$$\left(e^{\frac{0.015\sqrt{1-0.03q}}{q}} - e^{\frac{-0.005\sqrt{1-0.03q}}{q}} \right) \quad (72)$$

Using iteration method to solve Equation (72), this can be achieved by making one of the q's in Equation (72) the subject of the formula then taking initial conditions as well as rearranging Equation (72) by dividing through by $3.049\,e^{\left(\frac{-0.025}{q}\right)}$ gives

$$\frac{\left(1.7525\,e^{\frac{-0.03}{q}} + 3.9551\,e^{\left(\frac{-0.02}{q}\right)}\right)\left(1 - e^{\left(\frac{0.01\sqrt{1-0.03q}}{2q}\right)}\right)}{+\,3.049\,e^{\left(\frac{-0.025}{q}\right)}} = \left(e^{\frac{0.005\sqrt{1-0.03q}}{q}} - e^{\frac{0.015\sqrt{1-0.03q}}{q}} \right) \quad (73)$$

$$\frac{\left(1.5748\,e^{\frac{-0.005}{q}} + 1.2972\,e^{\left(\frac{-0.005}{q}\right)}\right)\left(1 - e^{\left(\frac{0.01\sqrt{1-0.03q}}{2q}\right)}\right)}{+\,3.049\,e^{\left(\frac{-0.025}{q}\right)}} = \left(e^{\frac{-0.005\sqrt{1-0.03q}}{q}} - e^{\frac{0.015\sqrt{1-0.03q}}{q}} \right) \quad (74)$$

Dividing Equation (74) by $\left(1 - e^{\left(\frac{0.01\sqrt{1-0.03q}}{2q}\right)}\right)$ gives

$$\left(0.5748\,e^{\frac{-0.005}{q}} + 1.2972\,e^{\left(\frac{0.005}{q}\right)}\right) = \frac{\left(e^{\frac{-0.005\sqrt{1-0.03q}}{q}} - e^{\frac{-0.015\sqrt{1-0.03q}}{q}} \right)}{\left(1 - e^{\left(\frac{0.01\sqrt{1-0.03q}}{2q}\right)}\right)} \quad (75)$$

Multiply all through by $e^{\frac{0.005\sqrt{1-0.03q}}{q}}$

$$e^{\frac{0.005\sqrt{1.0.03q}}{q}}\left(0.5748\,e^{\frac{-0.005}{q}} + 1.2972\,e^{\left(\frac{0.005}{q}\right)}\right) = e^{\frac{0.005\sqrt{1-0.03q}}{q}}\,\frac{\left(e^{\frac{-0.005\sqrt{1-0.03q}}{q}} - e^{\frac{-0.015\sqrt{1-0.03q}}{q}} \right)}{\left(1 - e^{\left(\frac{0.01\sqrt{1-0.03q}}{2q}\right)}\right)} \quad (76)$$

$$\left(0.5748\,e^{\frac{-0.005+0.005\sqrt{1-0.03q}}{q}} + 1.2972\,e^{\frac{0.005+0.005\sqrt{1-0.03q}}{q}}\right) = \frac{\left(1 - e^{\frac{-0.02\sqrt{1-0.03q}}{q}}\right)}{\left(1 - e^{\left(\frac{0.01\sqrt{1-0.03q}}{2q}\right)}\right)} \quad (77)$$

$$\left(0.5748e^{\frac{-0.005+0.005\sqrt{1-0.03q}}{q}}+1.2972e^{\frac{0.005+0.005\sqrt{1-0.03q}}{q}}\right)=\frac{\left(1-e^{\frac{-0.01\sqrt{1-0.03q}}{q}}\right)\left(1-e^{\frac{0.01\sqrt{1-0.03q}}{2q}}\right)}{\left(1-e^{\left(\frac{0.01\sqrt{1-0.03q}}{2q}\right)}\right)} \quad (78)$$

$$\left(0.5748e^{\frac{-0.005+0.005\sqrt{1-0.03q}}{q}}+1.2972e^{\frac{0.005+0.005\sqrt{1-0.03q}}{q}}\right)=\left(1+e\,\frac{0.01\sqrt{1-0.03q}}{q}\right) \quad (79)$$

Equation (79) can be written as shown in Equation (80) considering the RHS of Equation (79) results in

$$q=\frac{0.01\sqrt{1-0.03q}}{In\left[\left(0.5748\,e^{\frac{0.005+0.005\sqrt{1-0.03q}}{q}}+1.2972\,e^{\frac{0.005+0.005\sqrt{1-0.03q}}{q}}\right)-1\right]} \quad (80)$$

Considering the initial value of q=1, the iteration provides no solution for q and thus another q in Equation (80) is made the subject of the formula. Resolving Equation (80) algebraically for another expression for q,

$$q=\frac{1\frac{q^2}{0.0001}\left[In\left(0.5748\,e^{-0.005+0.005\sqrt{1-0.03q}}+1.2972\,e^{\frac{0.005+0.005\sqrt{1-0.03q}}{q}}\right)-1\right]^2}{0.03} \quad (81)$$

Considering the initial value of q=1, the iteration provides no solution for q and thus another q in Equation (80) is made the subject of the formula. Resolving Equation (80) algebraically for another expression for q,

$$q=In\,\frac{\frac{0.005+0.005\sqrt{1-0.03q}}{\left[\left(0.5748e^{\frac{-0.05+1-0.03q}{q}}+1.2972e^{\frac{0.005+0.005\sqrt{1-0.03q}}{q}}\right)+1\right]}}{1.2972} \quad (82)$$

Stating with initial value of q=1 and after 89 iterations the value of q calculated for is q = 0.008665668397.
Substituting the value of q = 0.008665668397 into Equation (80) gives

$$\left(0.5748\,e^{\frac{-0.005+0.005\sqrt{1-0.03q}}{q}}+1.2972\,e^{\frac{0.005+0.005}{q}}\right)=\left(1+e^{\frac{0.01\sqrt{1-0.03q}}{q}}\right)=4.170309018 \quad (83)$$

Substituting the value q = 0.008665668397 into equations (41), (44), (47), (48) and (49) to determine the values of a, b, A and B

Table 7. Experimental and theoretical values of the functional parameters for 1cm thickness of adsorbent.

Component	Values of functional parameters for adsorption of oil			
	Time (s)	Z (m)	Ma-model (g)	Ma-experiment (g)
Sample 1	30	0.01	2.42	2.09
Sample 2	60	0.01	3.03	2.62
Sample 3	90	0.01	3.79	3.28
Sample 4	120	0.01	4.75	4.10
Sample 5	150	0.01	5.95	5.12

Table 8. Experimental and theoretical values of the functional parameters for 2 cm thickness of adsorbent.

Component	Values of functional parameters for adsorption of oil			
	Time (s)	Z (m)	Ma-model (g)	Ma-experiment
Sample 1	30	0.02	4.32	7.11
Sample 2	60	0.02	5.41	4.84
Sample 3	90	0.02	6.77	5.31
Sample 4	120	0.02	8.48	7.01
Sample 5	150	0.02	10.615	5.69

$$a = \left(\frac{-1 + \sqrt{1 - 4\,(0.008665668397)(0.0075)}}{2\,(0.008665668397)} \right) = -0.007500487507$$

$$b = \left(\frac{-1 - \sqrt{1 - 4\,(0.008665668397)(0.0075)}}{2\,(0.008665668397)} \right) = -115.390407$$

$$A = \frac{3.9551\left(e^{-2.307883145} - e^{(-1.15405408)}\right) - \left(3.049e^{-3.46178215} - 1.7525\,e^{-3.4618622}\right)}{\left(e^{(-2.30810816)} - e^{-1.154279094}\right)}$$

A = 4.14410287

$$B = \frac{3.049e^{-0.0000750048507} - 1.7525e^{-0.000150009501}}{\left(e^{(-2.37883145)} - e^{(-1.15405408)}\right)} = -7.006344672$$

Substituting the determined functional parameters of q, a, b, A, and B into the developed model yielded the general solution for this project work.

$$M_a = e^{0.0075t}\left(4.14410287\,e^{-0.0075004875 7z} - 7.006344672\,e^{-115.390407z}\right) \quad (84)$$

Equation (84) is the Candy-Ukpaka's Model equation for the adsorption of crude oil using plantain pseudo-stem as adsorbent.

Testing the model

The following functional parameters were determined from the theoretical model developed in this project work.

The model used is known as Candy-Ukpaka's model for adsorption of crude oil using plantain pseudo stem as adsorbent. The results obtained are presented in the Tables 7, 8, 9 and Figure 6.

The graph presented in Figure 6 illustrates the comparison of experimental results with theoretical results in terms of mass of oil adsorbed for adsorbent of 1 cm thickness. Increase in the mass of oil adsorbed is observed with increase in contact time. Both

Table 9. Experimental and theoretical values of the functional parameters for 3 cm thickness of adsorbent.

Component	Values of functional parameters for adsorption of oil			
	Time (s)	Thickness (m)	Ma-experiment (g)	Ma-model (g)
Sample 1	30	0.03	4.19	4.72
Sample 2	60	0.03	6.15	12.01
Sample 3	90	0.03	7.71	2.91
Sample 4	120	0.03	9.65	8.09
Sample 5	150	0.03	12.08	11.11

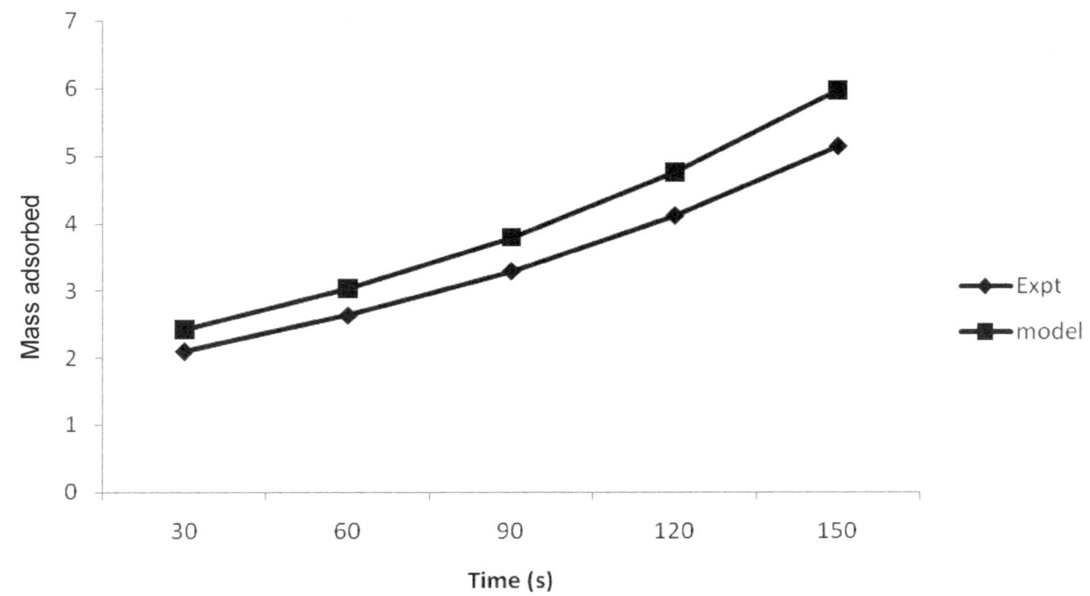

Figure 6. Comparison of experimental and theoretical values of mass of oil adsorbed versus time for 1 cm thickness of adsorbent.

experimental and theoretical results obey this rule as shown in Figure 6. The results illustrated in Figure 6 shows a good match for the comparison of experimental and theoretical result. Therefore the model developed can be found useful in monitoring and predicting the rate of adsorption using plantain pseudo-stem for effective remediation of polluted water environment.

Figure 7 illustrates the comparison of the experimental and theoretical result on the mass of adsorbed oil using plantain pseudo-stem as an adsorbent for 2 cm thickness. An intersection was obtained at time t=57 s, Ma=5.2 g as shown in Figure 7.

The results obtained for the model indicate increase in mass of adsorbed oil with increase in contact time, whereas for the experimental results, a decrease was observed with the contact time range of 30 to 60 s and sudden rise in the mass of oil adsorbed within 60 to 120 s as well as sudden decrease in the mass of oil adsorbed within time range of >120 s to <150 min as shown in Figure 6. The comparison of the experimental and theoretical results indicates a poor match.

In Figure 8, the experimental and theoretical results are presented for the rate of mass of oil adsorbed using adsorbent of 3 cm thickness. From the result presented in Figure 8 the mass adsorbed, based on the calculation from the developed model, increases with increase in contact time. On other hand, the experimental result for mass adsorbed increases and decreases in an irregular pattern, that is the mass of oil adsorbed increased within the time range of 30 to 60 s and later dropped within the time range of >60 to<90 s, and a sudden increase was observed within the time range of >90 s. This indicates that the experimental behavior of the rate of the mass of oil adsorbed did not initial obey the model but later followed the sequence of the model developed in this project.

General discussion

Figure 1 is a plot of the mass adsorbed by an adsorbent of 1cm thickness against time. The graph shows that an increase in the contact time increases the amount of oil

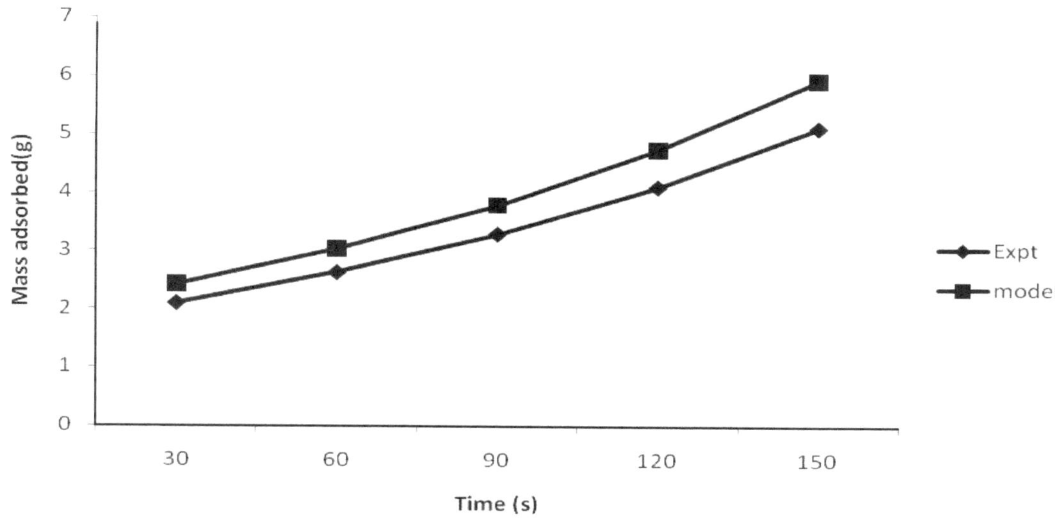

Figure 7. Comparison of experimental and theoretical values of mass of oil adsorbed versus time for 2 cm thickness of adsorbent.

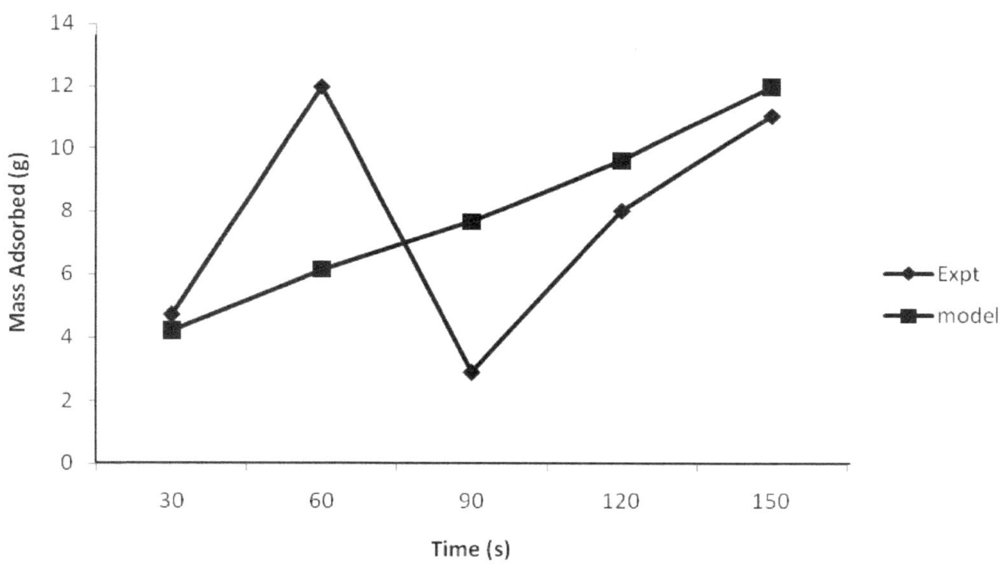

Figure 8. Comparison of experimental and theoretical values of mass of oil adsorbed versus time for 3 cm thickness of adsorbent.

been adsorbed by the adsorbent. This indicates that the adsorbent of 1 cm thickness is effective and thus can be used in facilitating remediation of an oil spilt site (in water environment). On the other hand, Figure 3 is more irregular, as the mass of oil adsorbed increases and decrease with time. This is expected, as it was previously stated in the literature review that plantain pseudo stem, as the name implies, is a false stem because it looks like a stem, but rather is made up of clustered cylindrical aggregation of the plantain leaf stalk. So for higher thicknesses say 2 cm and above, the possibility of having a adsorbent with pore spaces running without obstruction by an impermeable layer (which is a thin layer covering the front and back of the leaf stalk), from top to bottom is slim which is not only dependent on the nature of the leaf but also depends on the way the adsorbent was cut. Also it is true that the sample of 2 cm thickness with the highest amount oil adsorbed was probably hollow from top to bottom, as this is the only possible reason.

Likewise, Figure 4 shows a more irregular pattern than

the former, which proves that the higher the thickness of the adsorbent, the lower the possibility of having an adsorbent without impermeable layers. Figure 4 shows the relationship between mass of oil adsorbed by an adsorbent of 2 cm thickness with varying drying times against the drying time. From the plot, it is observed that mass adsorbed by the adsorbent decreases with an increase in the drying time of the adsorbent. This is anticipated as heating not only denatures the plant but also destroys the cellulose in the plant which is the main component responsible for the retention capability of plantain stem. That is, though the porosity of the stem aids adsorption by capillary action and diffusion, the adsorbent should have the ability to retain its adsorbed content which is the function of the cellulose, so by heating the adsorbent it becomes less or no longer effective.

Figures 6, 7 and 8 shows the relationship between experiment data and data estimated by the model for 1, 2 and 3 cm thickness of adsorbent respectively. In Figures 6, the deviation of model values from experiment is minimal as compared to the two other graphs. These errors arose as a result of assumption made in course of simplification, certain approximations and the method used for determining q. On the other hand, Figures 7 and 8 show the graphs estimated from the model for adsorbents which take into consideration the absence of a layer restricting adsorption in the direction of major concern (z-direction) based on this study. This implies that the model developed for this biosorption process with plantain stem as the adsorbent, characterizes to a great extent, oil adsorption process using adsorbent of 1 cm thickness and even smaller thicknesses, judging by the nature of the adsorbent. Deviations of a rather unacceptable magnitude are noted in the use of this particular model for adsorbents of 2 cm thickness. A rather outrageous set of results are got for adsorbents of 3 cm thickness. These errors can be attributed to the irregularity of the stem structure, which is of a variation from the model structure assumption. Also drying the adsorbent as a treatment step to enhance adsorption proved otherwise as drying reduces the effectiveness of the adsorbent.

Conclusion

Accidental spillage of crude is a major problem to mostly petroleum exploring nation and means of effectively remediating the polluted site at cheap rates is the target. On the bright side, plantain pseudo stem which is a major agricultural waste, as it is recorded in literature, has the desired specification required of every adsorbent, as it is clearly shown from this study that plantain pseudo stem can adsorb as much oil as its own weight. Also the problem associated with biosorbents adsorbing as much oil as water, is not the case for plantain pseudo stem as it has a high affinity for oil (oleophilic).

The Candy – Ukpaka's model,

$$M_a = e^{0.0075t}\left(4.14410287\, e^{-0.007500487657z} - 7.006344672\, e^{-115.390407z}\right)$$

On the other hand, it is useful for monitoring and predicting the amount of oil adsorbed for a given size of the adsorbent and time of contact required to remove a particular amount of oil spilt. It must be also be noted that plantain pseudo stem as an adsorbent is mainly a primary remediating material as it merely adsorbs the crude and removes it from the uncontrolled polluted site to a site where recovery or degradation of the crude oil can be deliberately carried out.

Nomenclature: c, solute concentration in the pore of the biomass (mg/L); **T,** solute concentration in the non biomass layer (mg/L); **c_b,** solute concentration in the bulk solution (mg/L); **r,** radial variable in the biomass core (cm); **R,** radius of the biomass core (cm); **KR,** radius of immobilized biomass particle (cm); **Q,** solute concentration in the biomass (mg/g); **A,** apparent biomass core porosity core (mg/L) (mg/g); **D,** effective solute diffusivity in the pore of the biomass core (cm^2/s); **P,** apparent biomass core density (g/cm^3); **D,** effective solute diffusivity in the non biomass layer (cm2/s); **δ,** non biomass layer thickness (KR - R)(cm); **k,** external fluid film mass transfer coefficient (cm/s).

REFERENCES

Acar Y, Alshawabkeh A (1993). Principles of Electrokinetic Remediation. Environ. Sci. Technol. 27:1638-1647.

Butler CS, Mason JR (1997). Structure Function Analysis of the Bacterial aromatic ring-hydroxylating dioxygenases. Adv. Microb. Physiol. 38:47-84.

Carberry J, Wik J (2001). Comparison of *Ex situ* and *In situ* Bioremediation of Unsaturated Soils Contaminated by Petroleum. J. Environ. Sci. Health, Part A-Toxic/ Hazard. Subst. Environ. Eng. 36:1491-1503.

Cerniglia CE (1997). Fungal metabolism of Polycyclic Aromatic Hydrocarbons: Past, Present and Future Applications in Bioremediation. J. Ind. Microbiol. Biotechnol. vol. 19, pp: 324-333.

Colwell RR, Mills AL, Walker JD Garcia-Tello P, Campos PV. (1978). Microbial Ecology of the Metula Spill in the Straits of Magellan. J. Fish. Res. Board Can. 35:573-580.

Deschamps G, Caruel H, Borredon ME, Bonnin C, Vignoles C (2003). Oil Removal From Water by Adsorption on Hydrophobic Cotton Fibers. Study of Adsorption Properties and Comparison With Other Cotton Fiber-based adsorbents. Environ. Sci. Technol. 37:1013-1015.

Federal Remediation Technologies Roundtable (2000). Remediation technologies Screening Matrix and Reference Guide, Version 4.0. Engineered Approaches to *in situ* Bioremediation of Chlorinated Solvents: Fundamentals and Field Applications, July 2000, EPA-542-R-00-008.

Goltz M, Bouwer E, Huang J (2001). Transport issues and Bioremediation Modeling for the *in-situ* Aerobic Co-metabolism of Chlorinated Solvents. Biodegradation, vol. 12, pp: 127-140.

Jackman S, Maini G, Sharman A, Sunderland G, Knowles C (2001). Electro-Kinetic Movement and Biodegradation of 2, 4-Dichlorophenoxyacetic acid in silt Soil. Biotechnol. 74:40-48.
Jain PK, Gupta VK, Gaur RK, Lowry M, Jaroli DP, Chauhan UK (2011). Bioremediation of Petroleum Oil Contaminated Soil and Water. Res. J. Environ. Toxicol. 5:1-26.
Mandelbaum R, Shati M, Ronen D (1997). In-situ Microcosms in Aquifer Bioremediation Studies. FEMS Microbiol. Rev. 20:489-502.
Morgan P, Watkinsons R (1992). Factors Limiting the Supply and Efficiency of Nutrient and Oxygen Supplements for the in-situ Biotreatment of Contaminated Soil and Groundwater. Water Res. vol. 26:73-78.
O.Connor C, Lepp N, Edwards R, Sundeland G (2003). The Combined use of Electrokinetic Remediation and Phytoremediation to Decontaminate Metal polluted soils: A Laboratory-Scale Feasibility Study. Environ. Monit. Assess. 84:141-158.
Page M, Page C (2002). Eelctroremediation of Contaminated Soils. J. Environ. Eng. ASCE 128:208-219.
Pamucku S (1997). Electro-Chemical Technologies for in-situ Restoration of Contaminated Subsurface Soils. Electronic J. Geotech. Eng. 37: 972-980.
Pamukcu S, Weeks A, Wittle J (2004). Enhanced reduction of CrVI by direct electric current in a contaminated clay. Environ. Sci. Technol. 38:1236-1241.
Prince R (1993). Petroleum Spill Bioremediation in Marine Environments. Crit. Rev. Microbiol. 19:217-242.
Reddy K, Chinthamreddy S (1999). Electrokinetic Remediation of Heavy Metal-Contaminated Soils Under Reducing Environments. Waste Manage. 19:269-282.
Reed B, Berg M, Thompson J, Hatfield J (1995). Chemical conditioning of Electrode Reservoirs During Electrokinetic Soil Flushing of Pb-Contaminated Silt Loam. ASCE J. Environ. Eng. 121:805-815.
Reynold JG, Coronado PR, Hrubesh LW (2001). Hydrophobic Aerogels for Oil Spill Cleanup-Intrinsic Absorbing Properties. Energy Source, 23:831-843.
Ribeiro A, Rodriguez-Maroto J, Mateus E, Gomes H (2005). Removal of Organic Contaminants From Soils by an Electrokinetic Process: The Case of Atrazine. Experimental and Modeling. Chemosphere 59:1229-1239.
Roszak DB, Colwell RR (1987). Survival Strategies of Bacteria in the Natural Environment. Microbiol. Rev. 51:365-379.
Safe Drinking Water Formulation, www.safewater.org., March 2000.
Shapiro A, Probstein R (1993). Removal of Contaminants from Saturated Clay by Electro-Osmosis. Environ. Sci. Technol. 27:283-291.
Teas C, Kalligeros S, Zanikos F, Stournas SL, Anastopoulos G (2001). Investigation of the Effectiveness of adsorbent Materials in oil Spills Clean up. Desalination 140:259-264.
Toyoda M, Inagaki M (2003). Adsorption and Recovery of Heavy Oils by Using Exfoliated Graphite. Spill Sci. Technol. Bull. 8:467-474.
Ukpaka CP (2004) Development of Model for Crude Oil Degradation in a Simplified Stream System. Int. J. Sci. Tech.ol. 3(2):34-37.
Ukpaka CP (2005). Investigation of Microbial Influenced Corrosion in Crude Oil Storage Tanks. J. Model. Simul. Control (AMSE) 66(4):1-22.
Ukpaka CP (2006). Microbial growth and Decay rate Kinetics on Biodegradation of crude oil." J. Model. Simul. Control (AMSE) 67(2): 59-70.
Ukpaka CP (2007). Modeling solid - Gas separation in a cyclone operating system. J. Sci. Ind. Stud. 5(1):39-45.
Ukpaka CP (2008). Smokeless Flare Modeling of an associated gas in a production oil flied. J. Model. Simul. Control (AMSE) 69(1):29-46.
Ukpaka CP (2010). Predictive Techniques to estimate the oxygen utilization by *Pseudomonas aeruginosa* in petroleum Hydrocarbon in a fluidized bed Reactor. ICASTOR J. Eng. 4(1):91-106.
USDOD (1994). Remediation Technologies Screening Matrix and Reference Guide. DOD Environmental Technology Transfer Committee. MK01\ RPT: 02281012.009\ compgde.fm, pp. 1-223.
USEPA (1999). Phytoremediation Resource Guide. United States Environmental Protection Agency, EPA/542/B-99/003. http://www.epa.gov/tio.
USEPA (2000). Introduction to Phytoremediation. United States Environmental Protection Agency, Washington, DC., USA, p. 80.
Van CL (1997). Electrokinetics: Ground-Water Remediation Technologies Analysis Center, Pittsbusrgh, PATO., pp: 97-103.
Virkutyte J, Sillanpaa M, Latostenmaa P (2002). Electrokinetic soil remediation - Critical overview. Sci. Total Environ. 289:97-121.
Wei QF, Mather RR, Fotheringham AF, Yang RD. (2003). Evaluation of Nonwoven Polyprophylene oil Adsorbents in Marine Oil Spill Recovery. Mar. Pollut. Bull. 46:780-783.

Performance of activated carbons in the catalytic wet peroxide oxidation (CWPO) of maleic acid

Lilian Daniel Kaale and Jamidu H. Y. Katima

Department of Chemical and Mining Engineering, College of Engineering and Technology, University of Dar es Salaam, P. O. Box 35131, Dar es Salaam, Tanzania.

Catalytic wet peroxide oxidation (CWPO) of maleic acid using palm fruit shell, peach stone and cashew nut shell as catalysts was investigated. Physical activation of pyrolised char with carbon dioxide was done at 900°C, 30 min holding time and heating rate of 20°C/min under nitrogen flowrate of 150 cm^3/min. The activated carbons had maximum Brunauer Emmett-Teller (BET) surface areas of 389, 428 and 692 m^2/g for palm fruit shell, peach stone and cashew nut shell, respectively. The morphology structures of the samples were determined using a scanning electron microscopy (SEM), while thermal stabilities were determined using thermal gravimetric analysis (TGA). The CWPO reactions were performed in a stirred batch reactor at an atmospheric pressure in the presence of H_2O_2 as an oxidant. The results showed that peach stone and palm fruit shell catalysts are more effective for liquid phase oxidation of maleic acid than cashew nut shell and can be applied for catalytic wastewater treatment systems.

Key words: Activated carbon, catalytic wet peroxide oxidation, activation, surface areas.

INTRODUCTION

Water pollution, particularly in developing countries, is a serious problem which requires vital consideration. It is a leading worldwide cause of death and diseases as it is reported to account for deaths of more than 14,000 people daily (Water pollution - http://en.wikipedia.org/wiki/water_pollution#_note-death2). Among the potential sources of ground and surface water pollutants are discharges from industrial processes and agricultural activities. Wastewaters such as those from chemical, petroleum, pharmaceutical, coke, pesticides, electroplating industries and paper manufacturing, are difficult to treat. Such pollutants include phenols and its derivatives. It is important to find out a suitable cost effective treatment technology for such pollutants. However, several technologies for degrading toxic wastes have been discovered by mankind overtime for the purpose of safeguarding human health, aquatic the desired removal efficiency, effectiveness, environmental conditions and cost (Masende et al., 2003; Taylor and Weygandt, 1974). However these methods are now losing popularity due to high operation cost and poor level of conversion of pollutants.

The catalytic wet oxidation (CWO) process is becoming more popular these days for oxidizing toxic organic wastes. It involves contacting of oxygen gas or air and organic compound like maleic acid and a catalyst in aqueous medium.

Application of catalysts in CWO has improved the old supercritical water oxidation and wet air oxidation processes (Debellefontaine et al., 1996; Masende et al., 2003; Mises, 2001; Santos et al., 2002).

However, one of the major drawbacks of catalysts is deactivation during liquid phase oxidation. The main causes for catalyst deactivation during CWAO are the leaching of the active metal phase (Fortuny et al., 1999) and/or the formation of carbonaceous deposits hampering access to the catalytic sites (Hamoudi et al., 1998). This problem could be avoided or minimized by employing activated carbon as catalyst or as catalyst support.

Activated carbon has broad range of applications in different fields such as food, chemical and pharmaceutical industries. Activated carbon is broadly defined to include a wide range of amorphous carbon based materials prepared in such a way that they exhibit a high degree of porosity and an extended surface area (Cecen, 2011). The primary use for activated carbon is in the treatment of water, including potable water; wastewater and groundwater remediation (Baker et al., 1992). They are also effective as catalysts and catalysts' supports (Jüntgen, 1986; Semikolenov, 1992). However, not every carbon is suitable for each use. The carbon skeleton products (after carbonization and/or activation), which result into porosity development inside the carbon particles, depend on the nature of the source materials and conditions of preparation. This need for task specificity among activated carbons necessitates careful selection of source material and conditions of preparation to suit the intended use.

Maleic acid was chosen as model compound in this study. Phenol reaction steps, in supercritical wet oxidation (SCWO), have been reviewed in which the formation of dimers and other intermediates like single-ring compound (e.g. hydroquinone) products, ring-opening products (e.g. maleic acid, glyoxylic acid, acetic acid and other organic acids) and gases (e.g. CO, CO_2) are reported (Devlin and Harris, 1984; Masende et al., 2003). Some of these partial oxidation products and intermediates, especially the dimmers are relatively more toxic than phenol. According to the reaction scheme proposed by Devlin and Harris (1984) for phenol oxidation, maleic acid found to be a crucial intermediate product as it appears almost at all reaction conditions, though in trace amounts (Mises, 2001; Santos et al. 2002). Furthermore, the oxidation of maleic acid over a number of catalysts showed a performance similar to phenol oxidation.

Nevertheless, many studies showing CWPO of phenolic compounds using activated carbons as catalysts or catalyst supports have been reported (Ayral et al., 2010; Missenden et al., 2003; Mohamed et al., 2013; Quesada-Peñatea et al., 2012). However, many of them are commercially too expensive especially for developing countries. There is a need to explore locally available and cost effective activated carbons and use them in waste water treatment. This paper presents characteristics of locally activated carbons prepared from cashew nut shell, palm fruit shell and peach stone and their performance in CWPO of maleic acid.

EXPERIMENTAL SECTION

Chemicals and materials used include: maleic acid 99% ($C_4H_4O_4$) (May & Baker), phosphoric acid 99% (H_3PO_4) distilled water, hydrogen peroxide (H_2O_2), nitrogen gas, carbon dioxide gas, palm fruit shells, peaches stones, and cashew nut shells. The equipments used are: CARBOLITE furnace type CTF 12/65, Autosorb-1 system from Quanta chrome Corp, thermo gravimetric analysis instrument 951, scanning electron microscopy and three-phase slurry batch tank reactor with a set of high performance liquid chromatography (HPLC).

Activation

Two-step CO_2 activation was performed in a horizontal tube furnace (CARBOLITE furnace type CTF 12/65). The equipment contains horizontal tubular heater and a EUROTHERM 808 temperature controller unit used for setting required temperatures. In each run, approximately 5 g of sample material was used and the system was then purged with 150 cm^3/min of N_2 for 15 min to ensure inert conditions in the reactor. The reactor was then heated at a rate of 20°C/min to a final temperature of 900°C, while the ACs wastes were thermally decomposed to porous carbonaceous materials. The sample was kept at 900°C for 90 min in the inert atmospheric of nitrogen. After carbonization while maintaining the temperature at 900°C the nitrogen stream was replaced by CO_2 for 30 min. The furnace was then switched off to allow it to cool to room temperature in the atmosphere of nitrogen. The samples were grinded and screened by using sieve of < 50 μm.

Characterization of activated carbons

Textural characterization

An Autosorb-1 system from Quanta chrome Corp. was used to determine the surface area and porosity characteristics of the activated carbons, according to IUPAC classification. Density Functional theory (DFT) was applied to investigate sample microporosity. DFT Kernel applied in this research was N_2 at 77K on carbon, (slit pores, NLFDFT equilibrium model). The BET surface area was calculated from N_2 adsorption isotherms by using the Brunauer–Emmett–Teller (BET). The total pore volumes were found from the amount of nitrogen adsorbed at a relative pressure of 0.99. A reference material, 8571 Alumina, (U.S. Department of Commerce National Institute of Standards and Technology (NIST)), was used to check the performance of the Autosorb-1. NIST reports the mean BET specific surface area for 8571 Alumina to be 158 m^2/g, but with a variation from 150 to 165 m^2/g. Autosorb-1 instrument therefore was tested and calibrated quite frequently to ascertain its performance. The surface areas of calibration material were measured to be 152 m^2/g, which is in total agreement with NIST published value.

The surface characteristics of samples were analysed using scanning electron microscopy (LEO 1500). The different powder samples were placed on aluminum SEM specimen mount stubs from Electron Microscopy Sciences (EMS) using carbon coated conductive adhesive tabs, with the accelerating voltage of 2500.

Table 1. Standard reactor operating condition.

Item	Standard	Range
Temperature, °C	50	50
Pressure, MPa *	1.2	1.2
Nitrogen flow rate (at room conditions)mil/min	90	10-120
Initial maleic acid concentration, mol/L	0.01	0.01
Volume of liquid in the reactor, ml	350	350
pH	Uncontrolled	2-7
Stirrer speed, rpm	350	350-1200
Activated carbons ACs, g/l	0.5	0.5

1.2 MPa * is a backward pressure which was used to facilitate easy withdrawal of samples from the reactor.

Thermogravimetric analysis

Thermal degradation characteristics of ACs waste were studied using a thermogravimetric method. Experiments were performed on TA instrument 951 using air (O_2, 21%). Each sample of 5–6 mg was heated from room temperature to a final temperature of 800°C Fixed carbon, volatile, and ash analyses were carried out using a thermogravimetric analyser. When the sample is heated under an inert atmosphere to 800°C, the weight loss during this step is a volatile component. The gas atmosphere is then switched to air to burn off fixed carbon, finally, any residue left after the system is cooled to room temperature is considered as ash.

Performance test

The experiments were conducted in a batch stirred Tank reactor (BSTR) in a temperature of 50°C and at an atmospheric pressure in the presence of H_2O_2 as an oxidant. Other experimental and reactor operating conditions were as listed in Table 1. The reactor used was a 500 ml volume autoclave (Autoclave Engineers, Zipperclave Hastelloy) with two internal baffles, thermowell, cooling coil, two separate liquid inlets, a liquid outlet with catalyst filters (filter type: Millipore HV 0.45 μm and a stainless steel 0.5 μm), immersion-pipe for level measurement, and a gas dispersion impeller. The reactor set-up is presented in Figure 1. The detailed procedure was similar to that described elsewhere (Masende et al., 2003).

RESULTS AND DISCUSSION

An effective use of activated carbon requires knowledge about the structure of its porosity obtained from equilibrium data namely: the pore-size distributions of the microporosity in particular, the pore-size distributions of the mesoporosity, the composition of the carbon surfaces onto which adsorption occurs and the knowledge of the dynamics of adsorption to indicate its effectiveness (Marsh et al., 2006). Traditional activated carbons possess a wide range of pore size, typically from 2 to 2000 Å (Martinez et al., 2003). The pores are divided into three groups, the micropores with diameter less than 20 Å, mesopores with diameters ranging between 20 and 50 Å and macropores with diameter greater than 50 Å. The micropores constitute a large surface area (about 95% of total surface area of activated carbon) and microvolume and therefore determine to a consideration extent of the adsorption capacity of a given activated carbon. However the molecular dimensions of the adsorbate are not too large to enter in micropores (Encyclopaedia of Surface and Colloid Science, Wikipedia). Macropores are not of considerable importance to the process of adsorption in activated carbon as their contribution to surface area does not exceed 0.5 m^2/g. Accordingly the surface areas and porosity characteristics obtained in this study are accepted as shown in Table 2.

The surface area of cashew nut shell was much higher compared to that found in Peach stone (PS) and palm fruit shells (PFS). In addition the cashew nut shell (CNS) had higher microporosity of 69.0 compared to PFS and PS, 68.2 and 67.3%, respectively. The total pore volume of cashew nut shell was 0.4637 cc/g, which was also higher than that found in peach stone and palm fruit shell, 0.3097 and 0.2654 cc/g, respectively. However, CNS showed poor performance in CWPO of maleic acid compared to other ACs. This might be due to thermal stability structure of CNS, from thermal stability data CNS showed poor thermal stability compared to other ACs.

From the BJH adsorption (Dv (d) in (cc/Å/g)) data of the activated carbons show that, Cashew nut shell, peach stone and palm fruit shell had diameters ranging between 13.00 to17.54 Å, 12.03 to16.36 Å and 12.02 to 16.38 Å respectively. It has been reported that micropores carbons like the one obtained in this study are more effective to retain small molecules (Martinez et al. 2003). On the other hand, the mesopores carbon are better for adsorbing large molecules. Table 2 clearly shows that CNS, PS and PFS have both micropores and mesopores carbon properties.

Figures 2 to 6 show micrographs of activated carbons from SEM analysis for the three materials at different magnification. Figures 2 to 5 are micrographs of palm fruit shell and peach stone ACs, respectively. It can be seen that the particles have irregular shapes and consist of pores.

Figure 6 shows the SEM for cashew nut shell. The micrograph clearly shows that there are physical difference

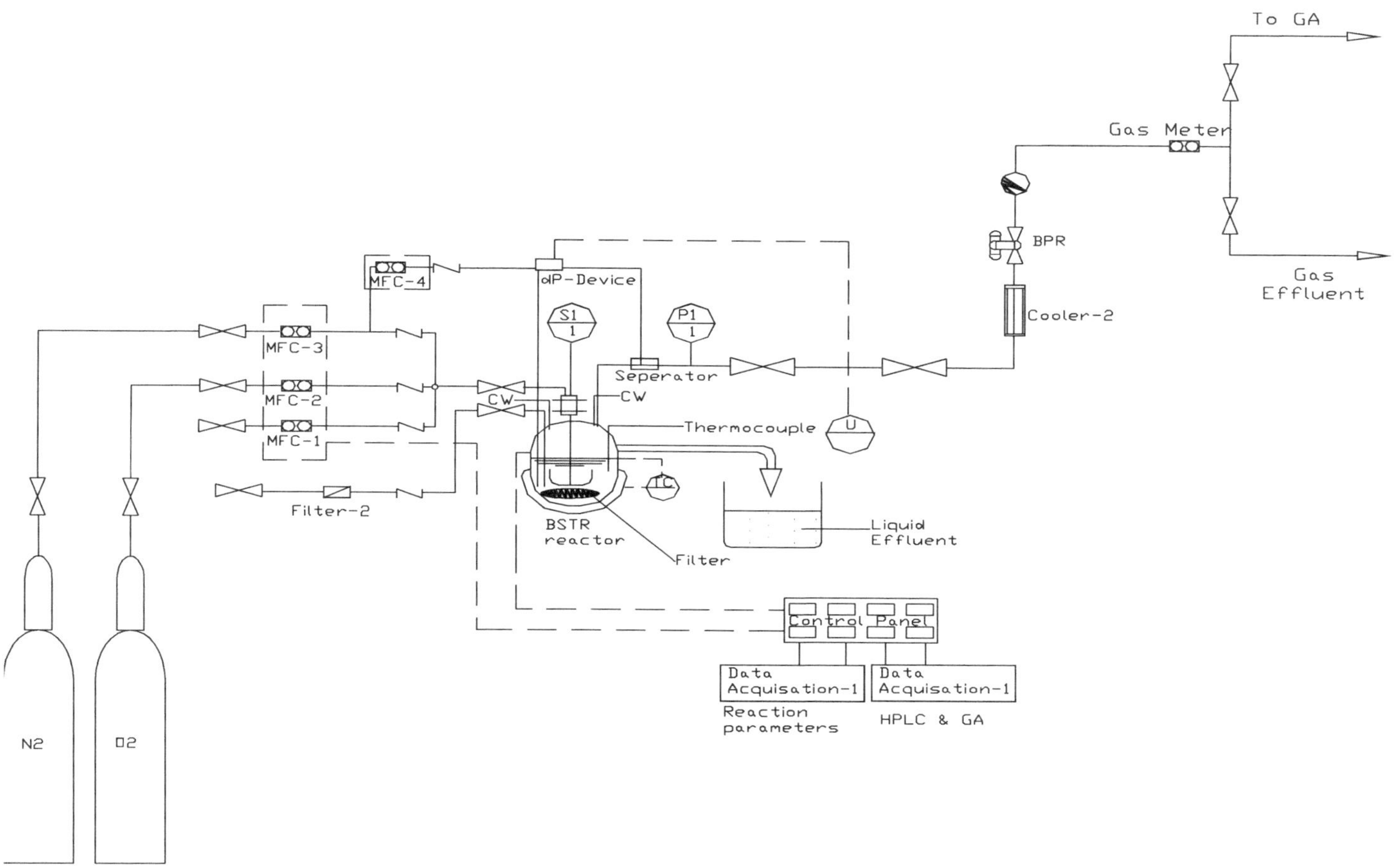

Figure 1. A three-phase batch reactor BTR.

Table 2. Surface area and porosity characteristics of the three activated carbons under study.

Sample name	Surface m^2/g	Micro-porosity (cc/g)	Micro-porosity (%)	Meso-porosity (cc/g)	Meso-porosity (%)	Macro-porosity (cc/g)	Macro-porosity %	Total porosity (cc/g)
PFS	389	0.1809	68.2	0.0643	24.2	0.0202	7.6	0.2654
PS	463	0.2138	69.0	0.0752	24.3	0.0207	6.7	0.3097
CNS	692	0.3120	67.3	0.1314	28.3	0.0203	4.4	0.4637

characteristics compared to PS and PFS. The micrograph also consists of irregular shapes.

Thermal stability

Thermogravimetric analysis (TGA) is an analytical technique for measuring the amount and rate of change in the weight of a material as a function of temperature or time. The changes in the mass can be caused by a variety of processes such as decomposition, degradation, sublimation, vaporization, adsorption, desorption, oxidation, and reduction (Vyazovkin, 2012). Figures 7 and 8 show results from thermogravimetric analyses (TGA) for the activated carbons. The maximum temperature of 800°C was used to ensure that the specimen weight was stable at the end of the experiment, implying that all chemical reactions were completed. This approach provided: oxidation temperature (To= dm/dtmax) (Figure 7) and ash content (residual mass, M_{res}) (Figure 8). In this context the oxidation temperature include temperature corresponding to the maximum weight loss rate (dm/dT_{max}) and the temperatures corresponding to onset weight loss onset (T_{onset}). The weight loss rate refers to the temperature of the maximum rate of oxidation, while the T_{onset} refers to the

Figure 2. Activated carbon from Palm fruit shell (low magnificatio 1.00X) SEM).

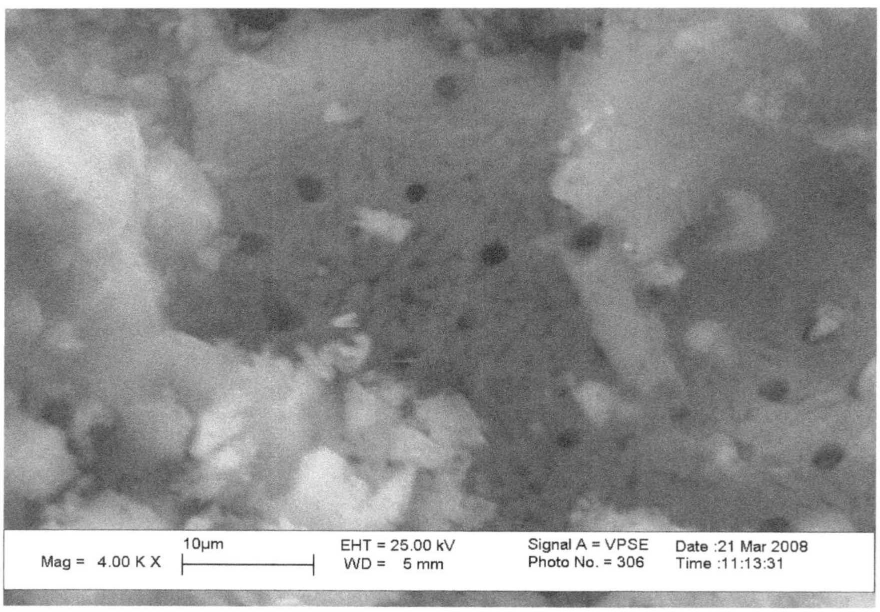

Figure 3. Activated carbon from Palm fruit shell (Modarately magnified (4.00X) SEM).

temperature when oxidation just begins.

All the curves indicated that the activated carbons underwent a single-step weight loss process due to pyrolysis (Figure 7). The derivative thermogravimetric (DTG) curves in clearly show that the peach stone and palm fruit shell are thermally stable up to 400°C while cashew nut shells were stable up to 300°C (Figure 7). It can also be seen that the mass loss graph vs. temperature/time for cashew nut shell is decreasing fast compared to other ACs Figure 8.

The thermogravimetric (TG) and derivative thermogravimetric (DTG) curves of devolatilisation revealed that two pyrolysis stages occurred. The main pyrolysis stage was completed up to 600°C, while the

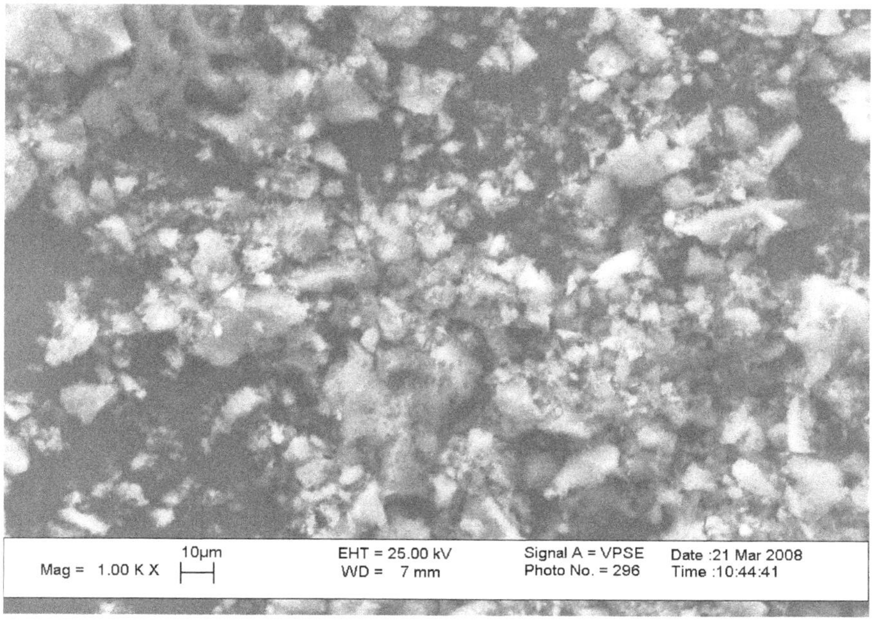

Figure 4. Activated carbon from peach stone (low magnification (1.00X) SEM).

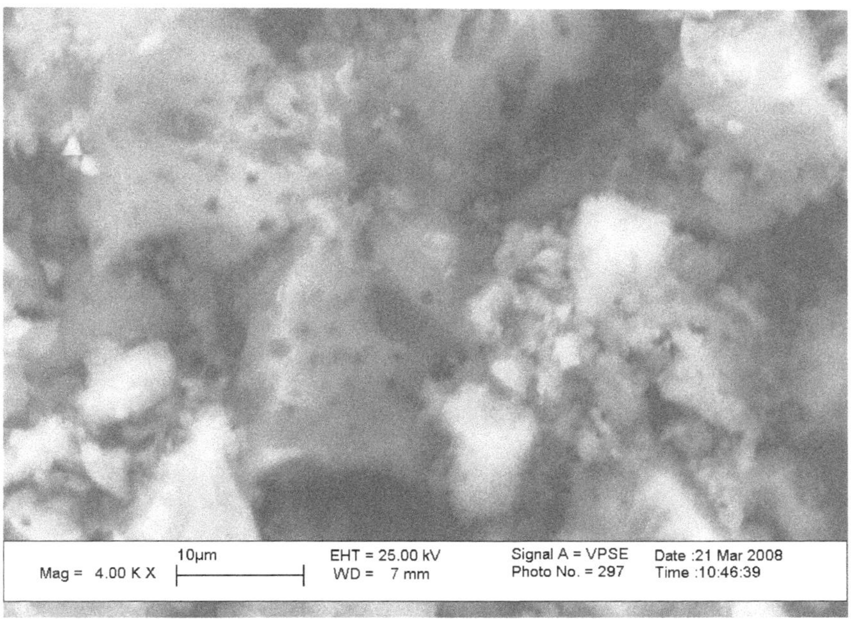

Figure 5. Activated carbon from peach stone.(Moderately magnified (4.00X) SEM).

second one continued until the test completion at 700°C. According to Skodras et al. (2005) and Tsamba et al. (2005) the pyrolysis for CNS was occurring in the same range. According to Fogler (1999) the DTG peak height is directly proportional to reactivity and the correspondent temperature is inversely proportional to this chemical activity parameter. Applying this theory, it can be concluded that reactivity of PFS is higher than that of CNS and PS.

Blank experiments

Blank experiments (that is with no activated carbons) were conducted under a nitrogen gas environment to determine whether a non-catalytic reaction, adsorption or vapour–liquid equilibrium effects were taking place under

Figure 6. Activated carbon from Cashew nut shell.

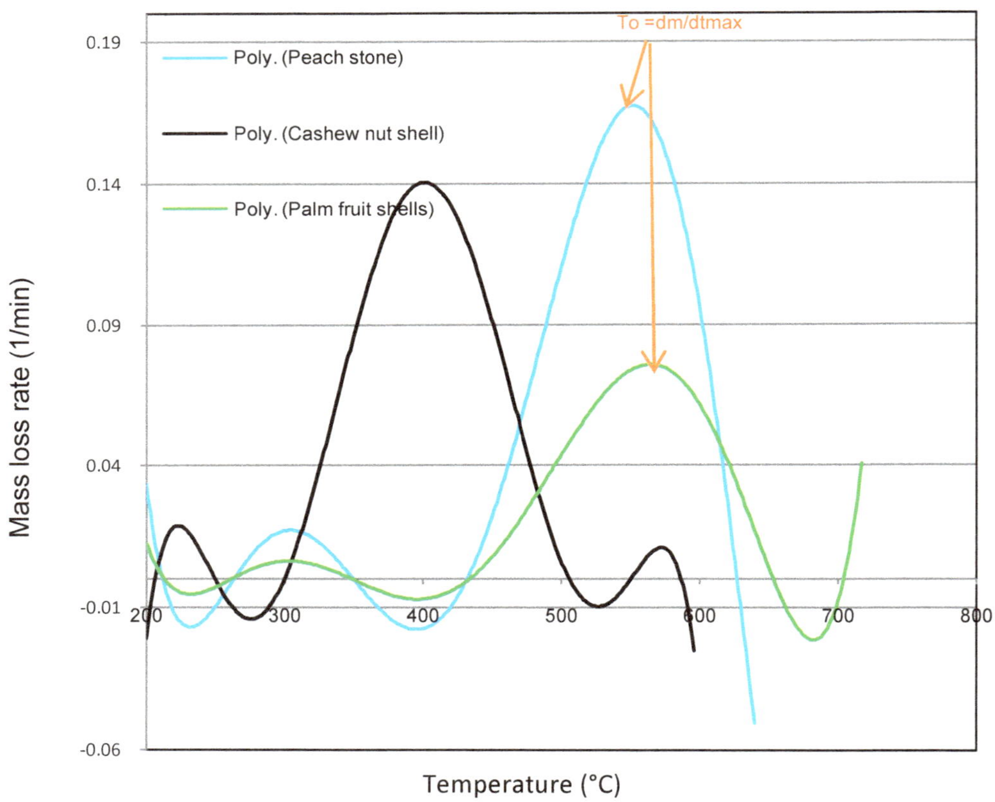

Figure 7. Mass loss rate (DTG) from the thermal decomposition of CNS, PS and PFS at 20°C.

experimental conditions. These experiments were performed at 50°C, atmospheric pressure and liquid residence time of 90 min. The results show that there was negligible maleic acid conversion (about 0.5% of the initial maleic acid concentration). Introduction of hydrogen peroxide in the reaction while maintaining the same reaction conditions slightly improved the conversion to about 2%. This could be attributed to either homogeneous reaction of maleic acid with hydrogen peroxide or reaction caused by presence of impurities in the solution or catalytic activity of the reactor material. The similar blank results were reported by Daniel and

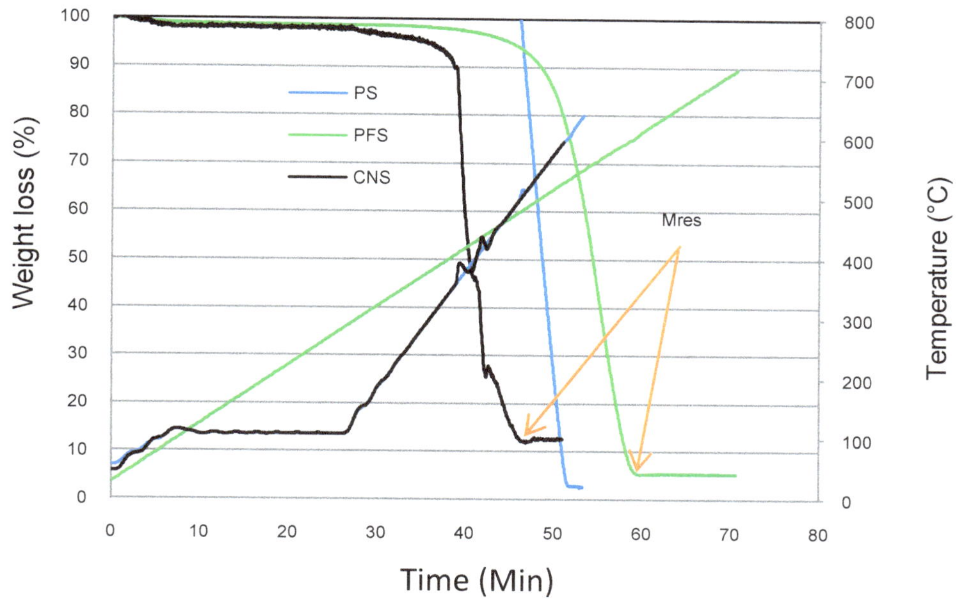

Figure 8. Weight loss %-Temperature with time for the PFS, PS and CNS.

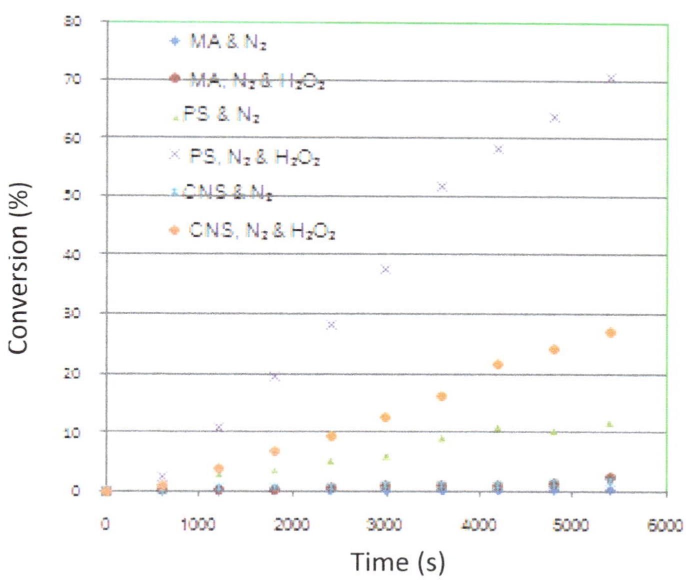

Figure 9. Blank experiments of conversion against time graph; Conditions: 50°C, C_o = 0.01 and effect of H_2O_2, activated carbons = 0.5 g/l.

Katima (2009). The effect of ACs loading on the CWPO of MA is shown in Figure 9. Blank experiments were used as reference for zero ACs loading. In general, the conversion of MA is enhanced in the presence of H_2O_2 and ACs. Two ACs among the three were tested in the blank experiments, the peach stone (the best performance in CWPO) and cashew nut shell (the least performance in CWPO). The introduction of peach stone or cashew nut shell with N_2 only in the reactor gave 11.9 and 1.8% of MA respectively, Figure 9. In the absence of H_2O_2, MA oxidation on ACs particularly CNS was comparable to that observed with H_2O_2 and N_2.

The reaction between maleic acid, hydrogen peroxide, activated carbons and nitrogen gas has improved the conversion of maleic acid to 70.7 and 27.3% for peach stone and cashew nut shell respectively. These results revealed that without H_2O_2 the conversion is low, suggesting that the performance of the peach stone or cashew nut shell (activated carbons) in the oxidation of maleic acid is highly influenced by the presence of H_2O_2. The catalytic oxidation using hydrogen peroxide is a more efficient process as compared to catalytic oxidation using air, since hydrogen peroxide is a stronger oxidant (Britto et al., 2008). Hydrogen peroxide is highly reactive and non-selective oxidants, which has been used to degrade many organic compounds (Hu, 2001). Rubalcaba et al. (2007) reported that, adding H_2O_2 to the CWAO process not only increases pollutant removal but also leads to higher mineralisation of the remaining oxidation products. This is because hydrogen peroxide acts as a free radical initiator, providing hydroxyl radicals that increase pollutant removal (Valkaj, 2011).

Performance of the three activated carbons in catalytic wet peroxide of maleic acid

Figure 10 shows the conversion/concentration of MA vs. time. The ACs were used as catalysts and hydrogen peroxide was used as oxidizing agent. The influence of the activated carbons on the degradation of maleic acid

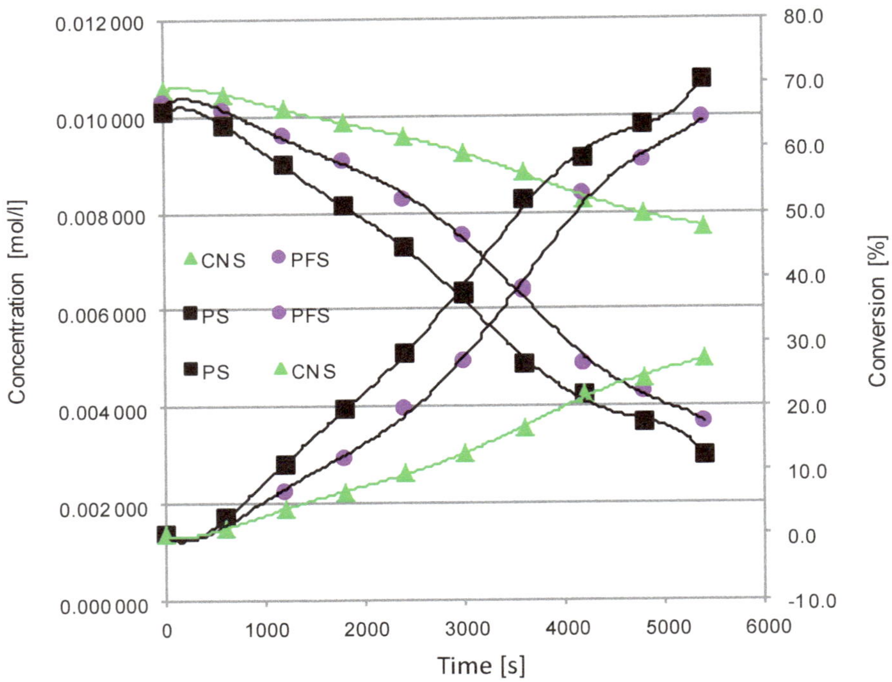

Figure 10. Influence of catalyst supports in CWPO of maleic acid. Conditions, Ca_o= 0.01 mol/L, Temperature = 50℃, activated carbons = 0.5 g/l.

was studied at 0.01 mol/L of maleic acid, 50°C and 0.5 g/l of activated carbons. It was found that cashew nut shell posses very low catalytic activity and had little adsorption capacity for maleic acid of about 27.3% after 90 min of reaction time. In contrast, due to suitable properties of the peach stone and palm fruit shell showed better catalytic activity in the degradation of maleic acid. The conversion of maleic acid using peach stone and palm fruit shell were 70.7 and 64.4%, respectively after 90 min of the reaction time. This might be due to their good porosity characteristics, thermal stability and porous structure of PFS and PS. Activated carbon alone is a good adsorbent for non-biodegradable organic waste pollutants (Stuber et al., 2005). Most forms of activated carbon are non-polar in nature, so they have the greatest affinity for other non-polar substances. As a result, they are most effective in the removal of a variety of organic pollutants. Activated carbon is an effective absorbent primarily due to its extensive porosity and very large available surface area (Jason, 1996). Ash et al. (2006) reported that the adsorption capacity of certain carbon is known to be a function of porous structure, chemical nature of the surface and pH of the aqueous solutions. Figures 11 and 12 show maleic acid chromatograph peaks for peach stone and cashew nut shell respectively.

It is clearly seen that the maleic acid peaks are decreasing when peach stone was employed as catalyst and after 90 min the maleic acid peak was almost a quarter of the original peak (Figure 11). Aguilar et al.

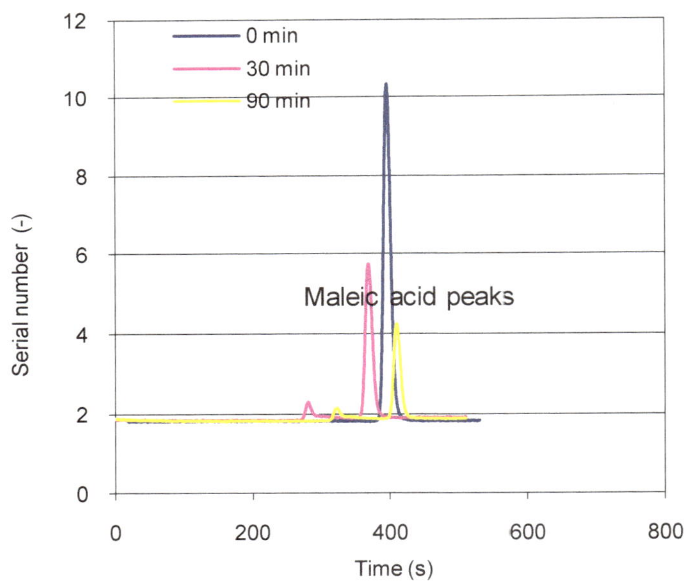

Figure 11. Chromatograph peak for peach stone; conditions Ca_o= 0.01 mol/L; Temperature = 50°C; activated carbons = 0.5 g/l.

(2005) used a commercial peach stone AC in the catalytic wet oxidation of aqueous methyl and dimethylamines and it was found to be a good catalyst, it has also been proven in this study that peach stone was the best catalyst. On the other hand, cashew nut shell

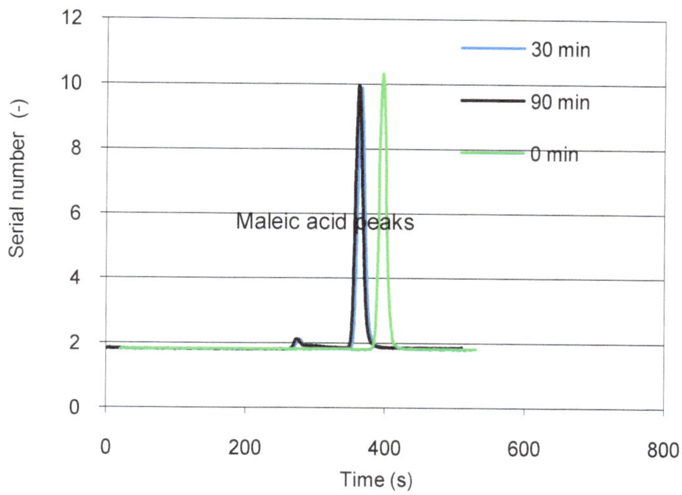

Figure 12. Chromatograph peak for cashew nut shell; conditions Ca_o= 0.01 mol/L; Temperature = 50°C; activated carbons = 0.5 g/l.

proved to perform poorly in the degradation of maleic acid (27.3% conversion of maleic acid). After 20 and 90 min, the maleic acid peaks (Figure 12) were almost the same and small difference was observed between 20/90 min and the original peak which explained that the conversion of maleic acid in the presence of CNS as catalyst was low though it had the best porosity characteristics compared to other ACs. This might be due to the stability structure of CNS. Thermal stability data for CNS showed less thermal stability compared to other ACs.

The catalytic wet peroxide oxidation of maleic acid using activated carbons has been reported elsewhere (Kaale, 2008). All these reasons and good preparation of the sample make the activated carbons to be used as catalyst.

Conclusions

Ps, PFS and CNs pyrolysis was successfully prepared, characterized and tested. Differences and similarities between these biomass wastes were found and discussed. The activated carbons had maximum Brunauer Emmett-Teller (BET) surface areas of 389, 428 and 692 m^2/g for palm fruit shell, peach stone and cashew nut shell, respectively. TG and DTG results showed that, the PS and PFS materials are thermally more stable than CNS. The catalytic wet peroxide oxidation (CWPO) of maleic acid was carried out using the three activated carbons. The PS and PFS showed better conversion of maleic acid, 70.7 and 64.4%, respectively as compared to CNS which had the conversion of 27.3%. The result demonstrated that, the PS and PFS are effective catalysts for liquid phase oxidation of maleic acid than cashew nut shell and can therefore be applied for catalytic wastewater treatment systems.

ACKNOWLEDGEMENT

The financial support provided by the Women Scholarship University of Dar es Salaam in the form of a University of Dar es Salaam Woman Postgraduate Scholars Award. Brown University-University of Dar es Salaam Fellowship Program (BUFP) is acknowledged, which financed part of the study at the Brown University, Rhode Island, USA.

REFERENCES

Aguilar C, García R, Soto-Garrido G, Arraigada R (2005). Catalytic oxidation of aqueous methyl and dimethylamines by activated carbon. Topics Catal. 33:201-206.

Ayral C, Lebigue CJ, Stuber F, Wilhelm A-M, Henri Delmas H (2010). Catalytic Wet Air Oxidation of Phenolic Compounds and Mixtures over Activated Carbon: Conversion, Mineralization, and Catalyst Stability. Ind. Eng. Chem. Res. 49:10707–10714.

Baker FS, Miller CE, Repik AJ, Tolles ED (1992). Activated carbon. Kirk- Othmer Encycl. Chem. Technol. 4:1015-1037.

Britto JM, Oliveira SBde, Rabelo D, Rangel MdC (2008). Catalytic wet peroxide oxidation of phenol from industrial wastewater on activated carbon. Catal. Today 133(135):582–587

Cecen F (2011). Water and Wastewater Treatment: Historical Perspective of Activated Carbon Adsorption and its Integration with Biological Processes: In Cecen, F and Aktasr, O (Eds.), Activated Carbon for Water and Wastewater Treatment: Integration of Adsorption and Biological Treatment 1st ed. Wiley-VCH Verlag GmbH & Co. KGaA, Weinheim Published.

Daniel L, Katima JHY (2009). Factors influencing catalytic wet peroxide oxidation of maleic acid in aqueous phase over copper/micelle templated silica-3-aminopropyltrimethoxysilane catalyst. Water Sci. Technol.-WST. 60:2621-2627.

Devlin HR, Harris IJ (1984). Mechanism of the oxidation of aqueous phenol with dissolved oxygen. J. Ind. Eng. Chem. Fundam. 23:387–392.

Encyclopaedia of Surface and Colloid Science. http://books.google.com/books?hl=en&lr=&id=n4luwcdQWv4C&oi=fnd&pg=PA4405&dq=Szabo,+and+Kallo,+1976&ots=NDzxh87G5&sig=HlAsY3ijTdaKGRwtXKtUwglgdlg#PPA4133,M1 [on 11/05 2008].

Hu X, Lam F, Cheung LM, Chan KF, Zhao XS, Lu GQ (2001). Copper/MCM-41 as catalyst for photo-chemically enhanced oxidation11 of phenol by hydrogen peroxide. Catal. Today 68:129–133.

Jason PP (1996). Activated Carbon and Some Applications for the Remediation of Soil and Groundwater Pollution. http://www.cee.vt.edu/ewr/environmental/teach/gwprimer/group23/webpage.htm [on 22/03/2008].

Jüntgen H (1986). Activated carbon as catalyst support: A review of new research results. *Funcat Cogas '86*: Int. Symp. 65:1436-1446.

Kaale LD (2008). Kinetic study of catalytic wet peroxide oxidation of maleic acid using heterogeneous catalysts. M.Sc. Thesis, University of Dar es Salaam.

Martinez ML, Moiraghi L, Agnese M, Guzman C (2003). Making and some properties of activated carbons produced from agricultural industrial residues. J. Argentine Soc. 91:103-108.

Masende ZPG, Kuster BFM, Plasinski KJ, Janssen FJJD, Katima JHY, Schouten JC (2003). Platinum catalysed wet oxidation of phenol in a stirred slurry reactor. Practical operation window. J. Appl. Catal. B: Environ. 41:247-267.

Mises MJM (2001). Liquid phase oxidation over platinum supported catalyst. M.Sc. graduation thesis, Eindhoven University of Technology. http://alexandria.tue.nl/extra2/ 200013654.pdf (retrieved on 21st June, 2007).

Mohamed WT, Ghani SA, Rasheed SM (2013). Fe/Activated Carbon as a Catalyst in Wet Oxidation of Phenolic Compounds in a Trickle Bed Reactor. Energ. Sources Part A. 35:299–311.

Quesada-Peñatea B, Julcour-Lebiguea C, Jáuregui-Hazac UJ, Wilhelma AM, Delmasa H (2012). Degradation of paracetamol by catalytic wet air oxidation and sequential adsorption – Catalytic wet air oxidation on activated carbons. J. Hazard. Mater. 221(222):131-138.

Rubalcaba A, Suárez-Ojeda ME, Carrera J, Font J, Stüber F, Bengoa C, Fortuny A, Fabregat A (2007). Biodegradability enhancement of phenolic compounds by Hydrogen Peroxide Promoted Catalytic Wet Air Oxidation. Catal. Today 124:191–197.

Semikolenov VA (1992). Modern approaches to the preparation of "palladium on charcoal" catalysts. Russ. Chem. Rev. 61:168-174.

Skodras G, Grammelis P, Basinas P (2005). Pyrolysis and combustion behavior of coal–MBM blends. J. Bioresour. Technol. 98:1-8.

Stuber F, Font J, Eftaxias A, Paradowska M, Suarez ME, Fortuny A, Bengoa C, Febregat A (2005). Chemical Wet Oxidation for the Abatement of Refractory Non-Biodegradable Organic Wastewater Pollutants. Proc. Saf. Environ. Prot. 83:371-380.

Surface Area Measurement. http://www.refiningonline.com/Engelhard KB/crep/TCR2_10.htm [on 04/05/2008].

Tsamba AJ, Yang W, Blasiak W (2005). Pyrolysis characteristics and global kinetics of coconut and cashew nut shells. J. Fuel Process. Technol. 87:523–530.

Valkaj KM, Wittine O, Margeta K, Granato T, Katoviæ A, Zrnèeviæ S (2011). Phenol oxidation with hydrogen peroxide using Cu/ZSM5 and Cu/Y5 catalysts. Pol. J. Chem. Technol. 13:28-36.

Water pollution http://en.wikipedia.org/wiki/Water_pollution#_note-death2 [on 08/06/2013].

Modeling the rate of biocorrosion and the effects of redox-reactions of metals in water environment

C. P. Ukpaka[1]*, S. A. Amadi[1], I. G. Ahuchogu[2] and J. Odharo[3]

[1]Department of Chemical/Petrochemical Engineering, Rivers State University of Science and Technology, Nkpolu, P. M. B. 5080, Port Harcourt, Nigeria.
[2]Department of /Petroleum and Natural Gas Processing, Petroleum Training Institute, P. M. B. 20 Effurun, Delta State, Nigeria.
[3]Engineering, Construction and Projects Division, Civil Works and Survey, Total E and P Nigeria Limited, Port Harcourt, Rivers State, Nigeria.

Laboratory studies were conducted to assess the effectiveness of biocorrosion through oxygen diffusion in zinc metal in water environment. The present investigation was undertaken to determine the corrosion rate of zinc metal in water environment with respect to Plank-Nernst Equation modified to monitor and predict the biocorrosion rate through oxygen diffusion which resulted to biofilm formation. The modified developed mathematical model of Plank-Nernst equations to evaluate the corrosion rate of zinc metal upon the influence of oxygen diffusion in a flowing system was given as $C_R = \frac{534w}{DAt_o e^{\lambda^2 t}}$ **where λ = 0.5 when compared with the normal corrosion rate equation,** $C_R = \frac{534w}{DAt}$ **then the results obtained showed a good match. The biocorrosion rate of zinc metal was tested in salt and fresh water environment and the results obtained showed that corrosion was faster on zinc metal immersed in salt water environment than the fresh water environment; this was attributed to the physicochemical parameters of the salt/fresh water environment.**

Key words: Modeling, corrosion, redox-reaction, biocorrosion, metal, water environment.

INTRODUCTION

Corrosion may be defined as the destruction of a material by the action of the surrounding environment. Material resistance to corrosion depends on many variables which include the material properties, and environmental characteristics. The most important aspects in the corrosion of a material are the mass transport characteristics, which occurs by convection, migration and diffusion. But among these characteristics, the aspect that is of emphasis is the diffusion. Before corrosion of metal occurs there are responsible factors, such as physical factor which consists of heat, temperature changes, water, wind and dust contributes greatly to the corrosion of metals. Another factor being the chemical which aids corrosion follows by the flare gases, acids, alkaline, salt and the biological factor which includes the microorganism, fungi, algae, worms and multi-cellular plants (Degarmo et al., 2003; Gardner, 2002; Johnson, 2000; Mars, 2005; Roy and Rofer, 1999).

The constant failure of metals due to biochemical process is of major concern to the engineers, in this study, suitable mathematical approach including experimental procedures in examining the biocorrosion of metal immersed in water environment will be clearly studied and presented on the findings of the research work. In this paper mathematical model was developed to simulate the corrosion rate of zinc in salt and fresh water

*Corresponding author. E-mail:chukwuemeka24@yahoo.com.

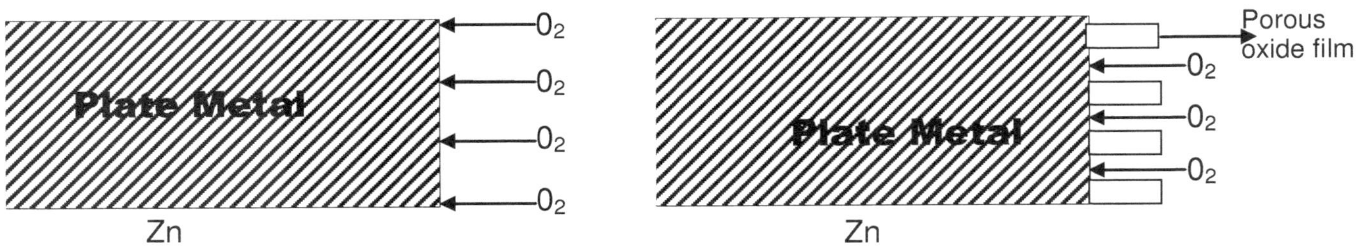

Figure 1. Area of oxide formation.

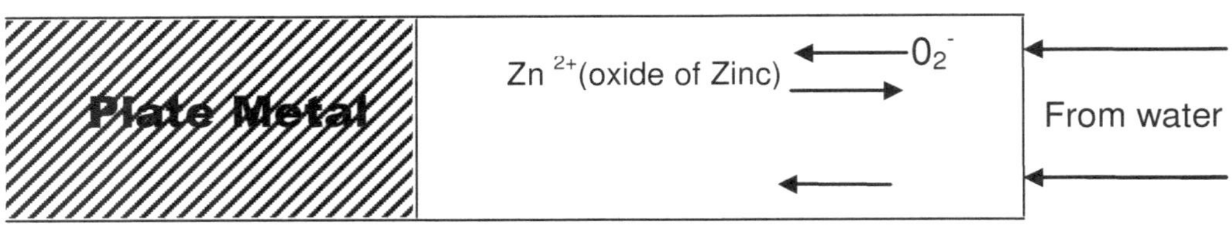

Figure 2. Oxidation reaction of zinc (Robinson and Holt, 1997).

media upon the influence of diffusion of the oxygen through the surface layer. The scope of the study is focused on the effect of redox-reactions on the bio-corrosion of zinc metal in both salt and fresh water medium.

Investigation carried out by various research groups revealed that corrosion is responsible for colossal loss in weight of materials occurring everywhere in the world (Ivor, 2004; Mar, 2005; Melchers, 2006). It is one of the most serious destructive agents and is one of the major technological problems in the modern society. Though a lot of scientific understandings of many phases of corrosion have been developed, and the results obtained are very complex phenomenon as there are a large number of complex variables responsible for it (Alkins, 1990; Ivor, 2004; Johnson, 2000; Melchers, 2006; Shifter, 2005; Sharma, 2003).

Redox reactions or oxidation-reduction reactions have a number of similarities to acid-base reactions. Fundamentally, redox reactions are family of reactions that are concerned with the transfer of electrons between species (metals) (Vijendra, 1999; Melchers, 2006; Mars, 2005; Liudvikas, 2003; John et al., 1999; Ivor, 2004). Oxidation reaction is referred to the loss of an electron, while reduction reaction refers to the gain of an electron. Thus,

$$Zn_{(s)} + 2HCl_{(l)} \rightarrow ZnCl_{2(l)} + H_{2\,(g)} \quad (1)$$

$$Zn_{(s)} + 2H_{(aq)} \rightarrow Zn^{2+} + H_{2(g)} \quad (2)$$

In equation 1 the hydrochloric acid reacts with the surface of the metal (Zinc) which resulted to the formation of aqueous zinc chloride and hydrogen gas librated. In equation 2 the solid zinc reacts with the aqueous hydrogen which yielded zinc ion and hydrogen gas liberated during the process. The chemical interaction between the solid zinc and aqueous hydrogen resulted to resulting and loss in weight of the metal (zinc).

MATERIALS AND METHODS

This monolayer of oxygen atoms which forms rapidly over the whole surface of the clean metal is said to be chemisorbed. The process involving the dissociation and ionization of oxygen molecules is called chemisorptions. Figure 1 illustrates the area of oxide formation in a metal. Figure 2 illustrate the corrosion oxidation reaction of zinc.

The chemical reaction is as stated as follows:

$$Zn \rightarrow Zn^{2+} + 2^{e-} \quad (3)$$

$$\tfrac{1}{2}O_2 + 2e^- \rightarrow O^{2-} \quad (4)$$

$$Zn^{2+} + O^{2-} \rightarrow ZnO \quad (5)$$

Investigation of sulphur reducing bacteria reveals that the microbes reduce the sulphate to sulphide.

$$4Zn \rightarrow 4Zn^{+2} + 8e^- \quad (6)$$

Anodic reaction: The anodic reaction involves dissociation of water and the chemical reaction is as shown in Equation (7).

$$8H_2O \rightarrow 8H^+ + 8OH \quad (7)$$

The depolarization process by SRBS yielded the chemical equation as presented in equation (8) as follows:

$$SO_4^{-2} + 8H^- \rightarrow S^{-2} + 4H_2O \quad (8)$$

$$Zn^{2+} + S^{2-} \rightarrow ZnS \quad (9)$$

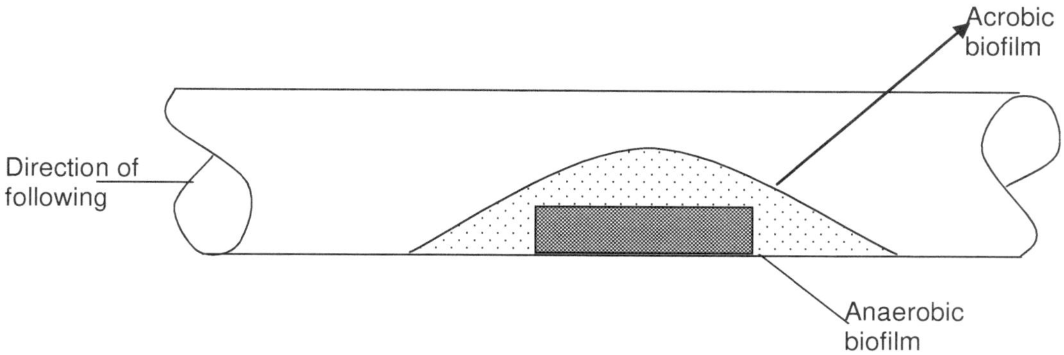

Figure 3. Typical biofilm formation on a pipe internal surface (Atkins, 1990).

Corrosion product are obtained as presented below in terms of the possible chemical reaction.

$3Ze^{2+} + 60H \rightarrow 3Zn(OH)_2$

Other corrosion product formed as a result of the chemical process is shown as follows.

$4Ze^{2+} + SO_4^{2-} + 4H_2O \rightarrow 3Zn\ (OH_2) + ZnS + 2OH$ (10)

Sulphur oxidizing bacteria are capable of oxidizing elemental sulphur or sulphur-bearing components to produce acidic compounds, which is corrosive to many metals. The typical reaction for this sulphur oxidizing bacteria (SOB) as:

$H_2S + 2O_2 \rightarrow H_2SO_4$ (11)

$2S + 3O_2 \rightarrow 2H_2O + 2H_2\ SO_4$ (12)

The instant, a metallic surface is immersed in water, a biofilm begins to form. The phenomenon can be described as a biomass composed of different aquatic microorganisms. Many species of these microorganism bind to various metals and form biofilm that develops in our stages. This is very common in pipes and has been associated to many pipe line failures (Figure 3).

The model

The model is restricted to that part of the corrosion process of steel subjected in aqueous medium which is controlled by the diffusion of the oxygen through the result layer neglecting the convective motion of species K (in our case, oxygen molecules) and considering by the plank – Nernst law as:

$$\overrightarrow{J_K} = - D_k^* \vec{\nabla}\ C_K + \left(\frac{-Z_K D^*{}_k f}{RT}\ C_K \right) \vec{\nabla}\ \phi \qquad (13)$$

The migration and diffusion are considered here. The transport of each specie satisfies the mass transport equation as follows:

$$\frac{\partial C_K}{\partial t} + \vec{\nabla}\ \overrightarrow{J_k} = S_R \qquad (14)$$

Where S_k is the source term that takes into account the production and/or consumption of oxygen. Assuming S_k equals zero which can be justified if we assume that oxygen consumption due to biochemical reaction, can only occur at the interface of the metallic surface and the rust, and considering only the 1 D problem on x-axis by replacing Equation (13) into Equation (14).

$$\frac{\partial c_K}{\partial t} + \frac{\partial}{\partial x}\left[\left(\frac{-z f D^*}{RT} \frac{\partial \phi}{\partial x} \right) C - D^* \frac{\partial c}{\partial x} \right] = 0 \qquad (15)$$

This is a migration diffusion equation. Now, let's only consider the diffusion process. Then,

$$\phi = \frac{\partial c}{\partial t} - \frac{\partial}{\partial x}\left[\frac{\partial c}{\partial x} \right] = 0 \qquad (16)$$

Let is suppose that D* = W (x) D which is a dimensionless mathematical function representing the rust layer porosity and D the diffusivity constant of oxygen in the water which is independent of the position (x). Therefore, the Equation (16) can be written as:

$$\frac{\partial c}{\partial t} - D \frac{\partial}{\partial x}\left[W\ (x) \frac{\partial c}{\partial x} \right] = 0 \qquad (17)$$

The Equation (17) can be expressed as:

$$\frac{\partial c}{\partial t} - DW\ (x) \frac{\partial^2 C}{\partial x^2} = 0 \qquad (18)$$

The Equation (18) is a second order differential

The mathematical application to equation (18)

The mathematical tool known as the separation of variables and Laplace transformation, was applied in resolving the problem as stated in equation (19)

$$\frac{\partial c}{\partial t} + D(x) \frac{\partial^2 C}{\partial x^2} \qquad (19)$$

Equation (19) was resolved by using Laplace approach. Thus,

$C = TX$ (20)

where T is the coefficient of t (time) which is dependent, x is the co-efficient of position or distance and C is the concentration of oxygen. From Equation (20), the following expressions can be obtained as:

$$\frac{\partial c}{\partial t} = T^1 X \quad (21)$$

$$\frac{\partial^2 C}{\partial_x^2} = TX^{II} \quad (22)$$

Substituting the Equations (21) and (22) into Equation (19) we have:

$T^I X = DW(x)\, TX^{II}$ (23)

Equation (23) can be written as shown in equation (24) in terms of a real constant value of λ^2.

$$\frac{T^1 X}{T} = DW(x)\frac{TX^{II}}{TX} = \lambda^2 \quad (24)$$

$$\frac{T^1}{T} = \lambda^2 \quad (25)$$

and

$$DW(x)\frac{11}{X} = \lambda^2 \quad (26)$$

Resolving the equation (25) mathematically using the necessary boundary conditions, it is noted that the equation (25) is a first order differential equation. That is:

$T^1 = \lambda^2 T$ (27)

$T^1 - \lambda^2 T = 0$ (28)

The quadratic equation obtained in equation (28) can be resolved using Laplace approach.

$T^1 = ST_{(s)} - T_{(0)}$ (29)

$T = T_{(s)}$

Substituting Equation (29) into Equation (28) we have:

$ST_{(s)} - T_{(0)} - \lambda^2 T_{(s)} = 0$

$ST_{(s)} - \lambda^2 T_{(s)} = T_{(o)}$

Considering $T_{(o)} = t_0$,

$T_{(s)}(S - \lambda^2) = t_o$, $T_{(s)} = {}^{t_o}/_{S} - \lambda^2$

$T_{(s)} = {}^{t_o}/_{S} - \lambda^2$ (30)

The inverse of equation (30) from S – domain to t – domain is given as:

$$T_{(t)} = t_o\, e^{\lambda^2 t} \quad (31)$$

From Equation (26) we have:

$$DW(x)\frac{X^{11}}{x} = \lambda^2$$

$$DW(x)\, X^{11} = \lambda^2 x$$

$$DW(x)\, S^{11} - \lambda^2 x = 0 \quad (32)$$

Applying the Laplace approach to the Equation (32) gives:

$X^{11} = S^2 X_{(s)} - SX1(0) - x\,0$ (33)

$X = X_{(s)}$

Putting the Equations (33) and (34) into Equation (32) gives the following expression:

$DW_{(x)}[S^2X_{(2)} - S^1 x(0) - x(0)] - \lambda^2 X_s = 0$

$DW(x)[S^2 - DW(x)]\, S^1 x(0) - DW(x)\, x(0)\ \lambda^2 X_s = 0$

Considering the boundary conditions;

$X^1(0) = =$ and $X(0) = X_0$

$DW(x)\, S^2X_{(s)} - DW(x)\, X_0 - \lambda^2 X_s = 0$

$DW(x)\, S2X_{(s)} - \lambda^2 X_s = DW_{(x)} X_o$

$Xs\,(DW(x)S^2 - \lambda^2) = DW(x)\, X_o$

Making $X_{(s)}$ the subject of the formula we have:

$$X_{(s)} = \frac{DW(x)X_o}{DW(x)S^2 - \lambda^2} \quad (34a)$$

Assuming DW(x) = 1 at equilibrium when the rate of diffusion of oxygen molecular is equal to the corrosion rate. Therefore Equation (34a) can be written as:

$$X_S = \frac{X_o}{S^2 - \lambda^2} \quad (34b)$$

Also Equation (34b) can be expressed as:

$$X_S = \frac{X_o}{(S-\lambda)(S+\lambda)} \quad (35)$$

Applying the law of partial fraction into Equation (35) yields;

$$\frac{X_0}{(S+\lambda)(S-\lambda)} = \frac{A}{(S-\lambda)} + \frac{B}{(S+\lambda)} \quad (35)$$

Multiplying Equation (35) by $(S - \lambda)(S + \lambda)$ gives,

$X_0 = A(S + (S - \lambda) + B(S - \lambda)$

$X_o = AS + A\lambda + BS - B\lambda$

By collecting the like terms we have

$X_o = AS + BS + A\lambda - B\lambda$

$$X_o = AS + BS \quad (36)$$

$$X_o = A\lambda - B\lambda \quad (37)$$

From equation (37), let S = λ, the equations becomes;

$$X_o = A\lambda + B\lambda \quad (38)$$

Solving Equation (37) and (38) simultaneously,

$$X_o = A\lambda + B\lambda \quad (39)$$

By adding these above equations give;

$2X_o = 2A\lambda$

$$\therefore A = X_o/\lambda \quad (40)$$

From Equation (37) let A = 0 to obtain the value of B

$X_o = A\lambda - B\lambda$, A = 0

$X_o = 0 - B\lambda$

$$\therefore B = -X_o/\lambda \quad (41)$$

$$\therefore X_s = \frac{A}{S-\lambda} + \frac{B}{S+\lambda} \quad (42)$$

As $A = X_o/\lambda$ and $B = -X_o/\lambda$ the Equation (42) becomes;

$$X_s = \frac{X_o}{\lambda(S-\lambda)} - \frac{X_o}{\lambda(S+\lambda)} = \frac{X_o}{\lambda}\left[\frac{1}{S-\lambda} - \frac{1}{S+\lambda}\right]$$

$$X_s = \frac{X_o}{\lambda}\left[\frac{1}{S-\lambda} - \frac{1}{S+\lambda}\right] \quad (43)$$

By transforming the equation (43) from S-domain to t-domain by the approach as Laplace transform gives;

$$X_t = \frac{X_o}{\lambda}\left(e^{\lambda t} - e^{-\lambda t}\right) \quad (44)$$

The Equation (44) obtained can be found useful in evaluating the functional parameters of the bio-corrosion reaction of the metal. From knowledge of corrosion engineering the general formula or mathematical expression to calculate the rate of corrosion is given as:

$$C_R = \frac{534W}{DAt} \quad (45)$$

Recall that $T_{(t)} = t_o e^{\lambda^2 t}$ in Equation (31)

where T(t) = T, and substitute into Equation (45) then CR becomes;

$$C_R = \frac{534W}{DA\, t_o e^{\lambda^2 t}} \quad (46)$$

Therefore, the Equation (46) becomes the developed model in terms of corrosion rate due to transport of oxygen diffusion in water environment. Since

C = TX

Therefore substituting equation (31) and (44) into equation (20) we have

$$C_{(t)} = t_o e^{\lambda^2 t}\left[\frac{X_o}{\lambda}\left(e^{\lambda t} - ^{-\lambda t}\right)\right] \quad (47)$$

Simplifying equation (47) yields

$$C_{(t)} = 0.679\,\frac{X_o}{\lambda} t_o e^{\lambda^2 t} \quad (48)$$

Equation (48) defines the rate of oxygen molecular diffusion and transport upon the metal per unit time in any given environment.

Experimental

The working specimens

The zinc metal plate was purchased in mile 3 market in Port Harcourt city of Rivers State in Nigeria. The specimens of zinc metal samples were transported to the Department of Chemical/Petrochemical Engineering Laboratory in Rivers State University of Science and Technology (RSUST) Nkpolu, Port Harcourt. The salt and fresh water samples were collected from Ntawogba and Orashi river all in Niger Delta area of Nigeria. The salt and fresh water samples were collected using plastic materials and all the samples collected were transported to the Chemical/Petrochemical Laboratory in the RSUST Port Harcourt for further investigation. The water samples were subjected to the physicochemical analysis to ascertain some of the functional parameters that influence biocorrosion of metals.

Experimental procedures

The zinc metal specimens was sub divided into two categories, group A of initial weight of 3230 mg and group B of 754 mg. Group A of 3230 mg by initial weight contains two specimens which were immersed in both salt and fresh water environment. Similarly, group B of 750 mg by initial weight contains two specimens which were immersed in both salt and fresh water.

The physiochemical parameters of the salt and fresh water environment were measured and the results obtained recorded as presented in this paper. The experimental work was monitored for a period 112 day at various days such as; 14, 28, 56 and 112days and level of corrosion was evaluated. The microbial species present in the water medium upon the surface of the zinc metal was isolated, identified and characterized as presented in this paper.

Microbial isolation and identification

The microbes from the water environment in counted with zinc metal interface on the water/zinc metal environment were isolated

Table 1. Analysis of some physicochemical parameters of salt and fresh water samples.

Samples	Unit (mg/l)	Salt water	Fresh water
pH	-	6.70	9.82
Chloride Ion	-	1560	3.8
Sulphate Ion	-	0.100	0.020
Total hardness	-	4.60	2.60
Iron content	-	1.08	0.24
Biochemical oxygen demand (BOD)	-	280	19.0
Chemical oxygen demand (COD)	-	500	135
Total suspended solids (TSS)	-	100.00	8.00
Temperature	-	35.0	29.5
Total dissolved solid (TDS)	-	530	186

Table 2. Evaluation of the weight loss of Zn in various days of sampling.

The weight loss of Zn in 14 Days (336 h)			
Aqueous solution	Zinc Initial (mg)	Zinc Final (mg)	Weight loss values of zinc specimen (mg)
Salt water A + Zn	8230	8198	32
Fresh water B + Zn	7450	7457	18
The weight loss of Zn in 28 Days (572h)			
Aqueous solution	Zinc Initial (mg)	Zinc Final (mg)	Weight loss values of zinc specimen (mg)
Salt water A + Zn	8198	8122	76
Fresh water B + Zn	7457	7411	46
The weight loss of Zn in 56 Days (1344h)			
Aqueous solution	Zinc Initial (mg)	Zinc Final (mg)	Weight loss values of zinc specimen (mg)
Salt water A + Zn	8122	8000	122
Fresh water B + Zn	7411	7323	88
The weight loss of Zn in 112 Days (2866-h)			
Aqueous solution	Zinc Initial (mg)	Zinc Final (mg)	Weight loss values of zinc specimen (mg)
Salt water A + Zn	8000	7823	177
Fresh water B + Zn	7323	7101	172

and identified according to the method of Buchana and Gibbons (1974) and Gerbardt et al. (1981).

Physicochemical parameter analysis

The physicochemical parameters of the salt and fresh water composition were analyzed using the APHA (1992) standard method.

RESULTS AND DISCUSSION

The analysis results obtained from the investigation were presented in Tables 1, 2 and 3 for both salt and fresh water environment.

The pH, chloride ion, sulphate ion, total hardness, iron content, biochemical oxygen demand (BOD), chemical oxygen demand (COD), total suspended solids (TSS), temperature and total dissolved solid of both the salt and fresh water samples were examined and the results obtained are presented in Table 1.

The results presented in Table 2 illustrate the weight loss by the metal at various days of exposure to the salt and fresh water environment. Increase in weight loss was observed with increase in period of the exposure (time). The weight loss was more in metal deposited in salt water environment than fresh water environment. The variation in weight loss of the metal in both salt and fresh water environment can be attributed to the physicochemical

Table 3. Comparison of existing model result with theoretical developed model result for corrosion rate of metal.

Aqueous solution	**Modeled equation** $C_R = \frac{534\ W}{D A t_o e^{\lambda^2 t}}$	**General formula** $C_R = \frac{534\ W}{D A T}$
For 14 days (336 h)		
Salt Water A + Zn	1.7 Mpy	1.7 Mpy
Fresh Water B + Zn	1.0 Mpy	1.0 Mpy
For 28 days (672h)		
Aqueous solution	Modeled equation $C_R = \frac{534\ W}{D A t_o e^{\lambda^2 t}}$	General formula $C_R = \frac{534\ W}{D A T}$
Salt Water A + Zn	2.0 Mpy	2.0 Mpy
Fresh Water B + Zn	1.2 Mpy	1.2 Mpy
For 56 Days (1344h)		
Aqueous solution	Modeled equation $C_R = \frac{534\ W}{D A t_o e^{\lambda^2 t}}$	General formula $C_R = \frac{534\ W}{D A T}$
Salt Water A + Zn	1.6 Mpy	1.6 Mpy
Fresh Water B + Zn	1.2 Mpy	1.2 Mpy
For 112 Days (2866 h)		
Aqueous solution	Modeled equation $C_R = \frac{534\ W}{D A t_o e^{\lambda^2 t}}$	General formula $C_R = \frac{534\ W}{D A T}$
Salt Water A + Zn	1.2 Mpy	1.2 Mpy
Fresh Water B + Zn	1.0 Mpy	1.0 Mpy

properties as well as period of exposure as presented in Table 2.

The results presented in Table 3 illustrate the corrosion rate of metal in both salt and fresh water environment. The results of the existing general corrosion model was compared with the developed model which shown a good match indicating that the developed model can be used in monitoring and predicting the corrosion rate of any metal.

The results presented in Table 1, illustrate the pH value of salt water and fresh water medium. It can be seen that the pH value of salt water is 6.70, which is acidic in nature while the fresh water medium is 9.82 alkaline in nature. From the characteristic of acid and base; acidic substance are more corrosive than alkaline. The microbial growth in salt water environment hereby increases the microbial activity as well facilitating corrosion.

All the physiochemical parameters considered are major factor's that influence corrosion, microbial activity is higher in the salt water medium than the fresh water medium as presented in Table 1. These parameters includes concentration of chloride ion, shulphate ion, total hardness, iron content, biochemical oxygen demand (BOD), chemical oxygen demand (COD) temperature, total suspended solids (TSS) and total dissolved solids (TDS).

The various microorganisms isolated and identified includes, *disiicforibrio, desulfobacter, gallionella, thiobacillus, pseudomomas sp, bacillus substillis, E-coli, desic formaculum and thooxidants*. From Table 2, it is seen that there is a decrease in weight of zinc metal with respect to sampling days as presented in the paper. The corrosion rate increases with increase in time. The weight loss values of zinc metal for salt water is given as 32 mg for 14days, 76 mg for 28day, 122 mg for 56days and 177 mg for 112days whereas for fresh water medium we have 18 mg for 14days, 46 mg for 28days, 58 mg for 56days and 172 mg for 112days.

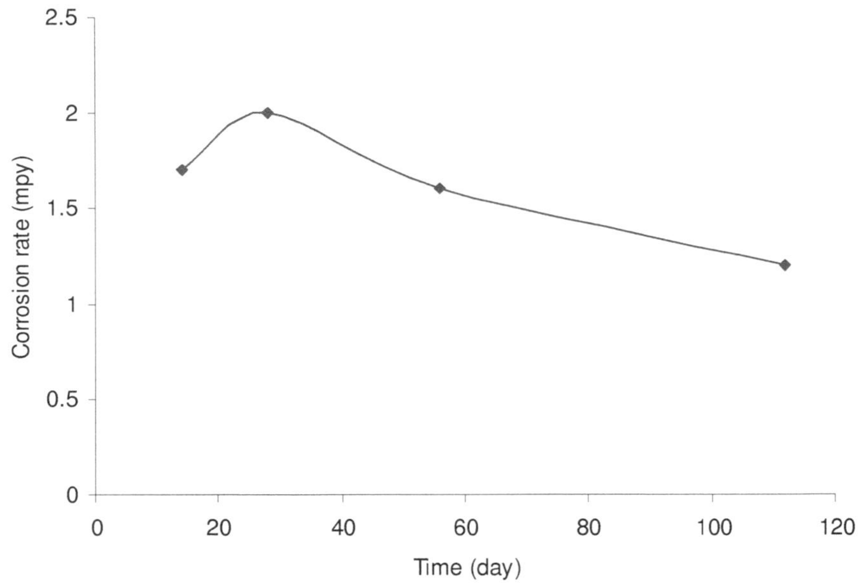

Figure 4. Corrosion rate of zinc metal in salt water medium versus time.

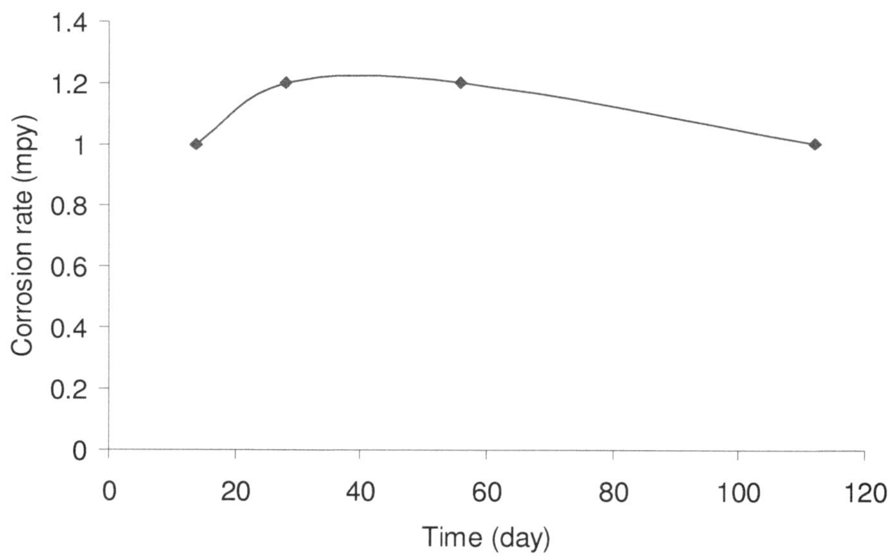

Figure 5. Corrosion rate of zinc metal in fresh water medium versus time.

Table 3 illustrates the corrosion rate of existing model with the modified model equation. The existing corrosion rate equation is given as $C_R = \frac{534w}{DAT}$ and the modified corrosion rat equation is given as $C_R = \frac{534w}{DAt_0e^{\lambda^2}t}$. The data obtained experimentally was feed into the existing and developed model and the same results were obtained as shown in Table 3. The corrosion rate of zinc metal increases with increase in time, from Figure 4, it is seen that the corrosion rate of zinc metal increase with increase in time for salt water environment. The variation in the corrosion rate of zinc metal can be attributed to variation in time as well as physicochemical parameters of the salt water medium. The variation was experienced for both equations (developed and general equation of corrosion). From Figure 5, it is observed that the corrosion rate of the zinc metal increase with increase in time. The increase in corrosion rate of zinc metal can be

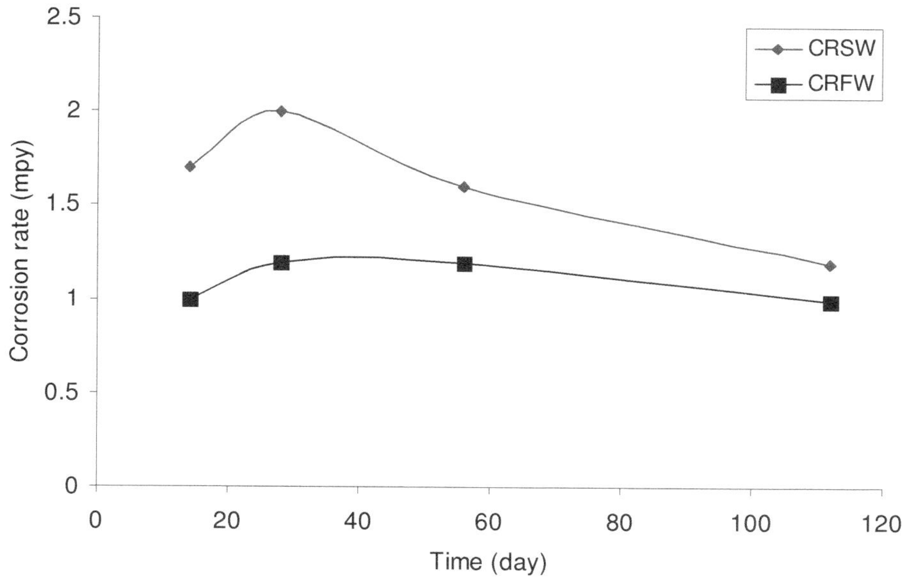

Figure 6. Comparison of corrosion rate of salt/fresh water medium versus time.

attributed to the microbial growth, physicochemical parameters and other environmental factors.

Figure 6 illustrates the comparison of the corrosion rate of zinc metal in salt and fresh water environment. The results obtained showed that corrosion of zinc is more effective in salt water medium than fresh water medium. This was attributed to the physicochemical properties of the salt and fresh water medium.

Conclusion

The following conclusions were drawn from the research work;

1) Redox reactions could contribute to the corrosion of metals
2) Physicochemical properties of the environment influence the rate of corrosion of any metal.
3) The rate of oxygen diffusion influence the corrosion rate of metal
4) Microbial growth influence the corrosion rate of any metal in the environment
5) Corrosion of metal is faster in salt water medium than fresh water medium.
6) The developed model in this paper as presented is found useful in monitoring and predicting the corrosion rate of any metal in any environment.

Based on the mathematical model developed on this paper, Ukpaka can write the mathematical formula of corrosion rate as $C_R = \frac{534w}{DAt_o e^{\lambda^2 t}}$, where λ = 0.5, thus, the value of t_o changes with change in corrosion rate.

Nomenclature: λ,Constant rate of corrosion which is assumed to be 0.5; $\overrightarrow{J_k}$ is the mass flux of species K; λ the constant rate of the reaction (dimensionless); ϕ the electric potential in the electrolyte; A area of metal (m^2); C_k the concentration of specie K (mg/l); C_R corrosion rate of specimen (mpy); D density of specimen in g/cm^3; D^*_k the diffusivity constant of oxygen in rust layer (mg/l); F is the faraday's constant (colomb); R the ideal gas constant; t constant corrosion time (day); w weight loss (mg); T the absolute temperature (℃); t_o changing time for corrosion; W weight loss of specimen (mg); $X_o$xt the original distance before the bio-corrosion reaction of the metal (m).

REFERENCES

APHA (1992). Standard Methods for the Examination of Water and Wastewater, 19th American Publication Health Association. Washington DC, p.18.
Atkins PW (1990). Physical chemistry. Fourth edition. Longman Express Ltd, pp. 15-81.
Buchana RE, Gibbons NE (1974). Bergey's Manual of Determination Bacteriology, 7th edition, Baltimore the William and Whisking Co. p. 274.
Degarmo EP, Ronald A, Kohser JT, Black E, Barley LX (2004). Materials and processes in manufacturing. Wandusoo Express, pp. 20-76.
Gardner CP (2002). Corrosion Science. Third edition. Crammer Hill Express, pp. 59-144.
Gerbardt P, Murrary RCE, Cstillon RW, Nester EW, Wood WA, Krig NR, Philips GB (1981). Manual of Methods for General Bacteriology.

Washington D.C American society of Microbial, pp. 142-240.
Ivor HS (2004). Building maintenance. Second edition. Macmillan Publishers Ltd, pp. 64-109.
John CX, Kotz L, Paul T (Jr.) (1999). Chemistry and chemical reactivity. Fourth edition. Saunders College Publishing, pp. 57-105.
Johnson JE (2000). Site control of materials. First edition. Butter Worths, London, pp. 78-90.
Liudvikas P (2003). Coating technology: ion beam deposition. Addison Wesley Longman Limited, pp. 17-30.
Mars GE (2005). Corrosion engineering. Third edition. McGraw Hill Publishing Company Limited, pp. 83-176.
Melchers CT (2006). Corrosion science. Second edition. Hilton Publishing Co. pp. 100-195.
Robinson O, Holt C (1997). The essentials of general chemistry. Houghton Mifflin Company Bosten, New York, pp. 303-382.
Roy C, Rofer C (1999). Construction Technology. Third edition. Addison Wesley Longman Limited, pp. 10-58.
Sharma SC (2003). Plant layout and materials handling. Winters Publishing Inc., pp. 24-40.
Shifter DA (2005). Corrosion Science. Fourth edition. Longman Express Limited, pp.183-274
Vijendra S (1999). Physical metallurgy. First edition. Academic Press Inc., pp. 74-161.

Permissions

All chapters in this book were first published in JETR, by Academic Journals; hereby published with permission under the Creative Commons Attribution License or equivalent. Every chapter published in this book has been scrutinized by our experts. Their significance has been extensively debated. The topics covered herein carry significant findings which will fuel the growth of the discipline. They may even be implemented as practical applications or may be referred to as a beginning point for another development.

The contributors of this book come from diverse backgrounds, making this book a truly international effort. This book will bring forth new frontiers with its revolutionizing research information and detailed analysis of the nascent developments around the world.

We would like to thank all the contributing authors for lending their expertise to make the book truly unique. They have played a crucial role in the development of this book. Without their invaluable contributions this book wouldn't have been possible. They have made vital efforts to compile up to date information on the varied aspects of this subject to make this book a valuable addition to the collection of many professionals and students.

This book was conceptualized with the vision of imparting up-to-date information and advanced data in this field. To ensure the same, a matchless editorial board was set up. Every individual on the board went through rigorous rounds of assessment to prove their worth. After which they invested a large part of their time researching and compiling the most relevant data for our readers.

The editorial board has been involved in producing this book since its inception. They have spent rigorous hours researching and exploring the diverse topics which have resulted in the successful publishing of this book. They have passed on their knowledge of decades through this book. To expedite this challenging task, the publisher supported the team at every step. A small team of assistant editors was also appointed to further simplify the editing procedure and attain best results for the readers.

Apart from the editorial board, the designing team has also invested a significant amount of their time in understanding the subject and creating the most relevant covers. They scrutinized every image to scout for the most suitable representation of the subject and create an appropriate cover for the book.

The publishing team has been an ardent support to the editorial, designing and production team. Their endless efforts to recruit the best for this project, has resulted in the accomplishment of this book. They are a veteran in the field of academics and their pool of knowledge is as vast as their experience in printing. Their expertise and guidance has proved useful at every step. Their uncompromising quality standards have made this book an exceptional effort. Their encouragement from time to time has been an inspiration for everyone.

The publisher and the editorial board hope that this book will prove to be a valuable piece of knowledge for researchers, students, practitioners and scholars across the globe.

List of Contributors

C. P. Ukpaka
Department of Chemical/Petrochemical Engineering, Rivers State University of Science and Technology, Nkpolu, P. M. B. 5080, Port Harcourt, Nigeria

E. A. M. Elshafei
Mechanical Power Department, Faculty of Engineering, Mansoura University, El-Mansoura 35516, Egypt

M. Safwat Mohamed
Mechanical Power Department, Faculty of Engineering, Mansoura University, El-Mansoura 35516, Egypt

H. Mansour
Mechanical Power Department, Faculty of Engineering, Mansoura University, El-Mansoura 35516, Egypt

M. Sakr
Mechanical Power Department, Faculty of Engineering, Mansoura University, El-Mansoura 35516, Egypt

C. P. Ukpaka
Department of Chemical/Petrochemical Engineering, Rivers State University of Science and Technology, Nkpolu, P. M. B. 5080, Port Harcourt, Nigeria

Mahmoud Ellouze
Laboratory of Analysis and Control Systems (LACS), ENIT, Tunis, C.U Belvedere 1060 Tunis, Tunisia
Superior Institute of Technology and Data processing, 2080, Ariana, Tunisia

Riadh Gamoudi
Laboratory of Analysis and Control Systems (LACS), ENIT, Tunis, C.U Belvedere 1060 Tunis, Tunisia

Abdelkader Mami
Laboratory of Analysis and Control Systems (LACS), ENIT, Tunis, C.U Belvedere 1060 Tunis, Tunisia
Department of Physics, Faculty of Sciences of Tunis, Electronic Laboratory, 2092 El Manar Tunis, Tunisia

S. Kennouche
Université de Boumerdès, Unité de recherche Matériaux-Procédé-Environnent, FSI, cite Frantz Fanon, 35000 Boumerdès, Algérie

A. Zerizer
Université de Boumerdès, Unité de recherche Matériaux-Procédé-Environnent, FSI, cite Frantz Fanon, 35000 Boumerdès, Algérie

A. Benmounah
Université de Boumerdès, Unité de recherche Matériaux-Procédé-Environnent, FSI, cite Frantz Fanon, 35000 Boumerdès, Algérie

B. Hami
Université de Boumerdès, Unité de recherche Matériaux-Procédé-Environnent, FSI, cite Frantz Fanon, 35000 Boumerdès, Algérie

M. Mahdad
(C.N.E.R.I.B) Centre National d'Etudes et de Recherches Intégrées du Bâtiment, Cité Nouvelle El Mokrani Souidania - Alger

H. Benouali
(C.N.E.R.I.B) Centre National d'Etudes et de Recherches Intégrées du Bâtiment, Cité Nouvelle El Mokrani Souidania - Alger

S. Bedjou
(C.N.E.R.I.B) Centre National d'Etudes et de Recherches Intégrées du Bâtiment, Cité Nouvelle El Mokrani Souidania - Alger

J. K. PATRA
Department of Biotechnology and Medical Engineering, National Institute of Technology, Rourkela, Odisha, India

S. GOUDA
Department of Biotechnology and Bioinformatics, AMIT, Bhubaneswar, Odisha, India

Nimain Charan Nayak
Department of Electronics Engineering (EEE), Sathyabama University, Chennai, India

C. Christober Asir Rajan
Department of Electronics Engineering (EEE), Pondicherry Engineering College, Puducherry, India

C. P. Ukpaka
Department of Chemical/Petrochemical Engineering, Rivers State University of Science and Technology, Nkpolu, P. M. B. 5080, Port Harcourt, Nigeria

Mohd Mustafa Al Bakri
School of Material Engineering, Universiti Malaysia Perlis (UniMAP), P. O. Box 77, d/a Pejabat, Pos Besar, 01007 Kangar, Perlis, Malaysia

H. Mohammed
King Abdul Aziz City Science and Technology (KACST), P. O. Box 6086, Riyadh 11442, Kingdom of Saudi Arabia

H. Kamarudin
School of Material Engineering, Universiti Malaysia Perlis (UniMAP), P. O. Box 77, d/a Pejabat, Pos Besar, 01007 Kangar, Perlis, Malaysia

I. Khairul Niza
School of Environmental Engineering, Universiti Malaysia Perlis (UniMAP), P. O. Box 77, d/a Pejabat, Pos Besar, 01007 Kangar, Perlis, Malaysia

Y. Zarina
School of Material Engineering, Universiti Malaysia Perlis (UniMAP), P. O. Box 77, d/a Pejabat, Pos Besar, 01007 Kangar, Perlis, Malaysia

C. P. Ukpaka
Department of Chemical/Petrochemical Engineering, Rivers State University of Science and Technology, Nkpolu, P.M.B. 5080, Port Harcourt, Nigeria

V. M. Ndor
Department of Mechanical Engineering, Rivers State University of Science and Technology, Nkpolu, P.M.B. 5080, Port Harcourt, Nigeria

N. K. Singh
Department of Physics, Veer Kunwar Singh (V. K. S) University, Arrah, Bhojpur, 802 301 Bihar, India

Pritam Kumar
Department of Physics, Veer Kunwar Singh (V. K. S) University, Arrah, Bhojpur, 802 301 Bihar, India

A. Kumar
Department of Physics, Veer Kunwar Singh (V. K. S) University, Arrah, Bhojpur, 802 301 Bihar, India

S. Sharma
Department of Physics, A. N. College, Patna, Bihar, India

Tadelech Atomssa Keyeta
Faculty of Physical and Chemical Sciences, Physics Department, Addis Ababa University, P.O.Box 1176, Addis Ababa, Ethiopia

A. V. Gholap
Faculty of Physical and Chemical Sciences, Physics Department, Addis Ababa University, P.O.Box 1176, Addis Ababa, Ethiopia

A. M Mustafa Al Bakri
Green Concrete@UniMAP, School of Material Engineering, Universiti Malaysia Perlis (UniMAP), P. O. Box 77, d/a Pejabat, Pos Besar, 01007 Kangar, Perlis Malaysia

H. Kamarudin
Green Concrete@UniMAP, School of Material Engineering, Universiti Malaysia Perlis (UniMAP), P. O. Box 77, d/a Pejabat, Pos Besar, 01007 Kangar, Perlis Malaysia

M. Bnhussain
King Abdul Aziz City Science and Technology (KACST), P.O. Box 6086, Riyadh 11442, Kingdom of Saudi Arabia

I. Khairul Nizar
King Abdul Aziz City Science and Technology (KACST), P.O. Box 6086, Riyadh 11442, Kingdom of Saudi Arabia

A. R Rafiza
Green Concrete@UniMAP, School of Material Engineering, Universiti Malaysia Perlis (UniMAP), P. O. Box 77, d/a Pejabat, Pos Besar, 01007 Kangar, Perlis Malaysia

Y. Zarina
Green Concrete@UniMAP, School of Material Engineering, Universiti Malaysia Perlis (UniMAP), P. O. Box 77, d/a Pejabat, Pos Besar, 01007 Kangar, Perlis Malaysia

G. R. Ebrahimi
Metallurgy and Materials Engineering Department, Sabzevar Tarbiat Moallem University, Sabzevar, Iran

H. Arabshahi
Physics Department, Ferdowsi University of Mashhad, Mashhad, Iran

N. G. Goki
Department of Geology and Mining, Nasarawa State University, P. M. B. 1022, Keffi, Nigeria

A. N. Amadi
Department of Geology, Federal University of Technology, P. M. B. 65, Minna, Nigeria

O. Amuneni
Department of Geology and Mining, Nasarawa State University, P. M. B. 1022, Keffi, Nigeria

P. Agbo
Department of Geology and Mining, Nasarawa State University, P. M. B. 1022, Keffi, Nigeria

N. O Okoye
Department of Geology, Federal University of Technology, P. M. B. 65, Minna, Nigeria

C. I. Unuevho
Department of Geology, Federal University of Technology, P. M. B. 65, Minna, Nigeria

Mao-Yu Wen
Department of Mechanical Engineering, Cheng Shiu University, Kaohsiung 833, ROC, Taiwan

Ching-Yen Ho
Department of Mechanical Engineering, Hwa Hsia Institute of Technology, Taipei 235, ROC, Taiwan

kuen-Jang Jang
Department of Mechanical Engineering, Cheng Shiu University, Kaohsiung 833, ROC, Taiwan

Asya S. Skal
P. O. Box 1836, Ariel 44837, Israel

A. Bahari
Department of Physics, University of Mazandaran, Babolsar, Iran

A. Anasari
Department of Physics, Islamic Azad University, Sari Branch, Sari, Iran

Z. Rahman
Department of Physic, Islamic Azad University, Centeral Tehran Branch, Iran

H. Arabshahi
Department of Physics, Ferdowsi University of Mashahd, Mashhad, Iran
Department of Physics, Payame Nour University of Fariman, Fariman, Iran

M. Y. Lim
Department of Physics, Faculty of Science, Universiti Putra Malaysia, 43400 UPM Serdang, Selangor, Malaysia

W. M. M. Yunus
Department of Physics, Faculty of Science, Universiti Putra Malaysia, 43400 UPM Serdang, Selangor, Malaysia

Z. A. Talib
Department of Physics, Faculty of Science, Universiti Putra Malaysia, 43400 UPM Serdang, Selangor, Malaysia

A. Kassim
Department of Chemistry, Faculty of Science, University Putra Malaysia, 43400 UPM Serdang, Selangor, Malaysia

J. G. Abuga
Department of Mathematics and Statistics, Jkuat, P.O.BOX 62000, Nairobi,Kenya

M. Kinyanjui
Department of Mathematics and Statistics, Jkuat, P.O.BOX 62000, Nairobi,Kenya

J. K. Sigey
Department of Mathematics and Statistics, Jkuat, P.O.BOX 62000, Nairobi,Kenya

A. Bahari
Department of Physics, University of Mazandaran, Babolsar, Iran

N. Noori
Department of Physics, Islamic Azad University, Sari Branch, Sari, Iran

N. P. Dharmadhikari
Department of Applied Physics, Sinhgad Academy of Engineering, Kondhwa (Bk.) Pune-48, (MS) India
Department of Civil Engineering, Maharashtra Institute of Technology, Pune-38, (MS), India

D. C. Meshram
Department of Geology, University of Pune, Pune-7, (MS), India

S. D. Kulkarni
Department of Environmental Science, New Arts Commerce and Sciences College, Ahmednagar-.414001, (MS), India

A. G. Kharat
Department of Civil Engineering, Sinhgad Academy of Engineering, Kondhwa (Bk.) Pune-48, (MS), India

S. S. Pimplikar
Department of Civil Engineering, Maharashtra Institute of Technology, Pune-38, (MS), India

Th. S. Dhahi
Institute of Nano Electronic Engineering, University Malaysia Perlis (UniMAP), Malaysia

U. Hashim
Institute of Nano Electronic Engineering, University Malaysia Perlis (UniMAP), Malaysia

N. M. Ahmed
School of Physics, University Sains Malaysia (USM), Malaysia

Md. Eaqub Ali
School of Physics, University Sains Malaysia (USM), Malaysia

T. Nazwa
School of Physics, University Sains Malaysia (USM), Malaysia

C. P. Ukpaka
Department of Chemical/Petrochemical Engineering, Rivers State University of Science and Technology, Nkpolu, P. M. B. 5080, Port Harcourt, Nigeria

S. A. Balogun
Department of Metallurgical and Materials Engineering, University of Lagos, Nigeria

G. I. Lawal
Department of Metallurgical and Materials Engineering, University of Lagos, Nigeria

O. I. Sekunowo
Department of Metallurgical and Materials Engineering, University of Lagos, Nigeria

S. O. Adeosun
Department of Metallurgical and Materials Engineering, University of Lagos, Nigeria.

Okoromah Candy Ozioma
Department of Chemical/Petrochemical Engineering, Rivers State University of Science and Technology, Nkpolu, P.M.B. 5080, Port Harcourt, Nigeria

Ukpaka C. Peter
Department of Chemical/Petrochemical Engineering, Rivers State University of Science and Technology, Nkpolu, P.M.B. 5080, Port Harcourt, Nigeria

Lilian Daniel Kaale
Department of Chemical and Mining Engineering, College of Engineering and Technology, University of Dar es Salaam, P. O. Box 35131, Dar es Salaam, Tanzania

Jamidu H. Y. Katima
Department of Chemical and Mining Engineering, College of Engineering and Technology, University of Dar es Salaam, P. O. Box 35131, Dar es Salaam, Tanzania

C. P. Ukpaka
Department of Chemical/Petrochemical Engineering, Rivers State University of Science and Technology, Nkpolu, P. M. B. 5080, Port Harcourt, Nigeria

S. A. Amadi
Department of Chemical/Petrochemical Engineering, Rivers State University of Science and Technology, Nkpolu, P. M. B. 5080, Port Harcourt, Nigeria

I. G. Ahuchogu
Department of /Petroleum and Natural Gas Processing, Petroleum Training Institute, P. M. B. 20 Effurun, Delta State, Nigeria

J. Odharo
Engineering, Construction and Projects Division, Civil Works and Survey, Total E and P Nigeria Limited, Port Harcourt, Rivers State, Nigeria